Computer Vision

Computer vision has made enormous progress in recent years, and its applications are multifaceted and growing quickly, while many challenges still remain. This book brings together a range of leading researchers to examine a wide variety of research directions, challenges, and prospects for computer vision and its applications.

This book highlights various core challenges as well as solutions by leading researchers in the field. It covers such important topics as data-driven AI, biometrics, digital forensics, healthcare, robotics, entertainment and XR, autonomous driving, sports analytics, and neuromorphic computing, covering both academic and industry R&D perspectives. Providing a mix of breadth and depth, this book will have an impact across the fields of computer vision, imaging, and AI.

Computer Vision: Challenges, Trends, and Opportunities covers timely and important aspects of computer vision and its applications, highlighting the challenges ahead and providing a range of perspectives from top researchers around the world. A substantial compilation of ideas and state-of-the-art solutions, it will be of great benefit to students, researchers, and industry practitioners.

Chapman & Hall/CRC Computer Vision Series

About the Series:
This series aims to capture the latest developments in Computer Vision (CV), as well as high-quality work summarizing or contributing to more established topics.

It welcomes contributions from across the spectrum of academic and industrial research and practice, and publishes a broad range of textbooks, reference books, handbooks, and primers.

Topics may include but are not limited to the following: CV applications and systems, 3D, action and behavior recognition, adversarial learning, autonomous vehicles, biometrics, bodies and faces, computational photography, detection and localization, efficient training and inference methods, explainable AI for CV, ethics and responsibility in CV, egocentric vision, image and video manipulation, machine learning for CV, medical and biological CV, motion and tracking, neural generative models, physics-based vision, retrieval and synthesis, robotics and embodied vision, scene analysis and understanding, text and document understanding, unsupervised learning, stereo, transfer learning, video analysis and understanding, vision and other modalities, visual reasoning and logical representation.

Computer Vision: Challenges, Trends, and Opportunities
First Edition
Atiqur Rahman Ahad, Upal Mahbub, Matthew Turk, Richard Hartley

Low-Power Computer Vision: Improve the Efficiency of Artificial Intelligence
First Edition
George K. Thiruvathukal, Yung-Hsiang Lu, Jaeyoun Kim, Yiran Chen, Bo Chen

For more information about this series please visit:
https://www.routledge.com/Chapman--HallCRC-Computer-Vision/book-series/CRCCV

Computer Vision
Challenges, Trends, and Opportunities

Edited by
Md Atiqur Rahman Ahad
Upal Mahbub
Matthew Turk
Richard Hartley

CRC Press
Taylor & Francis Group
Boca Raton London New York

CRC Press is an imprint of the
Taylor & Francis Group, an **informa** business

A CHAPMAN & HALL BOOK

Designed cover image: Shutterstock Images

First edition published 2025
by CRC Press
2385 NW Executive Center Drive, Suite 320, Boca Raton FL 33431

and by CRC Press
4 Park Square, Milton Park, Abingdon, Oxon, OX14 4RN

ISBN: 978-1-032-31705-2 (hbk)
ISBN: 978-1-032-35843-7 (pbk)
ISBN: 978-1-003-32895-7 (ebk)

DOI: 10.1201/9781003328957

Typeset in CMR10 font
by KnowledgeWorks Global Ltd.

Contents

Foreword

As a person who started working in the field of Computer Vision more than 50 years ago, recent advancements in the field have been nothing short of phenomenal.

Back then, digital cameras did not exist and I had to build my own digitizer for an NTSC analog video signal from a vidicon tube camera. The image size was a few hundred pixels in length and width at most. In my Japanese university's research laboratory, I hand-coded heuristic algorithms into a program, and ran it on a slow computer for testing with a small dataset. Computer Vision was a curiosity research field. Today, after five decades of research and development that include periods with both good and struggling progress, Computer Vision has found its applications everywhere from everyday cell phones to securities, entertainment, robotics, industrial, medical, and military domains. Advanced imaging sensors produce high-resolution images and videos at a high frame rate, even in 3D. High-performance low-power computers execute computation-hungry vision algorithms even in handheld devices. And, of course, Deep Neural Net Learning played the most critical role to turn Computer Vision into an amazingly impactful technology.

This book contains 15 chapters written by academic leaders and seasoned industrial practitioners of Computer Vision. They cover the diversity of today's application areas from face and forensic to medical and rehab, robots, media and XR, and sports, as well as the topic of neuro-morphic computing. This book, however, is more than just a collection of papers, each presenting particular work in various areas. Chapter 1 by Rama Chellappa, et al., presents insightful discussions on some challenges and their potential solutions in data-driven AI, in particular, domain shift, lack of robustness to adversarial attacks, and bias in decision; they all are the common issues in developing Computer Vision algorithms based on Deep Learning. Chapter 2 by Fatih Porikli gives an informal yet thoughtful story from an industrial perspective of how computer vision applications are conceptualized and developed into products. Each of the remaining chapters presents a comprehensive and informative overview of representative existing systems and state of the art methodologies in each area as well as discussions on future trends. Either skimming through or carefully reading those chapters, readers should be able to feel or understand the broad range of trends, challenges, and opportunities across Computer Vision.

This timely book will be enjoyed by today's researchers, practitioners, and students of Computer Vision.

Takeo Kanade
U. A. and Helen Whitaker University Professor
Carnegie Mellon University
Pittsburgh PA. USA
September 28, 2023

Preface

For many years, computer vision was seen by many as being a "promising" field, whose time had not yet come, but might do so at some point in the unspecified future. The first CVPR (Computer Vision and Pattern Recognition) conference took place in 1983, followed by ICCV (the International Conference on Computer Vision) in 1987 and ECCV (European Conference on Computer Vision) in 1990. Despite notable successes, the growth of the field was slow. In 1992, CVPR and ECCV had two or three hundred attendees, and the conference proceedings comprised a single volume.

People who have been involved in the field since that era might well view the current state of research in computer vision, and its importance in current world with astonishment, yet perhaps a vindication of their faith. Computer vision has evolved to become one of the most active and important disciplines in Computer Science and Artificial Intelligence. This volume provides a panorama of the ways computer vision is used in the present-day world, and the many ways it affects our lives.

The book comprises a collection of several chapters written by prominent researchers in the field, the authors giving their own perspective on some aspect of computer vision. Each set of authors outline the state of the art in their particular field, while at the same time enumerating outstanding research challenges. Thus, the book is not only a comprehensive survey of the present state of the art, but also serves as a guide for those wishing to further the field through their own research.

The book begins with a chapter by Rama Chellappa el al. Chapter 1, which sets the scene by introducing some of the main challenges, which are also recurring themes throughout the rest of the book, and summarizing current research addressing these challenges.

The first of these is *Domain Adaptation*. The early years of the era of Deep Learning were dominated by supervised learning, in which large-scale annotated images sets were used for training. Fortunately, such sets, such as ImageNet were available, and already widely used.

The use of data-sets and bench-marks, so prevalent these days was not common before the turn of the millennium. Robert Haralick was one who advocated their wide-spread use as a way of measuring and furthering progress in the field. Deep Learning for computer vision is still very dependent on such data-sets, and much effort goes into producing such data-sets. Many chapters of this book give extensive lists of available data-sets for different tasks, but as pointed out, for instance in Chapter 11, lack of suitable data-sets is still an obstacle to progress.

Annotated data-sets for every task are difficult and expensive to obtain, leading to the development of techniques such as Domain Adaptation, and the use of synthetic data for training, both of which are discussed in this volume.

Chapter 1 also highlights the challenges of **bias** in Deep Learning algorithms, and describes approaches to this problem. This theme is also developed in other chapters of the book, notably in Chapter 4, in relation to face or gender recognition. The problem is not as simple as balancing the mixture of types in the data-set.

Among the areas where computer vision is chiefly deployed in the present world are surveillance and security, social media, medical applications, environmental monitoring, robotics, autonomous driving and entertainment, and each of these is described in the chapters of this book.

Face and human recognition, as well as action recognition form an important part of **surveillance** and its attended security applications. However, they are also germane to domains such as online-security, social media, entertainment and health monitoring. These are discussed in detail in Chapter 9.

Medical applications are a major area of activity in computer vision research, and this field is represented here by Chapters 6 and 7, which give insight into some of the topics in this extensive area.

Environmental monitoring is one of the early fields in which computer vision was adopted, and motivated large part of computer vision research early on, such as remote sensing and the analysis of aerial and satellite imagery. In this book, Chapter 9 gives a view into the challenges of deploying vision systems underwater, and the use of undersea robots for a variety of tasks such as monitoring coral reefs, and fish detection.

The *entertainment industry* is a major area in which computer vision, and image-based graphics are used. The commercial importance of this area is highlighted in Chapter 10. Entertainment can be widely interpreted to include the topics of sport, content streaming and extended reality (XR). These topics are thoroughly surveyed in Chapters 10, 11, 12 and 14. These chapters describe the astonishing variety of ways that computer vision is used in Entertainment, and together provide current and prospective researchers with a comprehensive overview.

Another area in which computer vision has long been involved is that of *robotics*, where robotic vision has become a dominant direction of research, as vision becomes more reliable. Important applications of robotics include robot-assistance (Chapter 8), navigation and SLAM, and *autonomous driving*. This latter technology, which is finding its way into our lives, is represented in this book in Chapter 13, which provides just a sample of the challenges that face this field.

The success of computer vision in myriad aspects of our every-day life brings with it new problems of a technical and societal nature. Notable among these is the problems of deep fakes and fake news. As techniques for image and video synthesis improve, through the development of such systems as Deep-Fake, DALL-E, Stable Diffusion and Imagen, the issue arises of how one can ever be sure of what one sees online or in the news. As AI and computer vision create this problem, they are also crucial in mitigating its negative effects. This topic is discussed in detail in Chapter 5, and also in Chapter 10, where it is seen as as critical concern in trust of news sources, and of AI in general.

Another problem crucial for the trust in AI and vision systems concerns bias of algorithms and data-sets used in computer vision. Computer vision algorithms for tasks such as face recognition of gender recognition show different degrees of accuracy depending on gender or demographic type. This problem is highlighted as one of computer vision's core challenges in Chapter 1 and it is investigated in detail in Chapter 4. Is the problem one of balance of the dataset, intrinsic bias of the algorithms, or the difficulty of the task?

Multimodal applications have become a major topic in AI and computer vision research, particularly in the coming-together of image and text. This can be through image and video generation systems that generate content in response to textual descriptions, automatically seek to annotate image or videos, or use text as an aid to understanding images or video. This is an old topic in computer vision research, going back at least to work of Rohini Srihari, in the early 1990s[1]. An interesting aspect of this problem is human recognition based on textual descriptions, discussed in Chapter 3. It is also intrinsic to media content creation as discussed in Chapter 10.

One of the most dramatic changes in computer vision research over the last few decades is the move from an academic research area to one in which commercial companies play an increasingly dominant role. In the early 1990s, computer vision was largely an academic research area, with most computer vision research carried out in universities and research institutes. There were a handful of commercial computer vision groups, such as RCA Sarnoff Labs / RSI and General Electric. Larger scale research commercial computer vision research, such as at Microsoft Research, and MERL came along towards the end of the 1990s. There were also small start-up companies using computer vision methods. Contrast this with the present situation where much computer vision research is carried out at major companies with large resources and manpower. This reflects the success of the field. Some insights into the issues involved with carrying out research and development, particularly in computer vision, in a commercial environment are given by Fatih Porikli in Chapter 2. This chapter also contains an extensive list of open problems in computer vision, providing directions for future research, and emphasizing that the field still has a lot to do.

The trend in computer vision is for the use of larger and large amounts of data, larger networks and more and more computing power. The final chapter of the book, Chapter 15, provides a glimpse into possible ways the basic computing infrastructure may develop in the future, through the use of more biologically inspired computers.

With computer vision becoming more and more important in our lives, this book provides an extensive overview of the field of computer vision, past present and future, suggesting research directions and identifying resources to keep the field busy for many years to come.

Note

1. For instance, Rohini Srihari: "Use of captions and other collateral text in understanding photos", Artificial Intelligence Review, 1994.

Editor Biographies

Dr. Md Atiqur Rahman Ahad, (SMIEEE, SMOP-TICA) is at University of East London. He is a *Visiting Professor* at Kyushu Institute of Technology & UCSI University. He was Professor at University of Dhaka & Associate Professor at Osaka University. His research focuses on AI & healthcare, computer vision & sensors. He studied in Japan, Australia and Bangladesh. He authored 15 books, published over 220 papers/chapters, received 60+ awards/recognitions, & delivered 160+ keynotes/talks. He serves Editorial Board, *Scientific Reports, Nature; Chairs of* 6th *ABC,* 11th *ICIEV,* 6th *IVPR;* Guest Editor of PRL, JMUI; Reviewer of IEEE-TPAMI, IEEE-TBIOM, IJCV, PR, IEEE-Sensors, SR Nature, IEEE-TAC, ACM-IMWUT, etc. Through his research & activities, he strives to bridge the technological divide and empower all. More: http://ahadVisionLab.com

Dr. Upal Mahbub, a distinguished professional with a Ph.D. and Senior Membership in IEEE, currently holds the position of Staff Engineer at Qualcomm Technologies, Inc. His expertise extends to various facets of technology, and he serves as the Chair of the IEEE Computer Society San Diego Chapter, showcasing his commitment to advancing the field. Dr. Mahbub's scholarly contributions are notable, with a portfolio comprising more than 40 publications that have garnered more than 850 citations. Acknowledged for his expertise, he has taken on roles such as being a guest editor for Pattern Recognition Letters (PRL) Virtual Special Issues and serving as the Program Chair for the International Conference on Image and Vision Computing (IVPR) in 2021. To explore more about his work and insights, interested individuals can visit his website at https://umahbub.github.io/.

Dr. Matthew Turk holds a Ph.D. and is a Fellow of the ACM, IEEE, IAPR, AAIA, and AIAA. Currently serving as the President of the Toyota Technological Institute (TTIC) in Chicago, he also holds the title of Emeritus Professor in computer science at the University of California, Santa Barbara. Dr. Turk's contributions are reflected in his extensive portfolio of more than 240 publications, amassing citation count exceeding 44,000. Notably, he is widely recognized for introducing Eigenfaces, a groundbreaking concept co-authored with Prof. Alex Pentland. To delve deeper into his expertise and achievements, interested individuals can visit his web page at https://www.ttic.edu/mturk/.

Dr. Richard Hartley, an eminent scholar in the field, holds a Ph.D. and is recognized as a Fellow by IEEE, the Australian Academy of Science, and the Australian Mathematical Society and the Royal Society. Currently serving as an Emeritus Professor at the Australian National University, his academic legacy is characterized by an extensive body of work, with more than 400 publications that have garnered an impressive citation count surpassing 69,000. Dr. Hartley is most renowned for his significant contribution to the field of computer vision through the highly regarded book *Multiple View Geometry in Computer Vision* (Cambridge University Press, 2004), co-authored with Prof. Andrew Zisserman. To gain further insights into his contributions and expertise, interested individuals can visit his webpage at https://users.cecs.anu.edu.au/~hartley/.

1

Some Challenges and Solutions in Data-Driven AI

Rama Chellappa, Jiang Liu, Chun Pong Lau, and Prithviraj Dhar

Owing to the availability of massive amounts of multimodal data and the reemergence of neural network-based methods, data-driven artificial intelligence (AI) appears to be on the verge of becoming the dominant technology of the future. While this development has basically shaken many established fields such as computer vision, defense, medicine, natural language processing, robotics, and smart transportation, data-driven AI has also raised many concerns about robustness, vulnerability to adversarial attacks, and bias. In this chapter, we discuss the challenges posed by these issues and present a summary of recent research efforts that address them.

tions. In the case of images or videos, domain shift could arise due to changes in object pose, scene illumination, the type of sensors used to acquire the images, videos, day/night operation, etc. It has been known for almost a decade [100] that DL models can perform poorly, even when very small amounts of perturbations are added to the image, leading to misclassification. The literature is awash with numerous attacks and the consequences of such attacks. It has been shown that DL models, especially those designed for face recognition, exhibit different levels of performance for different subgroups of populations, leading to concerns such as bias or a lack of fairness [8]. In this chapter, we discuss the aforementioned three issues and present a summary of ongoing research efforts that address them.

1.1 Introduction

In recent years, data-driven artificial intelligence (AI) systems have been utilized extensively in a variety of applications. With their exceptional performance over the past decade, deep learning (DL) models have revolutionized the machine learning and AI fields. DL models have outperformed nearly all classical machine learning models, such as SVMs, naive Bayes classifier, k-means clustering, and nearest neighbor, in many applications, including computer vision [109], natural language processing [18], and speech processing [35], by a significant margin. In particular, with their outstanding capacity for feature extraction and generalizations, DL models have made significant advances in a wide range of computer vision problems, such as object detection [26, 62, 85], object tracking [111], image classification [40, 50], action recognition [97], image captioning [108], human pose estimation [76, 103], face recognition [20, 83], and semantic segmentation [11, 12, 127].

Despite their successes, DL models for AI have many weaknesses. The dominant issues that could affect the progress of data-driven AI include domain shift, lack of robustness to adversarial attacks, and bias in decision-making. Domain shift [33, 34, 82] refers to the situation when training and test data come from different distribu-

1.2 Domain Shift and Mitigation

With the availability of web-scale data obtained from different devices and varied acquisition conditions, we are often faced with scenarios where the data used to train a classifier have some properties that are different from what is presented during testing. Instances like this arise very naturally in object recognition applications where the training and testing data are captured under different lighting conditions, in speech processing where a speaker model trained in a noise-free indoor setting needs to be deployed in more realistic outdoor environments, or in multimedia indexing where tagged Flickr photos or YouTube videos are readily available using which a user would want to automatically index his/her own photo/video collection captured with a consumer camera. Domain adaptation (DA) refers to the class of techniques aimed to learn representations from a low-resource dataset by transferring knowledge from a related, yet shifted, resource-rich dataset. Shifts between datasets can exist in the form of changes in illumination, lighting, texture patterns, or any such biases inherent to the dataset. To transfer knowledge across domain-shifted

DOI: 10.1201/9781003328957-1

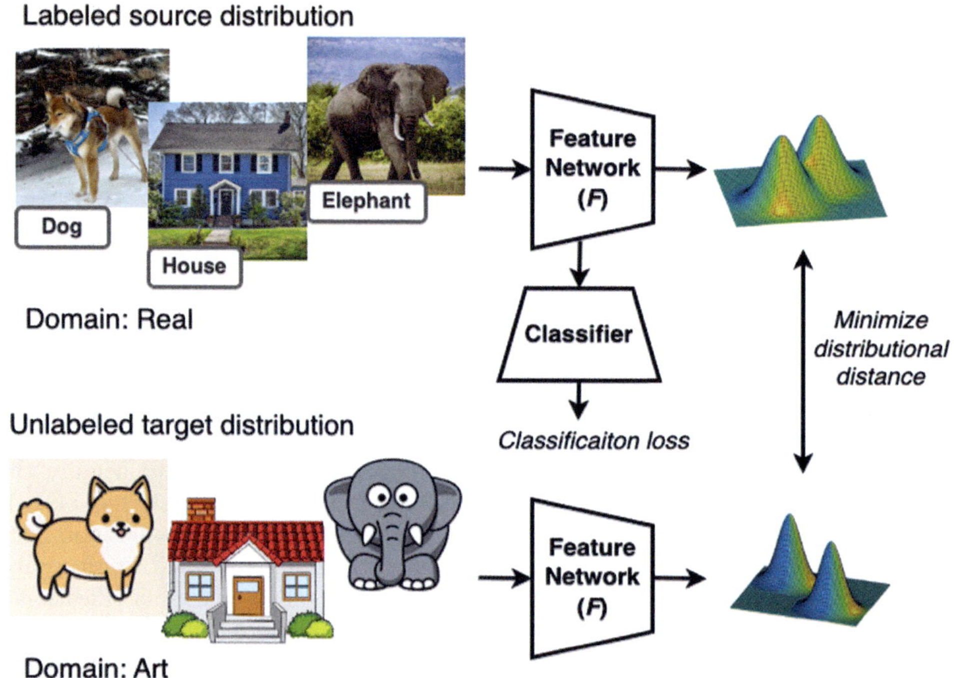

FIGURE 1.1
Domain adaptation using deep networks. Source and target feature spaces are aligned using a distributional distance minimization objective.

data distributions, a domain-invariant feature representation is sought that effectively phases out biases contributing to the domain shift.

There are two broad categories of DA techniques depending on whether the target domain test data either have partial labels (semi-supervised) or are completely unlabeled (unsupervised). While semi-supervised DA often uses correspondence from labeled target data to learn domain transformation [19, 88], unsupervised DA uses strategies that assume (1) a certain class of transformations between the domains [110], (2) the availability of discriminative features that are common to or invariant across both domains (namely "domain invariants") [3, 72], or (3) a latent space where the difference in distribution of source and target data is minimal [4]. In addition to adapting between a single source and a single target domain, there have been studies on multidomain adaptation (e.g., [71]) that consider more than one domain in source and/or target. While some of these approaches pursue "representation adaptation" by learning a domain shifting transformation, others [24] advocate a "classifier"-oriented approach that attempts to obtain target classifiers by manipulating or re-optimizing classifiers trained on the source domain. In the rest of this chapter, we focus mainly on the unsupervised setting.

Since 2011, extensive research has been done to address the challenge of unsupervised DA. Methods based on differential geometry [29, 33, 34, 43], sparse dictionaries [67, 77, 78, 95, 120], and, more recently, Generative Adversarial Networks (GANs) [92, 93] have been suggested. For more detailed expositions of DA research, the reader is referred to [82].

The framework for deep DA is shown in Figure 1.1. The source and the target feature distributions are first obtained by passing the corresponding images through a feature network F. Their feature distributions look dissimilar due to the domain shift. During DA, distributional distance between these feature spaces is minimized, while simultaneously training a classifier model on the labeled source data.

Several discriminative and generative approaches have been proposed for the DA problem [25, 44, 65, 66, 92]. In discriminative approaches, an additional loss function is typically used along with the classification objective to prevent the feature space drift. These functions typically take the form of a distributional distance minimization between source and target feature spaces [25, 65, 66]. In generative approaches, a generative model (typically a GAN) is trained to model the distribution of source and target images. This knowledge about the learned

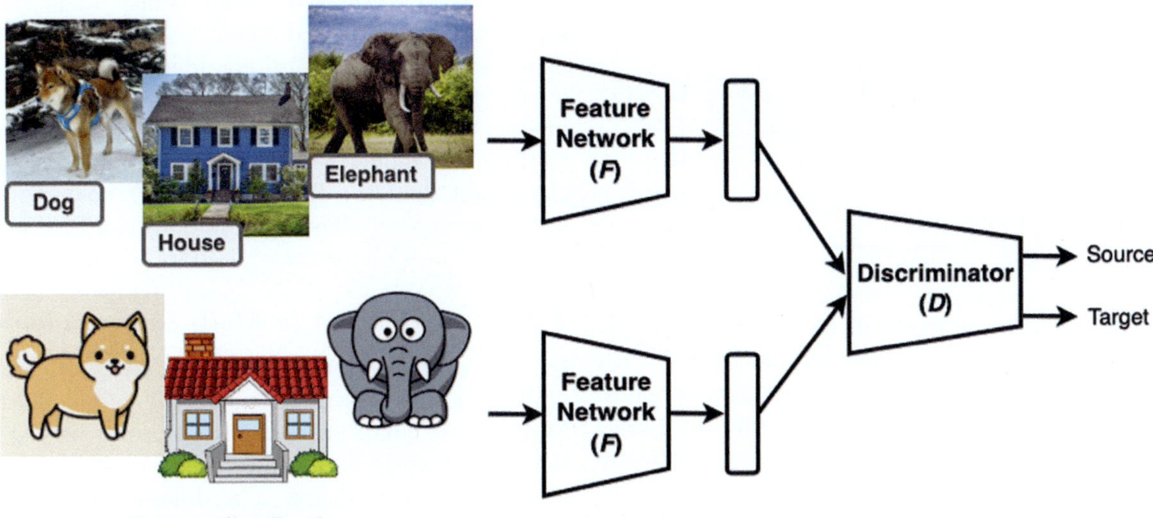

FIGURE 1.2
Domain adversarial training. The discriminator network predicts if the features come from the source or the target domain. The feature network is then trained adversarially to maximize the discriminator's loss. The training will converge when both source and target feature distributions are indistinguishable.

distributions is then used to induce domain invariance during training [44, 92]. A framework for such Domain Adversarial Training is shown inf Figure 1.2.

1.2.1 Discriminative Approaches for Domain Adaptation

Let F denote a deep neural network for extracting feature representations from images. The network F is typically a convolutional network that acts on input images and returns a vector as output. Let C denote a classification network that takes the feature representation as input and returns the classification logits. Let (x^s, y^s) denote input-output pairs of the source domain and (x^t) denote the target inputs. To train a classifier model in the source domain, we minimize a cross-entropy loss given by

$$\mathcal{L}_{cls} = \mathbf{E}[-y^s log(C(F(x^s)))] \qquad (1.1)$$

The resulting classifier would perform poorly on target domain due to the domain shift. The feature space drift is minimized using a domain adversarial loss that measures how different the source and target feature representations are. To implement the domain adversarial loss, we utilize an additional network called a discriminator (D), as shown in Figure 1.2. The inputs to the discriminator network are the feature representations obtained from the F network and the discriminator network predicts if the features come from the source or the target domain.

If there is sufficient mismatch between source and target features, the discriminator will obtain low loss value. The feature network is then trained adversarially to maximize the discriminator's loss. The training will converge when the discriminator fails at its task, that is, when both source and target feature distributions are indistinguishable (Figure 1.2). To accomplish the above objective, the discriminator network maximizes the following loss function:

$$\mathcal{L}_{disc} = log[D(F(x^s))] + log[1 - D(F(x^t))] \qquad (1.2)$$

During training, the models are optimized using a combination of classification loss and domain adversarial loss:

$$\min_{F,C} \max_{D} [\mathcal{L}_{cls} + \lambda\mathcal{L}_{disc}] \qquad (1.3)$$

The term λ controls the weight given to the domain adversarial term in the objective. It is a hyper-parameter that needs to be tuned while training the algorithm. A higher value of λ encourages the network to produce indistinguishable feature distributions of the source and target domains but may result in lower classification accuracy. The resulting algorithm is called domain adversarial training [25]. The networks F, C, and D are typically implemented as neural networks. In particular, the F network is a deep convolutional network (e.g., Resnet), and C and D networks are multilayer perceptrons. In [25], the adversarial loss is implemented using a gradient

TABLE 1.1
Domain adaptation performance (in %) on Office-31 dataset. A: Amazon, D: DSLR, W: Webcam [25]

Method	A→W	D→W	W→D
Source only	64.2	96.1	97.8
DDC [106]	61.8	95.0	98.5
DAN [64]	68.5	96.0	99.0
Domain Adversarial [25]	73.0	96.4	99.2

reversal layer in which the gradients coming from the discriminator network with $\mathcal{L}_{disc}$ loss are modulated with a $-\lambda$ factor before updating the feature network. Note that the feature network is also updated using the classification loss. All networks are optimized using stochastic gradient descent.

The performance of domain adversarial adaptation in comparison with other deep adaptation methods is shown in Table 1.1 for the Office-31 dataset, which has three domains–Amazon (A), DSLR (D), and Webcam (W). The task is to perform classification of 31 object categories. In Table 1.1, we observe that domain adversarial adaptation achieves significant gains in performance compared to source-only baseline, that is, baseline models which are trained only on source domain without any adaptation. Additionally, the model also performs better than Deep Domain Confusion (DDC) [106] and Deep Adaptation Networks (DAN) [64], the two other discriminative adaptation techniques, as shown in Table 1.1.

1.2.1.1 *Other Distance Measures*

While domain adversarial training is a popular choice for discriminative adaptation, other distance measures can also be used for measuring distributional mismatch between the source and the target distributions. Two such distance measures are Maximum Mean Discrepancy (MMD) and Wasserstein distance. In MMD, distributional distance is computed as distances between mean embeddings represented using a kernel from a Reproducing Kernel Hilbert Space (RKHS). Adaptation is then performed by minimizing the MMD between source and target feature distributions [64, 65]. In Wasserstein-based adaptation, Wasserstein distance between source and target feature distributions is used as a choice of distance measure. Dual form of Wasserstein distance is computed using a discriminator function. Adaptation is performed by minimizing the dual form of Wasserstein distance between source and target feature maps [96]. These approaches perform on par with adversarial adaptation.

1.2.1.2 *Pseudo-Labeling Approaches*

Different from adversarial approaches, in pseudo-labeling/self-training [90, 129, 130], pseudo-labels are generated on the unlabeled target samples using the current estimate of the model. Since pseudo-labels are typically noisy, most confident pseudo-labels are picked using some measures of confidence scores. One such measure is consistency across an ensemble of classifier models [90]. The mined pseudo-labels are then used to re-train the model to improve the predictive power on the target domain. This process is repeated iteratively until convergence. These techniques can also be used in combination with adversarial adaptation.

1.2.1.3 *Extension to Multi source Adaptation*

In multi source adaptation, the objective is to adapt from multiple source distributions to the target distribution. Multi source adaptation is a very useful setting in practice as real-world datasets typically contain a mixture of multiple latent domains. The domain adversarial training objective can be extended to the multi source setting using a k-way domain classifier as done in [121]. In [122], a curriculum-based weighting mechanism is used for selecting the best source domain samples to adapt for the given target domain. In cases where domain labels are available, latent domain discovery can be used for mining the unknown domain labels for multi-way adversarial adaptation [70].

1.2.2 **Generative Approaches for Domain Adaptation**

In generative approaches, the objective is to use generative models to estimate source and target distributions. The learned generative models are then used in the adaptation process to learn domain-invariant representations. Deep neural networks can be a popular choice for learning the generative models due to their high expressive power. Three popular choices of deep generative models are GANs [31], Variational Autoencoders (VAEs) [49], and normalizing flows [79]. GANs have been extensively used for DA as they have been very successful in generating high-fidelity samples. Training GANs is extremely challenging due to the min-max nature of the training objective. Several stabilization tricks are used in practice to make the GANs converge to good solutions [60]. Some of these tricks include use of Wasserstein [2] or hinge loss–based objective [74], using spectral normalization in network architecture [74], regularization techniques such as weight decay [60], gradient penalty [37], and feature matching losses [60].

1.2.2.1 *Distributional Distance Minimization with Generative Models*

An alternative approach to generative DA is by utilizing GANs for guiding distributional distance minimization. Recall that DA is posed as a distributional distance

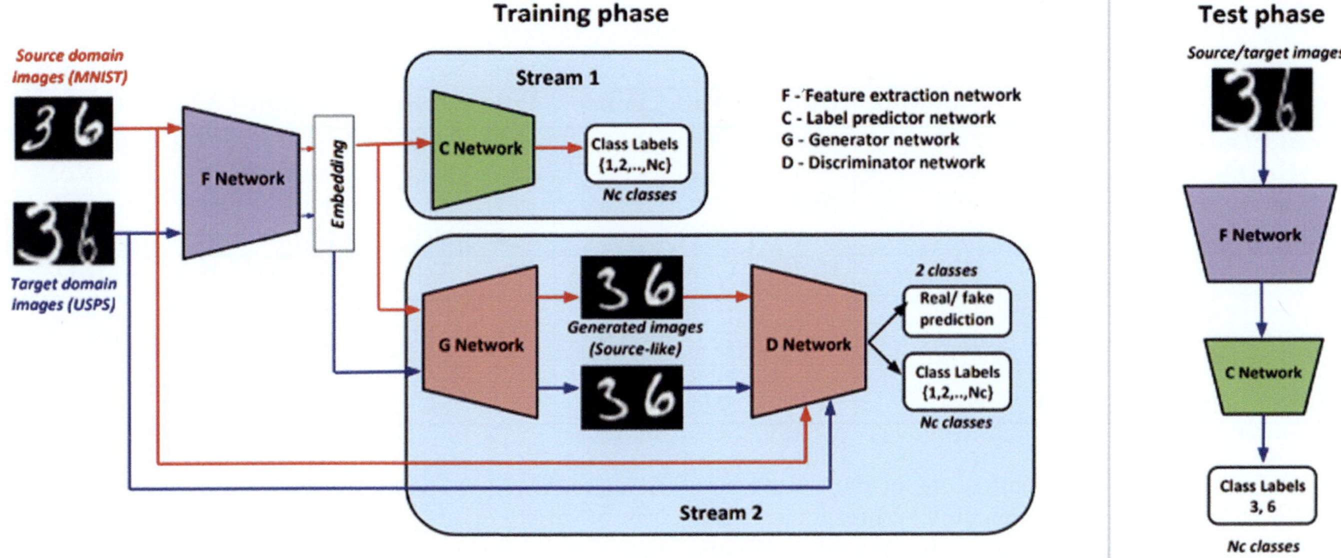

FIGURE 1.3
Framework of Generate to Adapt.

minimization problem between source and target feature distributions. Instead of directly performing distance minimization in the feature space, the idea is to project the feature back into pixel space using GANs and perform distributional distance minimization in this projected image space. This is referred to as "Generate to Adapt" [92].

Projecting features back into image space has two main benefits: First, the capacity of the discriminator network is effectively increased due to the projection step. Second, the projection step helps preserve semantic content in the generated features. That is, it prevents target features from one class to be mapped to a different class as the reconstruction network provides a supervisory signal in preserving the semantic content.

The framework of Generate to Adapt is shown in Figure 1.3. Features of source and target images are extracted using a feature network F. The obtained features are then passed through two streams: The first stream is a classification branch that is trained using cross-entropy loss on the source domain. The second stream is the distance minimization branch. In this stream, the source and the target features are first inverted back into image space using a generator network G. The generated samples are then passed through a discriminator network that discriminates if the samples come from real (source) or fake (target) domain. In addition, it also performs class label prediction on the source reconstructions using source labels as ground truth. This helps obtain class consistency in the reconstructed samples.

The feature network F is then trained adversarially to make the target reconstructions look like the source. This happens only when the source and target feature

TABLE 1.2
Domain adaptation performance using generative approaches on Digits dataset. Classification accuracy in % is reported. MN: MNIST, US: USPS, SV: SVHN [92]

Method	MN→US	US→MN	SV→MN
Source only	79.1	57.1	60.3
CoGAN [61]	61.8	95.0	98.5
PixelDA [6]	73.0	96.4	99.2
CyCADA [44]	95.6	96.5	90.4
Generate to Adapt [92]	92.8	95.3	92.4

distributions overlap. Additionally, the signal obtained from stream 1 helps features obtain class-consistent predictions.

1.2.2.2   Results

In Table 1.2, we show the results of generative approaches on cross-domain classification tasks using digits dataset. Three datasets are used for this purpose: MNIST, USPS, and SVHN. In each of the adaptation setting, we observe that source-only baseline achieves low performance. Both CyCADA [44] and Generate to Adapt [92] achieve significant performance gains compared to the baseline model. In addition, they also outperform CoGAN [61] and PixelDA [6], two other GAN-based adaptation approaches.

In Table 1.3, we report the performance on cross-domain semantic segmentation task. The source domain is GTA-5, which is a synthetic dataset of street scenes, while the target domain is the real dataset–Cityscapes.

TABLE 1.3

Semantic segmentation performance on GTA-5 $\rightarrow$ Cityscapes adaptation. IoU scores for each category and mean IoU scores (mIoU) are reported [93]

Method	Road	Sidewalk	Bldg	Wall	Fence	Pole	T light	T sign	Veg	Terrain
Source only	73.5	21.3	72.3	18.9	14.3	12.5	15.1	5.3	77.2	17.4
CyCADA	79.1	33.1	77.9	23.4	17.3	32.1	33.3	31.8	81.5	26.7
Generate to Adapt	88	30.5	78.6	25.2	23.5	16.7	23.5	11.6	78.7	27.2
Target only	96.5	74.6	86.1	37.1	33.2	30.2	39.7	51.6	87.3	52.6

Method	Sky	Person	Rider	Car	Truck	Bus	Train	Mbike	Bike	mIoU
Source only	64.3	43.7	12.8	75.4	24.8	7.8	0..0	4.9	1.8	29.6
CyCADA	69	62.8	14.7	74.5	20.9	25.6	6.9	18.8	20.4	39.5
Generate to Adapt	71.9	51.3	19.5	80.4	19.8	18.3	0.9	20.8	18.4	37.1
Target only	90.4	60.1	31.7	88.4	54.9	52.3	34.7	33.6	59.1	57.6

We use VGG-based FCN architecture for the feature network [92]. Intersection of Union (IoU) scores for each semantic class along with the mean IoU (mIoU) scores are reported. We observe that both CyCADA and Generate to Adapt achieve significant IoU gains compared to the source-only baseline model. However, there is a huge gap compared to the target-only model, which is the oracle model trained using true target labels.

1.3 Adversarial Attacks and Defenses

While DL models have shown promising performance on various computer vision tasks, maliciously crafted adversarial perturbations can severely degrade their performance. This serious security threat can easily fool almost any type of DL model by deliberately adding imperceptible perturbations to the inputs [87]. Consequently, this phenomenon, known as adversarial attacks, is regarded as a significant barrier to the widespread deployment of DL models for security-critical and large-scale systems. There are three categories of adversarial attacks: white-box attacks, gray-box attacks, and black-box attacks. White-box attacks occur when the attacker has complete access to the model's architecture, parameters, and training and testing routines. In the gray-box scenario, the attacker's knowledge is restricted to the model architecture and perhaps training and testing routines. The type of adversarial attack that is the most stringent is the black-box attack, in which the attacker has no knowledge of the target model. In recent years, a variety of methods have been proposed to undermine the performance of models, including Limited-memory Broyden–Fletcher–Goldfarb–Shanno (L-BFGS) [100], Fast Gradient Sign Method (FGSM) [32], Projected Gradient Descent (PGD) [68], Deep Fool (DF) [75], Carlini-Wagner (CW) [10], Jacobian-based Saliency Map

Attack (JSMA) [80], and Adversarial Patch [7, 48]. Moreover, a variety of adversarial attacks, particularly Elastic, Fog, Gabor, Snow, and JPEG, have been recently presented as differentiable attacks against deep models [46]. The details on the various adversarial attacks mentioned above can be found in the respective references. Figure 1.4 depicts some of the most popular adversarial examples against the ResNet-34 [40] network. It can be observed that the adversarial perturbations added to the images are not perceptible to human eyes.

In parallel, numerous defense strategies have been proposed to make DL models secure and resilient against these attacks. In this section, we introduce some recent research directions for adversarial defenses, including (1) on-manifold robustness [57, 91], which is about the adversarial samples perturbed within the latent space of some generators; (2) knowledge distillation-based defenses, including Mutual Adversarial Training (MAT) [58], which allows models to share their knowledge of adversarial robustness and teach each other to be more robust; and (3) defenses for object detectors, including Segment and Complete (SAC) [59], which is a general framework for defending object detectors against patch attacks through detection and removal of adversarial patches.

1.3.1 On-Manifold Robustness

Existing defenses consider properties of the trained classifier while ignoring the particular structure of the underlying image distribution. Recent advances in GANs and VAEs are shown to be successful in characterizing the underlying image manifold.[1]

Defense-GAN [91] is one of the early adversarial defense methods that leverages on-manifold images. Suppose we have a pre-trained GAN model G, which is trained with natural images. Then intuitively the manifold distribution from this GAN should be similar to the natural image distribution. To defend, Defense-GAN tries to find the most similar image from GAN to replace the perturbed

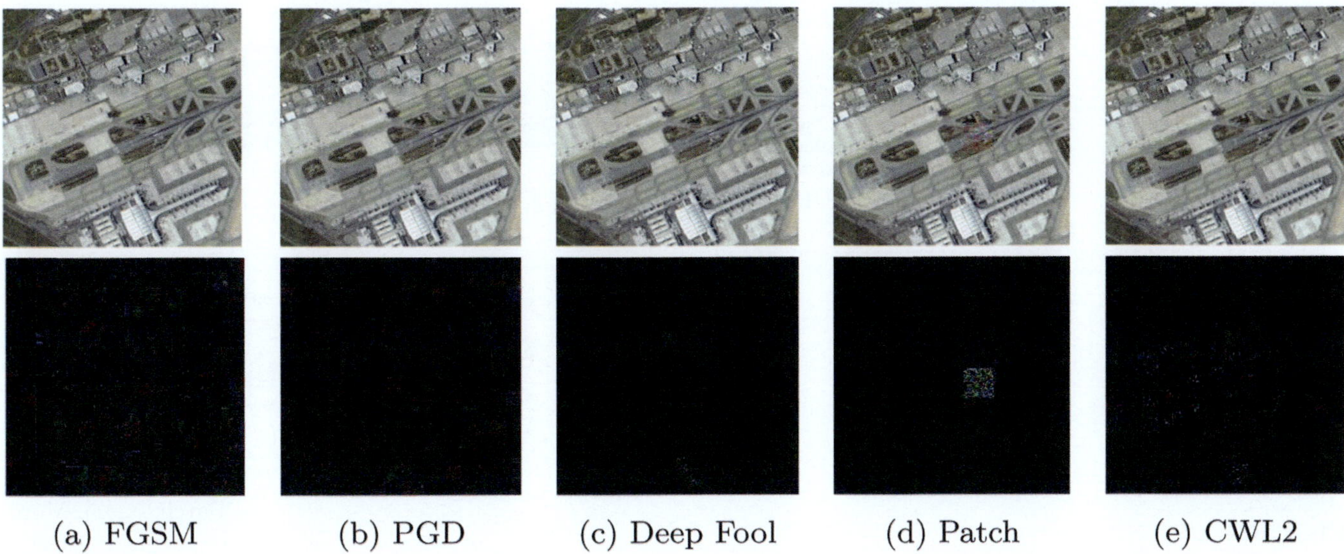

(a) FGSM (b) PGD (c) Deep Fool (d) Patch (e) CWL2

FIGURE 1.4
Adversarial samples (first row) and their respective relatively sparse perturbations (second row), generated based on a ResNet-34 network trained on the RESISC45 dataset (with courtesy from [15]) (a) Fast Gradient Sign Method (FGSM), (b)Projected Gradient Descent (PGD), (c) Deep Fool (DF), (d) Adversarial Patch, and (e) Carlini-Wagner (CW). Note that pixel values of perturbations are multiplied by a factor of 10 for the purpose of visualization.

image. Since the GAN model ideally has the natural image distribution, the image from GAN is unattacked. We call this process GAN projection. Therefore, this projected image is now used to be passed to the classifier, which can classify it correctly.

Lin *et al.* [57] investigate whether or not leveraging the underlying manifold information can boost model robustness, in particular against novel adversarial attacks. They consider the scenario when the manifold information of the underlying data is available. First, an "On-Manifold ImageNet" (OM-ImageNet) dataset where all the samples lie *exactly* on a low-dimensional manifold is constructed. This is achieved by first training a StyleGAN on a subset of ImageNet natural images, and then projecting the samples onto the learned manifold. With this dataset, they show that on-manifold adversarial training (AT) (i.e., AT in the latent space) could *not* defend against standard off-manifold attacks and vice versa. They proposed **D**ual **M**anifold **A**dversarial **T**raining (**DMAT**), which mixes both off-manifold and on-manifold AT (see Figure 1.5). AT in the image space (i.e., off-manifold AT) helps improve the robustness of the model against L_p attacks, while AT in the latent space (i.e., on-manifold AT) boosts the robustness of the model against unseen non-L_p attacks.

1.3.2 Knowledge Distillation based Defenses

AT [68] is considered to be one of the most effective algorithms for adversarial defenses. There have been many works for improving AT by using different loss functions, such as ALP [47], TRADES [125], and MART [114]. However, they only train one model without considering the synergy of a network cohort. In addition, most defense methods focus on a single perturbation type, which can be vulnerable to unseen perturbation types [69, 104]. For example, models adversarially trained on l_∞-bounded adversarial examples make them vulnerable to l_1 and l_2-bounded attacks.

Knowledge distillation (KD) [42] is a well-known method for transferring knowledge learned by one model to another. There are many forms of KD [112] including *offline* KD, where the teacher models are pretrained and the students learns from static teachers [42], and *online* KD, where a group of student models learn from peers' predictions [38, 99, 126]. Table 1.4 summarizes the differences among KD-based defenses. Defensive distillation (DD) trains a natural model, that is, a model trained with natural examples, with another natural model as the teacher, which cannot provide strong robustness as both the teacher and student are not adversarially trained; DD was found to be vulnerable, as demonstrated by Carlini and Wagner [9]. Adversarially robust distillation (ARD) [28] trains a robust model, that is, model trained with adversarial examples, with another robust model as the teacher to distill the robustness of a large network onto a smaller student. This strategy relies on the existence of strong teacher models, and the improvement of the student model is limited as the teacher is static. Adversarial concurrent training (ACT) [1] trains a natural model and a robust model jointly in an online KD manner to align the feature space of both. However, since

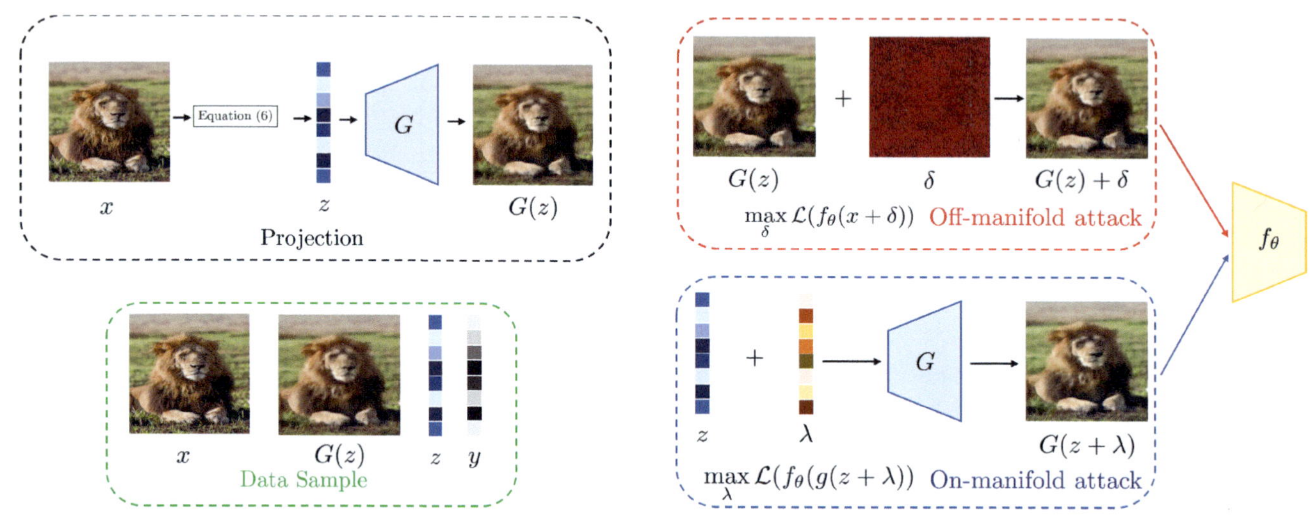

FIGURE 1.5
The overall pipeline of the proposed Dual Manifold Adversarial Training (DMAT) (with courtesy from [56].) DMAT considers the scenario when the information about the image manifold is available. This is achieved by projecting natural images x onto the range space of a trained generative model G. It is showed that either standard adversarial training or on-manifold adversarial training alone does not provide sufficient robustness, while DMAT achieves improved robustness against unseen attacks. During test time, images are directly passed to the adversarially trained classifier.

TABLE 1.4
Comparisons of KD-based defenses. "Robust" means the model is trained with adversarial examples, and "Natural" means the model is trained with natural examples.

Method	Teacher Model	Student Model	Form of KD	Multi-Perturbations
DD [81]	Natural	Natural	Offline	✗
ARD [28]	Robust	Robust	Offline	✗
ACT [1]	Natural	Robust	Online	✗
MAT [58]	Robust	Robust	Online	✓

natural and robust models learn fundamentally different features [105], aligning the feature space of a robust model with a natural model may result in degraded robustness. Liu *et al.* [58] proposed MAT that allows models to share their knowledge of adversarial robustness and teach each other to be more robust. Unlike previous KD-based defenses, MAT trains a group of robust models simultaneously, and each network not only learns from ground-truth labels as in standard AT, but also the soft labels from peer networks that encode peers' knowledge for defending against adversarial attacks to achieve stronger robustness.

1.3.3   Defenses for Object Detectors

Object detection is an important computer vision task that plays a key role in many security-critical systems including autonomous driving, security surveillance, including autonomous driving, security surveillance, identity verification, and robot manufacturing [107]. Despite the abundance [14, 51, 52, 55, 63, 94, 98, 102, 113, 117, 128] of adversarial patch attacks on object detectors, defenses against such attacks have not been extensively studied. Most existing defenses for patch attacks are restricted to image classification [27, 39, 53, 73, 84, 115, 118, 123]. Securing object detectors is more challenging due to the complexity of the task. Most existing defenses focus on global perturbations with the l_p norm constraint [13, 17, 124] and only a few defenses [16, 17, 45, 89, 119] for patch attacks have been proposed. Saha *et al.* [89] proposed Grad-defense and Out-of-Context defense for defending blindness attacks where the detector is blind to a specific object category chosen by the adversary. Ji *et al.* [45] proposed Ad-YOLO to defend against human detection patch attacks by adding a patch class on a YOLOv2 [86] detector such that it detects both the objects of interest and adversarial patches.

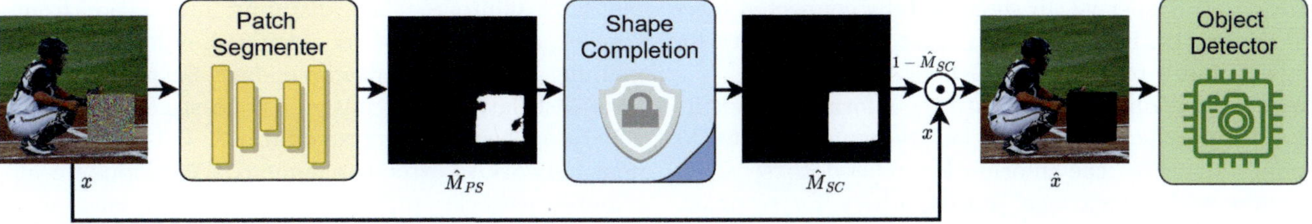

FIGURE 1.6
Pipeline of the SAC approach. SAC detects and removes adversarial patches on pixel-level through patch segmentation and shape completion and feeds the masked images into the base object detector for prediction.

DetectorGuard (DG) [119] is a provable defense against localized patch hiding attacks. These methods are designed for a specific type of patch attack or object detector.

SAC [59] is a general defense that can robustify any object detector against patch attacks without re-training the object detectors. SAC adopts a "detect and remove" strategy: It detects adversarial patches, removes the area from input images, and then feeds the masked images into the object detector. SAC first trains a patch segmenter to segment adversarial patches from the inputs and produce an initial patch mask. It uses a self-AT algorithm to enhance the robustness of the patch segmenter, which is efficient and object-detector agnostic. Since the attackers can potentially attack the segmenter and disturb its outputs under adaptive attacks, SAC further utilizes a robust shape completion algorithm that exploits the patch shape prior to ensure robust detection of adversarial patches. Shape completion takes the initial patch mask and generates a "completed patch mask" that is *guaranteed* to cover the entire adversarial patch, given that the initial patch mask is within a certain Hamming distance from the ground-truth patch mask. The overall pipeline of SAC is shown in Figure 1.6.

1.4 Bias Detection and Mitigation

Encoding of protected attributes such as gender or race in face descriptors results in bias with respect to these attributes when used for face recognition. A recent study from the National Institute of Standards and Technology [36] found evidence that characteristics such as gender and ethnicity impact verification and matching performance of face descriptors. Similarly, it has been shown that most face-based gender classifiers perform significantly better on male faces with light skintone than female faces with dark skintone [8]. One method of addressing privacy and bias issues is by producing face descriptors that are independent of the protected attribute(s). For instance, Debface proposed in [30] is an end-to-end method for producing face descriptors that are disentangled from protected attributes using an adversarial approach. Another common strategy for mitigating bias is to train face recognition systems using training datasets that are balanced in terms of sensitive attributes. However, building large datasets that are balanced in terms of the attributes we want to protect is difficult, expensive, and time-consuming. Moreover, once such a "fair" dataset is constructed, we still need to perform the costly operation of training a large recognition network from scratch.

End-to-end training of a large-scale network requires access to a large dataset and computing power and is time-consuming. Application of adversarial losses while training (as done in [30]) also slows down the training process. Several works [5, 54, 116] show that reducing the information of sensitive attributes while training a network results in a drop in overall performance. Even if a new network is trained to generate attribute-agnostic face descriptors, we need to replace the existing network and re-compute the descriptors for all the identities by feeding in the respective face images.

While face recognition networks are solely trained to recognize identities, it has been shown in [41, 101] that they also contain face-related information, such as gender, age, and pose, even when the networks are not trained to learn these attributes. A recent work [21] explores how these attributes are encoded in the internal layers of these networks, thus making the representations from face recognition networks more understandable. Expressivity is a measure of how much a feature vector informs us about an attribute, where a feature vector can be from internal or final layers of a network. Expressivity is computed by a second neural network whose inputs are features and attributes. The output of the second neural network approximates the mutual information between feature vectors and an attribute. Dhar *et al.* [21] investigate the expressivity for two different deep convolutional neural network (DCNN) architectures: A Resnet-101 and an Inception Resnet v2. In the final fully connected layer of the networks, it appears the order of expressivity for facial attributes to be Age > Gender > Yaw.

Relating this phenomenon with the information bottleneck theory, one can infer that face recognition networks encode and utilize the information related to sensitive attributes such as sex, age, among others. for recognizing faces, while being trained for identity classification. Such encoding has two major issues: (1) It makes the face representations susceptible to privacy leakage and (2) it appears to contribute to bias in face recognition. For a face recognition network to be lawful and respectful to the Nation's values, it must not demonstrate high bias against any specific gender, skintone, or age group.

A recent work [22] presents an adversarial de-biasing approach called Protected Attribute Suppression System (PASS). By reducing the encoding of sensitive attributes in face descriptors, PASS helps to make face descriptors more secure and resilient to privacy breach, thus aligning them with the principles for trustworthy AI. Additionally, the removal of sensitive attributes in face descriptors leads to a reduction in bias with respect to these attributes. Existing bias mitigation approaches generally require end-to-end training and are unable to achieve high verification accuracy. PASS is a descriptor-based framework that can be trained on top of descriptors obtained from any previously trained high-performing network to classify identities and simultaneously reduce encoding of sensitive attributes. This eliminates the need for end-to-end training. As a component of PASS, a novel discriminator training strategy that discourages a network from encoding protected attribute information is presented. PASS is able to reduce gender and skintone information in descriptors from SOTA face recognition networks like Arc Face. As a result, PASS descriptors outperform existing baselines in reducing gender and skintone bias on the IJB-C dataset, while maintaining a high verification accuracy. In [22], a new metric to better characterize bias/performance tradeoffs in face recognition is also presented.

Although adversarial de-biasing algorithms such as PASS are efficient in reducing bias in face recognition, these methods also demonstrate a drop in face verification performance. Therefore, in [23], a non-adversarial technique to reduce bias that minimizes the drop in face verification performance is presented. The proposed method is based on the observation that the regions of the face that a network attends to vary by the category of a given sensitive attribute. This might contribute to bias. Guided by this intuition, a novel distillation-based approach called Distill and De-bias (D&D) is proposed to enforce a network to attend to similar face regions, irrespective of the attribute category. In D&D, a teacher network is trained on images from one category of an attribute, for exam-

ple, light skintone. Then, distilling information from the teacher, a student network is trained on images of the remaining category, for example, dark skintone. A feature-level distillation loss constrains the student network to generate teacher-like representations. This allows the student network to attend to similar face regions for all attribute categories and enables it to reduce bias.

A second distillation step on top of D&D, called D&D++, is also proposed. For the D&D++ network, the "un-biasedness" of the D&D network is distilled into a new student network, the D&D++ network. The new network is trained on all attribute categories, for example, both light and dark skintones. This helps to train a network that is less biased for an attribute, while obtaining higher face verification performance than D&D. It is shown that D&D++ outperforms the existing baselines in reducing gender and skintone bias on the IJB-C dataset, while obtaining higher face verification performance than existing adversarial de-biasing methods. The effectiveness of the proposed methods is evaluated on two state-of-the-art face recognition networks: Crystalface and ArcFace.

1.5 Summary

Despite the success of data-driven AI systems, they have many weaknesses. In this chapter, we discuss some dominant issues that could affect the progress of data-driven AI, including domain shift, lack of robustness to adversarial attacks, and bias in decision-making. We also present a summary of recent research efforts that aim to address them.

Note

1. The term "manifold" is used to refer to the existence of lower-dimensional representations for natural images. This is a commonly used term in the generative model area. However, this definition may not satisfy the exact condition of manifolds defined in mathematics.

References

[1] Elahe Arani, Fahad Sarfraz, and Bahram Zonooz. Adversarial concurrent training: optimizing robustness and accuracy trade-off of deep neural networks. *arXiv preprint arXiv:2008.07015*, 2020.

[2] Martin Arjovsky, Soumith Chintala, and Léon Bottou. Wasserstein generative adversarial networks. In Doina Precup and Yee Whye Teh, editors, *Proceedings of the 34th International Conference on Machine Learning*, volume 70 of *Proceedings of Machine Learning Research*, pages 214–223. PMLR, 06–11 Aug 2017.

[3] John Blitzer, Koby Crammer, Alex Kulesza, Fernando Pereira, and Jenn Wortman. Learning bounds for domain adaptation. In *Advances in Neural Information Processing Systems 21*, Cambridge, MA, 2008. MIT Press.

[4] John Blitzer, Dean Foster, and Sham Kakade. Domain adaptation with coupled subspaces. In *Conference on Artificial Intelligence and Statistics*, Fort Lauderdale, 2011.

[5] Blaž Bortolato, Marija Ivanovska, Peter Rot, Janez Križaj, Philipp Terhörst, Naser Damer, Peter Peer, and Vitomir Štruc. Learning privacy-enhancing face representations through feature disentanglement. In *2020 15th IEEE International Conference on Automatic Face and Gesture Recognition (FG 2020)*, pages 495–502, 2020.

[6] Konstantinos Bousmalis, Nathan Silberman, David Dohan, Dumitru Erhan, and Dilip Krishnan. Unsupervised pixel-level domain adaptation with generative adversarial networks. In *Proceedings of the IEEE Conference on Computer Vision and Pattern Recognition*, pages 3722–3731, 2017.

[7] Tom B Brown, Dandelion Mané, Aurko Roy, Martín Abadi, and Justin Gilmer. Adversarial patch. *arXiv preprint arXiv:1712.09665*, 2017.

[8] Joy Buolamwini and Timnit Gebru. Gender shades: Intersectional accuracy disparities in commercial gender classification. In Sorelle A. Friedler and Christo Wilson, editors, *Proceedings of the 1st Conference on Fairness, Accountability and Transparency*, volume 81 of *Proceedings of Machine Learning Research*, pages 77–91. PMLR, 23–24 Feb 2018.

[9] Nicholas Carlini and David Wagner. Defensive distillation is not robust to adversarial examples. *arXiv preprint arXiv:1607.04311*, 2016.

[10] Nicholas Carlini and David Wagner. Towards evaluating the robustness of neural networks. In *2017 IEEE Symposium on Security and Privacy (sp)*, pages 39–57. IEEE, 2017.

[11] Liang-Chieh Chen, George Papandreou, Iasonas Kokkinos, Kevin Murphy, and Alan L Yuille. Deeplab: Semantic image segmentation with deep convolutional nets, atrous convolution, and fully connected crfs. *IEEE Transactions on Pattern Analysis and Machine Intelligence*, 40(4):834–848, 2017.

[12] Liang-Chieh Chen, George Papandreou, Florian Schroff, and Hartwig Adam. Rethinking atrous convolution for semantic image segmentation. *arXiv preprint arXiv:1706.05587*, 2017.

[13] Pin-Chun Chen, Bo-Han Kung, and Jun-Cheng Chen. Class-aware robust adversarial training for object detection. *arXiv preprint arXiv:2103.16148*, 2021.

[14] Shang-Tse Chen, Cory Cornelius, Jason Martin, and Duen Horng Polo Chau. Shapeshifter: Robust physical adversarial attack on faster R-CNN object detector. In *Joint European Conference on Machine Learning and Knowledge Discovery in Databases*, pages 52–68. Springer, 2018.

[15] Gong Cheng, Junwei Han, and Xiaoqiang Lu. Remote sensing image scene classification: Benchmark and state of the art. *Proceedings of the IEEE*, 105(10):1865–1883, 2017.

[16] Ping-Han Chiang, Chi-Shen Chan, and Shan-Hung Wu. *Adversarial Pixel Masking: A Defense against Physical Attacks for Pre-Trained Object Detectors*, pages 1856–1865. Association for Computing Machinery, New York, NY, USA, 2021.

[17] Ping-yeh Chiang, Michael J Curry, Ahmed Abdelkader, Aounon Kumar, John Dickerson, and Tom Goldstein. Detection as regression: Certified object detection by median smoothing. *arXiv preprint arXiv:2007.03730*, 2020.

[18] Kyunghyun Cho, Bart Van Merriënboer, Caglar Gulcehre, Dzmitry Bahdanau, Fethi Bougares, Holger Schwenk, and Yoshua Bengio. Learning phrase representations using RNN encoder-decoder for statistical machine translation. *arXiv preprint arXiv:1406.1078*, 2014.

[19] Hal Daume III and Daniel Marcu. Domain adaptation for statistical classifiers. *Journal of Artificial Intelligence Research*, 26:101–126, 2006.

[20] Jiankang Deng, Jia Guo, Niannan Xue, and Stefanos Zafeiriou. Arcface: Additive angular margin loss for deep face recognition. In *Proceedings of the IEEE/CVF Conference on Computer Vision and Pattern Recognition*, pages 4690–4699, 2019.

[21] Prithviraj Dhar, Ankan Bansal, Carlos D Castillo, Joshua Gleason, P Jonathon Phillips, and Rama Chellappa. How are attributes expressed in face DCNNs? In *2020 15th IEEE International Conference on Automatic Face and Gesture Recognition (FG 2020)*, pages 85–92. IEEE, 2020.

[22] Prithviraj Dhar, Joshua Gleason, Aniket Roy, Carlos D Castillo, and Rama Chellappa. PASS: Protected Attribute Suppression System for mitigating bias in face recognition. In *Proceedings of the IEEE/CVF International Conference on Computer Vision*, pages 15087–15096, 2021.

[23] Prithviraj Dhar, Joshua Gleason, Aniket Roy, Carlos D Castillo, P Jonathon Phillips, and Rama Chellappa. Distill and de-bias: Mitigating bias in face recognition using knowledge distillation. *arXiv preprint arXiv:2112.09786*, 2021.

[24] Lixin Duan, Dong Xu, and Shih-Fu Chang. Exploiting web images for event recognition in consumer videos: A multiple source domain adaptation approach. In *2012 IEEE Conference on Computer Vision and Pattern Recognition*, pages 1338–1345, 2012.

[25] Yaroslav Ganin, Evgeniya Ustinova, Hana Ajakan, Pascal Germain, Hugo Larochelle, François Laviolette, Mario Marchand, and Victor Lempitsky. Domain-adversarial training of neural networks. *The Journal of Machine Learning Research*, 17(1):2096–2030, 2016.

[26] Ross Girshick, Jeff Donahue, Trevor Darrell, and Jitendra Malik. Rich feature hierarchies for accurate object detection and semantic segmentation. In *Proceedings of the IEEE conference on computer vision and pattern recognition*, pages 580–587, 2014.

[27] Thomas Gittings, Steve Schneider, and John Collomosse. Vax-a-net: Training-time defence against adversarial patch attacks. In *Proceedings of the Asian Conference on Computer Vision*, 2020.

[28] Micah Goldblum, Liam Fowl, Soheil Feizi, and Tom Goldstein. Adversarially robust distillation. *arXiv preprint arXiv:1905.09747*, 2019.

[29] Boqing Gong, Yuan Shi, Fei Sha, and Kristen Grauman. Geodesic flow kernel for unsupervised domain adaptation. In *2012 IEEE Conference on Computer Vision and Pattern Recognition*, pages 2066–2073, 2012.

[30] Sixue Gong, Xiaoming Liu, and Anil K Jain. Jointly de-biasing face recognition and demographic attribute estimation. In *European Conference on Computer Vision*, pages 330–347. Springer, 2020.

[31] Ian Goodfellow, Jean Pouget-Abadie, Mehdi Mirza, Bing Xu, David Warde-Farley, Sherjil Ozair, Aaron Courville, and Yoshua Bengio. Generative adversarial nets. In Z. Ghahramani, M. Welling, C. Cortes, N. Lawrence, and K.Q. Weinberger, editors, *Advances in Neural Information Processing Systems*, volume 27. Curran Associates, Inc., 2014.

[32] Ian J Goodfellow, Jonathon Shlens, and Christian Szegedy. Explaining and harnessing adversarial examples. *arXiv preprint arXiv:1412.6572*, 2014.

[33] Raghuraman Gopalan, Ruonan Li, and Rama Chellappa. Domain adaptation for object recognition: An unsupervised approach. In *2011 IEEE International Conference on Computer Vision*, pages 999–1006, 2011.

[34] Raghuraman Gopalan, Ruonan Li, and Rama Chellappa. Unsupervised adaptation across domain shifts by generating intermediate data representations. In *IEEE Transactions on Pattern Analysis and Machine Intelligence*, 36(11):2288–2302, 2013.

[35] Alex Graves and Navdeep Jaitly. Towards end-to-end speech recognition with recurrent neural networks. In *PMLR International Conference on Machine Learning*, pages 1764–1772, 2014.

[36] Patrick J Grother, Patrick J Grother, Mei Ngan, and K Hanaoka. *Face Recognition Vendor Test (FRVT)*. US Department of Commerce, National Institute of Standards and Technology, 2014.

[37] Ishaan Gulrajani, Faruk Ahmed, Martin Arjovsky, Vincent Dumoulin, and Aaron C Courville. Improved training of wasserstein gans. *Advances in Neural Information Processing Systems*, 30, 2017.

[38] Qiushan Guo, Xinjiang Wang, Yichao Wu, Zhipeng Yu, Ding Liang, Xiaolin Hu, and Ping Luo. Online knowledge distillation via collaborative learning. In *Proceedings of the IEEE Conference on Computer Vision and Pattern Recognition*, pages 11020–11029, 2020.

[39] Jamie Hayes. On visible adversarial perturbations and digital watermarking. In *Proceedings of the IEEE Conference on Computer Vision and Pattern Recognition Workshops*, pages 1597–1604, 2018.

[40] Kaiming He, Xiangyu Zhang, Shaoqing Ren, and Jian Sun. Deep residual learning for image recognition. In *Proceedings of the IEEE Conference on Computer Vision and Pattern Recognition*, pages 770–778, 2016.

[41] Matthew Q Hill, Connor J Parde, Carlos D Castillo, Y Ivette Colon, Rajeev Ranjan, Jun-Cheng Chen, Volker Blanz, and Alice J O'Toole. Deep convolutional neural networks in the face of caricature. *Nature Machine Intelligence*, 1(11):522–529, 2019.

[42] Geoffrey Hinton, Oriol Vinyals, and Jeff Dean. Distilling the knowledge in a neural network. *arXiv preprint arXiv:1503.02531*, 2015.

[43] Huy Tho Ho and Raghuraman Gopalan. Model-driven domain adaptation on product manifolds for unconstrained face recognition. *International Journal of Computer Vision*, 109(1):110–125, 2014.

[44] Judy Hoffman, Eric Tzeng, Taesung Park, Jun-Yan Zhu, Phillip Isola, Kate Saenko, Alexei Efros, and Trevor Darrell. CyCADA: Cycle-consistent adversarial domain adaptation. In Jennifer Dy and Andreas Krause, editors, *Proceedings of the 35th International Conference on Machine Learning*, volume 80 of *Proceedings of Machine Learning Research*, pages 1989–1998. PMLR, 10–15 Jul 2018.

[45] Nan Ji, YanFei Feng, Haidong Xie, Xueshuang Xiang, and Naijin Liu. Adversarial YOLO: Defense Human detection patch attacks via detecting adversarial patches. *arXiv preprint arXiv:2103.08860*, 2021.

[46] Daniel Kang, Yi Sun, Dan Hendrycks, Tom Brown, and Jacob Steinhardt. Testing robustness against unforeseen adversaries. *arXiv preprint arXiv:1908.08016*, 2019.

[47] Harini Kannan, A. Kurakin, and Ian J. Goodfellow. Adversarial logit pairing. *ArXiv*, abs/1803.06373, 2018.

[48] Danny Karmon, Daniel Zoran, and Yoav Goldberg. Lavan: Localized and visible adversarial noise. In *International Conference on Machine Learning*, pages 2507–2515. PMLR, 2018.

[49] Diederik P Kingma and Max Welling. Auto-encoding variational bayes. *arXiv preprint arXiv:1312.6114*, 2013.

[50] Alex Krizhevsky, Ilya Sutskever, and Geoffrey E Hinton. ImageNet classification with deep convolutional neural networks. In *Advances in Neural Information Processing Systems*, pages 1097–1105, 2012.

[51] Dapeng Lang, Deyun Chen, Ran Shi, and Yongjun He. Attention-guided digital adversarial patches on visual detection. *Security and Communication Networks*, 2021, 2021.

[52] Mark Lee and Zico Kolter. On physical adversarial patches for object detection. *arXiv preprint arXiv:1906.11897*, 2019.

[53] Alexander Levine and Soheil Feizi. (De) Randomized smoothing for certifiable defense against patch attacks. *arXiv preprint arXiv:2002.10733*, 2020.

[54] Ang Li, Jiayi Guo, Huanrui Yang, Flora D Salim, and Yiran Chen. Deepobfuscator: Obfuscating intermediate representations with privacy-preserving adversarial learning on smartphones. In *Proceedings of the International Conference on Internet-of-Things Design and Implementation*, pages 28–39, 2021.

[55] Yuezun Li, Xiao Bian, Ming-Ching Chang, and Siwei Lyu. Exploring the vulnerability of single shot module in object detectors via imperceptible background patches. *arXiv preprint arXiv:1809.05966*, 2018.

[56] Tsung-Yi Lin, Michael Maire, Serge Belongie, James Hays, Pietro Perona, Deva Ramanan, Piotr Dollár, and C Lawrence Zitnick. Microsoft COCO: Common objects in context. In *European Conference on Computer Vision*, pages 740–755. Springer, 2014.

[57] Wei-An Lin, Chun Pong Lau, Alexander Levine, Rama Chellappa, and Soheil Feizi. Dual Manifold Adversarial Robustness: Defense against Lp and non-Lp Adversarial Attacks. In *Advances in Neural Information Processing Systems*, 2020.

[58] Jiang Liu, Chun Pong Lau, Hossein Souri, Soheil Feizi, and Rama Chellappa. Mutual adversarial training: Learning together is better than going alone. *arXiv preprint arXiv:2112.05005*, 2021.

[59] Jiang Liu, Alexander Levine, Chun Pong Lau, Rama Chellappa, and Soheil Feizi. Segment and complete: Defending object detectors against adversarial patch attacks with robust patch detection. *arXiv preprint arXiv:2112.04532*, 2021.

[60] Ming-Yu Liu, Xun Huang, Jiahui Yu, Ting-Chun Wang, and Arun Mallya. Generative adversarial networks for image and video synthesis: Algorithms and applications. *arXiv preprint arXiv:2008.02793*, 2020.

[61] Ming-Yu Liu and Oncel Tuzel. Coupled generative adversarial networks. *Advances in Neural Information Processing Systems*, 29, 2016.

[62] Wei Liu, Dragomir Anguelov, Dumitru Erhan, Christian Szegedy, Scott Reed, Cheng-Yang Fu, and Alexander C Berg. SSD: Single shot multibox detector. In *European Conference on Computer Vision*, pages 21–37. Springer, 2016.

[63] Xin Liu, Huanrui Yang, Ziwei Liu, Linghao Song, Hai Li, and Yiran Chen. DPatch: An adversarial patch attack on object detectors. *arXiv preprint arXiv:1806.02299*, 2018.

[64] Mingsheng Long, Yue Cao, Jianmin Wang, and Michael Jordan. Learning transferable features with deep adaptation networks. In *PMLR International Conference on Machine Learning*, pages 97–105, 2015.

[65] Mingsheng Long, Han Zhu, Jianmin Wang, and Michael I Jordan. Unsupervised domain adaptation with residual transfer networks. *Advances in Neural Information Processing Systems*, 29, 2016.

[66] Mingsheng Long, Han Zhu, Jianmin Wang, and Michael I Jordan. Deep transfer learning with joint adaptation networks. In *PMLR International Conference on Machine Learning*, pages 2208–2217, 2017.

[67] Boyu Lu, Rama Chellappa, and Nasser M Nasrabadi. Incremental dictionary learning for unsupervised domain adaptation. In *BMVC*, pages 108.1–108.12, 2015.

[68] Aleksander Madry, Aleksandar Makelov, Ludwig Schmidt, Dimitris Tsipras, and Adrian Vladu. Towards deep learning models resistant to adversarial attacks. *arXiv preprint arXiv:1706.06083*, 2017.

[69] Pratyush Maini, Eric Wong, and Zico Kolter. Adversarial robustness against the union of multiple perturbation models. In *International Conference on Machine Learning*, pages 6640–6650. PMLR, 2020.

[70] Massimiliano Mancini, Lorenzo Porzi, Samuel Rota Bulo, Barbara Caputo, and Elisa Ricci. Boosting domain adaptation by discovering latent domains. In *Proceedings of the IEEE Conference on Computer Vision and Pattern Recognition*, pages 3771–3780, 2018.

[71] Yishay Mansour, Mehryar Mohri, and Afshin Rostamizadeh. Domain adaptation with multiple sources. *Advances in Neural Information Processing Systems*, 21, 2008.

[72] Yishay Mansour, Mehryar Mohri, and Afshin Rostamizadeh. Domain adaptation: Learning bounds and algorithms. *arXiv preprint arXiv:0902.3430*, 2009.

[73] Michael McCoyd, Won Park, Steven Chen, Neil Shah, Ryan Roggenkemper, Minjune Hwang, Jason Xinyu Liu, and David Wagner. Minority reports defense: Defending against adversarial patches. In Jianying Zhou, Mauro Conti, Chuadhry Mujeeb Ahmed, Man Ho Au, Lejla Batina, Zhou Li, Jingqiang Lin, Eleonora Losiouk, Bo Luo, Suryadipta Majumdar, Weizhi Meng, Martín Ochoa, Stjepan Picek, Georgios Portokalidis, Cong Wang, and Kehuan Zhang, editors, *Applied Cryptography and Network Security Workshops*, pages 564–582, Cham, 2020. Springer International Publishing.

[74] Takeru Miyato, Toshiki Kataoka, Masanori Koyama, and Yuichi Yoshida. Spectral normalization for generative adversarial networks. In *International Conference on Learning Representations*, 2018.

[75] Seyed-Mohsen Moosavi-Dezfooli, Alhussein Fawzi, and Pascal Frossard. Deepfool: A simple and accurate method to fool deep neural networks. In *Proceedings of the IEEE Conference on Computer Vision and Pattern Recognition*, pages 2574–2582, 2016.

[76] Alejandro Newell, Kaiyu Yang, and Jia Deng. Stacked hourglass networks for human pose estimation. In *European Conference on Computer Vision*, pages 483–499. Springer, 2016.

[77] Hien V Nguyen, Huy Tho Ho, Vishal M Patel, and Rama Chellappa. Dash-n: Joint hierarchical domain adaptation and feature learning. In *IEEE Transactions on Image Processing*, 24(12):5479–5491, 2015.

[78] Hien V Nguyen, Vishal M Patel, Nasser M Nasrabadi, and Rama Chellappa. Sparse embedding: A framework for sparsity promoting dimensionality reduction. In *European Conference on Computer Vision*, pages 414–427. Springer, 2012.

[79] George Papamakarios, Eric Nalisnick, D Jimenez Rezende, Shakir Mohamed, and Balaji Lakshminarayanan. Normalizing flows for probabilistic modeling and inference, arxiv e-prints. *arXiv preprint arXiv:1912.02762*, 2019.

[80] Nicolas Papernot, Patrick McDaniel, Somesh Jha, Matt Fredrikson, Z Berkay Celik, and Ananthram

Swami. The limitations of deep learning in adversarial settings. In *2016 IEEE European Symposium on Security and Privacy (EuroS&P)*, pages 372–387, 2016.

[81] Nicolas Papernot, Patrick McDaniel, Xi Wu, Somesh Jha, and Ananthram Swami. Distillation as a defense to adversarial perturbations against deep neural networks. In *2016 IEEE Symposium on Security and Privacy (SP)*, pages 582–597, 2016.

[82] Vishal M Patel, Raghuraman Gopalan, Ruonan Li, and Rama Chellappa. Visual domain adaptation: A survey of recent advances. *IEEE Signal Processing Magazine*, 32(3):53–69, 2015.

[83] Rajeev Ranjan, Vishal M Patel, and Rama Chellappa. Hyperface: A deep multi-task learning framework for face detection, landmark localization, pose estimation, and gender recognition. *IEEE Transactions on Pattern Analysis and Machine Intelligence*, 41(1):121–135, 2017.

[84] Sukrut Rao, David Stutz, and Bernt Schiele. Adversarial training against location-optimized adversarial patches. *arXiv preprint arXiv:2005.02313*, 2020.

[85] Joseph Redmon, Santosh Divvala, Ross Girshick, and Ali Farhadi. You only look once: Unified, real-time object detection. In *Proceedings of the IEEE Conference on Computer Vision and Pattern Recognition*, pages 779–788, 2016.

[86] Joseph Redmon and Ali Farhadi. YOLO9000: Better, faster, stronger. In *Proceedings of the IEEE Conference on Computer Vision and Pattern Recognition*, pages 7263–7271, 2017.

[87] Kui Ren, Tianhang Zheng, Zhan Qin, and Xue Liu. Adversarial attacks and defenses in deep learning. *Engineering*, 6(3):346–360, 2020.

[88] Kate Saenko, Brian Kulis, Mario Fritz, and Trevor Darrell. Adapting visual category models to new domains. In *European Conference on Computer Vision*, pages 213–226. Springer, 2010.

[89] Aniruddha Saha, Akshayvarun Subramanya, Koninika Patil, and Hamed Pirsiavash. Role of spatial context in adversarial robustness for object detection. In *Proceedings of the IEEE/CVF Conference on Computer Vision and Pattern Recognition Workshops*, pages 784–785, 2020.

[90] Kuniaki Saito, Yoshitaka Ushiku, and Tatsuya Harada. Asymmetric tri-training for unsupervised domain adaptation. In Doina Precup and Yee Whye Teh, editors, *Proceedings of the 34th International Conference on Machine Learning*, volume 70 of *Proceedings of Machine Learning Research*, pages 2988–2997. PMLR, 06–11 Aug 2017.

[91] Pouya Samangouei, Maya Kabkab, and Rama Chellappa. Defense-gan: Protecting classifiers against adversarial attacks using generative models. In *International Conference on Learning Representations*, 2018.

[92] Swami Sankaranarayanan, Yogesh Balaji, Carlos D Castillo, and Rama Chellappa. Generate to adapt: Aligning domains using generative adversarial networks. In *Proceedings of the IEEE Conference on Computer Vision and Pattern Recognition*, pages 8503–8512, 2018.

[93] Swami Sankaranarayanan, Yogesh Balaji, Arpit Jain, Ser Nam Lim, and Rama Chellappa. Learning from synthetic data: Addressing domain shift for semantic segmentation. In *Proceedings of the IEEE Conference on Computer Vision and Pattern Recognition*, pages 3752–3761, 2018.

[94] Mahmood Sharif, Sruti Bhagavatula, Lujo Bauer, and Michael K. Reiter. Accessorize to a crime: Real and stealthy attacks on state-of-the-art face recognition. CCS '16, page 1528–1540, New York, NY, USA, 2016. Association for Computing Machinery.

[95] Sumit Shekhar, Vishal M Patel, Hien V Nguyen, and Rama Chellappa. Generalized domain-adaptive dictionaries. In *2013 IEEE Conference on Computer Vision and Pattern Recognition*, pages 361–368, 2013.

[96] Jian Shen, Yanru Qu, Weinan Zhang, and Yong Yu. Wasserstein distance guided representation learning for domain adaptation. In *Proceedings of the AAAI Conference on Artificial Intelligence*, volume 32, 2018.

[97] Karen Simonyan and Andrew Zisserman. Two-stream convolutional networks for action recognition in videos. In *Advances in Neural Information Processing Systems*, pages 568–576, 2014.

[98] Dawn Song, Kevin Eykholt, Ivan Evtimov, Earlence Fernandes, Bo Li, Amir Rahmati, Florian Tramer, Atul Prakash, and Tadayoshi Kohno. Physical adversarial examples for object detectors. In *12th USENIX Workshop on Offensive Technologies (WOOT 18)*, 2018.

[99] Guocong Song and Wei Chai. Collaborative learning for deep neural networks. In *Advances in Neural*

Information Processing Systems, pages 1832–1841, 2018.

[100] Christian Szegedy, Wojciech Zaremba, Ilya Sutskever, Joan Bruna, Dumitru Erhan, Ian Goodfellow, and Rob Fergus. Intriguing properties of neural networks. *arXiv preprint arXiv: 1312.6199*, 2013.

[101] Philipp Terhörst, Daniel Fährmann, Naser Damer, Florian Kirchbuchner, and Arjan Kuijper. Beyond identity: What information is stored in biometric face templates? In *2020 IEEE International Joint Conference on Biometrics (IJCB)*, pages 1–10, 2020.

[102] Simen Thys, Wiebe Van Ranst, and Toon Goedemé. Fooling automated surveillance cameras: Adversarial patches to attack person detection. In *2019 IEEE/CVF Conference on Computer Vision and Pattern Recognition Workshops (CVPRW)*, pages 49–55, Los Alamitos, CA, USA, 6 2019. IEEE Computer Society.

[103] Alexander Toshev and Christian Szegedy. Deeppose: Human pose estimation via deep neural networks. In *Proceedings of the IEEE Conference on Computer Vision and Pattern Recognition*, pages 1653–1660, 2014.

[104] Florian Tramèr and Dan Boneh. Adversarial training and robustness for multiple perturbations. In *Advances in Neural Information Processing Systems*, pages 5866–5876, 2019.

[105] Dimitris Tsipras, Shibani Santurkar, Logan Engstrom, Alexander Turner, and Aleksander Madry. Robustness may be at odds with accuracy. In *International Conference on Learning Representations*, 2019.

[106] Eric Tzeng, Judy Hoffman, Ning Zhang, Kate Saenko, and Trevor Darrell. Deep domain confusion: Maximizing for domain invariance. *arXiv preprint arXiv:1412.3474*, 2014.

[107] Abdul Vahab, Maruti S Naik, Prasanna G Raikar, and Prasad SR. Applications of object detection system. *International Research Journal of Engineering and Technology (IRJET)*, 6(4):4186–4192, 2019.

[108] Oriol Vinyals, Alexander Toshev, Samy Bengio, and Dumitru Erhan. Show and tell: A neural image caption generator. In *Proceedings of the IEEE Conference on Computer Vision and Pattern Recognition*, pages 3156–3164, 2015.

[109] Athanasios Voulodimos, Nikolaos Doulamis, Anastasios Doulamis, and Eftychios Protopapadakis. Deep learning for computer vision: A brief review. *Computational Intelligence and Neuroscience*, 2018, 2018.

[110] Chang Wang and Sridhar Mahadevan. Manifold alignment without correspondence. In *IJCAI*, volume 2, page 3, 2009.

[111] Lijun Wang, Wanli Ouyang, Xiaogang Wang, and Huchuan Lu. Visual tracking with fully convolutional networks. In *Proceedings of the IEEE International Conference on Computer Vision*, pages 3119–3127, 2015.

[112] Lin Wang and Kuk-Jin Yoon. Knowledge distillation and student-teacher learning for visual intelligence: A review and new outlooks. *arXiv preprint arXiv:2004.05937*, 2020.

[113] Yajie Wang, Haoran Lv, Xiaohui Kuang, Gang Zhao, Yu-an Tan, Quanxin Zhang, and Jingjing Hu. Towards a physical-world adversarial patch for blinding object detection models. *Information Sciences*, 556:459–471, 2021.

[114] Yisen Wang, Difan Zou, Jinfeng Yi, James Bailey, Xingjun Ma, and Quanquan Gu. Improving adversarial robustness requires revisiting misclassified examples. In *International Conference on Learning Representations*, 2019.

[115] Tong Wu, Liang Tong, and Yevgeniy Vorobeychik. Defending against physically realizable attacks on image classification. *arXiv preprint arXiv:1909.09552*, 2019.

[116] Zhenyu Wu, Zhangyang Wang, Zhaowen Wang, and Hailin Jin. Towards privacy-preserving visual recognition via adversarial training: A pilot study. In *European Conference on Computer Vision*, pages 606–624, 2018.

[117] Zuxuan Wu, Ser-Nam Lim, Larry S Davis, and Tom Goldstein. Making an invisibility cloak: Real world adversarial attacks on object detectors. In *European Conference on Computer Vision*, pages 1–17. Springer, 2020.

[118] Chong Xiang, Arjun Nitin Bhagoji, Vikash Sehwag, and Prateek Mittal. Patchguard: A provably robust defense against adversarial patches via small receptive fields and masking. In *30th USENIX Security Symposium (USENIX Security)*, 2021.

[119] Chong Xiang and Prateek Mittal. Detectorguard: Provably securing object detectors against localized patch hiding attacks. *arXiv preprint arXiv:2102.02956*, 2021.

[120] Hongyu Xu, Jingjing Zheng, and Rama Chellappa. Bridging the domain shift by domain adaptive dictionary learning. In *BMVC*, pages 96.1–96.12, 2015.

[121] Ruijia Xu, Ziliang Chen, Wangmeng Zuo, Junjie Yan, and Liang Lin. Deep cocktail network: Multi-source unsupervised domain adaptation with category shift. In *Proceedings of the IEEE Conference on Computer Vision and Pattern Recognition*, pages 3964–3973, 2018.

[122] Luyu Yang, Yogesh Balaji, Ser-Nam Lim, and Abhinav Shrivastava. Curriculum manager for source selection in multi-source domain adaptation. In *European Conference on Computer Vision*, pages 608–624. Springer, 2020.

[123] Ping yeh Chiang, Renkun Ni, Ahmed Abdelkader, Chen Zhu, Christoph Studor, and Tom Goldstein. Certified defenses for adversarial patches. In *International Conference on Learning Representations*, 2020.

[124] Haichao Zhang and Jianyu Wang. Towards adversarially robust object detection. In *Proceedings of the IEEE/CVF International Conference on Computer Vision*, pages 421–430, 2019.

[125] Hongyang Zhang, Yaodong Yu, Jiantao Jiao, Eric Xing, Laurent El Ghaoui, and Michael Jordan. Theoretically principled trade-off between robustness and accuracy. In *International Conference on Machine Learning*, pages 7472–7482, 2019.

[126] Ying Zhang, Tao Xiang, Timothy M Hospedales, and Huchuan Lu. Deep mutual learning. In *Proceedings of the IEEE Conference on Computer Vision and Pattern Recognition*, pages 4320–4328, 2018.

[127] Hengshuang Zhao, Jianping Shi, Xiaojuan Qi, Xiaogang Wang, and Jiaya Jia. Pyramid scene parsing network. In *Proceedings of the IEEE Conference on Computer Vision and Pattern Recognition*, pages 2881–2890, 2017.

[128] Yusheng Zhao, Huanqian Yan, and Xingxing Wei. Object hider: Adversarial patch attack against object detectors. *arXiv preprint arXiv:2010.14974*, 2020.

[129] Yang Zou, Zhiding Yu, BVK Kumar, and Jinsong Wang. Unsupervised domain adaptation for semantic segmentation via class-balanced self-training. In *Proceedings of the European Conference on Computer Vision (ECCV)*, pages 289–305, 2018.

[130] Yang Zou, Zhiding Yu, Xiaofeng Liu, BVK Kumar, and Jinsong Wang. Confidence regularized self-training. In *Proceedings of the IEEE/CVF International Conference on Computer Vision*, pages 5982–5991, 2019.

2

Challenges of Computer Vision Research from an Industry Perspective

Fatih Porikli

2.1 Introduction

Computer vision aims at understanding the physical world through visual perception. It is based on processing, analyzing, and interpreting sensory information acquired by single or multiple cameras in the form of images and videos into consumable intelligence for humans and machines. This could be through detecting, tracking, classifying, recognizing, reconstructing, enhancing, encoding, and decoding the underlying and constituting patterns, motions, semantics, objects, events, and concepts in the surrounding space. As an interdisciplinary field, it often leverages other signals and modalities, including text, depth, wireless, haptic, sound, and speech.

Thanks to recent advances driven by machine learning, computer vision has been successfully applied to numerous industrial use cases, from autonomous driving to augmented reality, camera pipelines, visual commerce, robotics, visual surveillance, biometrics, remote sensing, computer graphics, human-computer interfaces, video compression, medical imaging, digital pathology, computer-assisted intervention, and many more.

Chances are already familiar with the above since you are reading this book. In this chapter, let's take a journey together as two computer vision researchers in an industrial setting, galvanized with excitement and prospects of creating a real impact and improving lives.

Imagine you just commenced your new computer vision expert role with a great research lab. Congratulations! You attended the orientation that talked about the company, its culture, code of conduct, missions, policies, procedures, and everything except what you will actually be doing. Your team welcomed you, and your immediate manager tried to explain your first assignment. Now, you are sitting in front of your computer, ready for takeoff.

Your career feels like a new boat in a harbor on a sunny day, poised to set sail toward uncharted horizons soon. Yet, you have lingering apprehensions in your mind as the captain of your future career. How much do you know about your destination port? Who has the maps, and if you have one, does your compass point in the right direction? How long will your first voyage take? Will it be smooth navigation? Where are passageways and ship-wrecking reefs? What storm is waiting for you in open waters? How good a swimmer are you? You will need a strong boat and an accurate compass to navigate your course effectively. Likewise, you will need solid research competence and impactful product contribution in an industrial research setting. We will talk about these in the following sections.

Below, we arrange what it would take to work as a computer vision researcher in the industry into two intercoupled phases: conception of ideas and execution. These phases are not independent of each other, and they may not even be on a linear time scale. Depending on your findings, the project timeline, and the obstacles confronted, you often switch back and forth or cycle over them as the project progresses. We will build helpful guidelines without going into technical aspects.

2.2 Conception Phase

Hard work forges solutions, yet problems ignite ideas. The more challenging and demanding the problems are, the more creative and resourceful the ideas become.

Computer vision has undeniably confronted and worked on the most demanding perception problems. Within the last four decades, it has built upon inventive concepts emanating from major undertakings aiming to solve formidable challenges. One such challenge involves automatically learning descriptive image representations, eliminating the need for handcrafting features, which is fragile and cumbersome. To this end, neural networks–and some may also recall dictionary learning— have been thoroughly investigated, allowing the data and learning objectives to facilitate the creation of the best

DOI: 10.1201/9781003328957-2

representations. Another example is domain shift, which arises when the source data has a different distribution from the target data, creating a gap between the training assumptions and test settings. To tackle this challenge, extensive research has been conducted from meta-learning to few-shot learning to ensure solutions are adaptable and generalizable across domains.

Apart from these two examples, there is a long list of open computer vision problems that continually fuel innovative ideas. We will not delve into these specific challenges here, as esteemed conferences and journals offer valuable insights into them. Rather, we will encapsulate several aspects of idea generation within industrial research labs.

2.2.1 Ideas and Innovative Process

Much like in other domains, computer vision research centers around devising novel methods to enhance performance, a pursuit commonly recognized as innovation. This may manifest as, for example, a power-efficient motion deblurring framework for a camera pipeline or a highly precise panoptic segmentation model for a visual search engine. Innovations can take the form of fresh solutions to existing challenges or the introduction of new functionalities within established applications.

Creativity and innovation in computer vision involve intricate processes. As we often seek to simplify and structure complex concepts, let's endeavor to outline how ideas can be nurtured in our research, as illustrated in Figure 2.1.

1. **One hammer for all nails:** Consider a powerful technique, such as convolutional networks, generative adversarial networks, self-

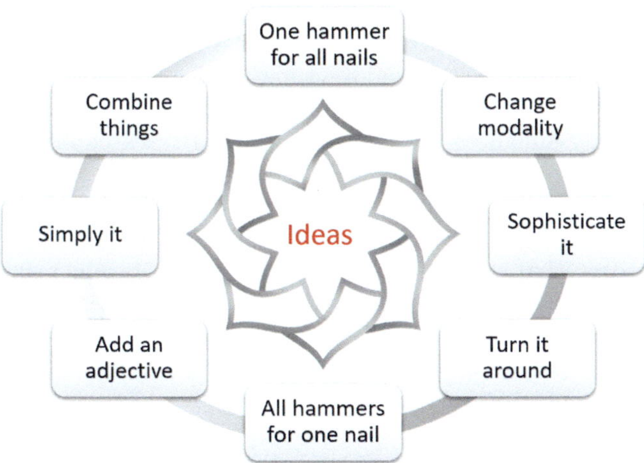

FIGURE 2.1
Common strategies of cultivating ideas.

attention, neural reconstruction, diffusion, and more. One strategy for innovation is to wield this technique as a hammer and then pound on any possible computer vision tasks as our nails. The premise is that if a hammer proves effective for one task, it is likely to be suitable for similar tasks. Not surprisingly, many publications adopt this strategy.

2. **All hammers for one nail:** We might have a computer vision task at hand, which is the nail, and we bring forth a range of methodologies to address it, that is, all the hammers we can think of to put it down. For instance, when tackling image segmentation, we can employ techniques like spectral clustering, level sets, graph-cut, mean-shift, message passing, and dense labeling networks, each acting as a distinct hammer.

3. **Simply it:** Streamline and untangle your solution. For instance, neural network pruning reduces model complexity while maintaining accuracy. Foundation models apply to various tasks instead of requiring separate designs for each. Language models can generate actionable skills and motion, eliminating the need for laborious, task-specific robot programming.

4. **Sophisticate it:** Go crazy, push the boundaries, and make it larger, more expansive, and more powerful. This is exemplified in inception networks and, even more remarkably, in large generative artificial intelligence(AI) models boasting billions of parameters.

5. **Change modality:** When seeking inspiration, we can broaden the core approach by exploring different dimensions or modalities. This could, for instance, involve transitioning from 2D convolutions to 3D convolutions or from image diffusion models to video diffusion models, among other possibilities.

6. **Add an adjective:** Build upon a base technique to make it real-time, lightweight, energy-efficient, memory-efficient, distributed, domain-adaptive, robust, generalizable, self-supervised, and so on. This is another typical strategy found in publications.

7. **Turn it around:** We have the option to pivot and consider the opposite approach. For example, instead of utilizing semantic segmentation for 3D depth estimation, we can employ 3D depth estimation to enhance segmentation. This strategy is often used for trading off latency with accuracy, memory with computing,

spatial coverage with channel depth, and vice versa.

8. **Combine things:** We can fuse seemingly disparate techniques, such as combining sound signatures with image representations to learn audible object classes, or employing large language models alongside vision transformers for text-to-image generation, among other innovative approaches.

Regarding the 'hammers' mentioned earlier, one avenue for cross-pollination and sparking fresh ideas is through interdisciplinary research. For instance, concurrent advancements in applied mathematics, differentiable topology, optimization, natural language processing, information theory, signal processing, stochastic processes, computer graphics, data analytics, robotics, control, and neuroscience have instigated innovative and effective solutions for computer vision pipelines. This underscores that there are valuable insights and growth opportunities, whether through collaboration with experts from other disciplines or by keeping a close watch on progress in diverse fields.

For additional perspectives, you can consult the talks by Ramesh Raskar [41] and Tom Dietterich [11], the tutorial by Serge Demeyer [10], the lecture notes by Jan van Gemert [45], and the presentation by Kilian Weinberger [47].

2.2.2 Need-Oriented Research

Having briefly reviewed the innovation strategies, let's examine the roots and dynamic forces propelling innovation forward. In an industrial research setting, success is fundamentally tied to financial gains. Consequently, research endeavors are geared toward enhancing business profitability. Predictably, the sources of current or potential revenue dictate the allocation of limited resources among various projects, ultimately impacting the sustainability and growth prospects of research teams. This dynamic creates a competitive environment encompassing research, development, and business departments.

Now, let's contemplate what underpins profitability. Among many other factors, profitability depends on how a product is positioned in the market compared to its competitors. Rarely does a tech company hold a monopoly in its sector, resulting in an ongoing competition that requires a constant effort to offer features that provide unique value, ultimately persuading customers to choose your product over alternatives. This competition fuels an unending quest for product advancements and improvements, which is one reason why there are dedicated research teams within companies. So, there is a crucial demand for research in the industry. This is even more pronounced in computer vision due to its capacity to catalyze disruptive technologies that are essential for competitiveness.

As a researcher, one of your key missions is to understand what innovations and advancements are needed for your products. This will steer your creative thinking to the right research areas. By cultivating a broader perspective that takes considers product competitiveness, you can align your research trajectory well with the company's objectives. To achieve this, foster open and constructive channels of communication with the people who can provide valuable feedback and articulate product demands, ideally sourced from users and customers of your products. Often, the first point of contact for such insights would be those interfacing directly with customers—namely, the product teams and development managers. They are best positioned to identify and communicate priorities, areas needing improvement, pain points, competitive offerings, and target timelines, in addition to customer requirements. This is precisely the information you want to know in order to orient and justify your research efforts.

Furthermore, this would provide you a clear sense of purpose and clarity on why you are working on a research challenge. Enhanced clarity enables you to zero in on the most promising research challenges. Taking proactive steps to engage with product teams and initiate discussions proves invaluable for you and your research team. This is sometimes seen as a leadership responsibility in research. Nevertheless, your immediate manager may not always have the most up-to-date information. In fact, astute research leaders actively seek and value input from the entire team to refine and redirect research efforts.

Another vital consideration in industrial research is timing. While numerous research ideas hold the potential to contribute to product development, it's important to discern which improvements carry the most weight. Understanding priorities and devising a feasible timeline for addressing them is paramount. This strategic approach not only enhances the efficiency of your research efforts but also garners greater recognition internally.

In essence, the primary challenge in the realm of industrial research lies in cultivating a comprehensive perspective that encompasses expectations for product enhancement. It is crucial to establish the rationale behind a research endeavor before diving deep into it. First comes understanding the need from the company's perspective and finding realistic answers to questions like why you should work on a specific research problem and why now. This involves identifying and engaging with individuals who possess valuable insights to strategize and prioritize your research efforts.

2.2.3　Exploratory Research and Culture

The points outlined above do not necessarily confine research activities solely to existing products. Industrial research has also prospects of instigating novel products and services. Research inherently is a risk-taking endeavor; more often than not, groundbreaking innovations come from outside the fence by venturing into unexplored possibilities. For this reason, research labs oftentimes encourage looking beyond shorter-term agendas and exploring opportunities in complementary directions by allowing specific efforts to diverge from current product lines.

Exploration comes at a cost, and it requires a delicate balance between addressing immediate product needs and exploring new directions. Going all-in on uncharted territory could jeopardize the competitiveness of your existing products. Moreover, investing heavily in new ideas does not guarantee a significant impact, even after allocating substantial resources.

Nevertheless, there is a bigger danger of tunnel visioning if all attention is dedicated to existing products or present ways of solving problems. Investing your energy only in product needs and ignoring explorative research may isolate you from what is emerging on the horizon. Besides, such a myopic focus may cause research to get stale and fall behind the competition.

Exploratory research is particularly critical in computer vision since the field is moving forward quickly. The most recent advancements published on platforms like arXiv can swiftly surpass findings from just a few months or years prior, rendering them obsolete. As a computer vision researcher, you cannot afford not to know the latest in our field. It is crucial for your success to stay abreast of the latest developments in computer vision and supporting areas. Research beyond your comfort zone will also facilitate the expansion and refinement of your expertise.

Many would argue that the optimal approach to computer vision research lies at the intersection of need-driven and exploratory endeavors, often aligning with the prevailing company culture. The proximity to the product and customer dictates the emphasis on meeting immediate product needs. Certain research labs may allocate minimal resources to research activity that deviates from established pipelines, prioritizing existing products. Meanwhile, others can afford to dedicate resources to speculative, blue-sky research.

Additionally, there are cyclical shifts in focus, oscillating from needs to exploration, then back to needs, and so on. It is not fruitful to fixate on mismatched objectives at the wrong time; it is your responsibility to anticipate the timing.

The company culture plays a significant role in determining whether the research work takes a leading or following position in the field. Choosing to follow can result in safer bets and quicker technology adoption, as you can leverage existing developments from external sources, bypassing some initial steps. However, what is already publicly available also means that others might have even better solutions that are not yet accessible to you as they are not published. This can result in a vicious cycle where your ideas lag behind, as others might have already pursued similar innovations while you work to refine an existing solution. Thus, a decision lies in whether you can operate within established technologies without pursuing the most cutting-edge innovations, or if you're positioned to lead the field and are willing to accept the potential for failure.

2.3　Execution Phase

The next stage is a dynamic process that advances ideas from conceptualization to tangible proofs of concept and beyond. This encompasses a series of crucial steps: implementation, rigorous testing, iterative refinement, and ultimately, presenting the refined concepts to key decision-makers. Further enhancements are made based on their feedback, and with successful outcomes, these ideas are seamlessly integrated into our product line.

In the forthcoming sections, we will explore the multifaceted journey of executing a research project, shedding light on its various facets and navigating through the underlying challenges.

2.3.1　Performance Objectives

Research efforts are typically geared toward improving existing solutions. But, what does constitute 'improvement' in this context, and how can we gauge it through benchmarks and performance indicators?

In industrial computer vision, two pivotal yet conflicting drivers are at play, as exemplified in Figure 2.2. The first revolves around accuracy-related objectives, which focus on the solution's quality within the constraints of the target platforms. These objectives encompass a spectrum of requirements: reliability in meeting product specifications, resilience against input variations, consistency across diverse settings, generalizability to unseen data, adaptability to different domains, and scalability across platform, compute, and volume variations.

To quantify these accuracy-related objectives, various metrics come into play. For instance, F-score and average precision (AP) are pivotal for tasks like detection and classification. Intersection over union (IoU or Jaccard index) stands paramount for segmentation, while

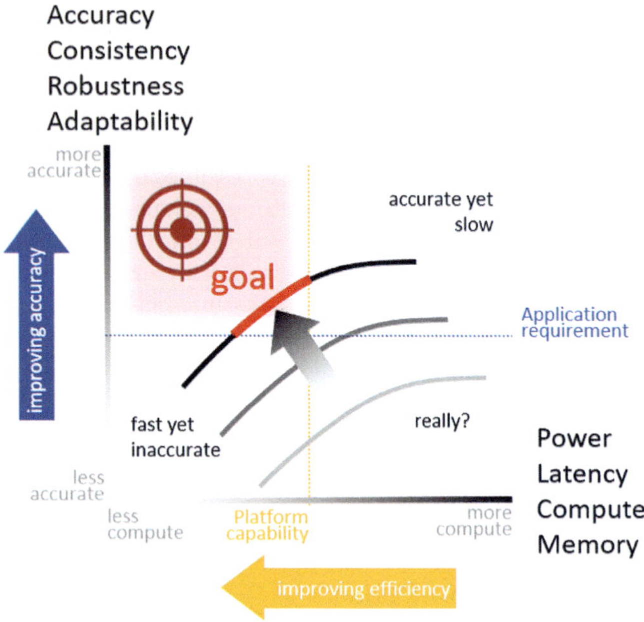

FIGURE 2.2

An illustration of the performance objectives in computer vision research. Accuracy and actualization factors counteract. Better solutions improve both accuracy and efficiency. There is a limit on the platform capability, while applications require minimum accuracy and efficiency levels.

precision-recall and area under the curve (AUC) come into play for recognition. Other metrics, like peak signal-to-noise-ratio (PSNR) for image enhancement and end-point error (EPE) for motion estimation, find their niche in fundamental computer vision tasks. The choice of accuracy metric itself is a pivotal decision, as it directly shapes the research objective.

The other driving force concerns actualization-related efficiency objectives linked to practical implementation. These encompass factors like power consumption, latency, computational load, memory usage, parameter count, bandwidth, and throughput. For instance, latency is measured by milliseconds, inference per second (IPS), and frame per second (FPS) on hardware, computational load is quantified by the number of multiplications and additions (MAC), the number of binary operations, and the number of floating-point operations (FLOP). Power consumption is measured in milliwatts.

Efficiency is closely tied to the capabilities of the platform and hardware. While the number of computations can influence power consumption, it may not necessarily scale latency, as memory access can often be the primary bottleneck. Metrics like computations, model size, and memory footprint can be derived in a hardware-agnostic manner. In contrast, latency, power consumption, and throughput depend on specific hardware specifications.

There exists an inherent trade-off between accuracy and actualization objectives. For instance, larger neural networks, assuming they are trained on extensive and diverse datasets, often yield higher accuracy scores compared to their more compact counterparts. However, such larger networks also demand increased computational resources and, consequently, more power. A parallel analogy can be drawn with input image quality. Higher resolution and dynamic range images, or utilizing multiple images, invariably boost object detection accuracy. Yet, this improvement comes at the cost of a proportional increase in computational load.

For mobile computer vision platforms like XR devices, mobile phones, and aerial drones, how much power is depleted is a critical consideration, particularly given such devices run on battery. This concern extends even to electric vehicles, as higher power consumption rates of computer vision algorithms, especially those processing video and data streams from multiple cameras and sensors, can significantly impact their operational range. Furthermore, specific tasks necessitate execution at real-time speeds. Examples include hand tracking for mixed reality experiences and vehicle tracking for autonomous driving applications.

While some of the mentioned metrics can be readily assessed on validation datasets, there are inherent challenges. For instance, obtaining power measurements often requires using complex instruments or relying on Pareto optimal schemes. Additionally, the real-world performance might not always align with benchmark results, owing to the long-tailed nature of task and scenario distributions.

2.3.2 Evolving Requirements

The primary objective is evidently to enhance accuracy while adhering to the specified power and latency thresholds set by both the hardware specifications and the application's requirements. Nevertheless, it's crucial to recognize that this represents just the surface of industrial research. Each project introduces a unique set of conditions and considerations across its various stages. Performance objectives evolve over the course of a project. The reason is that industrial computer vision research is expected to yield tangible product-related enhancements and create measurable value aligned with the company's bottom line. Such advancements must be transparent across the organization to promote acknowledgment and garner support.

This dynamic drives our research efforts toward short and evolving timelines and priorities. For instance, the initial focus may center on enhancing accuracy in line with competitive technology benchmarks set by research directors. Later, the emphasis might shift toward efficiency, aligning with power requirements stipulated by product managers.

Consider another scenario: many computer vision applications require the concurrent execution of multiple tasks. In such cases, efficiency may focus on minimizing redundant computations across tasks that operate at varying framerates and often on multimodal data, all while utilizing a unified neural network backbone. This entails a research effort aimed at accurately training the same backbone for multiple tasks without compromising performance. However, having a unified dataset annotated with separate task ground truths is highly unlikely. This leads to the adoption of iterative training schemes, where the backbone is optimized for a single task with each update while still maintaining performance in other tasks. This iterative approach allows for various strategies, such as starting with a pretrained backbone and fine-tuning it on multiple tasks or designing a completely new backbone and training it. The choice of training objectives for these options sets different timelines.

After initial proof-of-concept demos that establish the solution's adherence to the expected accuracy goals, attention turns toward adapting the model for target platforms. This involves transferring the model onto on-device systems and optimizing additional objectives, such as latency and power, to align with design requirements. Often, this requires revisiting the initial design, fine-tuning various components to reduce computational load, and exploring alternative designs to compensate for potential losses resulting from reduced computing.

This distinction is particularly noteworthy when comparing academic and industrial research. In academia, resources for such a transfer to on-device platforms may not always be readily available, whereas in industrial research, this process is inherent to the nature of the business.

As a computer vision researcher closely collaborating with dedicated engineering and product teams, you will come to understand that they operate on different timelines. Their availability for supporting your research can vary, necessitating synchronization efforts. These varying timelines will establish distinct milestones and fluctuating performance objectives for your research outcomes.

Furthermore, essential elements like authentic training and test data, along with customer feedback on pain points, are sourced from the product teams. These directly influence the performance targets. The quality and quantity of data provided by the product teams or relayed from customers may fluctuate throughout the project, affecting research directions and evolving performance goals for computer vision solutions. Thus, aligning your timeline and performance objectives with the availability of data and customer feedback is pivotal for your research success.

2.3.3 Dataset Challenges

Computer vision stands as one of the most extensive research disciplines in computer science owing to its task diversity. Each task comes with its own long list of benchmark datasets and specific ways of evaluating performance, called protocols. For a comprehensive reference on such benchmarks, please consult [1, 8, 37, 48]. Despite these resources, datasets remain one of the most significant challenges in computer vision research, particularly within the realm of industrial research.

Driven by data-centric AI approaches like deep learning, computer vision research heavily relies on extensive datasets, which play a pivotal role in validating model performance and, even more crucially, in training such models. A rough rule of thumb suggests that doubling the volume of training data often leads to a halving of performance errors. The success of deep learning can be directly attributed to the growing availability of these large datasets.

At one end of the spectrum, certain computer vision applications like autonomous driving generate petabytes of raw data daily. The task of sifting through, curating, and selecting the right subsets for labeling while also ensuring the quality of annotations is costly and difficult to scale. Such applications require data that accurately represents the underlying population and encompasses real-world scenarios. This means the data must be diverse and comprehensive, covering all edge and typical cases to accurately depict the problem at hand. Moreover, it should maintain consistency in labeling to extract meaningful insights. Nevertheless, dealing with these extensive datasets demands increased workloads when we access, transfer, process, and run evaluations.

At the other end, not all computer vision tasks have the advantage of accessing such extensive, diverse datasets since acquiring accurately annotated datasets tailored for specific tasks is a formidable challenge. For example, generating optical flow ground truth annotations for unconstrained real-world scenarios requires intricate post-processing or calibrated data from an alternate modality, such as 3D, with the assumption of feature point correspondences across frames. Similarly, tasks like super-resolution demand a splitting lens setup with multiple sensors or, ideally, an imaging system capable of zooming in and out to capture both low-resolution and high-resolution images of the same scene. However, the latter necessitates a static scene. Similarly, obtaining a noise-free image ground truth for denoising is no straightforward feat. In the case of 3D depth estimation, ground truth is sourced from another sensor, such as a time-of-flight camera, which introduces calibration and limited resolution challenges. In the realm of medical computer vision, as well as in other applications to some extent, privacy concerns can handicap research from accessing large datasets. These factors collectively curtail the availability of annotated data and introduce elements of bias and uncertainty in data quality.

As you may have observed, in industrial research, releasing products with models trained on public datasets is often prohibited due to licensing constraints. Consequently, for product integration, it becomes imperative to retrain your models using proprietary data after initial experimentation with public or synthetically generated datasets. This transition may induce and require resolving domain shift challenges.

Let us emphasize the importance of avoiding assessments or performance evaluations based on just a few samples. This can lead to overfitting on unrepresentative data, hindering a comprehensive understanding of the actual challenges. Additionally, it may lead to duplicated efforts when applied to target datasets. Engaging in such narrowly focused experimentation may not effectively demonstrate your findings, nor should it be solely convincing for your own evaluations.

In addition, there are other factors that can introduce errors and affect dataset quality, such as lack of diversity, representation bias, domain mismatch, mislabeling, inadequate data maintenance, ambiguous protocols, disk space requirements, and access costs. If left undetected, these issues, along with the ones mentioned earlier, can lead to devastating repercussions on cost, research timeline, and model performance.

2.3.4 Role of Baselines

Beyond the dataset and performance objectives, choosing the right baseline methods for evaluating your improvements is pivotal. To a considerable degree, these baselines represent your competition by establishing tangible performance benchmarks. They also provide a clear definition of the problem at hand.

It is worth noting that many computer vision tasks in the industry have a well-established history, often spanning decades. Not only conventional and handcrafted solutions but also machine learning approaches have matured significantly in recent years. This evolution is evident in premier computer vision publications, where proposed methods are routinely benchmarked against multiple baselines, sometimes even against those published within the same year. Each accepted paper and successful leaderboard entry pushes the state of the art further, effectively becoming a new baseline itself.

Each computer vision task comes with a multitude of baselines, yet this chapter is not the ideal location for an exhaustive listing. Nonetheless, let's explore some insights on how to select the appropriate baselines for benchmarking industrial research.

- **Customer's model:** If available, start with what the customer already has. It is vital to ensure that your solution offers clear advantages and improvements over the customer solution. This comparison is something the customer will likely request, and it will also provide valuable insights into areas for further development in their existing model.

- **Top ranking models at leaderboards:** Sooner or later, questions about how your solution compares to top models will arise. Outperforming these top-ranking models is a significant milestone, as it builds confidence and may open doors to additional resources and time for your project. Even if you have not reached this point yet, familiarizing yourself with these leading solutions can provide valuable insights and potentially inspire ideas for further refining your model.

- **Most recent models:** Studying the most recent models can be a source of inspiration and prevent redundant efforts of inventing the wheel twice. Additionally, since these models are up-to-date, they likely boast respectable performance results.

- **Accessible models:** Given limited research time, you will need to carefully select which ideas to pursue for a new project. Thankfully, it is common in both academia and occasionally industry for researchers to provide public implementations of their models and sometimes even training pipelines for validation. Utilizing such repositories, often found on platforms like GitHub, will expedite your testing process and provide a solid foundation to build upon.

- **Well-known and extensively reported models:** Although popular papers may not always house the best solutions, they do tend to be more widely recognized. Including them as baselines can bolster confidence when presenting your own advancements to stakeholders.

2.3.5 Platform Factors

One striking difference between academic and industrial computer vision research is the emphasis on specific target hardware platforms in industry. Industrial computer vision platforms typically encompass chipsets that incorporate CPUs, GPUs, FPGAs, DSPs, specialized hardware engines, and, increasingly, neural processing units (NPUs) that are often designed for mobile and low-power applications. These components vary in terms of configurability, power consumption, fixed or mixed-point functions, supported neural operations, memory architectures, storage capacities, and clock speeds. Certain platforms are specialized for neural processing, while others execute meticulously engineered and hardened solutions.

Most platforms, especially mobile ones, are resource-constrained. Recent progress in neural networks has demanded rethinking of how computer vision models are designed, trained, and deployed. The efforts in this area can be broadly categorized as:

- **Knowledge Distillation:** This process refers to model compression by instructing a smaller student network using a pre-trained larger teacher network, all while maintaining comparable performance with the teacher model.

- **Neural Architecture Search:** This aims to automate the architectural design of neural networks to enhance performance in terms of either accuracy or latency on the specified task and dataset.

- **Model Pruning:** This technique reduces the model's memory footprint and computational cost. It can result in sparse computational graphs or changes in the input and output configuration of layers and weight matrices.

- **Quantization:** This entails converting original floating-point representations learned in training to low-precision fixed-bit or mixed-bit integer values. This reduces memory usage, power consumption, and, in many cases, latency.

- **Co-design of Models and Hardware:** This involves customizing neural network models to leverage specific target hardware platforms, while also redesigning hardware processing to efficiently handle core neural components.

As a computer vision researcher in the industry, you will confront challenges in porting your solution to such platforms, for instance, developing efficient models robust to quantization, quantizing weights and activations while maintaining the accuracy objectives, and then optimizing the resulting compute graphs of neural models by proper tiling and scheduling for Neural Signal Processors (NSPs).

Nevertheless, software tools for lower-bit quantization may have limited availability and support on target platforms, as these platforms are constantly evolving. Quantization-aware training can prove to be unstable and cumbersome due to hyperparameter tuning, and neural architecture search is still in its early stages for multi-frame and video use cases.

A valuable tip is to be mindful of the target platform right from the beginning. This allows you to design around its limitations and requirements, saving you a significant amount of time and effort that might otherwise be spent trying to rectify issues later.

2.3.6　Agile Development

Another distinction between academic and industrial research lies in the necessity for practical, rather than purely theoretical, work in industrial research, where timelines are considerably shorter compared to academia, driven by the rapid pace of development and the need to stay ahead of constant competition. Moreover, resources are often stretched thin due to a multitude of ongoing projects. In industrial research, a research timeframe extending beyond a year is considered a luxury, and even those longer projections are typically more hypothetical than concrete. There is a constant pressure to demonstrate tangible progress every three to six months, in order to maintain momentum and ensure continued allocation of resources. This requires a delicate balance between agile development and achieving periodic milestones, all while exploring new ideas.

As you can see, it's not just about generating innovative ideas, but also about how quickly you can implement, test, and evaluate them. While many recognize that a significant portion of our time in industrial computer vision research is spent implementing and debugging code, we often overlook how crucial having strong software development skills is for our work. Our research follows a cyclic process of learning, conceptualizing ideas, designing solutions, implementing them, testing, identifying shortcomings, and reevaluating options, which is repeated throughout the project. The more efficient you are at each step of this process, the more productive your research will be.

To achieve this, plan ahead to prevent redundancies, reuse and expand on your prior work and implementations, and establish a uniform codebase to facilitate software development. Make sure you're proficient in the programming environments and can effectively navigate large software frameworks. Strive to be not only an exceptional researcher but also an excellent software developer.

Adopting the best software development principles and practices will greatly streamline your work. While a detailed exploration of these principles is beyond the scope of this chapter, here are some high-level pointers to bear in mind:

- Document and comment on your code to provide comprehensive explanations of the logic. Additionally, implement version control to track changes, collaborate effectively with team members, and ensure that you can revert to earlier versions if needed.

- Use standard naming conventions to ensure that your code is not only readable to you but also to your collaborators, making it easier for everyone to understand the purpose and functionality of each component.

- Create a comprehensive blueprint design before implementation, outlining the architecture, data structures, and key functionalities. This proactive approach ensures that you adhere to quality coding practices consistently, resulting in a well-structured and maintainable codebase.

- Familiarize yourself with AI software development tools and standardized containers. Utilize Docker and other OS-level virtualization techniques for consistent code execution across different environments, minimizing the risk of unintended code breaks during updates.

- Simplicity is key. Ensure your code is clear and easy to understand, allowing for quick bug identification. Begin with the simplest possible code. Introduce complexity only when necessary.

- Break complex sections into smaller, focused blocks, which makes future modifications easier. Prioritize incremental addition of functionalities rather than attempting to implement everything at once. This leads to leaner, more manageable code.

- Refactor common logic into functions instead of copy-pasting code in different places. Avoid directly copying repositories from GitHub. If you must, consider creating API wrappers around them.

- Strive for independence between software components and minimize communication between different classes. Organize related object classes in the same package to achieve cohesion.

Staying current with the latest AI trends, your research might entail coding neural network models using leading deep learning platforms and accelerators (such as PyTorch or TensorFlow). This involves setting up tools to efficiently manage your data, training models, storing checkpoints, validating models, extracting results, and evaluating performance. Rather than starting from square one with each project, focus on building a robust codebase that allows for seamless reuse and swift development.

A notable challenge in this context is allocating time to refine and expand your software development proficiency.

2.3.7 Wrap up, Follow up, Pivot

In a computer vision research project, envision yourself as a relay race athlete. Yo need to know exactly when to pass the baton to the next team member and to keep the momentum going until you reach the finish line.

There are several ways to pass the baton. This could be through a prototype, a showcase demo, or a codebase, through which you transfer the technology to an engineering or product team. Your models and code might be engineered into a feature, integrated into an application pipeline, or incorporated into a product framework. This is a process where you interact with teams outside core research, enabling you to understand how to revise and polish your overall research and communication to make the technology transfer more efficient next time. Moreover, this interaction may spark improvements and collaborations, potentially leading to new project directions.

Once you pass the baton, it is essential not to retreat to the dressing room. A critical follow-up process, sometimes overlooked even by senior researchers, involves staying in touch with the receiving teams. Inquire about the integration of your solutions into products or their impact on the company. Periodic catch-ups allow them to keep your expertise in mind, making it more likely for them to seek your assistance in the future.

While we strive for success in our research efforts, it is imperative to acknowledge that achieving targeted objectives is not always guaranteed, even when all other factors align. The very essence of research lies in seeking answers to questions we do not have at the outset, and with this comes inherent risk. It is perfectly acceptable for a research project to not entirely meet its initial goals. Embracing this reality can be a process, especially for individuals deeply driven to solve complex problems—the very reason many of us choose a career in computer vision research.

In our pursuit of excellence, it is equally vital to learn from our missteps and gracefully transition our focus to new projects. Every project has a lifespan and a runway, but it is not boundless. Recognizing when to conclude a project is a key skill to prevent getting trapped in a prolonged pursuit with diminishing returns.

Furthermore, consider publishing your work in reputable venues or submitting your results to leaderboards. This practice is pivotal for several reasons. Firstly, it provides invaluable validation for the novelty and robustness of your work, offering an impartial seal of approval devoid of conflicts of interest. It enables you to benchmark against state-of-the-art techniques, showcasing the competence of your method within and beyond the company. Additionally, it brings recognition and visibility to both your research and the company, enhancing the recruitment of top candidates and potentially securing valuable product contracts.

Perhaps humbly, you need to actively promote your achievements and seize opportunities to highlight the innovative and pivotal nature of your research. Often, we ignore this aspect, assuming that success will automatically garner recognition. However, even within research labs, the definition of success can be subjective and contingent on diplomatic skills. Researchers who are adept at presenting and articulating their work tend to find it easier to gain recognition for their efforts.

Industrial research companies safeguard intellectual property through an internal review process. This ensures that all ideas meet legal requirements before external submission or publication. During this phase, experts may offer valuable feedback on your ideas for further refinement. Legal checkpoints extend to datasets, pre-trained models, and software tools and can take weeks to months for approval. You need to plan for these reviews ahead of publication deadlines. Some innovations, especially those slated for product integration, may not be cleared for external disclosure. Yet, they may lead to significant contributions and recognition within the company.

2.3.8 Teamwork

You might have noticed that high-quality computer vision papers from the industry often have multiple authors, even a lengthy list. This reflects not only the extensive effort required to produce them but also the depth of collaboration and collective contributions of many.

Although project teams often operate within established management structures, it is still possible to forge connections and secure additional support. A valuable strategy for this is to offer assistance to others, creating a reciprocal environment for collaboration. This proactive approach, in turn, fosters a network of collaborators, ultimately enhancing the productivity of your research efforts. Whether you're an individual contributor or a research leader, success hinges on effective communication, seamless information sharing, and adept time management. Mastering these skills ensures seamless teamwork and productive outcomes.

Keeping track of your progress and maintaining an up-to-date project document is essential for effective communication. This document, preferably in a simple format like a slide deck or confluence page, records your progress, detailing what you've attempted and the milestones you have achieved. Such a working document serves as a medium to share observations and align goals. It helps you reporting your efforts, results, and critical issues regularly and timely to others and management, which is an element of success in research.

2.3.9 Self-Improvement

Learning never ends in computer vision. It stands as one of the fastest-growing and evolving research fields, with new methods emerging at an unprecedented pace. To stay ahead, it is essential to invest in your own development and consistently refresh your knowledge in your research area.

To stay at the forefront of computer vision, take advantage of technical training opportunities such as computer vision summer schools, tutorials at flagship conferences, and local university events. Participate in coding boot camps and use survey papers as your entry point into new areas or to refresh your existing knowledge. Commit to specific reading goals, like one paper per week or even per day. Volunteer to present papers in your discussion groups. Make sure you are well-informed about the benchmarks and competitions in your specific computer vision tasks and keep a close eye on related papers released on platforms like arXiv. Additionally, consider serving as a reviewer at computer vision and machine learning conferences and journals. These activities not only keep you connected with the latest research and trends but also broaden your perspective and help you avoid duplicating existing work.

2.4 Open Problems

A chapter dedicated to computer vision R&D challenges would be lacking without a discussion of active research areas and outstanding questions. Below, some of those are categorized into core computer vision, machine learning, and application domains.

2.4.1 Core computer vision tasks

- *Long-term multi-object tracking* [29]. One big challenge is maintaining accurate correspondence and minimizing identity switches when objects have similar appearances and undergo scale variations, deformations, and occlusions while the background changes.

- *Occlusion handling* [16, 50]. While many core computer vision tasks, from segmentation to object detection, can be solved when the target objects and semantic classes are adequately visible, there are often self, partial, or full occlusions in real-life settings.

- *Depth estimation* for specular, transparent, and featureless objects [6, 27]. In addition to these difficulties, direct estimation of downstream representations such as surface meshes and planar compositions instead of point clouds still requires further research. Moreover, scale-consistent depth estimation [34] in monocular vision is a challenge since the predicted depth is obtained up to a scale factor to the real world.

- *Semantically consistent optical flow* [53]. Further performance improvement can be accomplished by incorporating semantic and contextual cues in optical flow estimation.

- *Small object detection* in cluttered scenes [21]. The true detection vs. false alarm operating points for small targets, such as vehicles in the far field, are unsuitable for robust applications. There exist hierarchical and multi-resolution schemes for recursive inference of fine-grained features, still, there is more to be done to progressively refine their estimations from more salient and larger objects to more intricate and smaller objects in an effective way.

- *Contextual-aware panoptic segmentation in uncontrolled settings* [33]. Even though segmentation has been one of the oldest undertakings in computer vision, it remains as a challenge.

- *Temporally consistent video perception.* Not only is this task computationally demanding, but proper temporal consistency measures also need to be developed.

- *Inverse rendering* including physically based scene decomposition [56] and scene lighting estimation, *inverse kinematics* from visual data, and *image vectorization*, that is, encoding and image as a composition of basic parts (e.g., brush strokes) and learning to compose such primitives, components, and concepts back into visual data. See [43] for compositional computing.

- Open-world *visual question answering* grounded in an image and video is still in its infancy [5]. This task requires resolving language ambiguity and understanding the interaction of visual context and question at linguistic levels, semantics, syntax, and pragmatics.

- *Action recognition and localization* [25]. These video-understanding tasks still demand more research efforts to attain acceptable levels of accuracy and robustness.

2.4.2   Machine learning for computer vision

- *Reliance of large data.* Recent learning strategies, for example, deep learning networks, adversarial training, and generative models, heavily rely on large datasets. Yet, data is often incomplete, irregular, and many times, it lacks diversity. The annotated data is limited and annotation quality is inconsistent. Besides, not all research labs can access large datasets due to license restrictions. Even if it is accessible, training cost limits only a few to be able to leverage such large datasets, which creates a skewed and disadvantaged competition in industrial and academic research.

- *Uncertainty quantification* for risk-aware computer vision assesses confidence by unsupervised validation on target data [23]. AI models can interpolate from the data that trained them. However, they may fail to extrapolate if the test samples lie outside the training data distribution. Existing models do not extend to high-order reasoning to assess confidence and connect the dots the way a human intuition can do. It is critical for a system, such as, an autonomous vehicle or a robotic surgeon, to communicate its uncertainty along with an explanation of the source of the uncertainty to the human decision-maker.

- Related to above, *model explainability* aims to understand machine learning algorithms' decisions and underlying factors in human-interpretable terms and mental models [3, 4, 19]. Transparency in the representation and processing of knowledge is required to trust an AI model's decisions.

- How to defend against *adversarial attacks* [2], in particular, understanding the existence of adversarial examples, intrinsically robust models, robustness-accuracy trade-offs, adversarial training, and certified defenses are longstanding problems of this area.

- Lifelong and *continual learning* without catastrophic forgetting with efficient models [31]. Continual learning enables updating models with new data. However, incrementally updating a model only with new data when previous data is unavailable due to storage, computing, or privacy issues may degrade the ability of the model to perform well on previously seen data, causing catastrophic forgetting. Industrial-grade computer vision solutions, however, have to strike a balance between adaptation, learning plasticity, and performance assurance.

- Effective unsupervised or self-supervised *personalization* [38, 55]. This also relates to how to go beyond the gradualist models that rely on backpropagation and advocate for gradual or incremental changes rather than radical ones, in order to learn usable knowledge from a single sample, for domain adaptation and personalization.

- *Imitation learning* by observing an oracle, for example, human [20, 40]. This has the potential to pave the way for efficiently learning highly complex tasks to be executed by robots imitating how people perform them. Another related challenge is the ability to learn by receiving explicit instructions and interactions (see [9] for retrospectives from embodied AI perspective), in particular, sample-efficient exploration for reinforcement learning [24, 49]. Deep reinforcement learning methods require a large number of training iterations to achieve reasonable performance; otherwise, they can be trapped in local minima because of insufficient exploration. However, they are impractically expensive due to their lack of sample efficiency.

- Expanding conventional learning to *commonsense reasoning* [51, 52], where models contain core formalizations that can be reused to solve a wide range of variants of the original task and other problems. Commonsense reasoning is a huge undertaking itself, with research lines in causal representation learning, probabilistic reasoning, intuitive physics, and psychology. This also relates to interpreting data and locating hidden aspects (as in intuition) based on the gestalt laws and the principle of totality that argues the whole is more than the sum of its parts [12].

- Data-driven *learning conditioned on analytical manifolds* [14], in particular, differentiable proxies or direct solutions that can retain and take advantage of the topology.

- Merging the latest data-driven learning methods and *symbolic AI* for visual understanding. See [15, 26, 42] for a long list of potential directions.

2.4.3   Systems and applications

- *Open set recognition*, improving the robustness of perception models in open-world settings and for long-tail distributions [22, 39]. Computer vision algorithms are often developed inside a closed-world paradigm, for example, recognizing objects from a fixed set of categories. In real-world computer vision tasks, our visual world is naturally open, containing dynamic, vast, and unpredictable situations. Collecting training samples for all classes is not always possible. Still, product quality solutions require efficiently adapting, robustly generalizing, and accurately classifying the seen and unseen classes.

- The above also relates to practical *on-device learning* [32], which is one of the holy trails in mobile computer vision productization that promises further performance improvements.

- *Multi-task learning* with multiple incomplete and noisy datasets [46, 54]. Like multimodal perception, multi-task learning is intrinsic to computer vision applications. Any headway here can open up opportunities for better products. Seamless integration and efficient execution of diverse multimodal perception tasks. Major applications from autonomous vehicles to XR rely on multimodal data. Related to this, another major challenge is how to optime *scheduling of multiple computer vision tasks* that need to run in parallel to reduce power while attaining the desired level of performance.

- How to optimize and balance model inference across *heterogeneous computing* platforms including multiple GPUs, NPUs, CPU cores, as well as, edge and cloud balancing while optimizing for latency, memory, bandwidth, and power requirements. The complexity and range of hardware platforms and computer vision tasks make this problem undeniably a topmost challenge in industrial research.

- Memory and compute *efficient generative AI* models for visual data creation, manipulation, and enhancement [7]. Porting large models with billions of parameters to mobile platforms and running them efficiently is an active research area. From GPT variants to diffusion-based generative AI models, the widespread public interest and potential product and service opportunities around them make this challenge a high priority. There is a demand for real-time generative AI solutions.

- Relating to above, hardware support for *attention mechanisms* including self and cross-attention [17]. Exciting approaches have been proposed to improve the computational load of the attention mechanisms and transformers. However, more work must still be done to adapt them on hardware for pixel-wise dense estimation tasks as well as for attention across frames for video tasks.

- Compute and memory-*efficient perception models* that can be easily customized for different hardware. For instance, efficient camera essentials such as super-resolution, denoising, and deblurring [28], and efficient pixel-wise correspondence and cost-volume computation [44]. Such dense estimation tasks are among the most compute-intensive processes in computer vision; thus, any optimizations here would boost the product adaptation of such solutions. Even though it may rank among the most accurate approaches to gathering evidence from limited data (e.g., from a single image for denoising), executing non-local means in a compute-efficient manner is still a big challenge. Among other similar challenging tasks, efficient visual SLAM [30] for large input image resolutions, and adaptation for camera intrinsics [13] are notable. There are initial studies, yet for better systems, more work is needed to take full advantage of camera intrinsics. These and similar perception models have to satisfy very low power and latency. Yet, there is still a substantial gap between the most accurate solutions and what can run on current mobile platforms.

- The gap between theoretical and practical efficiency of dynamic neural models [18]. Many exciting studies demonstrate that *conditional computing* and dynamic routing can improve the number of computations. Yet, hardware only supports such methods partially.

- Efficient training and neuromorphic architectures for *spiking neural networks* [36] that can do visual perception at extremely low power levels, like a biological brain. Although spike neural networks have been around for a decade, solving the fundamental issues in their productization still promises a big reward.

- *Learning manipulation skills* to interact with seen and unseen objects [35], bridging the gap between visual perception and robotics. Dexterous manipulation of a robot is an integral part of embodied intelligence. Yet, existing techniques can only perform simple tasks in structured environments. Object manipulation from 3D data poses significant challenges in building generalizable visual perception and policy models; for instance, what are the primitives of robotic manipulation?

2.5 Final Words

The research challenges discussed in this chapter often resonate with those found in academic settings, blurring the once distinct lines between computer vision in industry and academia. It is hardly surprising that there has been a significant flow of talent from academia to industrial labs, accompanied by a surge of pioneering research initiatives from industrial labs to academia in recent years. It is common to see faculty members spearheading start-up ventures while industrial researchers claim best paper awards. Rather than viewing these realms as opposing forces demanding a bridge, it is more apt to see them as interlocking components of the mechanism. As people propelled by the excitement of conquering monumental challenges and enriching lives, we are integral to the vibrant computer vision research community.

Good luck and bon voyage in your research!

References

[1] AI-benchmark. https://ai-benchmark.com

[2] Naveed Akhtar, Ajmal Mian, Navid Kardan, and Mubarak Shah. Advances in adversarial attacks and defenses in computer vision: A survey. *IEEE Access*, 9, 2021.

[3] Alejandro Barredo Arrieta, Natalia Díaz-Rodríguez, Javier Del Ser, Adrien Bennetot, Siham Tabik, Alberto Barbado, Salvador García, Sergio Gil-López, Daniel Molina, Richard Benjamins, et al. Explainable artificial intelligence (xai): Concepts, taxonomies, opportunities and challenges toward responsible AI. *Information Fusion*, 58, 2020.

[4] Adrien Bennetot, Jean-Luc Laurent, Raja Chatila, and Natalia Díaz-Rodríguez. Towards explainable neural-symbolic visual reasoning. *arXiv preprint arXiv:1909.09065*, 2019.

[5] Raffaella Bernardi and Sandro Pezzelle. Linguistic issues behind visual question answering. *Language and Linguistics Compass*, 15(6), 2021.

[6] Xiaotong Chen, Huijie Zhang, Zeren Yu, Anthony Opipari, and Odest Chadwicke Jenkins. Clearpose: Large-scale transparent object dataset and benchmark. In *European Conference on Computer Vision*, 2022.

[7] Florinel-Alin Croitoru, Vlad Hondru, Radu Tudor Ionescu, and Mubarak Shah. Diffusion models in vision: A survey. *IEEE Transactions on Pattern Analysis and Machine Intelligence*, 2022.

[8] DatasetList. https://www.datasetlist.com

[9] Matt Deitke, Dhruv Batra, Yonatan Bisk, Tommaso Campari, Angel X Chang, Devendra Singh Chaplot, Changan Chen, Claudia Pérez D'Arpino, Kiana Ehsani, Ali Farhadi, et al. Retrospectives on the embodied ai workshop. *arXiv preprint arXiv:2210.06849*, 2022.

[10] Serge Demeyer. How to perform research? (and get "empirical" results): https://jvgemert.github.io/HowToDoResearchInDL-slides.pdf

[11] Tom Dietterich. Research methods in machine learning: https://web.engr.oregonstate.edu/~tgd/talks/new-in-ml-2019.pdf

[12] Willis Ellis. *Chapter: Laws of Organization in Perceptual Forms by Max Wertheimer, originally published in 1923)*. 1938.

[13] Jose M Facil, Benjamin Ummenhofer, Huizhong Zhou, Luis Montesano, Thomas Brox, and Javier Civera. Cam-convs: Camera-aware multi-scale convolutions for single-view depth. In *Proceedings of the IEEE/CVF Conference on Computer Vision and Pattern Recognition*, 2019.

[14] Zhi Gao, Yuwei Wu, Xiaomeng Fan, Mehrtash Harandi, and Yunde Jia. Learning to optimize on riemannian manifolds. *IEEE Transactions on Pattern Analysis and Machine Intelligence*, 2022.

[15] Marta Garnelo and Murray Shanahan. Reconciling deep learning with symbolic artificial intelligence: Representing objects and relations. *Current Opinion in Behavioral Sciences*, 29, 2019.

[16] Shane Gilroy, Edward Jones, and Martin Glavin. Overcoming occlusion in the automotive environment—A review. *IEEE Transactions on Intelligent Transportation Systems*, 22(1), 2019.

[17] Kai Han, Yunhe Wang, Hanting Chen, Xinghao Chen, Jianyuan Guo, Zhenhua Liu, Yehui Tang, An Xiao, Chunjing Xu, Yixing Xu, et al. A survey on vision transformer. *IEEE Transactions on Pattern Analysis and Machine Intelligence*, 45(1):87–110, 2023.

[18] Yizeng Han, Gao Huang, Shiji Song, Le Yang, Honghui Wang, and Yulin Wang. Dynamic neural networks: A survey. *IEEE Transactions on Pattern Analysis and Machine Intelligence*, 2021.

[19] Andreas Holzinger, Anna Saranti, Christoph Molnar, Przemyslaw Biecek, and Wojciech Samek. *Explainable AI Methods - A Brief Overview*. 2020.

[20] Jiang Hua, Liangcai Zeng, Gongfa Li, and Zhaojie Ju. Learning for a robot: Deep reinforcement learning, imitation learning, transfer learning. *Sensors*, 21(4), 2021.

[21] Licheng Jiao, Ruohan Zhang, Fang Liu, Shuyuan Yang, Biao Hou, Lingling Li, and Xu Tang. New generation deep learning for video object detection: A survey. *IEEE Transactions on Neural Networks and Learning Systems*, 2021.

[22] KJ Joseph, Salman Khan, Fahad Shahbaz Khan, and Vineeth N Balasubramanian. Towards open world object detection. In *Proceedings of the IEEE/CVF Conference on Computer Vision and Pattern Recognition*, 2021.

[23] Lance Kaplan, Federico Cerutti, Murat Sensoy, Alun Preece, and Paul Sullivan. Uncertainty aware AI ML: Why and how, 2018.

[24] Robert Kirk, Amy Zhang, Edward Grefenstette, and Tim Rocktäschel. A survey of generalisation in deep reinforcement learning. *arXiv preprint arXiv:2111.09794*, 2021.

[25] Yu Kong and Yun Fu. Human action recognition and prediction: A survey. *International Journal of Computer Vision*, 130(5), 2022.

[26] Luis C Lamb, Artur Garcez, Marco Gori, Marcelo Prates, Pedro Avelar, and Moshe Vardi. Graph neural networks meet neural-symbolic computing: A survey and perspective. *arXiv preprint arXiv:2003.00330*, 2020.

[27] Zhengqin Li, Yu-Ying Yeh, and Manmohan Chandraker. Through the looking glass: Neural 3d reconstruction of transparent shapes. In *Proceedings of the IEEE/CVF Conference on Computer Vision and Pattern Recognition*, 2020.

[28] Anran Liu, Yihao Liu, Jinjin Gu, Yu Qiao, and Chao Dong. Blind image super-resolution: A survey and beyond. *IEEE Transactions on Pattern Analysis and Machine Intelligence*, 2022.

[29] Wenhan Luo, Junliang Xing, Anton Milan, Xiaoqin Zhang, Wei Liu, and Tae-Kyun Kim. Multiple object tracking: A literature review. *Artificial Intelligence*, 293, 2021.

[30] Andréa Macario Barros, Maugan Michel, Yoann Moline, Gwenolé Corre, and Frédérick Carrel. A comprehensive survey of visual slam algorithms. *Robotics*, 11(1):24, 2022.

[31] Zheda Mai, Ruiwen Li, Jihwan Jeong, David Quispe, Hyunwoo Kim, and Scott Sanner. Online continual learning in image classification: An empirical survey. *Neurocomputing*, 469, 2022.

[32] Arnab Neelim Mazumder, Jian Meng, Hasib-Al Rashid, Utteja Kallakuri, Xin Zhang, Jae-Sun Seo, and Tinoosh Mohsenin. A survey on the optimization of neural network accelerators for micro-AI on-device inference. *IEEE Journal on Emerging and Selected Topics in Circuits and Systems*, 11(4), 2021.

[33] Shervin Minaee, Yuri Y Boykov, Fatih Porikli, Antonio J Plaza, Nasser Kehtarnavaz, and Demetri Terzopoulos. Image segmentation using deep learning: A survey. *IEEE Transactions on Pattern Analysis and Machine Intelligence*, 2021.

[34] Yue Ming, Xuyang Meng, Chunxiao Fan, and Hui Yu. Deep learning for monocular depth estimation: A review. *Neurocomputing*, 438, 2021.

[35] Tongzhou Mu, Zhan Ling, Fanbo Xiang, Derek Yang, Xuanlin Li, Stone Tao, Zhiao Huang, Zhiwei Jia, and Hao Su. Maniskill: Generalizable manipulation skill benchmark with large-scale demonstrations. *arXiv preprint arXiv:2107.14483*, 2021.

[36] João D Nunes, Marcelo Carvalho, Diogo Carneiro, and Jaime S Cardoso. Spiking neural networks: A survey. *IEEE Access*, 10, 2022.

[37] Computer Vision Online.

[38] Poojan Oza, Vishwanath A Sindagi, Vibashan VS, and Vishal M Patel. Unsupervised domain adaptation of object detectors: A survey. *arXiv preprint arXiv:2105.13502*, 2021.

[39] Pramuditha Perera, Poojan Oza, and Vishal M Patel. One-class classification: A survey. *arXiv preprint arXiv:2101.03064*, 2021.

[40] Yuzhe Qin, Yueh-Hua Wu, Shaowei Liu, Hanwen Jiang, Ruihan Yang, Yang Fu, and Xiaolong Wang. DexMV: Imitation learning for dexterous manipulation from human videos. In *European Conference on Computer Vision*, 2022.

[41] Ramesh Raskar. How to think like an MIT Media Lab inventor: https://www.youtube.com/watch?v=fYnJPtEJj4s, 2013

[42] Md Kamruzzaman Sarker, Lu Zhou, Aaron Eberhart, and Pascal Hitzler. Neuro-symbolic artificial intelligence: Current trends. *arXiv preprint arXiv:2105.05330*, 2021.

[43] Paul Smolensky, R. Thomas McCoy, Roland Fernandez, Matthew Goldrick, and Jianfeng Gao. Neurocompositional computing: From the central paradox of cognition to a new generation of AI systems, 2022.

[44] Zachary Teed and Jia Deng. RAFT: Recurrent all-pairs field transforms for optical flow. In *European Conference on Computer Vision*, pages 402–419. Springer, 2020.

[45] Jan van Gemert. How to do research with me in deep learning: https://jvgemert.github.io/HowToDoResearchInDL-slides.pdf

[46] Simon Vandenhende, Stamatios Georgoulis, Wouter Van Gansbeke, Marc Proesmans, Dengxin Dai, and Luc Van Gool. Multi-task learning for dense prediction tasks: A survey. *IEEE Transactions on Pattern Analysis and Machine Intelligence*, 2021.

[47] Kilian Weinberger. The importance of deconstruction: https://slideslive.com/38938218/the-importance-of-deconstruction

[48] Papers with code. Computer vision benchmarks and tasks: https://paperswithcode.com/area/computer-vision

[49] Tengyang Xie, Nan Jiang, Huan Wang, Caiming Xiong, and Yu Bai. Policy finetuning: Bridging sample-efficient offline and online reinforcement learning. *Advances in Neural Information Processing Systems*, 34, 2021.

[50] Tingman Yan, Yangzhou Gan, Zeyang Xia, and Qunfei Zhao. Segment-based disparity refinement with occlusion handling for stereo matching. *IEEE Transactions on Image Processing*, 28(8), 2019.

[51] Keren Ye and Adriana Kovashka. A case study of the shortcut effects in visual commonsense reasoning. In *Proceedings of the AAAI Conference on Artificial Intelligence*, volume 35, 2021.

[52] Rowan Zellers, Yonatan Bisk, Ali Farhadi, and Yejin Choi. From recognition to cognition: Visual commonsense reasoning. In *Proceedings of the IEEE/CVF Conference on Computer Vision and Pattern Recognition*, 2019.

[53] Mingliang Zhai, Xuezhi Xiang, Ning Lv, and Xiangdong Kong. Optical flow and scene flow estimation: A survey. *Pattern Recognition*, 114, 2021.

[54] Yu Zhang and Qiang Yang. A survey on multi-task learning. *IEEE Transactions on Knowledge and Data Engineering*, 2021.

[55] Kaiyang Zhou, Ziwei Liu, Yu Qiao, Tao Xiang, and Chen Change Loy. Domain generalization: A survey. *IEEE Transactions on Pattern Analysis and Machine Intelligence*, 2022.

[56] Rui Zhu, Zhengqin Li, Janarbek Matai, Fatih Porikli, and Manmohan Chandraker. Irisformer: Dense vision transformers for single-image inverse rendering in indoor scenes. In *Proceedings of the IEEE/CVF Conference on Computer Vision and Pattern Recognition*, 2022.

3

Soft Biometrics for Human Identification

Mark S. Nixon and Emad Sami Jaha

3.1 What Are Soft Biometrics for Identification?

We describe the use of a form of eyewitness description to recognise a known subject. The description concerns biometrics that can be perceived from a distance: it concerns the broad aspects of a subject's face, their body or their clothing. A victim can often provide a description of a subject. As illustrated in Figure 3.1, given a database of subjects who have been described using similar terms, it might be possible to match a subject to one of the subjects in the database. Alternatively, there might be an image of the crime scene and the descriptions can be derived by computer vision and the same matching process can be used. This allows for database search: if we have an image of a person and can derive the descriptions automatically, then we can generate the descriptions for each subject in the database. In more formal terms, this is zero-shot identification as we proceed from description to recognition. A major question here is what can be learned from human vision. The procedure is not to determine whether the image matches the chosen categories: that is computer vision and identification. Here, identification is achieved by using fine-grained semantic categories, which it might be possible to learn from images. We can then search videos using human descriptions of people: that is the aim of soft biometrics for human recognition. As we shall find, the approach needs ways to gather descriptions and computer vision techniques to generate the descriptions.

3.1.1 Recognition by Humans

Identification by eyewitness statements must have a history as long as society itself. The Bible includes descriptions of people and their biometrics, long before a notion that automated identification might be achieved. The design of subject identification forms, the nature and makeup of an identification parade, the accuracy of reconstruction from a verbal description or its assemblage by a

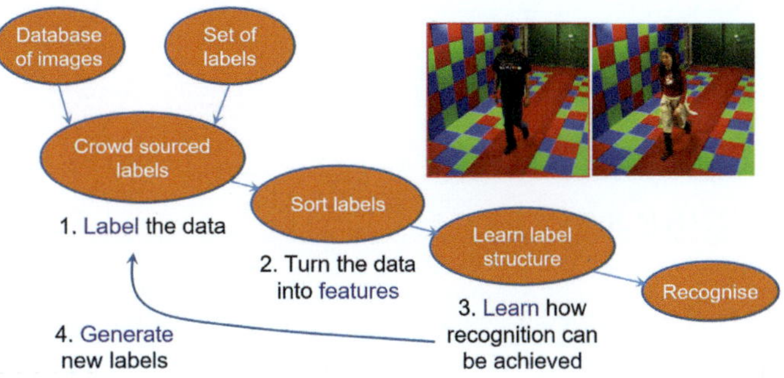

FIGURE 3.1
Procedure for identification by soft biometrics.

photofit procedure, together with the accuracy of human descriptions have all received careful consideration. It is now common to broadcast (on public television) images of suspects derived from crime scenes, largely in the hope that a member of the public will see the footage and can identify the target. We shall not dwell on these processes long since they are well-established and not without debate. The terminology used in computer vision is shown in Figure 3.2 where a subject is recognised from descriptions of their biometric attributes: the intersection between human analysis and the attributes is how they are defined for recognition purposes.

The newer aspects and technologies for eyewitness identification have been covered [1, 59] and some of the more established material on eyewitnesses has been previously well described [35]. The completeness and accuracy of eyewitnesses' descriptions have been studied [60] using 2299 statements of 1313 eyewitnesses for 582 different robbers from official police records and investigated the factors that could affect the accuracy and completeness of the statements. It was found that eyewitnesses tended to describe more general attributes (e.g. age and gender). In addition, the study showed that completeness (of the descriptions) did not necessarily imply accuracy, so even if an eyewitness provided sparse information, it tended to be accurate. Naturally, there are many more studies of this ilk: Burton et al. [8] focused on subjects' ability to identify target people from surveillance videos; Fahsing

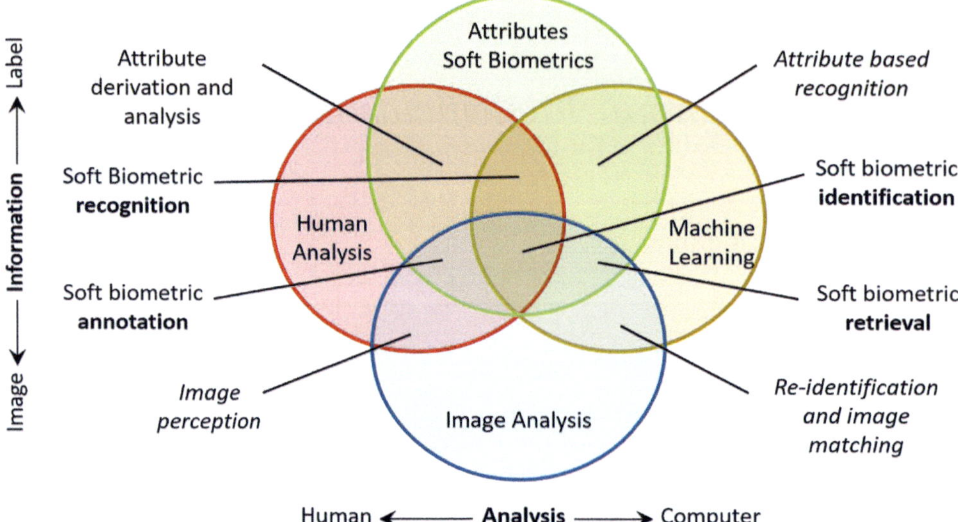

FIGURE 3.2
Approaches to identification.

et al. [17] studied 250 offender descriptions by witnesses of armed bank robberies and there are systems aimed to create images of criminals by using reconstruction techniques [20].

Eyewitness forms do suggest descriptive terms that are used in identification, though there is no exact consensus. One modern development concerns super-recognisers [13], people who have an above-average ability for face recognition. The lower range of ability in face recognition is prosopagnosia where subjects have no ability to recognise faces (some cannot even recognise their own mother) and prosopamnesia which is the lack of ability to learn new faces (thus forgetting them). At the other end of the scale, super-recognisers can pick people from crowds, even when their previous glimpse of the target was very brief. Clearly, the need for robust identification drives much research in this area; our interest here is to benefit from the descriptions that are already used for identification purposes. We have explored the linkage of our approaches to eyewitness statements and phrased in psychology terms [45].

3.1.2　Biometrics and Computer Vision for Recognition

As well as human recognition, Figure 3.2 also describes the technology areas in automated recognition. The approaches include image analysis/computer vision and machine learning/pattern recognition and when it is just the intersection of these two areas, the topic is re-identification. When attributes are used with these two technologies, the approach is used to recover or retrieve subjects from image data. When the technologies concern human analysis of image data, the concern is how to attach the attributes to describe the image data. The topic of this chapter is to be found at the intersection of human analysis, soft biometrics, image analysis and machine learning.

3.2　Biometric Attributes for Recognition

3.2.1　Categorical Attributes

The earliest approach to soft biometrics by describing the human body [52] obtained the descriptions, the soft biometric labels, for each subject in the Southampton gait database (SGDB) [56]. The process aimed to mimic and was informed by eyewitness statements for describing the appearance of people. The gait database comprises subjects walking unsupervised along a straight track, in a green-screen laboratory and viewed from the side. The majority classes of the walking subjects were young white males (aged around 22) and slightly older Chinese females. Labellers were presented with a single image of 10 subjects selected from the SGDB, and annotations from 38 annotators were collected using a web-based interface. The original soft biometric labels were categorical and used a five-point Likert scale. The categorical labels for the body were inspired by Macleod's earlier study [37]. The attributes from the 13 most significant attributes identified by a statistical analysis were selected as categorical soft biometrics for the human body. The selection was based on suitability for use when a subject was at a

TABLE 3.1
Categorical Biometric Attributes

Attribute	Term/Label	Attribute	Term/Label
0. Arm Length	(0.1) Very Short	12. Weight	(12.1) Very Thin
	(0.2) Short		(12.2) Thin
	(0.3) Average		(12.3) Average
	(0.4) Long		(12.4) Big
	(0.5) Very Long		(12.5) Very Big
2. Chest	(2.1) Very Slim	13. Age	(13.1) Infant
	(2.2) Slim		(13.2) Pre Adolescence
	(2.3) Average		(13.3) Adolescence
	(2.4) Large		(13.4) Young Adult
	(2.5) Very Large		(13.5) Adult
3. Figure	(3.1) Very Small		(13.6) Middle Aged
	(3.2) Small		(13.7) Senior
	(3.3) Average	18. Facial Hair	(18.1) None
	(3.4) Large		(18.2) Stubble
	(3.5) Very Large		(18.3) Moustache

distance (or low resolution) and detail could not be perceived. Note that in 2008, this long preceded the availability of the sophisticated crowdsourcing tools that are now available, and these will be found later.

Examples of some of the attributes used in this study and their descriptions (terms) are given in Table 3.1. Each of the 10 subjects selected from the SGDB had annotations from 38 annotators [52]. The traditional soft biometrics: age, ethnicity and sex were collected. For the body descriptions, the body and face features were usually described using a five-point scale, though age was given a larger set of descriptions (especially for younger subjects since they change faster in appearance). The body labels collected were those suited to use at a distance and the fine-grained identifications often used for facial image identification were not collected relying instead on those features available at low resolution.

Euclidean distance was used to evaluate the similarity between the probe and gallery feature vectors – this was possible due to the ordinal nature of the labels. The subjects were ordered based on their similarity to the probe. The position of the suspect's gallery biometric signature within the ordered list shows the retrieval performance of the system. A 90% Correct Classification Rate (CCR) – those subjects at rank 1 – could be achieved when descriptions were used alone and when more than 40% of the features were used ('just annotation' in Figure 3.3). Though not perfect, this was indeed an encouraging start. When combined with a gait biometrics approach (a system using computer vision-derived measures from video, used for identification) which could itself reach 98.1% CCR ('just visual' in Figure 3.3), the recognition performance rose to 99.5% CCR ('score fused' in Figure 3.3). This demonstrated that the semantic features could be

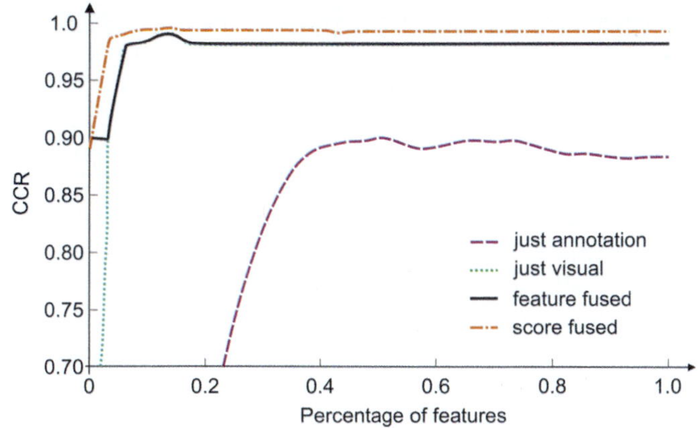

FIGURE 3.3
Performance of categorical biometric attributes.

used alone (though imperfectly) or to reinforce gait biometrics. This was indeed an encouraging start to using human descriptions of attributes for recognition.

3.2.2 Mitigating the Psychology Effects

As the process uses humans to describe other humans, the labels and the way they were presented aimed to mitigate known psychology effects as far as possible:

1. **Memory:** Labellers were allowed to view images for as long as they found necessary so that poor memory would not affect the labels provided;

2. **Defaulting:** No default labels were provided, and labellers were explicitly required to state values for labels;

3. **Anchoring:** This was initially addressed by the use of 'unsure' as opposed to 'average'. The order in which subjects were presented to the labellers was randomised to avoid anchoring;

4. **Categorisation:** Labellers were provided with five distinct categories for each label; and

5. **Owner variables:** A labeller's description of themselves was also collected, so aiming to be able to reduce this effect.

The only known and pertinent psychological factor that was explicitly omitted was the cross-race effect, since ethnicity is notoriously unstable – especially in modern society. It is hard to imagine how a factor which more affects the recognition procedure, rather than affecting a labelling structure, could be accommodated in this scenario. 'Politically correct' labels were avoided to ensure labellers understood the precise meaning of their instructions: a subject's figure was simply labelled as Thinner or Fatter; categorising Height avoided any notion of vertical challenge.

3.2.3 Comparative Attributes

An extension to the labelling approach was motivated by concerns about the accuracy of the perception of height [49]. In categorical labels, Height for a single (walking) subject was labelled as Very Short, Short, Medium, Tall or Very Tall. In Samangooei's study, the labels depended on the labellers' perception and on their impression of scene geometry (only the ceiling/floor and the position and size of a lamp gave any height suggestion; the distance of the subject from the camera was not known by the labellers). As shown in Figure 3.4, it was found that a subject who was actually 310 pixels height in the video

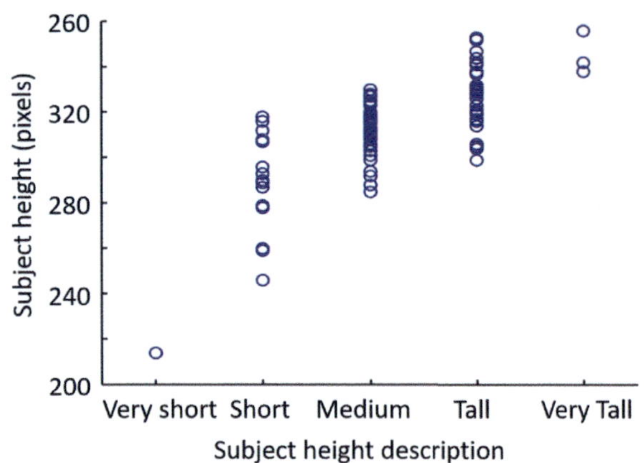

FIGURE 3.4

Relationship between subjects' height and categorical labels of height.

could be given labels of Short, Medium or Tall; in contrast, the labels Very Tall and Very Short were rarely confused. The problem with categorical labels is that there is no grounding. The collection of a labeller's perception of themselves proved to be of little use since many preferred to label themselves as medium for many attributes (one is perhaps disinclined to label oneself as fat though will readily supply that label for others).

The problem was solved by changing the structure of the labels: to derive labels by comparison. The new labels were derived by comparing the attributes of one subject to those of another subject using a new Web interface, shown in Figure 3.5. This then avoids the problem with observer variables since labellers' judgements were not influenced by their perception of themselves, especially for more personal terms such as those describing a subject's figure. The comparative labels of the body from two subjects were Much Shorter, Shorter, Same, Taller or Much Taller. Initially, 19 comparative attributes of 100 subjects from the SGDB were labelled by 57 annotators, together with three additional categorical attributes [49]. The initial development of comparative attributes was contemporaneous with relative attributes [46]. These were also formulated to have greater descriptive capability than categorical statements and are well known for such capability in the computer vision community. The only other work as yet on comparative labelling in biometrics has been to crowd-source the relative quantity of information for pairwise comparisons of latent fingerprint image quality [11].

3.2.4 Deriving rank-ordered measures from comparative labels

Comparative labels are just that: is one attribute different from another? To change this into a form usable for recognition, the comparisons need to be transformed into a rank-ordered list and there are many ways to transform a set of comparative labels. The Elo rating system [16] was originally aimed to quantify the ranks of chess players so the results of their matches could be used to derive rankings. There is no opportunity for chess players to play all others but the ranks can be inferred from the results by using the comparisons between them. Similarly, in soft biometrics, the comparisons are sorted into a ranked list per attribute and the position in the ranked list becomes the value of that attribute for a particular target. The ranking process can then be considered as one that transforms comparisons into categorical values.

In Elo, a 'match' is a comparison between two players, i and j. The match outcome reflects status, in chess whether one player is superior to another, or not. The outcome of a match is used to update the players' ratings. At iteration n for two players i and j, the ratings

FIGURE 3.5
Website developed for deriving comparative labels. (With courtesy from [48].)

R_i^n and R_j^n are updated to be the value at iteration $n+1$ according to the results of a match between them. The result of a match S is 1 for winning, 0.5 for a tie, and 0 for losing $S_i^{<n>} = 1$ when player i wins at iteration n. The ratings for two players at iteration $n+1$ are updated as

$$R_i^{<n+1>} = R_i^{<n>} + k(S_i^{<n>} - E_i)$$
$$R_j^{<n+1>} = R_j^{<n>} + k(S_j^{<n>} - E_j) \qquad (3.1)$$

where E denotes the expected outcome given the current ratings. The rating is updated by the difference between what has been achieved and what was expected. The parameter k controls the maximum rating adjustment possible. In the case of soft biometric labels, k depends on the available number of comparisons N_C. The maximum rating, M, is used to define $k = MN_C$ allowing M to be fully explored by any number of comparisons. Values for E are then calculated by

$$Q_i = 10^{R_i U}$$
$$Q_j = 10^{R_j U}$$
$$E_i = Q_i(Q_i + Q_j) \qquad (3.2)$$
$$E_j = Q_j(Q_i + Q_j)$$

where U is chosen to reflect how a player's current rating can affect the expected result. Here, the terms describing soft biometric attributes were assigned a number ranging from -2 to $+2$ based on their order. We shall find later that this was reduced to use -1, 0 or 1 representing lower, equal or higher, respectively, since this gave better performance. The 'score' from a comparison is determined by normalising the given label's value to within 0 and 1. If the actual result is the same as that which is expected, then the rankings do not change since there is no difference to change them. On the other hand, if the actual result differs from the expected one, the targets' rankings are increased or decreased according to the comparison. The magnitude of adjustment depends on the difference between the actual and the expected results. In this way, we determine feature vectors for different subjects that are vectors of the labels associated with the categorical attributes and the rankings of the comparative attributes.

When this was applied to the comparative measures for height, the correlation between subjects' height in pixels and their height derived by Elo ranking comparisons was statistically significant (Pearson's correlation $= 0.87$, $p < .0001$), showing that the relative measurements inferred from human comparisons strongly represent the physical attributes. This suggests that the Elo rating system resulted in, from visual comparisons, an accurate ordering of the subjects based on height. The correlation between pixel height and the absolute height labels used previously (Figure 3.4) was weaker (Pearson's correlation $= 0.71$, $p < .0001$) than relative measurements, mainly due to the highly subjective and categorical nature of

FIGURE 3.6
Recognition performance by comparative labels and by categorical labels.

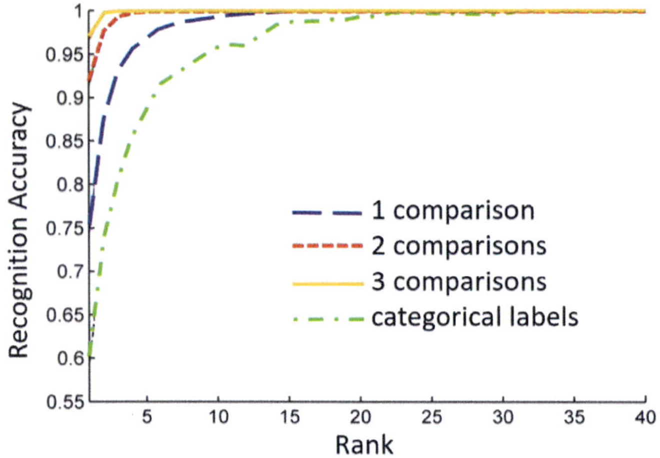

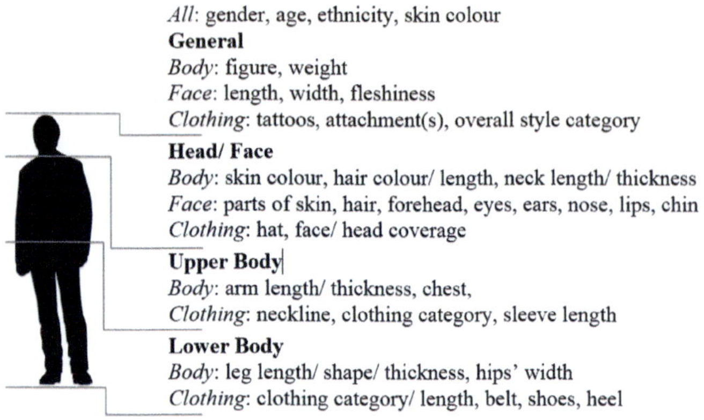

FIGURE 3.7
Varieties of labels used for describing people.

the absolute labels. Note that the comparative basis and ranking approach essentially projected a single attribute along a chosen axis. This is the underlying reason for the increase in discriminatory ability.

3.2.5 Recognition Using Comparative Attributes

Using comparative labels led to a considerable improvement in recognition, illustrated in Figure 3.6 in comparison with recognition by the categorical labels. When only a single comparison was made to derive the attributes, recognition performance was close to that of the original categorical labels. When each subject was compared with more subjects, a remarkable increase in performance was observed. The recognition performance at rank 1 rose to over 90% compared with around 50% by categorical labels. By rank 10, 100% recognition could be achieved using comparative labels, whereas categorical labels were close to peak performance first at rank 40. Clearly, the comparative labels can improve recognition considerably. This is likely due to the removal of bias: by comparing one attribute with that of another, one removes the bias and subjectivity of the human observations. This was seen in the order of the resulting height attributes; here, we see the effect in a gamut of attributes.

3.3 Body Soft Biometrics for Identification

As we now have the methodology and the confirmation that labels derived from humans can be used to identify people, we shall move to approaches that use more data and newer techniques. The techniques are largely focused on particular means of identification: the body, the face and the subject's clothing, as shown in the overview in Figure 3.7. All techniques use the traditional soft biometrics: gender, age, ethnicity and skin colour. There are some other terms that are used between modalities: the body features concentrate on those features available at a distance and those features include some facial features; the clothing features relate not only to the body but also to the head. The labels are collected by crowdsourcing, since this had advanced rapidly following Mechanical Turk. Also, there are new classification techniques, particularly deep learning. As we shall find, these have all been used to advance the state-of-the-art in soft biometrics for human identification.

3.3.1 On Body Attributes for Use at a Distance

There has been some previous interest in psychology in identifying people by their body, though most attention focuses on the use of the face. Perhaps stimulated by a plethora of scene of crime pictures which are now broadcast where the face was concealed, recently, there has been a study on the potency of body measurements vs those of the face, for identification, in some ways redeeming Bertillon's original approach [36]. The study was based on using anthropometric measurements of 3982 individuals from the US Army Anthropometry Survey (ANSUR) database and concluded that 'The body is more variable than the face and should be used in identification', also stating practical advantage.

More recently, a study has considered classifying gender, whole-body imagery, from anthropometric ratios mitigating noise from computer vision-based measurements [29]. The use of the ANSUR database [36] mitigates Bertillon's previous assertion that anthropometric

FIGURE 3.8
Southampton tunnel data samples.

measures could be used for recognition and previous studies [50, 52] have shown that human descriptions can be used, and how they can best be deployed. The next stage was to translate their work with more subjects and labellers using a more contemporary machine learning and computer vision framework and to benefit from improved technology.

3.3.2 Crowdsourcing Body Attributes

The new approach to large-scale use of soft biometrics for the body Martinho-Corbishley et al. [38, 39] aimed to investigate binary, categorical, relative binary and relative continuous labelling techniques. These were derived for 12 attributes of 100 subjects captured from eight camera viewpoints, comprising 1600 images. The 1000 subjects were chosen to be gender-balanced comprising 50 female and 50 male subjects. The images were captured in a controlled environment, Figure 3.8, from up to 12 camera viewpoints in a synchronised video. The background and the lighting were controlled so as for efficient walking subject extraction and were aimed to provide a 3D reconstruction of the walking subject for recognition purposes [54]. The controlled nature of the dataset allowed for a more principled study of the topic of interest, avoiding reliance on extraneous factors consistent with real-world images.

By collectively reviewing the most significant, prevalent and stable traits from [37, 50, 52, 57], the final soft biometric attribute lexicon included two global (Gender, Age) and ten body soft attributes (Height, Weight, Figure, Chest size, Arm thickness, Leg thickness, Skin colour, Hair colour, Hair length, Muscle build), and the terms for three of these can be seen in Table 3.3. Ethnicity was excluded as a global soft trait, as separation by ethnicity has been challenged previously [36] as it can be misinterpreted when describing low-quality imagery, and the notion of ethnicity in modern society is very complex. Given the intention to avoid owner variables, gender and skin colour were now collected as comparative attributes. This appeared to be the first notion that gender could be ascribed as a comparative attribute (as a continuous measure) rather than as categorical (as a binary measure).

The Crowdflower platform was used to build and acquire the body-attribute annotation for identification. As labellers were drawn from a wide pool, it was necessary to provide a satisfactory environment in order to determine the best set of labels. Equally, it was necessary to reject some labellers on grounds of quality. Several subsets of questions were tested to measure the acceptability of the predefined test questions. Some overall considerations that led to the layout were

1. Respondents were rejected (and not paid) if their response distribution varied largely from the average response distribution formed during the initial trials.

2. 'Can't see' was capped at a maximum 20% per respondent, rejecting those who exceeded this value.

TABLE 3.2
Example Attributes and Labels for Identification by Body Descriptions

Attribute	Response label					
	5	4	3	2	1	0
Gender	Much more feminine	More feminine	Same	More masculine	Much more masculine	Cannot see
Age	Much more old	More old	Same	More young	Much more young	Cannot see
Height	Much more tall	More tall	Same	More short	Much more short	Cannot see

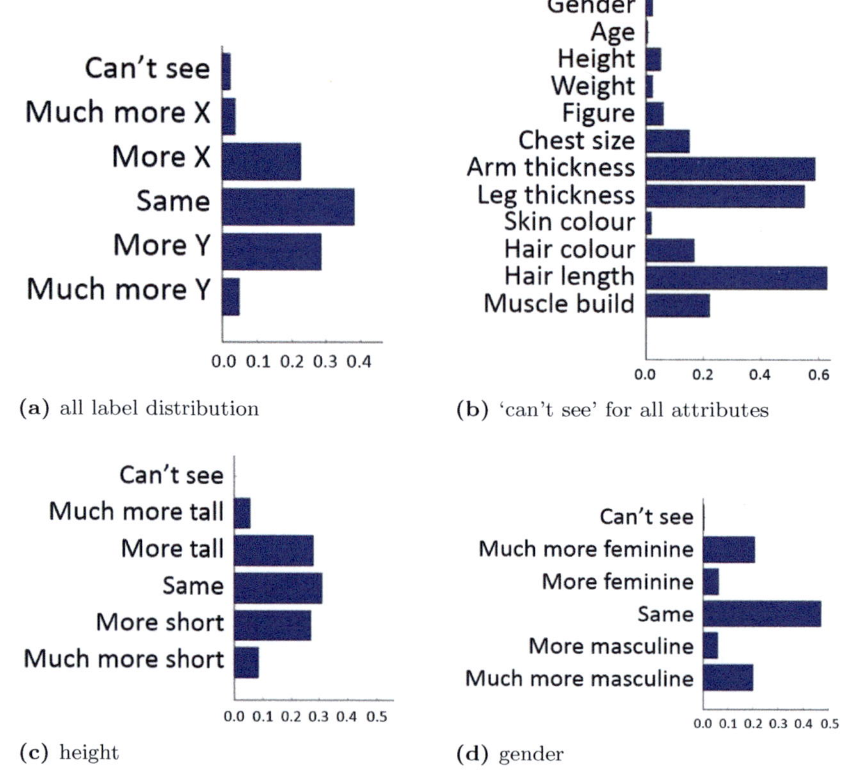

(a) all label distribution

(b) 'can't see' for all attributes

(c) height

(d) gender

FIGURE 3.9
Analysing crowdsourced response distributions: (a) distribution of all labels, (b) distribution of "can't see" labels across attributes, (c) distribution of heights, and (d) distribution of gender.

3. To make test questions fair, they were sampled from what appeared to be the more obvious comparisons and only the most fundamentally incorrect responses were rejected. However, respondents must score at least 80% to proceed.

4. In addition to a large number of introductory examples, each question included text and highlighting, intentionally reiterating the task 'compare the person on the left, to the person on the right'. The response form was formatted using vertically aligned radio buttons, enabling quick and instinctive responses. Psychological considerations were also involved: labellers could view pages for as long as required to avoid memory effects; initial answers were left blank to avoid anchoring (Section 3.2.2), and comparison reduces the effect of owner variables (as Section 3.2.5).

3.3.3   Analysing Body Attribute Labels

Via crowdsourcing 59,400 unique attribute comparisons were collected from 892 trusted labellers (124 untrusted labellers were unused, and 4383 responses were rejected). The distributions of the labels appear close to what might be expected when labelling a large population. People are generally similar in appearance as shown in Figure 3.9(a) and the attributes were chosen to be visible and few cannot be seen. Of the labels that had the highest proportion of 'can't see' in Figure 3.9(b) most are concerned with width and length. It is interesting that the hair length can be difficult to perceive in a viewpoint that was chosen to mimic surveillance video. Evidently, annotation uncertainty is trait-specific and independent and gender and skin colour are two of the least uncertain traits by this comparative attribute collection. Compared with previously the distribution for height in Figure 3.9(c) is now much closer to that of a normal distribution, as would be expected, and that comparative height could always be labelled. The distribution for gender labels in Figure 3.9(d) shows that for most subjects the gender is clear, but there are some for whom it is not as clear as for others. There was actually one subject whose gender could not be attributed even by expert labellers and 'can't see' was the label used there.

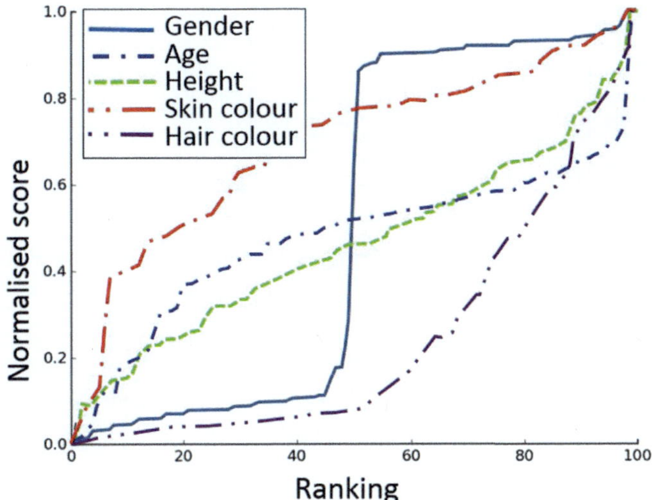

FIGURE 3.10

Normalised relative scores vs ranks for each attribute. (With courtesy from [38].)

3.3.4 Ranking the Comparative Body Attributes

The body attribute comparisons were ranked using a soft-margin RankSVM formulation to infer relative attributes from the set of ordered pairwise comparisons per trait, as previously applied to texture attributes [41]. Essentially, this derives the relative continuous scores for each comparison of each attribute and this essentially projects each attribute along a single axis. When plotting the normalised relative scores vs ranks for each attribute, Figure 3.10, an evenly distributed attribute would be a diagonal line from the origin to the point with rank 100 and score 1.0. The closest to that are Age and Height. Others

which are not shown (Weight, Figure, Chest size, Arm thickness, Leg thickness, Hair length and Muscle build) have a similar distribution to Age and Height. Of the other three, Hair colour appears to verge towards the darker (ranks (0–50), whereas lighter shades are more easily distinguished (ranks 50–100). Skin colour veers towards lighter (largely reflecting the composition of the population, though illumination might also be a factor). Gender should be more binary distributed (and not close to an even – or normal – distribution). This appears to be the case, though it is clearly not an absolutely binary categorisation, reflecting the previously observed uncertainty in the gender label (Figure 3.9(d)). The correlations between the attribute ranks were largely similar to those observed found for categorical labels. Since the labels are now closer to a continuous variable, the attributes are called super-fine to differentiate them from the coarser binary and fine attributes are generally to be found in computer vision [46].

3.3.5 Recognition by Body Descriptions

The question remains: can the descriptions be used to identify people? The recognition performance by crowdsourced body attributes is shown in Figure 3.11(a) for using signatures built both from relative normalised scores and from the ranked position of each attribute. Evidently, performance was not as potent as found previously, achieving the best performance (close to [50]) only with multiple comparisons (the match rate by crowdsourced descriptions reaches 95% CCR, whereas the previous ones reached 95%). Note that the crowdsourced labels include no inferences and thus avoid self-correlation and are drawn from a deeper and wider pool of

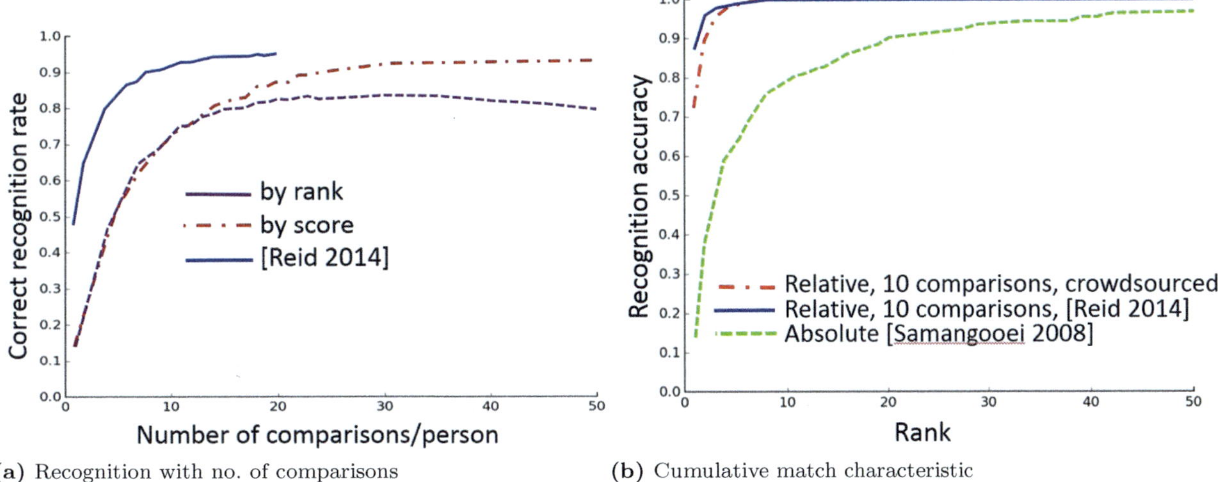

(a) Recognition with no. of comparisons   (b) Cumulative match characteristic

FIGURE 3.11

Recognition by crowdsourced body attributes: (a) Recognition with no. of comparisons, (b) Cumulative match characteristic.

labellers. By this, the results of crowdsourcing appear more representative of a realistic application of these techniques.

The CMC analysis for 10 comparisons, Figure 3.11(b), confirms that crowdsourced comparative labels are slightly inferior to their previous (selected and including inference) versions, and with considerably greater potency than the categorical versions. Clearly, the body can be used for recognition as Bertillon originally observed, as was confirmed over a century later [36].

3.3.6 Closing the Semantic Gap: Estimating Human Body Descriptions by Computer Vision

The ResNet-152 CNN was selected as an architecture for estimating the super-fine attributes [40]. First, labels were crowdsourced from human labellers to determine human capability on the (PETA) dataset. This used similarity comparisons, given their established performance, which were embedded and clustered to find prototypes representing the most characteristic images. Then the superfine labels were fed to the deep learning network for regression to match the labels to subjects. This is then using deep learning to understand the performance of human-derived similarity comparisons. A more conventional approach is to classify images by using deep learning to relate the structure of the categorical labels to the image dataset. This provides a mechanism by which the comparative labels can be estimated for a much larger dataset by extending the classifications learned from a smaller dataset and their correlation to image data, and thus to be able to label unseen image data.

The new approach was used on the PETA dataset which has categorical 'ground truth' labels. Note that there is actually uncertainty in these ground truth labels and an investigation into gender [39] confirmed the need for continuous labels (and the existence of uncertainty in the ground truth). The architecture was evaluated in comparison with a selection of contemporaneous studies on PETA: multi-attribute learning for attribute recognition DeepMAR [34], and a multi-attribute residual network MAResNet [6]. A subset of the performance for attribute estimation is shown in Table 3.3, which shows that in multi-shot classification (re-identification), the performance is generally good and can be excellent (for all three architectures). On average multi-attribute learning could exploit the scenario best and the residual network least. The main point here is that the attributes can indeed be learned from the data, and with confidence. One would expect that to be the case for multi-shot identification, since it is a matter of associating labels with extant data. More interestingly, it is possible to extrapolate the learning process to estimate labels in previously unseen

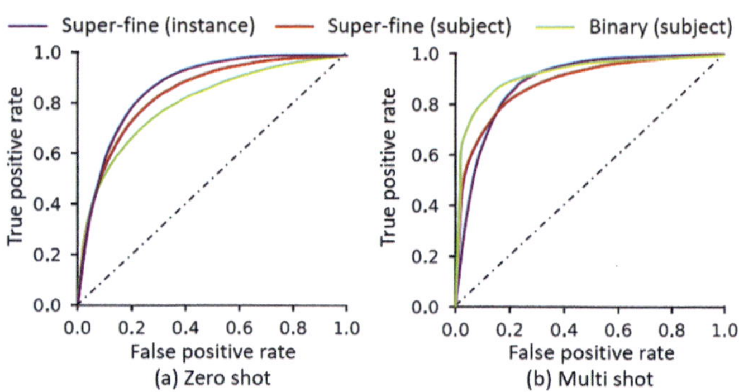

FIGURE 3.12

Classification by superfine attributes: (a) Zero shot, (b) Multi shot. (With courtesy from [40].)

data. This is critical for soft biometrics to be deployed in surveillance image analysis. The performance shows that it can indeed be done, and the performance of this first approach was encouraging, but far from perfect.

Beyond estimation, the next question is whether the approach can be used to classify subjects correctly. As it is machine learning, the arrangement of data is very important and here the training, validation and test sets are split at the instance level, potentially containing multiple instances of a subject across all sets. In multi-shot evaluation, both instance and subject-level super-fine labels consistently outperform binary labels by up to 11.2% mAP and 7.2% mAP for gender and age, respectively, though all 35 binary attributes comfortably surpass all three (gender, age and ethnicity) super-fine traits in this scenario (Figure 3.12(b)). This is, of course, the usual testing scenario for computer vision algorithms. For soft biometric identification, the performance on previously unseen data also needs to be established. The zero-shot retrieval result emulates a real-world surveillance scenario where estimators are evaluated without prior exposure to similar instances during training; a subject can be classified from within new and unseen video material (Figure 3.12(a)). Learned descriptors that overfit (classify noise) to subject identities are now penalised when the architecture is at the test stage, as they do not generalise to unseen subject instances. In zero-shot, the binary labels perform particularly poorly; conversely, super-fine attributes gender and age to improve mAPs up to 6.8 % and 14.8%, respectively, in the zero-shot scenario. The classification by just three combined superfine attributes (gender, age and ethnicity) outperforms 35 binary attributes by 6.5% in zero-shot identification. Clearly, the use of comparative labelling has led to a major performance advantage. Equally clearly, subjects can also be recognised in surveillance footage, and with the advance in computational architectures and software, one can only anticipate that the performance will improve.

TABLE 3.3
Superfine Attribute Estimation

| Attribute | Proportion | DeepMAR | Multi Shot | | Zero Shot |
			MAResNet	ResNet-152	ResNet-152
Gender Male	54.9	89.90	76.60	**93.06 ± 0.33**	83.96 ± 0.56
Age 15–30	49.7	85.80	78.38	**87.28 ± 0.48**	74.39 ± 1.24
Age 30–45	32.9	81.80	75.55	**82.98 ± 0.47**	60.60 ± 3.10
Age 45–60	10.2	**86.30**	80.87	83.95 ± 0.97	55.10 ± 0.71
Age >60	6.2	**94.80**	86.29	92.74 ± 0.53	56.59 ± 4.08
Acs Hat	10.2	**91.80**	81.69	91.64 ± 0.26	59.45 ± 3.09
Acs Nothing	74.9	85.80	74.65	**87.16 ± 0.44**	67.66 ± 1.37
Cry M.Bag	29.6	82.00	71.99	**83.06 ± 0.21**	64.80 ± 1.38
Cry Nothing	27.6	83.10	71.31	**84.45 ± 0.38**	68.34 ± 0.56
Upr Casual	85.3	84.40	75.20	**86.58 ± 0.37**	62.98 ± 1.34
Upr S.Sleeve	14.2	**87.50**	77.35	86.62 ± 0.69	85.06 ± 0.77
Lwr Casual	86.1	84.90	77.39	**86.90 ± 0.96**	63.94 ± 2.16
Lwr Trousers	51.5	84.30	71.51	**85.75 ± 0.70**	50.00 ± 0.83
Average		**82.60**	75.43	81.65 ± 0.83	66.45 ± 1.49

3.4 Face Soft Biometrics

3.4.1 On Face Attributes for Identification

People are used to verbally describing the appearance of another person's face. One of the earliest approaches to use attributes in computer vision was for faces [31] focused on attributes and simile attributes for identification within the Labelled Faces in the Wild (LFW) database [32]. Klare was the first to pioneer the interaction between automatically extracted soft biometrics and human-generated soft biometrics [30]. The study was largely concerned with using proposed categorical facial attributes for identification in criminal investigations. A set of 46 facial attributes was defined aimed to encapsulate persistent characteristics (performance over time was a particular issue), and an SVM regressor was trained to estimate the attributes automatically from face images. Identification experiments were performed using the FERET database with all possible combinations of the probe-gallery (i.e. human vs machine). Identification using an automatic probe in an automatic gallery resulted in the best recognition accuracy as compared with the other three identification scenarios in which human annotations are used (i.e. for a probe, gallery, or both). An interesting methodology has also been proposed for constructing biometric signatures that are based on the relationships between categorical facial attributes. Their method has demonstrated its effectiveness for face verification using the LFW and the PubFig [31] databases. Parikh and Grauman first explored the estimation of

relative facial attributes [46], where a RankSVM with similarity constraints was trained to predict the relative strengths of facial attributes from pairs of face images. Using relative facial attributes, their approach showed better accuracy on a large database when predicting unseen categories as compared with the categorical (binary) approach.

By capability and performance, it appeared natural to translate the methods used in soft biometrics for identification to the human face. The first approach for face identification using comparative soft biometrics [49] used a subset of attributes from the Aberdeen University Face Rating Schedule (FRS) [60], which offered a comprehensive selection of attributes that has been used in other studies [20, 64]. A later study aimed to investigate more deeply the accuracy and nature of soft biometrics for face recognition using crowdsourced comparative labels [2].

Eyebrow shape was the only feature for which categorical descriptions were used since its description was too complex to form in a comparative manner. Also, traditional soft biometrics (gender, skin colour and age) were included to aid the identification performance. Gender was in a comparative format as introduced in [39] to achieve a level of consistency among the attributes. Table 3.4 shows the attributes that were selected and the terms used to describe them. Initially, the comparisons were based on a five-point bipolar scale that ranged from 2 to +2 according to the label assigned to that attribute. For example, the attribute Eyebrow Thickness was described as [Much Thinner, More Thin, Same, More Thick, Much Thicker] and Skin Colour as [Much Lighter, More

TABLE 3.4

Comparative Face Attributes and Labels [2]

Attribute Class	No.	Attribute	Labels
Traditional soft biometrics	1	Age	[More Young, Same, More Old]
	2	Figure	[More Thin, Same, More Thick]
	3	Skin Colour	[More Light, Same, More Dark]
	4	Gender	[More Feminine, Same, More Masculine]
Eyes	5	Eye Shape	[More Tilted Inward, Same, More Tilted Outward]
	6	Eye Size	[More Small, Same, More Large]
	7	Inter- pupil Distance	[More Close, Same, More Wide]
	8	Eye-to-Eyebrow Distance	[More Small, Same, More Large]
	9	Eyebrow Thickness	[More Thin, Same, More Thick]
	10	Eyebrow Length	[More Short, Same, More Long]
	11	Inter Eyebrow Distance	[More Close, Same, More Wide]
Jaw	12	Chin Height	[More Small, Same, More Large]
	13	Jaw Shape	[More Chiseled, Same, More Lantern]
	14	Cheek Shape	[More Flat, Same, More Prominent]
	15	Cheek Size	[More Small, Same, More Large,]
Hair	16	Forehead Hair	[Less Hair, Same, More Hair]
Mouth	17	Mouth Width	[More Narrow, Same, More Wide]
	18	Lips Thickness	[More Thin, Same, More Thick]
Face	19	Face Width	[More Narrow, Same, More Wide]
	20	Face Length	[More Short, Same, More Long]
	21	Nose Length	[More Short, Same, More Long]
	22	Nose Width	[More Narrow, Same, More Wide]
	23	Nose – Mouth Distance	[More Short, Same, More Long]
	24	Face Shape	[More Oval, Same, More Round]

Light, Same, More Dark, Much Darker]. All the attributes had a 'Do not know' label to address the cases in which an attribute was occluded. Nevertheless, the recognition performance showed that compression of the comparison levels improves the identification performance. Therefore, the five-point scale was compressed to a three-point scale. Using Crowdflower for crowdsourcing ensured a wide population coverage within a professional labelling system. The attribute Eyebrow Thickness was described as [More Thin, Same, More Thick] and Skin Colour as [More Light, Same, More Dark]. The full set of face attributes and their compressed descriptions are shown in Table 3.4.

As previously, in Figure 3.13, recognition was good and more comparisons improved recognition. Interestingly, compressing the labels further improved recognition capability. There are several possible reasons for this. Perhaps the most straightforward reason is that labellers found difficulty or were inconsistent when discriminating 'Much more' from 'More'. It is likely that there is a much deeper reason: comparative attributes reduce the cognitive bias associated with labelling by categorical labels and yet using a five-point scale could actually perpetuate

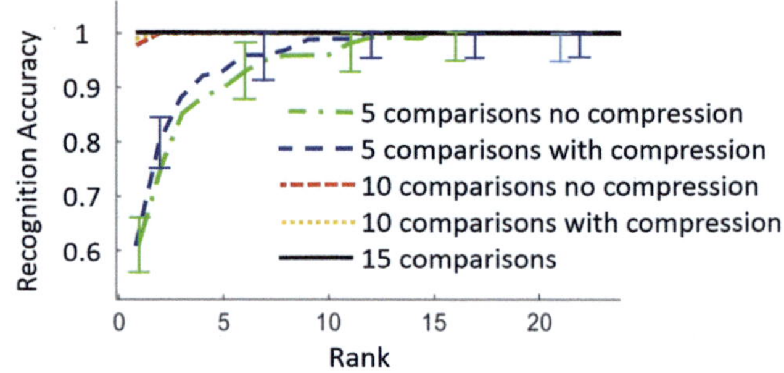

FIGURE 3.13

Effect of label scale compression on recognition.

this problem. Compressing the scale then removed any form of categorical basis and so it is perhaps unsurprising that good eventual performance was observed since the human ability for comparative judgement is used in its most pure form. Note that this effect is observed little when more comparisons are made, as the error reduces. This is consistent with the reduced variance of the

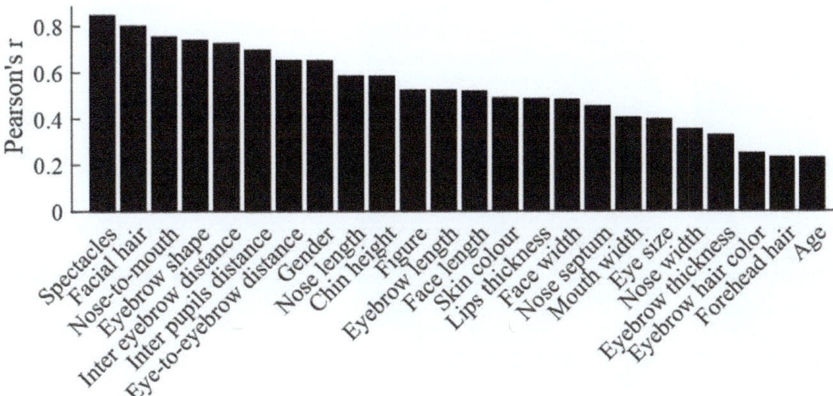

FIGURE 3.14
Face attribute stability based on Pearson's *r*.

recognition accuracy (computed according to a random sampling of the database) with larger numbers of comparisons. Given that we are aiming for an application scenario of eyewitness identification, any scenario that targets fewer comparisons will always be preferred.

3.4.2 Performance of Crowdsourced Face Descriptions

It is not just sufficient to analyse the performance of the system for recognition, for it is also important to know how recognition is achieved and how well descriptions fit features themselves. This might later influence the selection of particular features for identification or to give weight to their likely importance in recognition. It is possible to understand the capability of features by statistical analysis (e.g. by correlation and by ANOVA), as well as by automated learning algorithms. The trait stability reflects the level of agreement that has been achieved on particular face features by different groups of labellers. This is shown in Figure 3.14 for crowdsourced face attribute comparisons on the 4038 subjects within the View 1 subset of the LFW database where all images were normalised so that subjects had the same inter-pupil distance of 50 pixels (to remove the influence of overall size/distance from the camera [3]. The analysis shows that binary and clearly evident facial attributes such as spectacles and facial hair are labelled consistently. Interestingly, the separation between nose and mouth precedes a group of features which label the eyes. This reflects the importance of such analysis: what is consistently understood and thus labelled is not necessarily that which would be identified by human intuition. The role of the eyes in human analysis is to be expected, but it might not be anticipated that humans often agree on nose-to-mouth spacing or on face length, as this analysis shows.

3.4.3 Recognition by Human Descriptions

As we say in England, the proof of the pudding is in the eating: can these descriptions be used for recognition/ identification? The evaluation used six-fold cross-validation wherein the 4038 subjects in LFW View 1 were split into six sets; one set was used as a test and the other five for training (the gallery). The number of comparisons corresponds to each different face attribute. Clearly, the more comparisons, the higher the recognition rate. The identification approaches recognition quickly. Given the relevance of human descriptions, it is worth noting that by using 24 attributes with 10 comparisons, a match will be found in the top ten results with a probability of 97%. That the number of comparisons is 10 is pertinent since that is the average size of an ideal identity parade. By comparison with another study of soft biometrics on the FERET dataset, Klare et al. [30] achieved a rank-1 accuracy of 22.5% using 46 features (19 binary and 27 categorical), whereas her performance is 30.2% at rank-1 with 10 comparisons. One suspects the advantage in performance is due to the nature of the comparative labels and is further confirmation of their superior performance.

3.4.4 Exploring Comparative Attributes with Computer Vision

Having established that semantic face signatures could be used to recognise faces, the question then becomes the same as for the body: is it possible to estimate the same labels by computer vision? Should the performance be sufficient for recognition, then such an approach can be used to extend the labels to much more data and derive semantic-based recognition on a database allowing for the search of a subject in a large set of images. Thus, an approach was derived, which combined the human labelled attributes with established computer vision techniques

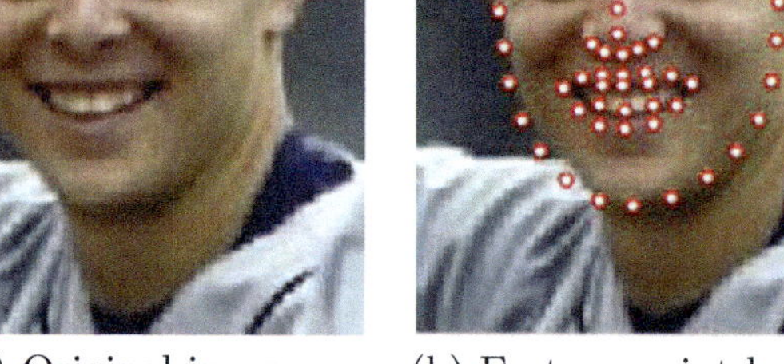

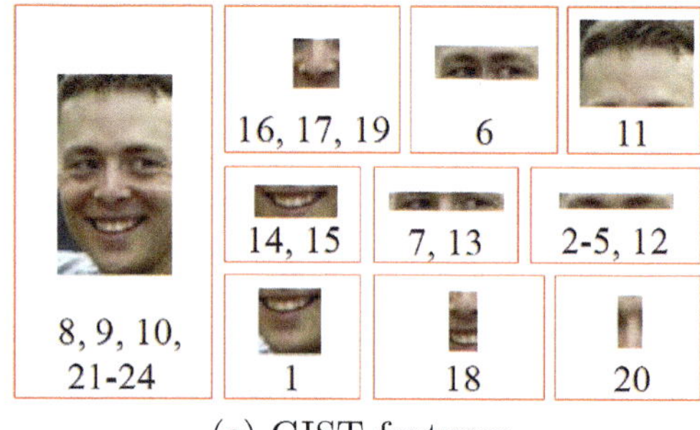

(a) Original image   (b) Feature point locations   (c) GIST features

FIGURE 3.15
Labelling face features automatically: (a) Original image, (b) Feature point locations, (c) GIST features. (With courtesy from [4].)

[4]. The LFW- 199 MS4 database was used as it is well known and often used in automatic face recognition. This subset contains 1720 images of 430, as the subjects with at least four sample face images were used. The sample faces were aligned and then normalised so that the inter-pupil distance was 50 pixels. This removes from consideration the effect of subject size and distance from the camera. The attributes were labelled using Crowdflower, as in the previous section. Nearly 250,000 attribute comparisons were gathered for around 10,000 subject comparisons by around 10,000 labellers. The attributes used the compressed comparative scale (a Likert scale of 3).

Then to derive the labels by computer vision, a shape model via the Active Appearance Model (AAM) approach was combined with response maps that are generated by trained detectors using Constrained Local Models (CLMs) for face alignment. The AAM combines shape and texture to iteratively fit a shape model – of the labelled points derived from training images – to a new image. A framework trained for face alignment in-the-wild [39] was used to locate the facial landmarks. The extraction process is illustrated in Figure 3.15. Figure 3.15(a) shows an image from the LFW dataset, Figure 3.15(b) shows the points extracted by AAM/ CLM and Figure 3.15(c) shows the regions extracted automatically as face features.

A Support Vector Regressor (SVR) was trained using the GIST features along with the ranks from the attributes by Elo so that the relative rank of an attribute for a new or unseen subject can be determined. The 430 subjects in the LFW-MS4 369 were randomly allocated to ten folds of the same size. Ten-fold cross-validation was used to derive an average rank by Elo and by SVR across ten folds of four subject samples. The SVR system was trained using the GIST features from four sam-

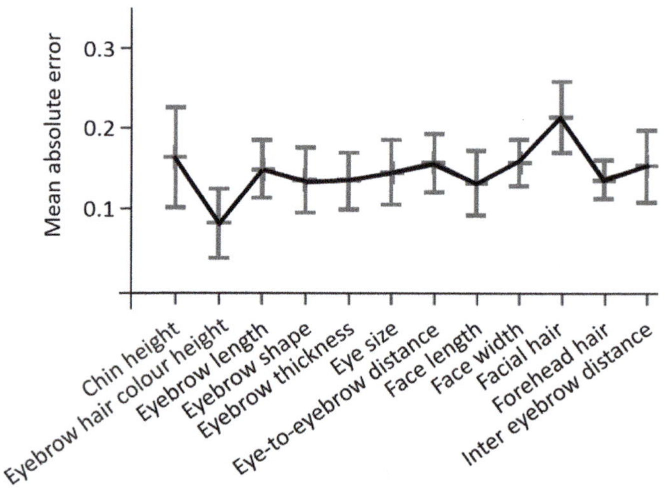

FIGURE 3.16
Mean absolute error between visual and semantic features.

ples per subject together with the crowdsourced relative comparisons. The ranks for each attribute were inferred, and results are derived from the average over ten folds and four samples per subject. The mean absolute error was derived between the ground truth and the predictions and is shown in Figure 3.16 for a subset of the attributes. The errors in the attributes were largely similar, and the exceptions were spectacles, facial hair and eyebrow hair colour. The first two of these are binary attributes, and consistency was witnessed in the other attributes. These errors and their consistency suggest that the labels predicted by the automated system are sufficient for this purpose. Of course, accuracy could be improved as this is invariably the case. The study shows that it is possible

(a) Gender estimation   (b) Age estimation

FIGURE 3.17
Matching faces to automatically derived labels: (a) Gender estimation, (b) Age estimation. (With courtesy from [4].)

to use computer vision to predict attribute values. The training process is guided by computer vision and ensures that the generalisation to larger datasets by automated means not only avoids the cost (time and financial) in the labelling process but also achieves good accuracy.

Another interpretation is the ranks of images which are retrieved according to the automatically derived labels. Here, the analysis is by intuition: do the images appear to match the labels? Figure 3.17 shows the highest- and lowest-ranked subjects for gender and for age. To human vision and knowledge, the visual features V match the semantic features S. Age by human vision is generally more challenging, though the results in Figure 3.17(b) appear correct. Note that this analysis rather contravenes the general premise that comparative labels have greater discriminatory power than categorical ones. It is invariably tempting to revert to categorical analysis as it is easier to explain and understand, as here, though the performance of categorical descriptions is actually impaired.

Again the proof of the approach is in classification capability and accuracy, and the CMC is shown in Figure 3.18. There are three different scenarios where the probe P and the database (gallery) DB are derived from the semantic S and the visual features V [4]. The best performance is observed when the test semantic features P_s are compared with a gallery of semantic features DB_v with a rank-5 identification rate of 98.4%. Using ten comparisons only, the search range can be narrowed down to four subjects with a high probability that a correct match

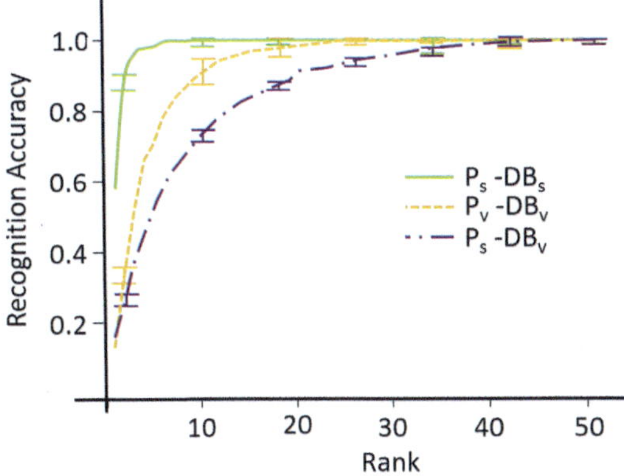

FIGURE 3.18
Recognition performance by face semantic description combined with computer vision.

with the subject will be found. By way of contrast, a set of 46 attributes (19 binary and 27 categorical) resulted in an accuracy of 22.5% at rank-1 using 1196 subjects from the FERET database [30].

Performance decreases when the vision-based probe, P_v, is a visual face signature that is used to search a visual database, DB_v, imitating a scenario in which a sample facial image is used to identify an unknown subject from a database of facial images. Recognition is low at rank-1 and increases to 72.8% at rank-10, reaching 90.7% at rank-20. One advantage of the approach is that it can be used to narrow the search range for the probe subject in the database to 10.2%. Face recognition is well known to be more popular now compared with when Mark Nixon published his first paper on automated face recognition in 1985 [43]. Many of the sophisticated approaches train machine learning on face databases like LFW to derive very high recognition rates but without reference to the structure of the data studied. It is clear that if a database of potatoes was labelled as faces and then recognised by these approaches, then a 'face' recognition rate would ensue (potatoes do have eyes, after all). Here, these early works have shown that categorical attributes [30] and comparative attributes can be used in the recognition process, similar to eyewitness statements. These attributes are explicitly and intimately linked to face structure and could be used within an evidential framework (whereas a black box cannot). This allows recognition and reduction of the search space, which should be welcome to future criminal procedures. There is clearly room for improvement, and these pioneering works have shown that recognition can be achieved in this way, and improving performance is an interesting and fertile area for development.

3.5 Clothing Soft Biometrics for Identification

3.5.1 Can Clothing Be Used for Subject Recognition?

It is easily observable that clothing is often highly unique. In some scenarios, such as the armed forces, the clothing is largely the same, though it might be worn slightly differently. Despite this, clothing can be used as an identifier, even if it is short term, since it is easy to change one's clothes. The first deployment of clothing for recognition as a soft biometric occurred, as the computer vision approaches for clothing identification were not sufficiently advanced at the time. The UK Crown Prosecution Service section on Identification: Is the Witness Identifying a Person? UKCrownProsecution [58] notes that in one case [15] 'the Divisional Court analysed one set of circumstances, in which the witness described the clothing and approximate ages of two young men which led to the arrest of two young men'. Stating also that '...there was nothing in Mr. Challice's description which bore upon the facial appearance, colouring, build, height, or manner of walking or moving of those he had seen ... an identification parade could have served no useful purpose'. In another case [58] 'the Court of Appeal stated that no parades were necessary if a witness purported to identify a distinguishing feature such as clothing only rather than the

offender'. This is an example of precedence guiding the UK legal system. It is common for public appeals for suspect identification to mention clothing. As Shakespeare observed:

"For the apparel oft proclaims the man" (Hamlet: Act 1, Scene 3); and
"Know'st me not by my clothes?" (Cymbeline Act 4 Scene 2)

In that spirit, we shall move to how clothing can be described to recognise individuals.

3.5.2 Describing and Analysing Clothing

There has been little work on clothing for biometric purposes, with more focus on attribute estimation in surveillance, particularly for re-identification [33, 42, 61, 63, 68]. Many of the current approaches employ computer-vision algorithms and machine-learning techniques to extract and use clothing descriptions in a variety of applications. In the past, approaches used attribute classifiers such as Support Vector Machines; the currently favoured approach is deep learning. The applications include soft attributes for re-identification [55], online person recognition [42, 63]; along with attribute-based people search [14, 55]; detecting and analysing semantic descriptions (labels) of clothing colours and types to supplement other bodily and soft facial attributes in automatic search and retrieval [61]. Even with images captured on different days, there remains sufficient information to compare and establish identity, since clothes are often worn again or a particular individual may prefer a specific clothing style or colour [21].

Clearly, there are problems with clothing analysis, as there are with body and face analysis, and these can again use the rich power of human description. This led to the first approaches to using clothing for biometric purposes. In many ways, the analysis is similar as the approaches achieve a semantic description which is used for recognition [24] and then leads to a way by which the labels can be estimated automatically [26]. There have been no attempts as yet to determine the style or wear (with time) of clothing as that might buttress identity.

Unlike most existing vision-based approaches concerning clothing attributes for fashion search/retrieval or clothes recognition/classification, the first approach to using clothes as a biometric proposed a novel set of soft clothing attributes for human identification and retrieval [26]. Each soft attribute was derived automatically and described at a high level using a small group of semantic categorical/comparative labels for biometric purposes. There has previously been a semantic description of clothing [41], though this was not for identification purposes.

TABLE 3.5

Semantic Clothing Attributes

Body Zone	Semantic Attribute	Categorical Labels	Comparative Labels
Head (4)	1. Head clothing category	[None, Hat, Scarf, Mask, Cap]	
	2. Head coverage	[None, Slight, Fair, Most, All]	[Much Less, Less, Same, More, Much more]
	4. Hat	[Yes, No, Don't know]	
Upper body (4)	5. Upper body clothing category	[Jacket, Jumper, T-shirt, Shirt, Blouse, Sweater, Coat, Other]	
	6. Neckline shape	[Strapless, V-shape, Round, Shirt collar, Don't know]	
	7. Neckline size	[Very Small, Small, Medium, Large, Very Large]	[Much Smaller, Smaller, Same, Larger, Much Larger]
Lower body (4)	9. Lower body clothing category	[Trouser, Skirt, Dress]	
	10. Shape	[Straight, Skinny, Wide, Tight, Loose]	
	11. Leg length (of lower clothing)	[Very Short, Short, Medium, Long, Very Long]	[Much Shorter, Shorter, Same, Longer, Much Longer]
Foot (2)	13. Shoes category	[Heels, Flip flops, Boot, Trainer, Shoe]	
	14. Heel level	[Flat/low, Medium, High, Very high]	[Much Lower, Lower, Same, Higher, Much higher]
Attached to body (4)	16. Bag (size)	[None, Side-bag, Cross-bag, Handbag, Backpack, Satchel]	[Much Smaller, Smaller, Same, Larger, Much Larger]
	19. Gloves	[Yes, No, Don't know]	
General style (1)	20. Style category	[Well-dressed, Business, Sporty, Fashionable, Casual, Nerd, Bibes, Hippy, Religious, Gangsta, Tramp, Other]	
Permanent (1)	21. Tattoos	[Yes, No, Don't know]	

As with all human descriptions, the most intuitive is to use categorical attributes. The high-level categories were given binary labels and the terms at the lower level included distinct categories. In this way, the presence of a hat is termed either Yes, No or Don't know, whereas the head clothing category could be None, Hat, Scarf, Mask or Cap. Footwear did not have a high-level category, for it is unusual in the data used for people to go barefoot, and the footwear categories included Heels, Flip flops, Boots, Trainers or Shoes. As this was an initial study, it was not intended for the list to be an all-embracing ontology of items and categories, but more to demonstrate whether clothing could be used for identification or not.

Comparative attributes were also used where appropriate such as (Neckline size: 'Much smaller', 'Smaller', 'Same', 'Larger' and 'Much larger'). Twenty-one semantic attributes were proposed, and for each attribute, a suitable group of categorical labels is specified to be used for describing these attributes. Furthermore, 7 of the 21

attributes are both categorical and relative, whereas the remaining 14 are unsuited for comparison because they are binary or multi-class attributes that can be described using only categorical (absolute) labels. Thus, the categorical labels of these seven relative attributes are extended to their corresponding comparative labels. Table 3.5 shows the list of proposed semantic attributes with their assigned categorical and comparative labels.

The SGDB [56] is a standard database used for the initial analysis of recognition by clothing (Figure 3.19). Beyond the gait sequences, the database also comprises a subset of full-body front- and side-view still images of each subject in the database with soft biometrics' labels available for body and face [50, 53]. The front-view images were used to collect the clothing descriptions. This subset consists of 115 individuals with a total of 128 front samples. There are multiple samples of some subjects, and multiple samples of a single individual were considered as different and independent entities by which each

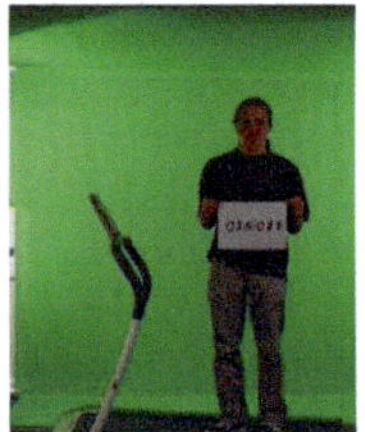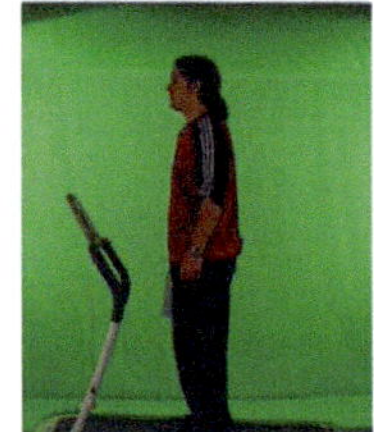

FIGURE 3.19
Front- and side-view (static) images from the Southampton Gait dataset.

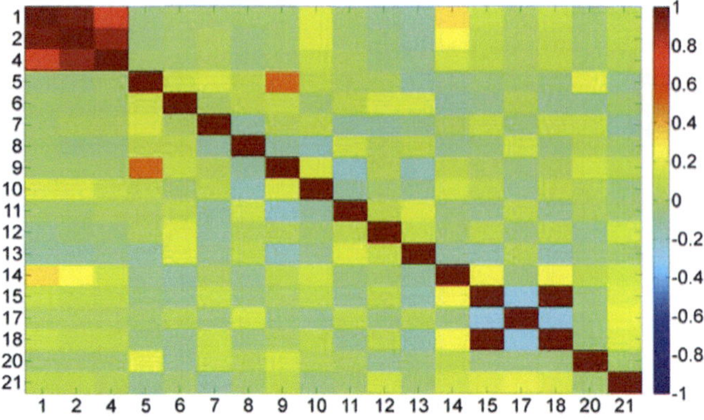

FIGURE 3.20
Clothing label correlation.

entity represents that individual only if wearing exactly the same clothing. If the subject was wearing different clothing, their image is considered as another entity, even though it belongs to the same individual. This is because the study focused on the use of clothing for identification. Had the focus differed, say to recognise individuals, then the sample where subjects wore different clothes would have been considered as from the same subject, but that was not the case here. Note that many of the subjects in the database wear similar clothing (particularly jeans and T-shirts) since the majority of the subjects were students. This data appeared to be sufficiently challenging for the first study to show the uniqueness of clothing. It is interesting that the subject who was difficult to individuate by body shape was also difficult to individuate by clothing. Vice versa the subject who was easy to individuate by body shape was also easy to individuate by clothing.

The Clothing Attribute Dataset [10] was developed by harvesting images from Sartorialist 1 and Flickr, by applying an upper body detector to select pictures with people; a total of 1856 images were collected that contain clothed people (mostly pedestrians on the streets). Amazon Mechanical Turk (AMT) was used to collect ground truth labels. The approach largely concerned learning 26 clothing attributes and presented a fully automatic system that learned semantic attributes for clothing on the human upper body. Projected applications include analysis of clothing style and determination of gender.

Using a custom-developed website, 27 labellers provided categorical and comparative labels of 128 subjects from the Southampton Gait dataset. All 128 samples were labelled by multiple users, with one or more separate user annotations per subject describing the 21 categorical attributes. All subjects were compared using the seven relative attributes by multiple users. To enrich the comparison data from the available number of collected comparisons, additional comparisons were inferred when two subjects were both compared with another same subject. A soft-margin Ranking SVM [28], along with a supporting formulation of similarity constraints [46], was used to rank the comparative attributes.

3.5.3   Analysing the Potency of Soft Clothing Attributes

The correlation matrices were calculated for collected labels, comparisons and inferred attributes using Pearson's r coefficient. The calculated linear correlation coefficient was considered significant with respect to the resulting p-value when $p \leq 0.05$. A low correlation between two different attributes did not suggest that there was no relationship between the two attributes but can suggest that this correlation is not prevalent within the clothing dataset currently used such as the attributes 'Face covered' and 'Bag'.

Figure 3.20 shows the correlation between all labels of 18 clothing attributes (see Table 3.5); attributes without correlation are not shown. High correlation is symbolized by orange, and low by blue/ green. As such, attributes relating to head coverage can be observed to be highly correlated, as are the attributes (15) and (18) relating to the description of items attached to the body. Clothing was well correlated for upper (5) and lower (9) body. The matrix structure suggests that the desired uniqueness has been achieved.

The distributions of terms for the comparative data, Figure 3.21, show close to normality for some and clearly non-normality for others. This is largely inherent in the attributes and their descriptions. For example, attribute 7 Neckline size shows, to visual inspection, a near normal distribution, whereas attribute 14 shows that it is unusual to wear high-heeled shoes in a university laboratory (especially when one will walk on a treadmill). This is not unexpected. It is, of course, possible to test distributions for normality, but it was doubted that this would be informative here. It was sufficient that the distributions showed anticipated variation. It is possible to show the most significant soft clothing attributes and to suggest which attributes might offer better recognition performance than others using ANOVA analysis to derive

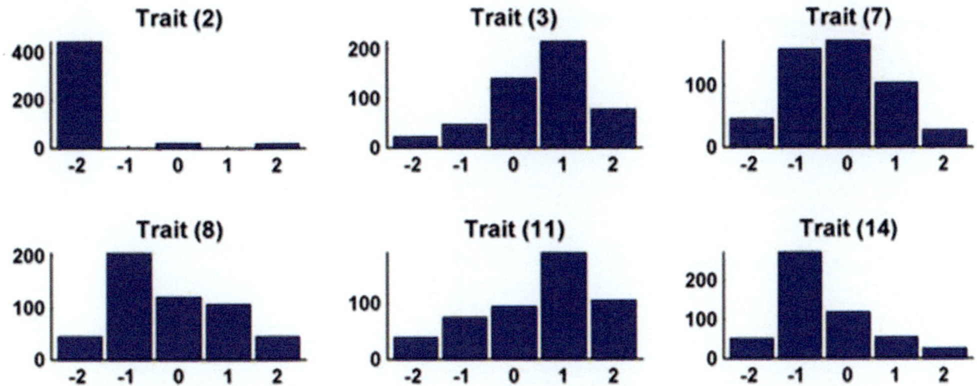

FIGURE 3.21
Distributions of labels for comparative soft clothing attributes.

F-ratios with their corresponding p-values, with degrees of freedom df. Table 3.6 lists examples of the resulting ANOVA values for categorical and comparative clothing attributes. Attribute 9, lower body clothing, was observed to be the most discriminative attribute. Head coverage (2) is highly discriminative since few subjects covered their heads. It is perhaps surprising that sleeve length (8) is so discriminative, especially compared with the length of the trousers (11), but that is what this analysis reveals, and no summary analysis is possible by human vision. Multivariate analysis of variance (MANOVA) was also applied to the categorical and the comparative attributes to explore interactions between multiple attributes and measures significance of multivariate discrimination capability via different standard statistics (Pillai's Trace, Wilks Lambda etc.), and all measures provided significant p-values<0.0001 [44]. It is also possible to estimate the Statistical Dependency to examine whether the values of a feature are dependent on the associated class labels and whether those feature values and class labels are likely to co-occur by chance. That analysis showed that the comparative attributes are more statistically dependent by receiving higher SD values than the categorical attributes, which appears intuitive. Sleeve length (8) ap-

peared to be the highest dependent comparative attribute and among the highest categorical attributes, emphasising the potential usefulness of that attribute.

As we are using human labelling, it appears prudent to investigate the labellers' performance as well as that of the attributes. A framework for such analysis was designed and applied to annotators participating in a relatively similar task, such as word sense annotation in Natural Language Processing (NLP) [7]. This allows for analysis of singular performance (statistical outliers), clustering by the similarity of behaviour and to recognise potential confusion. It might be more statistics, but it does appear to be more balanced to analyse labellers as well as the attributes that are labelled. Human performance is imperfect, though there is often much consistency. Since each labeller was asked to annotate only a subset of ten randomly selected subjects in the database to achieve wide coverage, all subjects were not labelled by all labellers. This was, after all, the first study aiming to show whether or not it could be achieved. A more comprehensive study would aim for completeness of the attributes their labels, and by all labellers.

Compared with word sense annotation NLP mechanisms [7, 47], the structure of clothing annotation is likely

TABLE 3.6
ANOVA of Labels for Soft Clothing Attributes. The tables below show (a) Categorical Attributes, and (b) Comparative attributes.

(a) Categorical attributes			
Clothing attribute		F-ratio ($df = 443$)	p-value ($p \leq 0.05$)
2.	Head coverage	19.651	3.90e-99
13.	Shoes category	4.282	3.64e-26
6.	Neckline shape	2.413	1.68e-10
14.	Heel level	1.863	5.58e-6
7.	Neckline size	1.688	1.15e-4
18.	Object in hand	0.721	9.84e-1

(b) Comparative attributes			
Clothing attribute		F-ratio ($df = 443$)	p-value ($p \leq 0.05$)
2.	Head coverage	15.670	3.68e-86
8.	Sleeve length	12.021	1.02e-71
11.	Leg length	4.819	2.49e-30
16.	Bag	4.689	2.42e-29
3.	Face covered	2.559	9.38e-12
7.	Neckline size	2.515	2.26e-11
14.	Heel level	1.636	2.73e-4

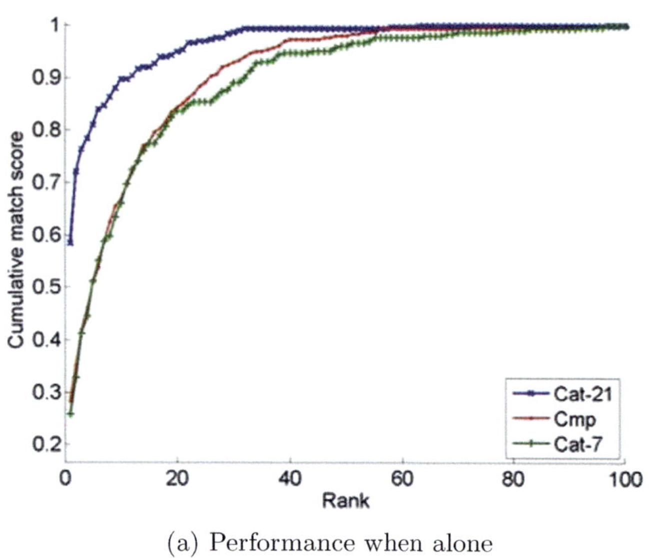

(a) Performance when alone

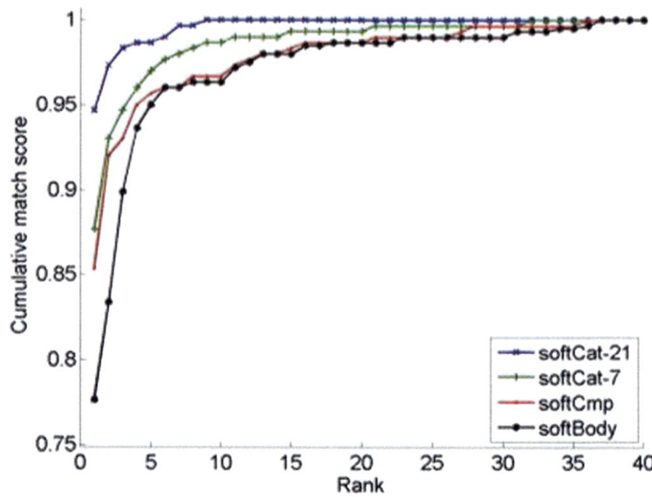

(b) Performance when fused with body soft biometrics

FIGURE 3.22
Performance of soft clothing biometrics: (a) Performance when alone, (b) Performance when fused with body soft biometrics. (With courtesy from [26].)

to have a smaller room of subjectivity and systematic differences among a subset of annotators. That is due to two major characteristics. Firstly, a limited set of labels in multiple-choice wise is provided for an annotator to describe an attribute. Secondly, the labelling task exclusively depends on a single displayed subject or two displayed subjects (in the comparison task), as an annotator is asked to exactly describe what they see, not what they infer, which tends to minimise the variation of judgement affected by their natural differences in opinion and behaviour. Two groups of user-annotators provided categorical labels for the same group of subjects. Thereby, two subsets of annotation data describing a number of subjects, where each subject has multiple annotations provided by exactly the same group of user-annotators. Group 1 contained three labellers of ten subjects, and Group 2 contained four labellers of five subjects (different from those studied by Group 1).

To analyse inter-labeller agreement in the context of soft clothing attribute annotation, Jaha used a statistical (consensus) measurement of Cohen's kappa [9, 12] and was previously used in the context of NLP annotation tasks [27]. The Kappa agreement (κ) score can be calculated as:

$$\kappa = \frac{P(a) - P(e)}{1 - P(e)} \quad (3.3)$$

where $P(a)$ represents the proportion of observed agreement among annotators. $P(e)$ is the proportion of agreement on a certain label by chance, which is equivalent to the probability of selecting a dominant agreed label

from a set of N available labels, so it can be calculated as $1/N$ since each label has an equal chance among a set of designated labels per attribute. The inter-annotator agreement varied in attribute level ranging from 0.5 to 1.0, while the received total average scores of κ (0.82 in Group-1) and (0.79 in Group-2) indicated a high overall consensus among inter-annotator labelling.

Surprisingly, the least agreement between the two groups was for neckline size and leg length. This might be in part due to potential obscuration of the lower leg, or difference in interpretation (e.g. length from floor to hip vs length from floor to crotch). The more evident characteristics, such as head coverage and heel height, show much greater agreement, and this is to be expected.

3.5.4 Recognising People by Human Descriptions of Their Clothing

There is ever a possibility of feature set selection (using the earlier ANOVA) and so reduced sets are denoted $-n$, where n is the number of best-performing features used. The initial dataset used for clothing analysis had 128 subjects viewed from the side and the front. When retrieving an image given a labeller's description of the subject, one is essentially testing the uniqueness of the labels. When the soft clothing biometrics are used alone, it is possible to use the 21 categorical labels alone, a set of the seven most potent categorical labels derived by statistical analysis (Section 3.5.3). Figure 3.22 shows the match and recognition performances just by using these sets of clothing measures alone and when fused with body soft biometrics (Section 3.3). In Figure 3.22(a), the 21 categorical measures, in general, appear to offer the best recognition capability. Clearly, clothing can be used alone, but it can also be used in fusion with body soft biometrics and their performance increases considerably (Figure 3.22(b)). In Figure 3.22(a), the correct recognition rate at rank 1 increases by around 50% from when used alone and increases the body recognition rates too. Evidently, clothes can be used for recognition on their own, and when fused with other biometrics.

It is also possible to achieve viewpoint invariant retrieval since the clothing descriptions are immune to change in pose of the subject relative to the viewer [25]. The problem is then to retrieve the image that most closely relates to the descriptions provided. This implies a need to generate descriptions of both sets of images, in this case the side view and front view. In this case, it would be impossible to match the face, as the pose change is too extreme for most recognition approaches. One could not use other biometrics, such as the ear, since it cannot be seen in the front view. The body can be seen, though some descriptions are possible only in one view (measures of width for the front view and of depth for the side view).

3.5.5 Closing the Gap: Estimating Human Clothing Descriptions via Computer Vision

As noted in Section 3.5.4, there is now much interest in automatic clothing analysis. The imagery analysed so far to demonstrate that clothing can be used for identification has been constrained to laboratory data from the SGDB. The nature of this data readily allows for automatic analysis. First, the images were collected using a chromakey background. Together with controlled illumination, this provided for a controlled background and this eases the separation of the object from their background. Body detection was achieved using a deformable model-based detector [18], and a cascade classifier based on Haar-like features was used for face detection [62]. In both cases, reassessment and realignment were applied exploiting the intersection and relative mutual positioning between the two estimated bounding boxes of body and face, with a higher priority placed on the detected face.

Afterwards, the final estimated region containing the face/head (if there was no detectable face) was used to initialise a starting region of a foreground highlighting method [19], which used the Grabcut technique for foreground object extraction [51] by learning foreground and background colour models that initially suggest subregions where the person is likely to be present or absent. Implicit in these is the use of colour models to exploit the chromakey background and attain accurate full-body seg-

mentation. The colour models were more conservative in extracting shape and details of the full appearance of the person, including the lower body and feet regions, which were also to be considered in the foreground model as likely candidate regions. Finally, shadows were removed from around the feet. As can be seen, the resulting silhouette is sufficient for automated clothing analysis, and that was the aim of the whole procedure.

The amount of skin revealed by a person can indicate style as well as season. 'Skin exposure' and 'Clothing season' attributes for overall, upper and lower body clothing were inferred from derived vision-based relative measurements for each of the three body parts. These relative measurements are computed with respect to the proportion of exposed skin and then compared with a pre-defined scale with five/three ranges of the five/three categorical labels, leading to the assignment of the best representative label for each soft attribute. Subsequently, a sliding window was used to detect and examine the region containing the feet to annotate the 'Footwear category'. This was performed by utilising skin detection in the feet zone to investigate whether the footwear is closed-toed, covering the skin of the entire feet, or open-toed, exposing some of the skin. Eventually, the detected skin was omitted from all body segments and these segments are refined to prepare a pure clothing representation to be passed and used in the analysis of clothing colour and patterns, and that was the aim of the whole procedure.

Clothing colour was used to determine attributes describing: the brightness of upper and lower garment pieces (A7 and A13); overall contrast between upper and lower clothing (A3); quantitative description of dominant colours of upper and lower parts (A9 and A15); and colour-scheme category of overall, upper and lower clothing (A4, A8 and A14). A five-scale colour-map model was used to classify all possible grayscales (from 0 to 255) into five groups of brightness levels which correspond to categorical labels. Clothing segments were processed to extract and analyse visual clothing patterns in terms of simplicity and complexity, which led to a high-level description of the soft attributes (A10 and A16) of upper and lower clothing patterns. Uniform LBP was used to analyse texture as it is now the predominant approach in computer vision. A threshold was used to increase noise resilience, compared with the usual binary comparison in ULBP. Clothing segment boundary points were removed by using edge detection since the major concern was embedded patterns of pieces of clothing, not the colour transitions between them and the background. Example automatic generated attributes are described in Table 3.7, and the augmentation of season and pattern can be observed.

The vision-based clothing annotations were used to compose two categorical feature vectors of automatic

TABLE 3.7

Example Automatic Soft Clothing Attributes, Ordered by p-Value. The tables below show (a) Categorical Attributes, and (b) Comparative attributes [26]

(a) Categorical attributes		
Clothing Attribute	F-ratio ($df = 255$)	p-value ($p \leq 0.05$)
A12 Lower Clothing season	214.69	1.36e-111
A4 Overall Colour-scheme	8.90	6.33e-30
A1 Overall Skin exposure	6.74	5.00e-24
A2 Overall Clothing season	4.05	1.77e-14
A9 Upper Dominant colour	3.21	8.58e-11
A6 Upper Clothing season	3.12	2.33e-10
A15 Lower Dominant colour	0.82	8.62e-01

(b) Comparative Attributes		
Clothing Attribute	F-ratio ($df = 255$)	p-value ($p \leq 0.05$)
A7 Lower Clothing season	86.23	5.32e-87
A3 Lower Brightness	25.17	8.98e-55
A16 Lower Pattern	15.45	1.09e-42
A10 Lower Colour-scheme	7.25	1.56e-25
A5 Upper Colour-scheme	3.69	6.07e-13
A9 Upper Pattern	3.21	8.57e-11
A15 Overall Colour-scheme	0.82	0.86

clothing descriptions. The first feature vector was composed of the full set of 17 automatic categorical clothing attributes describing a single subject. Then, the 14 most effective and discriminative categorical clothing attributes were derived by using feature subset selection. These 14 attributes are the most effective, and discriminative attributes are selected based on ANOVA, and a number of similarly structured feature vectors for all subjects are gathered to construct the first categorical gallery of 14 automatic clothing attributes (referred to as AutoCat-14). A ranking SVM method was used to derive a comparative form of the soft clothing biometrics. The automatic categorical feature vectors of all subjects, where each comprises only ten values describing the ten automatic relative attributes, are used as a training dataset to learn ten optimal ranking functions for the ten automatic relative soft attributes (A1, A3, A5, A7, A9, A10, A11, A13, A15, A16), All obtained similarly structured comparative feature vectors were gathered to build the first comparative gallery of five automatic clothing attributes (referred to as AutoCmp-5).

The same analysis of performance was used for the automatically derived versions of the attributes. The ANOVA was included with the list of automatic attributes (Table 3.7). The correlation between automatically derived attributes and the distributions of each were analysed. The correlation analysis showed similar aspects to the correlation performance observed for human labellers . Note that the upper body attributes show a consistent correlation. There is no correlation between season and clothing category, which again is expected.

There are now several sets of attributes available for recognition by clothing. Essentially, there are manual and automatic labels, being those derived from human labelling and those derived by computer vision. There are also categorical and comparative attributes.

The clothing-based soft attributes were used in isolation, aiming to achieve successful retrieval and to investigate the max potency of pure soft clothing attributes when used alone in person retrieval, which is deemed to be a mimic of challenging real-case forensic scenarios when soft clothing attributes are the only observable soft attributes. The retrieval evaluation results and performance analysis for all five clothing-based approaches are shown in Figure 3.23. Despite the modest retrieval rate of all approaches at rank 1 that do not exceed 53%, the retrieval rate can be observed to increase sharply to score more than 77% on average at up to rank 10, especially in the three automatic approaches.

The consequent full CMC analysis of all approaches is illustrated in Figure 3.23. By all measures, the automated soft biometrics (softAuto - Cat-14 & Cmp-5) attains the highest retrieval performance receiving the best scores in all evaluation measurements. The same set without the comparative measures (softAutoCat-14) offered the next highest performance approach with very similar scores.

As Shakespeare knew, clothing can be used to describe a person. He would not have imagined our automatic methods, but one suspects he might have pointed out that the results from human labels were within expectations.

3.6   Fusing Body, Face and Clothing for Identification at Differing Distance

The focus of all these approaches to soft biometrics for identification was to recognise people by human descriptions. An early analysis showed the advantages of fusion [45]. It then behoved to determine which modality was best when the subject was at some distance from the camera, and thus at low resolution. Intuitively, the face would be most likely to perform best at short distances, but which would be best at longer ones: body or clothing?

The approach [22] used the comparative labelling process on image data synthesised so that not only was the background removed but also it was constant over image sequences as shown in Figure 3.24. Images were then

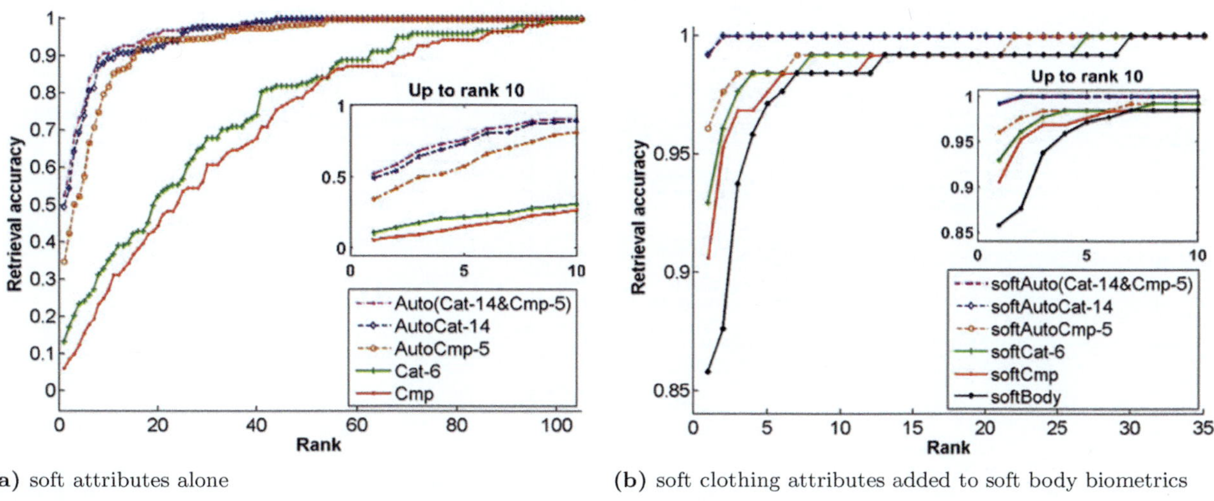

(a) soft attributes alone   **(b)** soft clothing attributes added to soft body biometrics

FIGURE 3.23
CMC performance of soft clothing attributes: (a) soft attributes alone, (b) soft clothing attributes added to soft body biometrics.

(a) close   (b) far

FIGURE 3.24
Synthesised images for evaluating soft biometrics at a distance: (a) close, (b) far. (With courtesy from [22].)

constructed by inserting subjects at the same controlled distances: close, medium and far. The labels used were those defined in previous studies, removing some that are not appropriate at long distance, such as eyebrow shape and distance, and were acquired using a three-point Likert scale. The face and body were labelled by the comparison between each of the 200 subjects and 20 randomly chosen subjects. The total number of comparisons is 4000, each one labelled by 20 people; the categorical clothing attributes were labelled for 200 subjects by 20 people. The Elo system was used to rank the labels, and so a rank-score fusion method was proposed to combine the complementary information in rank and similarity scores.

The normalised joint density of similarity score and rank order was used as to weigh the match similarity score. Similar scores, as obtained in previous studies, were observed for all three modalities at close distance, as shown in the CMC curves Figure 3.25. As expected, at far distance, the face performed worst, largely with random performance. Clothing had the least variability and was of comparable performance across the three distances. This is reflected similarly in the rank-1 performances (Figure 3.25).

When fused at a close distance 98.5% recognition accuracy was achieved by fusion (and it is better to use the face alone). At medium distance and at far distance, the

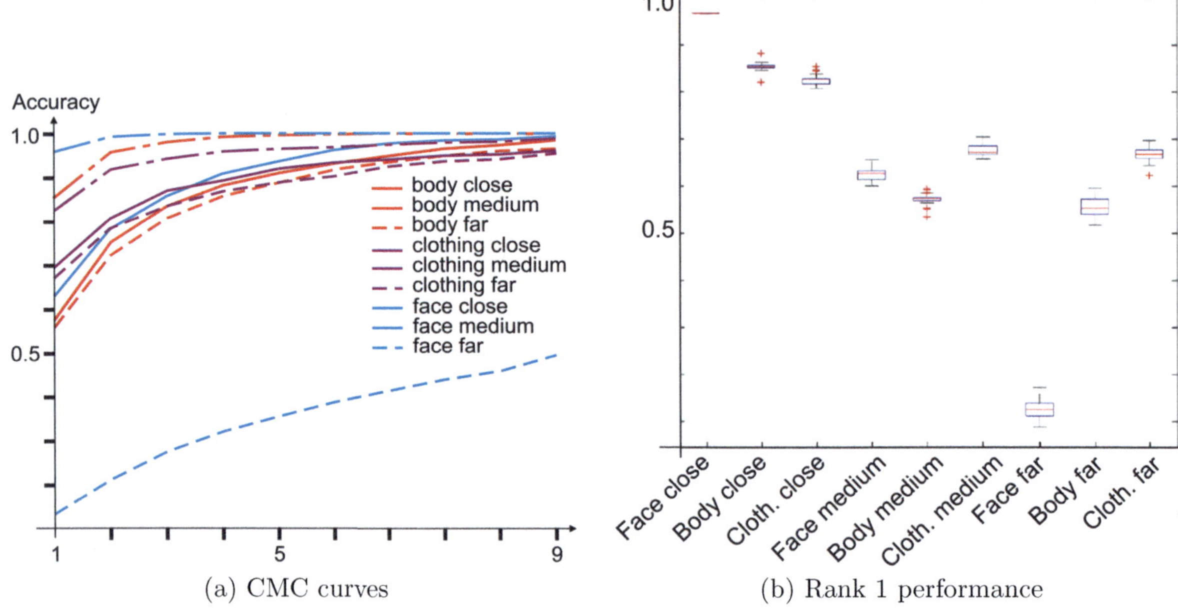

(a) CMC curves   (b) Rank 1 performance

FIGURE 3.25
Soft biometrics performance at different distances: (a) CMC curves, (b) Rank 1 performance.

rates were 92.5% and 82.6%, respectively, which is better than any modality alone and showing the advantages that clothing brings to the table. Other fusion approaches were also evaluated but did not perform as well as ranked score fusion, perhaps since the ranks were integral to the evaluation process.

3.7   Discussion and Future Work

This chapter has focused on using attributes for recognition by soft biometrics. Fine-grained descriptions were derived by using ranked comparative attributes. Using deep learning for attribute estimation continues to improve (e.g. [65]), especially for person re-identification, though this continues largely to focus on categorical labels (e.g. 'Hat: no; boots: yes; gender: female'), which have less discriminatory capability than fine-grained descriptions. A more recent survey on face attribute estimation [67] describes attribute estimation and manipulation. The estimation approaches reside in parts-based and holistic methods; the analysis approaches in the manipulation stage use generative adversarial networks and autoencoders as backbones for their approaches. Again, the approaches are very sophisticated and concentrate largely on categorical descriptions. A clear advantage of the deep learning approaches is their need for and ability to handle large data, and learning structures across highly represen-

tational data. There is also more interest of late in the analysis of clothing in images: as an example of the current state of the art in clothing analysis by automated image analysis [66]. There have also been considerations of beauty estimation by computer vision and an approach has suggested that attractiveness can be added to the panoply of soft biometrics [5] with discriminatory power appearing to be similar to that of gender. No doubt other soft biometrics will be added to the mix in time. A recent survey on soft biometrics [23] analysed soft biometrics approaches, particularly using the data described herein for face, clothing and body, showing advantages and properties of this data. The study concluded that 'the development of a robust and highly accurate soft biometrics recognition system remains a challenging task', and is one that will benefit society greatly indeed.

3.8   Conclusions

Using human descriptions allows for a different recognition approach. It is one which requires knowledge of the limitations and advantages of human vision. When these are addressed, measures are described that are sufficient for recognition purposes. These can be used to describe a human body in a way which would be difficult for hand-crafted approaches to address and hard to explain for deep-learning approaches based purely on images. The approaches can be used to describe all characteristics that are available at a lower resolution, as in surveillance images acquired at a distance. The measures appear to have properties that are similar to those of the conventionally-acquired 'hard' biometric descriptions. This allows the new approaches to be fused for recognition with different advantages at differing distances/resolutions. In all, these powerful new techniques incorporate the power of human description and even provide a basis by which the labels can be generated anew from unseen images. As such, soft biometrics for identification provide a new radically different approach to the difficult problem of identification within surveillance imagery.

References

[1] Agency, N. P. I. (2012). *The PNC User Manual, Version 12.01*. UK National Archives.

[2] Almudhahka, N., Nixon, M., and Hare, J. (2016). Human face identification via comparative soft biometrics. In *Identity, Security and Behavior Analysis (ISBA), 2016 IEEE International Conference on*, pages 1–6. IEEE.

[3] Almudhahka, N. Y., Nixon, M. S., and Hare, J. S. (2016). Unconstrained human identification using comparative facial soft biometrics. In *Biometrics Theory, Applications and Systems (BTAS), 2016 IEEE 8th International Conference on*, pages 1–6. IEEE.

[4] Almudhahka, N. Y., Nixon, M. S., and Hare, J. S. (2018). Semantic face signatures: Recognizing and retrieving faces by verbal descriptions. *IEEE Transactions on Information Forensics and Security*, 13(3):706–716.

[5] Alnamnakani, M., Mahmoodi, S., and Nixon, M. (2019). On the potential for facial attractiveness as a soft biometric. In *International Symposium on Visual Computing*, pages 516–528. Springer.

[6] Bekele, E., Narber, C., and Lawson, W. (2017). Multi-attribute residual network (maresnet) for soft-biometrics recognition in surveillance scenarios. In *Automatic Face & Gesture Recognition (FG 2017), 2017 12th IEEE International Conference on*, pages 386–393. IEEE.

[7] Bhardwaj, V., Passonneau, R. J., Salleb-Aouissi, A., and Ide, N. (2010). Anveshan: A framework for analysis of multiple annotators' labeling behavior. In *Proceedings of the Fourth Linguistic Annotation Workshop*, pages 47–55. Association for Computational Linguistics.

[8] Burton, A. M., Wilson, S., Cowan, M., and Bruce, V. (1999). Face recognition in poor-quality video: Evidence from security surveillance. *Psychological Science*, 10(3):243–248.

[9] Carletta, J. (1996). Assessing agreement on classification tasks: The kappa statistic. *Computational linguistics*, 22(2):249–254.

[10] Chen, H., Gallagher, A., and Girod, B. (2012). Describing clothing by semantic attributes. *Computer Vision–ECCV 2012*, pages 609–623.

[11] Chugh, T., Cao, K., Zhou, J., Tabassi, E., and Jain, A. K. (2017). Latent fingerprint value prediction: Crowd-based learning. *IEEE Transactions on Information Forensics and Security*.

[12] Cohen, J. (1960). A coefficient of agreement for nominal scales. *Educational and psychological measurement*, 20(1):37–46.

[13] Davis, J. P., Jansari, A., Lander, K., et al. (2013). 'i never forget a face!'. *The Psychologist*, 26(10):726–729.

[14] Denman, S., Fookes, C., Bialkowski, A., and Sridharan, S. (2009). Soft-biometrics: unconstrained authentication in a surveillance environment. In *Digital Image Computing: Techniques and Applications, 2009. DICTA '09.*, pages 196–203. IEEE.

[15] Div.Court (1998). Dance v dpp.

[16] Elo, A. E. (1978). *The rating of chessplayers, past and present*. Arco Pub.

[17] Fahsing, I. A., Ask, K., and Granhag, P. A. (2004). The man behind the mask: Accuracy and predictors of eyewitness offender descriptions. *Journal of Applied Psychology*, 89(4):722.

[18] Felzenszwalb, P. F., Girshick, R. B., McAllester, D., and Ramanan, D. (2010). Object detection with discriminatively trained part-based models. *IEEE Transactions on Pattern Analysis and Machine Intelligence*, 32(9):1627–1645.

[19] Ferrari, V., Marin-Jimenez, M., and Zisserman, A. (2008). Progressive search space reduction for human pose estimation. In *IEEE Conference on Computer Vision and Pattern Recognition, 2008. CVPR 2008.* pages 1–8. IEEE.

[20] Frowd, C. D., Hancock, P. J., and Carson, D. (2004). Evofit: A holistic, evolutionary facial imaging technique for creating composites. *ACM Transactions on Applied Perception (TAP)*, 1(1):19–39.

[21] Gallagher, A. C. and Chen, T. (2008). Clothing cosegmentation for recognizing people. In *IEEE Conference on Computer Vision and Pattern Recognition, 2008 CVPR 2008.*, pages 1–8. IEEE.

[22] Guo, B. H., Nixon, M. S., and Carter, J. N. (2019). Soft biometric fusion for subject recognition at a distance. *IEEE Transactions on Biometrics, Behavior, and Identity Science*, 1(4):292–301.

[23] Hassan, B., Izquierdo, E., and Piatrik, T. (2021). Soft biometrics: A survey. *Multimedia Tools and Applications*, pages 1–44.

[24] Jaha, E. S. and Nixon, M. S. (2014). Soft biometrics for subject identification using clothing attributes. In *2014 IEEE International Joint Conference on Biometrics (IJCB)*, pages 1–6. IEEE.

[25] Jaha, E. S. and Nixon, M. S. (2015). Viewpoint invariant subject retrieval via soft clothing biometrics. In *2015 International Conference on Biometrics (ICB)*, pages 73–78. IEEE.

[26] Jaha, E. S. and Nixon, M. S. (2016). From clothing to identity: Manual and automatic soft biometrics. *IEEE Transactions on Information Forensics and Security*, 11(10):2377–2390.

[27] Jena, I., Bhat, R. A., Jain, S., and Sharma, D. M. (2013). Animacy annotation in the hindi treebank. In *LAW@ ACL*, pages 159–167.

[28] Joachims, T. (2002). Optimizing search engines using clickthrough data. In *Proceedings of the Eighth ACM SIGKDD International Conference on Knowledge Discovery and Data Mining*, pages 133–142. ACM.

[29] Kakadiaris, I. A., Sarafianos, N., and Nikou, C. (2016). Show me your body: Gender classification from still images. In *2016 IEEE International Conference on Image Processing (ICIP)*, pages 3156–3160. IEEE.

[30] Klare, B. F., Klum, S., Klontz, J. C., Taborsky, E., Akgul, T., and Jain, A. K. (2014). Suspect identification based on descriptive facial attributes. In *2014 IEEE International Joint Conference on Biometrics (IJCB)*, pages 1–8. IEEE.

[31] Kumar, N., Berg, A. C., Belhumeur, P. N., and Nayar, S. K. (2009). Attribute and simile classifiers for face verification. In *2009 IEEE 12th International Conference on Computer Vision*, pages 365–372. IEEE.

[32] Learned-Miller, E., Huang, G. B., RoyChowdhury, A., Li, H., and Hua, G. (2016). Labeled faces in the wild: A survey. In *Advances in Face Detection and Facial Image Analysis*, pages 189–248. Springer.

[33] Li, A., Liu, L., Wang, K., Liu, S., and Yan, S. (2015). Clothing attributes assisted person reidentification. *IEEE Transactions on Circuits and Systems for Video Technology*, 25(5):869–878.

[34] Li, D., Chen, X., and Huang, K. (2015). Multiattribute learning for pedestrian attribute recognition in surveillance scenarios. In *2015 3rd IAPR Asian Conference on Pattern Recognition (ACPR)*, pages 111–115. IEEE.

[35] Loftus, E. F. and Palmer, J. (1996). Eyewitness testimony. In *Introducing Psychological Research*, pages 305–309. Springer.

[36] Lucas, T. and Henneberg, M. (2016). Comparing the face to the body, which is better for identification? *International Journal of Legal Medicine*, 130(2):533–540.

[37] MacLeod, M. D., Frowley, J. N., and Shepherd, J. W. (1994). Whole body information: Its relevance to eyewitnesses. *Adult Eyewitness Testimony: Current Trends and Developments*, pages 125–143.

[38] Martinho-Corbishley, D., Nixon, M. S., and Carter, J. N. (2016). Analysing comparative soft biometrics from crowdsourced annotations. *IET Biometrics*, 5(4):276–283.

[39] Martinho-Corbishley, D., Nixon, M. S., and Carter, J. N. (2016). On categorising gender in surveillance imagery. In *2016 IEEE 8th International Conference on Biometrics Theory, Applications and Systems (BTAS)*, pages 1–6. IEEE.

[40] Martinho-Corbishley, D., Nixon, M. S., and Carter, J. N. (2018). Super-fine attributes with crowd prototyping. *IEEE Transactions on Pattern Analysis and Machine Intelligence*, 41(6):1486–1500.

[41] Matthews, T., Nixon, M. S., and Niranjan, M. (2013). Enriching texture analysis with semantic data. In *Proceedings of the IEEE Conference on Computer Vision and Pattern Recognition*, pages 1248–1255.

[42] Niinuma, K., Park, U., and Jain, A. K. (2010). Soft biometric traits for continuous user authentication.

IEEE Transactions on Information Forensics and Security, 5(4):771–780.

[43] Nixon, M. (1985). Eye spacing measurement for facial recognition. In *Applications of Digital Image Processing VIII*, volume 575, pages 279–285. SPIE.

[44] Nixon, M. S., Correia, P. L., Nasrollahi, K., Moeslund, T. B., Hadid, A., and Tistarelli, M. (2015). On soft biometrics. *Pattern Recognition Letters*, 68:218–230.

[45] Nixon, M. S., Guo, B. H., Stevenage, S. V., Jaha, E. S., Almudhahka, N., and Martinho-Corbishley, D. (2017). Towards automated eyewitness descriptions: Describing the face, body and clothing for recognition. *Visual Cognition*, 25(4-6):524–538.

[46] Parikh, D. and Grauman, K. (2011). Relative attributes. In *2011 IEEE International Conference on Computer Vision (ICCV)*, pages 503–510. IEEE.

[47] Passonneau, R. J. and Carpenter, B. (2014). The benefits of a model of annotation. *Transactions of the Association for Computational Linguistics*, 2:311–326.

[48] Reid, D. (2013). *Human identification using soft biometrics*. PhD thesis, University of Southampton, 2013.

[49] Reid, D. A. and Nixon, M. S. (2011). Using comparative human descriptions for soft biometrics. In *2011 International Joint Conference on Biometrics (IJCB)*, pages 1–6. IEEE.

[50] Reid, D. A., Nixon, M. S., and Stevenage, S. V. (2014). Soft biometrics; human identification using comparative descriptions. *IEEE Transactions on Pattern Analysis and Machine Intelligence*, 36(6):1216–1228.

[51] Rother, C., Kolmogorov, V., and Blake, A. (2004). Grabcut: Interactive foreground extraction using iterated graph cuts. In *ACM Transactions on Graphics (TOG)*, volume 23, pages 309–314. ACM.

[52] Samangooei, S., Guo, B., and Nixon, M. S. (2008). The use of semantic human description as a soft biometric. In *2008 IEEE Second International Conference on Biometrics: Theory, Applications and Systems*, pages 1–7. IEEE.

[53] Samangooei, S. and Nixon, M. S. (2010). Performing content-based retrieval of humans using gait biometrics. *Multimedia Tools and Applications*, 49(1):195–212.

[54] Seely, R. D., Samangooei, S., Lee, M., Carter, J. N., and Nixon, M. S. (2008). The university of southampton multi-biometric tunnel and introducing a novel 3d gait dataset. In *2nd IEEE International Conference on Biometrics: Theory, Applications and Systems, 2008. BTAS 2008.* pages 1–6. IEEE.

[55] Shi, Z., Hospedales, T. M., and Xiang, T. (2015). Transferring a semantic representation for person re-identification and search. In *Proceedings of the IEEE Conference on Computer Vision and Pattern Recognition*, pages 4184–4193.

[56] Shutler, J. D., Grant, M. G., Nixon, M. S., and Carter, J. N. (2004). On a large sequence-based human gait database. In *Applications and Science in Soft Computing*, pages 339–346. Springer.

[57] Tome, P., Fierrez, J., Vera-Rodriguez, R., and Nixon, M. S. (2014). Soft biometrics and their application in person recognition at a distance. *IEEE Transactions on Information Forensics and Security*, 9(3):464–475.

[58] UKCrownProsecution (2000). R v doldur 2000.

[59] Valentine, T. and Davis, J. P. (2015). Forensic facial identification. *Forensic Facial Identification: Theory and Practice of Identification from Eyewitnesses, Composites and CCTV*, pages 323–347.

[60] Van Koppen, P. J. and Lochun, S. K. (1997). Portraying perpetrators: The validity of offender descriptions by witnesses. *Law and Human Behavior*, 21(6):661.

[61] Vaquero, D. A., Feris, R. S., Tran, D., Brown, L., Hampapur, A., and Turk, M. (2009). Attribute-based people search in surveillance environments. In *2009 Workshop on Applications of Computer Vision (WACV)*, pages 1–8. IEEE.

[62] Viola, P. and Jones, M. (2001). Rapid object detection using a boosted cascade of simple features. In *Proceedings of the 2001 IEEE Computer Society Conference on Computer Vision and Pattern Recognition, 2001. CVPR 2001.* volume 1, pages I–I. IEEE.

[63] Wang, H., Bao, X., Choudhury, R. R., and Nelakuditi, S. (2013). Insight: Recognizing humans without face recognition. In *Proceedings of the 14th Workshop on Mobile Computing Systems and Applications*, page 7. ACM.

[64] Wogalter, M. S. (1991). Effects of post-exposure description and imaging on subsequent face recognition performance. In *Proceedings of the Human Factors Society Annual Meeting*, volume 35, pages 575–579. SAGE Publications Sage CA: Los Angeles, CA.

[65] Wu, J., Liu, H., Jiang, J., Qi, M., Ren, B., Li, X., and Wang, Y. (2020). Person attribute recognition by sequence contextual relation learning. *IEEE Transactions on Circuits and Systems for Video Technology*, 30(10):3398–3412.

[66] Zhang, Y., Zhang, P., Yuan, C., and Wang, Z. (2020). Texture and shape biased two-stream networks for clothing classification and attribute recognition. In *Proceedings of the IEEE/CVF Conference on Computer Vision and Pattern Recognition*, pages 13538–13547.

[67] Zheng, X., Guo, Y., Huang, H., Li, Y., and He, R. (2020). A survey of deep facial attribute analysis. *International Journal of Computer Vision*, 128(8):2002–2034.

[68] Zhu, J., Liao, S., Lei, Z., and Li, S. Z. (2017). Multi-label convolutional neural network based pedestrian attribute classification. *Image and Vision Computing*, 58:224–229.

Exploring Causes of Demographic Variations in Face Recognition Accuracy

Gabriella Pangelinan, K.S. Krishnapriya, Vitor Albiero, Grace Bezold, Kai Zhang, Kushal Vangara, Michael C. King, and Kevin W. Bowyer

4.1 Introduction

Automated facial recognition (FR) technology dates back to the early 1970s, with Takeo Kanade's 1973 Ph.D. thesis *Picture Processing System by Computer Complex and Recognition of Human Faces* [1] often cited as an early landmark work. However, it was not until the late 2010s that increased availability and power of FR technology increased its routine usage. In 2017, Apple introduced the iPhone X as the "smartphone industry's benchmark" [2], with their new facial identification system, Face ID, as a primary innovation and selling point [3, 4] (Figure 4.1). The corresponding security guide [5] even claimed "the probability that a random person in the population could look at your iPhone X and unlock it using Face ID is approximately 1 in 1,000,000 (versus 1 in 50,000 for Touch ID)." Yet a security flaw was quickly publicized: some Chinese users reported that their iPhones opened for other, non-authorized individuals [6, 7]. The underpinnings of these incidents would soon be explored by the research community, leading to a major question: does face recognition perform equally across all demographics?

FIGURE 4.1
Apple's Face ID ad campaign touted ease of use and improved security for face recognition over the prior fingerprint standard, and made bold promises.

In 2018, Buolamwini and Gebru explored a related question in *Gender Shades: Intersectional Accuracy Disparities in Commercial Gender Classification* [8]. Their work evaluated the accuracy of commercial gender classification software. For one classifier, they reported that error rates for lighter-skinned males were only 0.8%, while error rates for darker-skinned females were dramatically higher, at 34.7%. Provocative headlines like "Facial Recognition Is Accurate, if You're a White Guy" [9] sparked public interest in Buolamwini and Gebru's work. Media coverage generally failed to make any distinction between gender classification, the task of assigning a gender label to one face image, and face recognition, the task of deciding whether or not two face images are from the same person.

Government research organizations quickly addressed the growing public concern around possible "bias" in face recognition accuracy. As part of their 2018 biometric technology rally, the Department of Homeland Security assessed the effect of demographic factors on performance of commercial face biometric systems, as measured by transaction times and by similarity scores of pairs of images of the same person [10]. In 2019, the National Institute of Standards and Technology (NIST) released a Face Recognition Vendor Test (FRVT) report on demographic accuracy variations in commercial face recognition tools [11]. The report provides extensive results across multiple datasets and matchers. However, NIST did not explore why the accuracy variations occur and listed "analyze cause and effect" as an item under the heading "what we did not do" [11].

As public and private use of FR technology expands, it is more important than ever to assess how the technology interacts with people across gender and race. Current FR research relies on demographic labels that are either self-reported, annotated by manual reviewers, or generated by automated classifiers. Each possible source

DOI: 10.1201/9781003328957-4

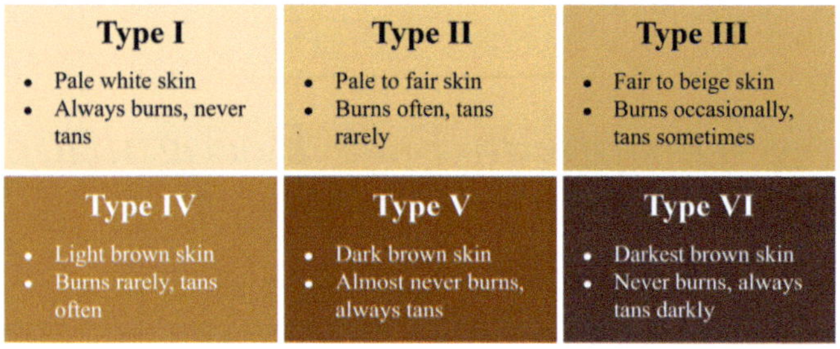

Type I
- Pale white skin
- Always burns, never tans

Type II
- Pale to fair skin
- Burns often, tans rarely

Type III
- Fair to beige skin
- Burns occasionally, tans sometimes

Type IV
- Light brown skin
- Burns rarely, tans often

Type V
- Dark brown skin
- Almost never burns, always tans

Type VI
- Darkest brown skin
- Never burns, always tans darkly

FIGURE 4.2
The Fitzpatrick skin type (FST) scale is used in dermatology. (With courtesy from [13].) FST-inspired six-tone scales are often used in face recognition research.

of demographic labels has weaknesses. Gender is typically treated as a binary male/female label, and race is typically treated in terms of a small number of discrete categories (e.g. Caucasian, African American, and Asian). Works focusing on skin tone often use a six-tone, light-to-dark scale like the Fitzpatrick skin type (FST) scale (shown in Figure 4.2). However, researchers have noted that true FST assignments are meant to be provided in person by a trained practitioner. Google AI's recent introduction of the ten-shade Monk Skin Tone (MST) scale [12] speaks to the growing desire, in industry and research alike, for greater inclusivity of the "spectrum of skin tones we see in our society" [12].

The following sections highlight the ongoing collaboration between research teams at the Florida Institute of Technology and the University of Notre Dame to better understand how, and why, face recognition accuracy varies across demographics. Our work has provided insight into the complexity of this multi-faceted problem. There is still much work to be done – by researchers and developers alike – in the quest to create truly equitable and fair face recognition systems.

4.2 Demographic Variations in Accuracy

There is consensus in the research literature that face recognition accuracy is lower for females, who often have both a higher false match rate and a higher false non-match rate. In 2021, Albiero et al. [14] presented the first experimental analysis to identify the major causes of this "gender gap," which is quantifiable in terms of separation between female and male impostors and genuine distributions. Albiero et al. analyzed these distributions for Caucasian, African American, and Asian demographics using two curated datasets and a top-performing open-source matcher (ArcFace [15] trained on MS1Mv2 [16]).

The Caucasian and African American demographics in [14] come from the MORPH 3 dataset [17], a large-scale collection of mugshot-style images acquired with controlled lighting and an 18% gray background. Subjects generally have a frontal pose and neutral facial expression. Twins and mislabeled or duplicate images were removed from the curated version of MORPH, yielding the demographic distribution given in Table 4.1.

There is not yet a large generally available dataset of Asian face images acquired in a controlled manner similar to that of MORPH. For this reason, Albiero et al. curated a subset of the web-scraped Asian-Celeb dataset. Curation of a MORPH-comparable subset included removal of mislabeled or duplicate images and pose constraints, with the final composition given in Table 4.1.

Albiero et al. [14] used ArcFace, a state-of-the-art deep convolutional neural network (CNN), as the matcher, providing similarity scores between pairs of images. The experimental instance of ArcFace corresponds to a set of publicly available weights, trained on the publicly available MS1Mv2 dataset. ArcFace takes as input aligned face images resized to 112×112 pixels, extracts 512-d features, and matches using cosine similarity. (In recent work, we have found that a newer instance of ArcFace trained on the Glint-360K dataset [18] gives higher overall accuracy than the version trained on MS1Mv2. However, the same demographic variations in accuracy are still seen with the Glint-trained ArcFace.)

Impostor and genuine distributions for each demographic group are given in Figure 4.3. The impostor distribution contains match scores resulting from comparisons between images of two different individuals, or *non-mated* image pairs. The genuine distribution's match scores result from comparisons between different images of the same individual, or a *mated* pair.

For all three demographic groups, the female impostor distribution is shifted toward higher similarity scores

TABLE 4.1
Number of Images and Subjects in the Curated Datasets Used in [14]. (*Source:* Table adapted from [14].)

	Female		Male	
	Images	Subjects	Images	Subjects
Caucasian	10,941	2798	35,276	8835
African American	24,857	5929	56,245	8839
Asian	43,356	6083	73,376	12,673

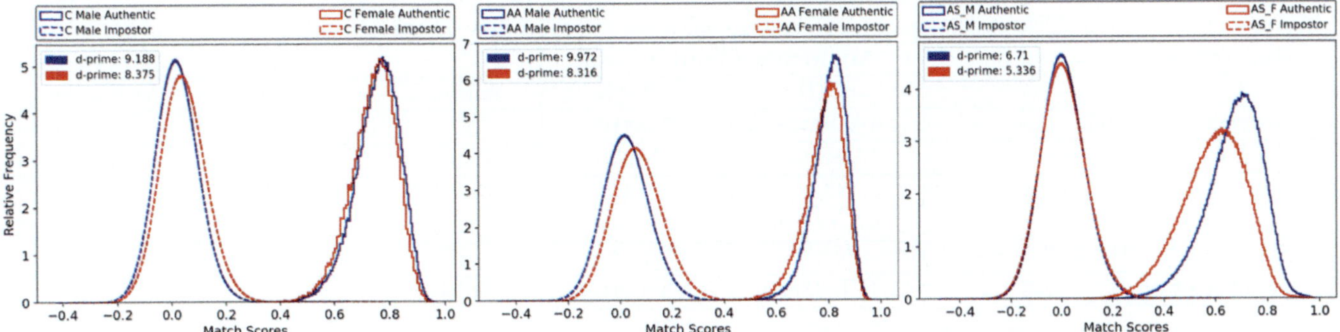

FIGURE 4.3
ArcFace impostor and genuine (authentic) distributions for Caucasian, African American, and Asian groups (from left to right). (Figure adapted from [14].)

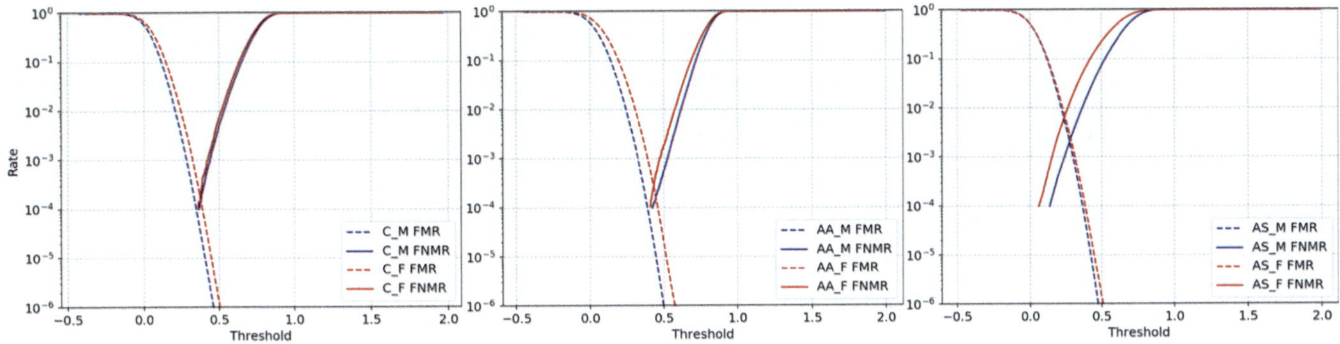

FIGURE 4.4
ArcFace FMR and FNMR for Caucasian, African American, and Asian groups (from left to right). (Figure adapted from [14].)

than its male counterpart. This shift indicates a higher false match rate (FMR): that is, a higher rate at which a given image instance is classified as an incorrect identity (a "false accept"). Each female genuine distribution ranges over lower similarity scores, indicating a higher false non-match rate (FNMR): a higher rate at which a given image instance is not correctly classified with its true identity (a "false reject"). For each group, the d-prime value – a statistical measure of a recognition system's ability to distinguish between samples – is lower for females than males. The lower d-prime value indicates that the matcher is less effective in differentiating between images of females than males.

The FMR and FNMR curves for each group are given in Figure 4.4. The FNMR is consistently higher for females than males across all datasets, though the corresponding gender gap is larger for Asian-Celeb and MORPH African American groups than MORPH Caucasian. For both MORPH datasets, FMR is significantly higher for females than males; for Asian-Celeb, the cross-gender FMRs are more similar. Albiero et al. speculate that this difference can be attributed to the fact that Asian-Celeb is a web-scraped dataset – despite curation, impostor pairs may still have factors other than gender affecting the FMR.

With this foundational understanding of the gender gap, we proceed to review analyses of its various speculated causes. Namely, we explore how cross-demographic recognition performance varies with skin tone, face geometry, representation in training data, and face pixel information.

4.3 Does Darker Skin Tone Cause Increased False Match Rate?

Media coverage has hyped darker skin tone as a primary driver of recognition accuracy varying across race.

TABLE 4.2

Previous Comparisons of African American (AA) versus Caucasian (C) Accuracy. (*Source:* Table adapted from [24].)

Reference	Accuracy Comparison		Dataset		
	FMR	FNMR	Subjects	Images	Available
Grother [20], 2017	AA better	C better	N/A	300K	No
Cook et al. [10], 2019	C better	N/A	363+199	6K	No
Howard et al. [22], 2019	N/A	C better	363+199	6K	No
Krishnapriya et al. [25], 2019	AA better	C better	11K	45K	Yes
Wang et al. [21], 2019	N/A	C better	6K	20K	Yes
Grother et al. [11], 2019	C better	AA better	Multiple large datasets		No

Articles in prominent news sources have stated that "face recognition tech is less accurate the darker your skin tone" (BBC) [19] and "the darker the skin tone, the more errors arise" (New York Times) [9].

Research studies also seem to have supported this notion. Cook et al. [10] reported that darker skin tone is associated with lower similarity scores for mated pairs. Grother [20] found that "African Americans give slightly lower FNMR than Whites" and a "much higher FMR." Wang et al. [21] reported that Caucasians have higher verification accuracy measured on in-the-wild imagery with intentionally difficult pairs. Grother et al. [11] observed that Caucasians have more false negatives in mugshot-quality images, while African Americans have more false negatives in lower-quality images. Interestingly, Howard et al. [22] found that for males younger than 40, African American FMR is greater than Caucasian, with the opposite effect observed for males aged 40 or older.

Table 4.2 provides an overview of previous results regarding African American versus Caucasian FR accuracy. (We exclude analyses based on ROC curves, as they can hide the fact that different demographics achieve the same FMR at different similarity thresholds.) The general consensus is that, for a fixed score threshold, African Americans experience a higher FMR and Caucasians experience a higher FNMR. Many different matchers have been tested across the highlighted works. The only dataset that is generally available to researchers and contains mugshot-style images is MORPH. The only other dataset that has been used and is generally available to researchers is IJB-C [23]. The datasets that are available are typically not balanced on skin tone, race, or other factors.

4.3.1 Skin Tone and False Match Pairs

As concern around possible racial bias in face recognition continued to grow, Krishnapriya et al. [25] observed that existing comparisons across African American and Caucasian images had confounding effects of skin tone and face morphology. It was impossible to ascertain which of the two effects was the driving factor. (In light of the recent work of Albiero [14], we might now add that differing social conventions for hairstyle present another confounding effect.) In *Issues Related to Face Recognition Accuracy Varying Based on Race and Skin Tone*, Krishnapriya et al. [24] published the first experiment intended to isolate the effect of skin tone alone on accuracy. To directly test the premise that face recognition is less accurate for darker skin tone, they examined a range of tones within the single demographic of African American males (AAMs) using the MORPH dataset.

They used two matchers to produce AAM impostor distributions: ArcFace (as described in Section 4.2) and a publicly available VGGFace2 model [26]. VGGFace2 is representative of the state-of-the-art in CNN matchers prior to ArcFace. It is based on the popular ResNet-50 network structure and trained on the VGGFace2 dataset [27] with standard softmax loss. Faces are detected, aligned, and resized to 224×224 pixels, and a 2048-d feature vector is taken from the next-to-last layer. As with ArcFace, cosine similarity is measured between feature vectors.

Krishnapriya et al. compared the frequency of images with darker skin tone in two regions of the AAM impostor distribution. The high-similarity tail (HST) is the region containing the non-mated image pairs that are most likely to cause false matches. To represent the HST, image pairs were sampled from just above the commonly used 1-in-10,000 (1-in-10k) FMR threshold. This threshold corresponds to the non-mated similarity score at which one error occurs for 10,000 image-pair comparisons. The center of the AAM impostor distribution was sampled as the "no-false-match region." Image pairs are selected from just above the 1-in-2 FMR threshold. The two regions are visualized in Figure 4.5.

Each image was assigned a skin tone rating from I (lightest) to VI (darkest), inspired by the FST scale, by three independent observers. The observers were shown exemplar images (Figure 4.6) selected from the well-known and manually annotated IJB-C dataset [23] to

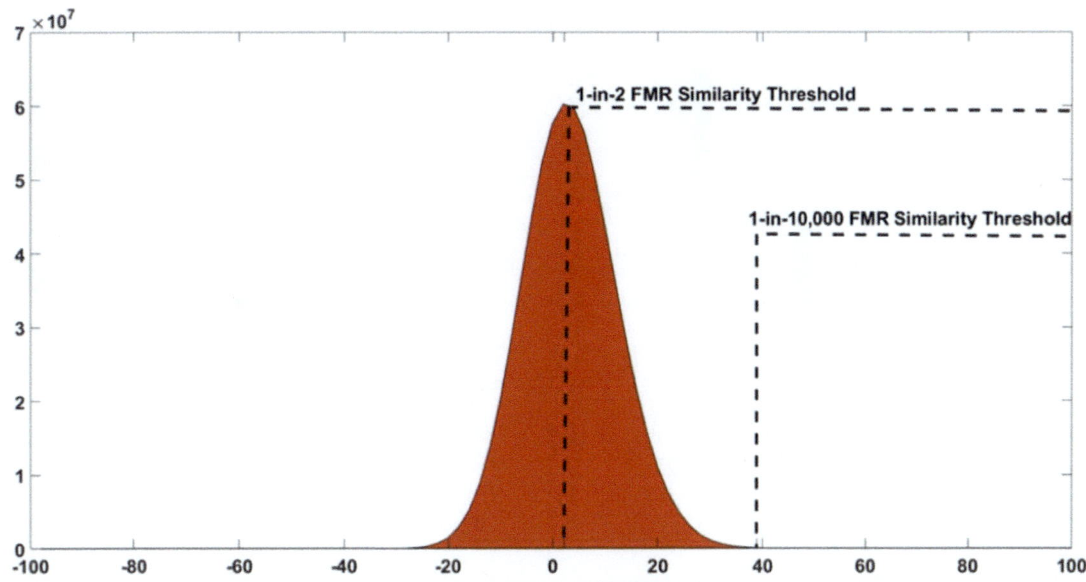

FIGURE 4.5
The ArcFace impostor distribution for the African American male cohort of MORPH, with the center and HST indicated. (Figure adapted from [24].)

FIGURE 4.6
Exemplar images of skin tone ratings I–VI used by manual raters for reference. (Figure adapted from [28].)

guide their ratings. If two or three of the raters agreed on a given skin tone rating, that rating was used. If all three gave different ratings, the middle rating of the three was used. Across the four total samplings (two samples from each of the two matchers), two or three viewers agreed on the skin tone rating for 93%–96% of images.

The two sets of manual ratings are visualized in Figure 4.7. If darker skin tone *is* a cause of increased false matches, we would expect to see a significantly higher

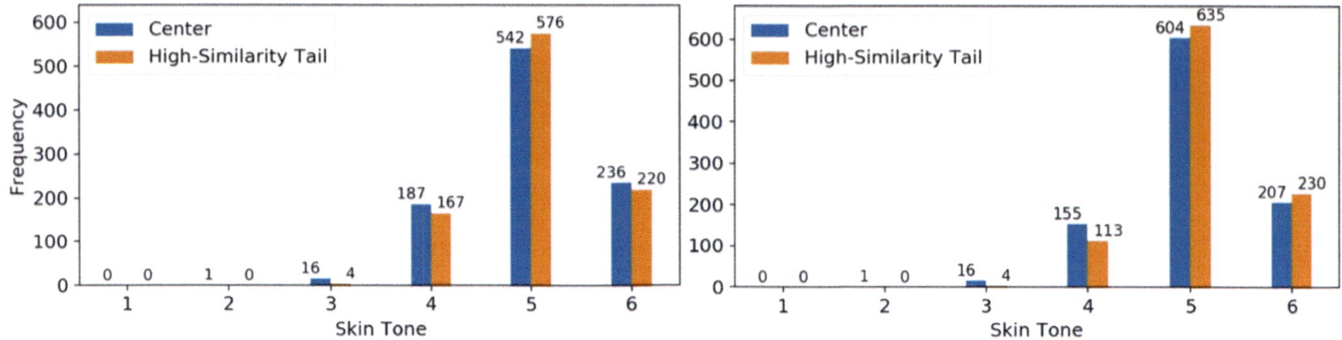

FIGURE 4.7
Comparison of skin tone ratings from the center and HST of the AAM impostor distribution from ArcFace (left) and VGGFace2 (right). (Figure adapted from [24].)

frequency of skin tones IV–VI in HST images. However, for both matchers, the distribution of skin tone ratings varies only slightly between center and HST images. For both sets of images, the most frequent skin tone rating is V, followed by IV, then VI, then III, with almost no images rated I or II. There is no clear shift toward higher or lower skin tone ratings in either region of the impostor distribution.

4.3.2 Standardized Skin Tone and False Match Pairs

Extending the aforementioned work, Krishnapriya et al. [28] conducted a second experiment, adding color correction, an automated skin tone rating system, and additional manual raters. In agreement with Howard et al. [29], they determined that a color correction step is crucial to the accurate estimation of apparent skin tone to account for inconsistent image lighting or coloration. The controlled capture environment of the MORPH dataset allows for color correction of the images based on the standard 18% gray background. The original face images in the top row of Figure 4.8 (adapted from [28]) feature a notable range of variation in the background color. After color correction, the images show a more consistent gray background.

Even a *single* subject's skin tone may vary significantly, both by visual and automated estimation, across their image set. Figure 4.9 gives an example of the automated skin type ratings associated with 150 face image instances of a single subject. The two pie charts give the distribution of skin type ratings based on the subject's original, uncorrected images (left) and the color-corrected versions (right). Four sample images are shown before and after correction in the top and bottom rows, respectively.

FIGURE 4.8
Noticeable variations exist in the gray background of the original MORPH images (top row), while their color-corrected versions (bottom) more accurately display the standard 18% gray background.

For an African American subject, we expect higher-valued skin type ratings. However, the ratings of the original images in Figure 4.9 primarily span over I–III (lighter) values. Inspection reveals overexposure: spectral highlights are present across the subject's forehead, nose, and cheek regions, interpreted as artificially lightened pixels. The majority of the corrected image ratings range from III to V, which is likely a more accurate estimation of the subject's true skin tone.

After color correction, manual and automated skin tone ratings were recorded for the center and HST images. Six manual raters independently assigned Fitzpatrick-inspired skin tone ratings from I to VI. For each image, the average of the six ratings was used as its consensus manual rating. Automated ratings were generated using six-toned, light-to-dark individual typology angle

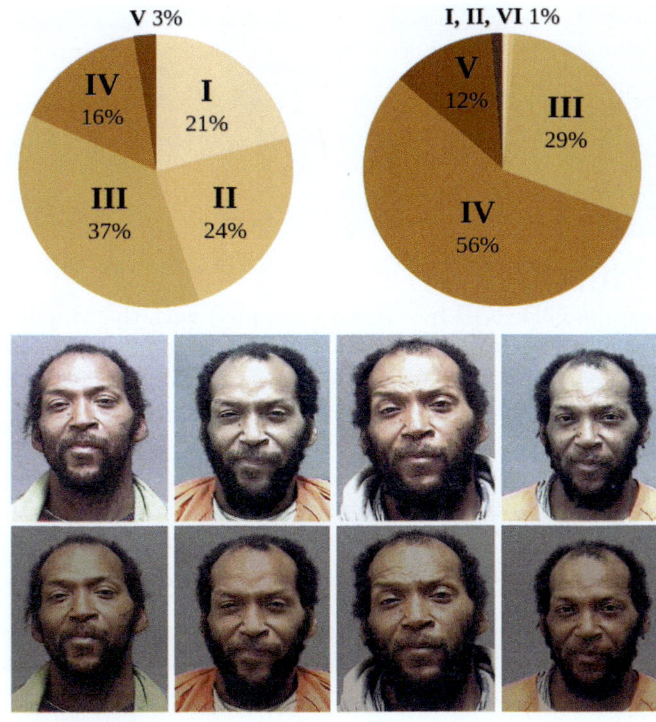

FIGURE 4.9
Automated ratings assigned to a single subject, before and after color correction.

(ITA) measurements, which have been used in previous research [30, 31] to directly assess skin tone from imagea. Krishnapriya et al. produced automated ratings for each image by mapping ITA values onto the FST-like scale.

Figure 4.10 gives the distribution of automated and consensus manual skin tone ratings for color-corrected images from the center and HST of the AAM impostor distribution. The two rating sets are generally consistent across center and HST images, reinforcing the previous finding that region of impostor distribution has little effect on skin tone rating.

4.3.3 Summary of Skin Tone Results

In their two experiments designed to directly assess the impact of skin tone on FMR, Krishnapriya et al. find no clear evidence to support the idea that skin tone is a driving factor behind the higher false match rate typically reported for African American individuals. That is, their results do not support a general conclusion that darker skin tone, in and of itself, causes an increased FMR. They note that the premise "face recognition is less accurate for darker skin tones" appears to be an over-simplification of previous research results on the ROC curves or FMRs for African Americans versus Caucasians.

Though the same FMR is typically achieved for different demographics at different thresholds, real-world recognition systems often use the same threshold across demographics. This practice has likely contributed to the common conclusion that the higher FMR for darker-skinned individuals is driven by skin tone. Krishnapriya et al. conclude that while the issue certainly merits

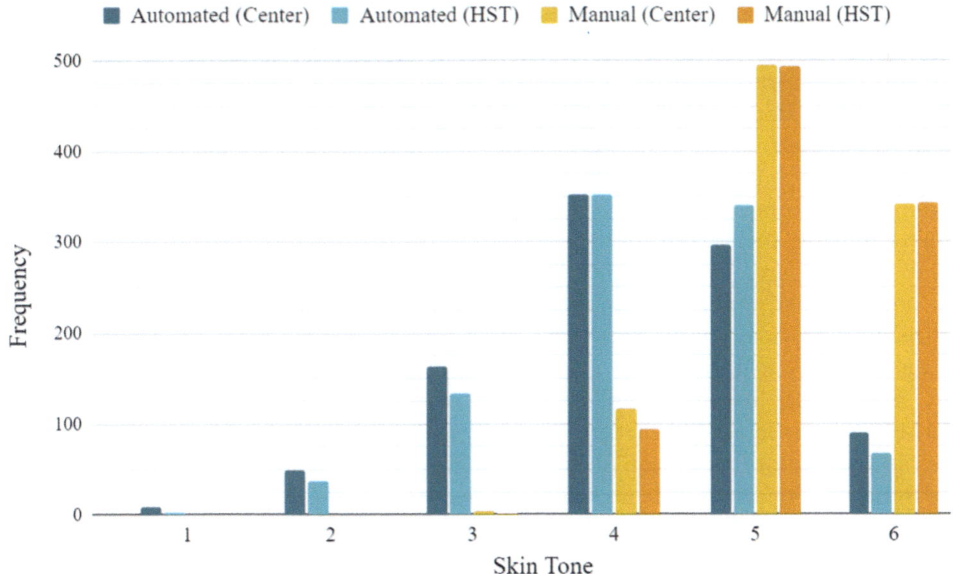

FIGURE 4.10
Distribution of ratings on color-corrected center versus HST images. (Figure adapted from [28].)

TABLE 4.3
Select Mean and Standard Deviation (SD) Values and
Male–Female SD Differences, as Reported in [32].

	Male		Female		Male–Female
Distance	Average	SD	Average	SD	SD
n-gn	121.3	6.8	111.8	5.2	+1.6
sn-gn	71.9	6	65.5	4.5	+1.5
n-sn	53	3.5	48.9	2.6	+0.9
ex-ex	89.4	3.6	86.8	4	-0.4

further study, their results suggest that factors *other* than
skin tone should be considered in looking for the cause of
race-asymmetric recognition accuracy.

4.4 Do Gender Differences in Face Size and Shape Cause Accuracy Differences?

The study of facial landmark points by Farkas et al. [32]
is well-known in the face recognition research community.
Demographic differences in face size and shape have been
studied in other fields as well. For example, these factors
impact the design of personal protective equipment [33].
In this section, we investigate several female/male differences in face size and shape to assess their impact on
the gender gap. To collect face measurements, we use a
3D+2D face image dataset originally collected at the University of Notre Dame and distributed as part of the Face
Recognition Grand Challenge (FRGC) program [34].

Table 4.3 from [32] gives average face size and shape
measurements for individuals representing a "North
American White Young Adult Population." We are particularly interested in two measurements. The **ex-ex** distance is measured between the outer eye corners ("exo-
canthion" points) and relates to face width. The average
ex-ex distance is 89.4 mm for males and 86.8 mm for
females. The **n-gn** distance relates to face height. It is
measured from the point of high curvature on the cen-
terline of the nasal bridge ("nasion," or **n**) to the point
on the midline of the chin where face curvature turns
("gnathion," or **gn**). The average **n-gn** distance is 121.3
mm for males and 111.8 mm for females.

Table 4.3 also gives the standard deviation values for
each distance measurement. The standard deviation of
ex-ex distance is 3.6 mm for males and 4 mm for females.
The standard deviation of **n-gn** distance is 6.8 mm for
males and 5.2 mm for females. These differences in face
width and height and their standard deviations suggest
that there is a gender-based difference in the average aspect ratio of the face.

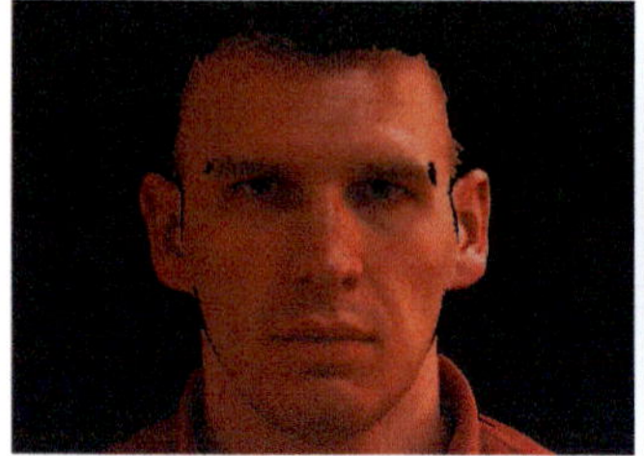

FIGURE 4.11
Example 2D RGB face images acquired with the Minolta
Vivid 910 3D scanner. Pixels are mapped to black if there
is no corresponding (x,y,z) location.

In this section, we examine whether the female/male
differences in face measurements impact the gender gap
in recognition accuracy. We design experiments to assess
the following factors:

- **Face width:** Does the larger average **ex-ex** distance
  for males result in a greater amount of face information and better accuracy?

- **Variation in face height:** Does the larger standard
  deviation in **n-gn** distance for males lead to lower
  similarity scores for impostor (non-mated) pairs of
  images?

- **Face aspect ratio:** Does balancing the aspect ratio
  of male and female faces decrease the gender gap?

4.4.1 Dataset, Landmark Points, and Distances

The Minolta Vivid 900/910 sensor captures a series of 2D
color images (480×640 pixels) and uses triangulation to
obtain 3D (x,y,z) coordinates. The resulting data points
are registered, associating pixels in the 2D image with
their 3D (x,y,z) locations in the scene. Sample 2D images
from the FRGC dataset are shown in Figure 4.11. The
3D data points are missing in areas where the face surface
is too far from the sensor, or where the projected light
beam is not seen clearly in the 2D image, for example, the
pupils and some hair regions. The pixels shown as black
in Figure 4.11 have no corresponding 3D point.

We begin by measuring average face width. Two independent viewers mark **ex** points in the 2D images. The
corresponding 3D coordinates are used to compute **ex-
ex** distance. Images with more than 0.5 cm difference in
the independently computed **ex-ex** distance are marked
again. If this difference persists in a second round of
marking, the image is dropped from analysis. From the
final set of **ex-ex** distances, the median value is taken as
the size of the face for a given image. The two rounds
of **ex-ex** markings resulted in a dataset of 1618 images

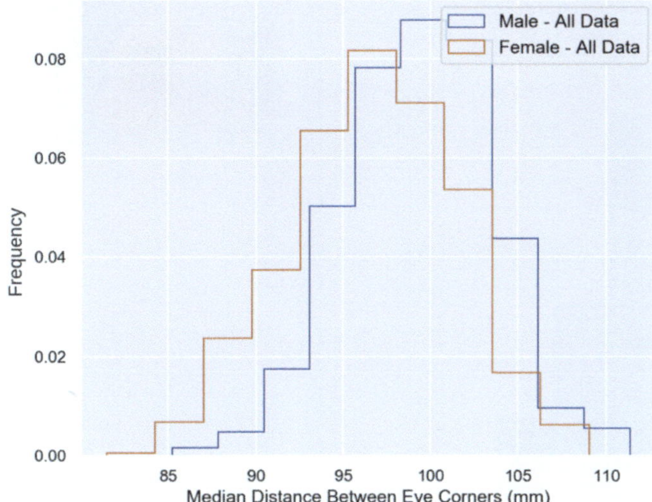

FIGURE 4.12
Distribution of female/male face width (**ex-ex**) values.

Finally, we compute face aspect ratio, dividing face width (**ex-ex** distance) by height (**n-gn**). The aspect ratio indicates whether overall face shape is more elongated or more rounded. Figure 4.15 shows the distributions for the female and male aspect ratio measurements. Females tend to have rounder faces, while males tend to have more elongated faces.

4.4.2 Do Size and Shape Differences Drive Accuracy Differences?

Our first experiment analyzes the impact of face width. We divide both male and female 2D images into two **ex-ex** distance groups. Faces in the first group of images ("Lower Half") have smaller-than-median **ex-ex** distance. Faces in the second group ("Upper Half") have larger-than-median **ex-ex** distance. We then use ArcFace to generate impostor and genuine distributions for each group and gender. These distributions are shown in Figure 4.16.

For both genders and distribution types, there is no clear difference in lower- and upper-half scores. In both plots of Figure 4.16, there is significant overlap between the blue and red curves (corresponding to the lower- and upper-half groups, respectively). Thus, we find no evidence that differences in average face width contribute to the observed differences in female and male impostor or genuine distributions.

The next experiment considers face height variation. We are interested in whether faces with greater height variation yield lower non-mated scores. Lower non-mated scores correspond to an improved impostor distribution and imply a lower false match rate. Given that males have a larger average **n-gn** distance and standard deviation (Table 4.4), we have designed an experiment to balance **n-gn** distance across male and female face images.

representing 583 females and 480 males. The resulting distribution of face width is compared for females and males in Figure 4.12.

We follow the same process to measure face height variation, marking **n** and **gn** points on each image and calculating **n-gn** distances. In some images, **gn** landmark points could not be accurately marked due to issues like facial hair or the chin being cropped. Sample removed images are shown in Figure 4.13. The distributions of **n-gn** distances for females and males are given in Figure 4.14. The means and standard deviations of the **n-gn** distances for females and males, given in Table 4.4, are similar to those reported by Farkas [32].

We create subsets of male and female images that have a nearly equal standard deviation in the distribution of **n-gn** distances using an iterative two-step process. In the first step, we remove the two male face images with the maximum and minimum **n-gn** distances. Then, the standard deviation of the remaining male set is compared to the female standard deviation (5.6 mm). After 21 iterations, the remaining subset of male images had an **n-gn** standard deviation of 5.9 mm. While this balancing approach does not perfectly balance the overall distributions, we find it sufficient to test the impact of the **n-gn** distribution on the impostor distribution.

The resulting impostor and genuine distributions are shown in Figure 4.17. Balancing the standard deviation of the female and male **n-gn** distances has no noticeable effect on the difference between the female and

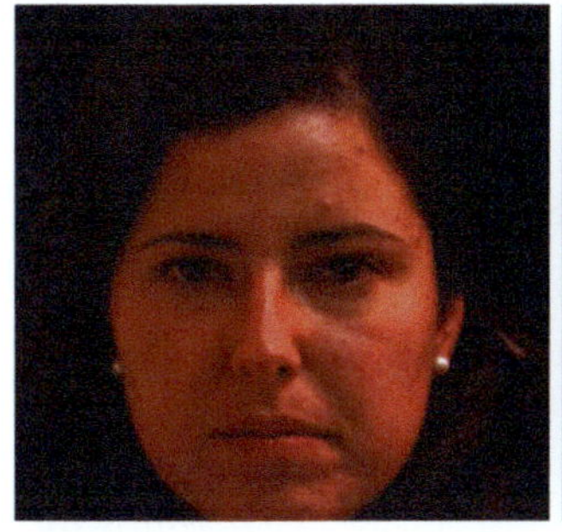

FIGURE 4.13
Example problematic images for marking face height landmark points. *Left:* The **gn** point is cut off at the bottom of the image. *Right:* The **gn** point is obscured by facial hair.

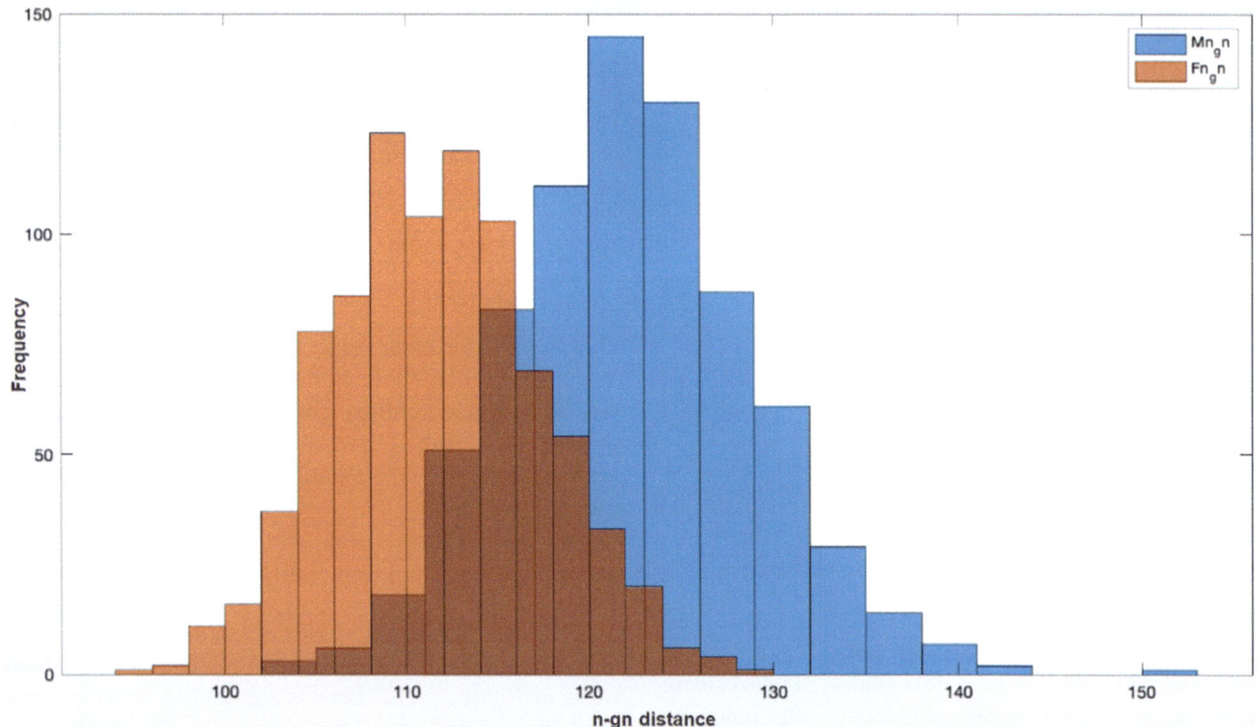

FIGURE 4.14
Distributions of **n-gn** distances for females and males.

TABLE 4.4
Measured Distances for Landmark Points (Left), Compared to Those Reported in [32] (right).

Distance	Male (SD)	Female (SD)	Farkas Male	Farkas Female
n-gn	122.3 (6.6)	111.6 (5.6)	121.3 (6.8)	111.8 (5.2)
sn-gn	70 (5.6)	63.1 (4.8)	71.9 (6)	65.5 (4.5)
n-sn	54.3 (3.6)	50.1 (3.3)	53 (3.5)	48.9 (2.6)

male impostor distributions. The female impostor distribution (indicated by the red dashed line) remains skewed toward higher similarity scores than the male distribution. That is, we find no clear evidence that controlling for face height decreases the gender gap seen in impostor distributions.

As a final experiment, we combine the previous width and height measurements to study cross-gender aspect ratio. We begin by balancing female/male images with respect to face aspect ratio. Each male image is paired with a female image that has an aspect ratio difference less than or equal to 0.025. This balancing yielded a set of 464 male images and 464 female images with approximately matched aspect ratios. The resultant image distributions are given in Figure 4.18.

The impostor and genuine distributions for the subsets balanced on aspect ratio are given in Figure 4.19. Again, we find no observable impact on the distributions.

The female distributions remain closer together than male distributions, with impostor scores skewed toward higher values and genuine scores toward lower values.

4.4.3 Summary of Face Size and Shape Results

In this section's experiments, we measure facial landmark points and distances using 3D+2D face images. Our measurements and results align with those previously reported by Farkas [32]. The measurements reveal consistent and notable differences in face size and shape for males versus females. In general, males have larger average values for facial measurements. Their faces are usually longer and have larger aspect ratios. Additionally, men tend to have larger standard deviations in face measurements than women, a hypothesized reason why male faces may be easier to differentiate than female faces.

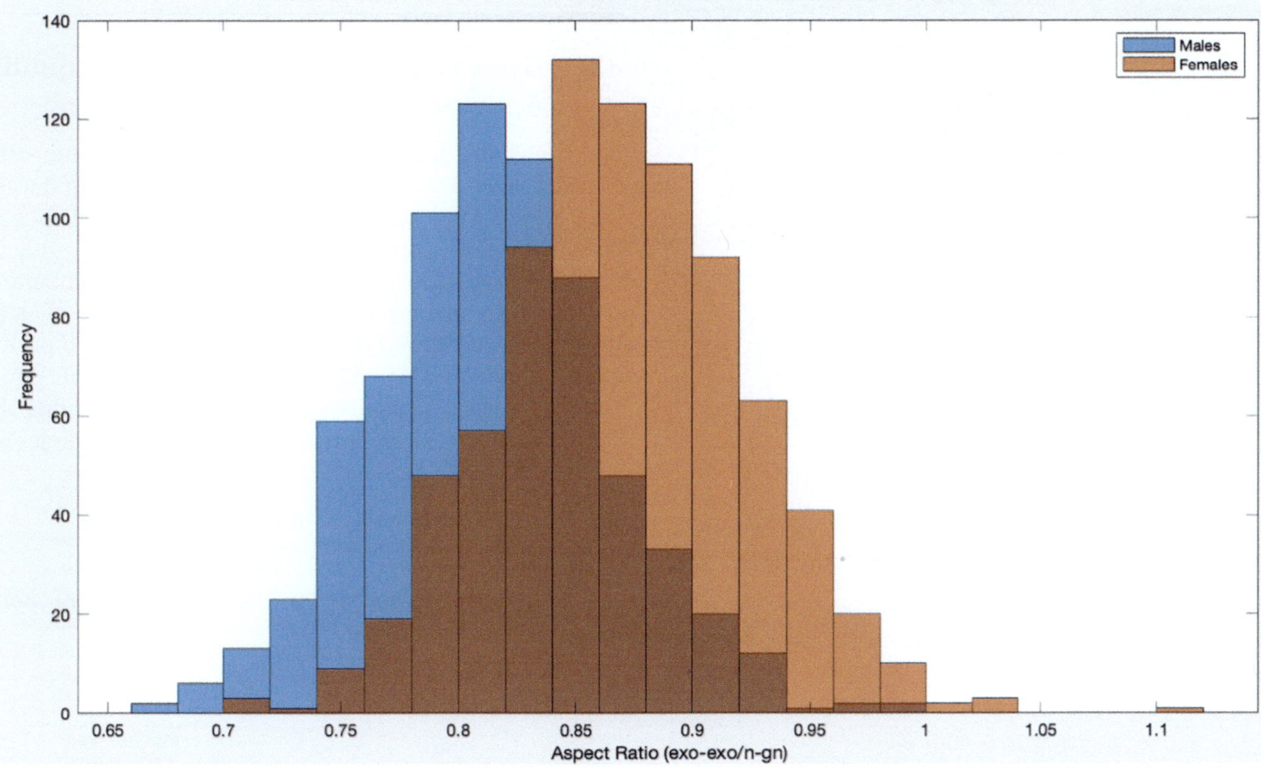

FIGURE 4.15
Distributions of female and male aspect ratios.

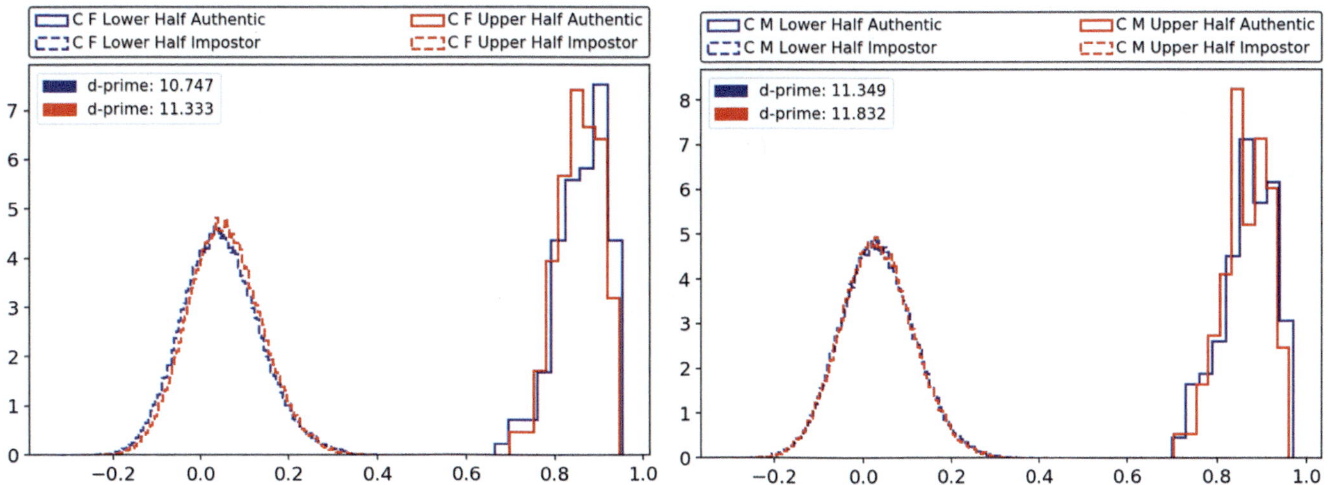

FIGURE 4.16
Impostor and genuine (authentic) distributions of face width groups, with the Caucasian female plot on the left and male on the right.

In the three experiments, we analyze how facial width, height variation, and aspect ratio impact male and female impostor and genuine score distributions. In each case, balancing female/male images with respect to the given measurement has no significant effect on the resulting distributions. We thus find no clear evidence that the observed facial differences contribute to the gender gap.

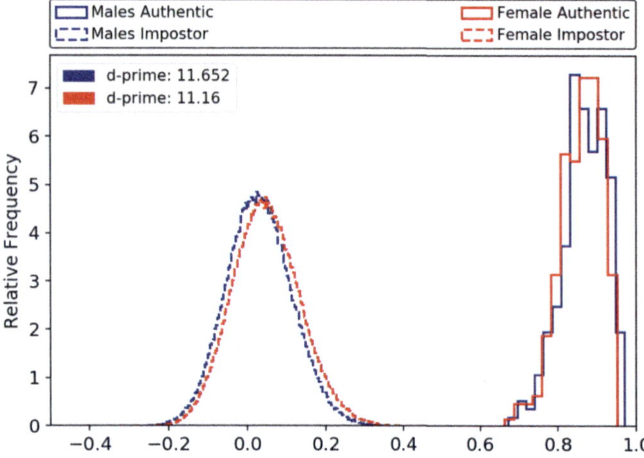

FIGURE 4.17

Impostor and genuine distributions of the male and female subsets balanced on n-gn standard deviation (height).

4.5 Does Balanced Training Data Equalize Accuracy in Test Data?

Some authors have suggested that demographic differences in accuracy are due to imbalanced training datasets [35], which tend to have "a higher number of face images from [the] Caucasian ethnicity and males" [36]. Researchers note that biases resulting from such imbalances do not seem to be intentional [37] but assume that systems built on skewed data are simply "bound to produce biased models" [31]. The speculation from this line of thought is that explicitly gender-balanced training data should eliminate or at least greatly reduce the gender gap.

4.5.1 Training with Balanced Data

Albiero et al. first addressed the issue in *Analysis of Gender Inequality In Face Recognition Accuracy* [38]. For an

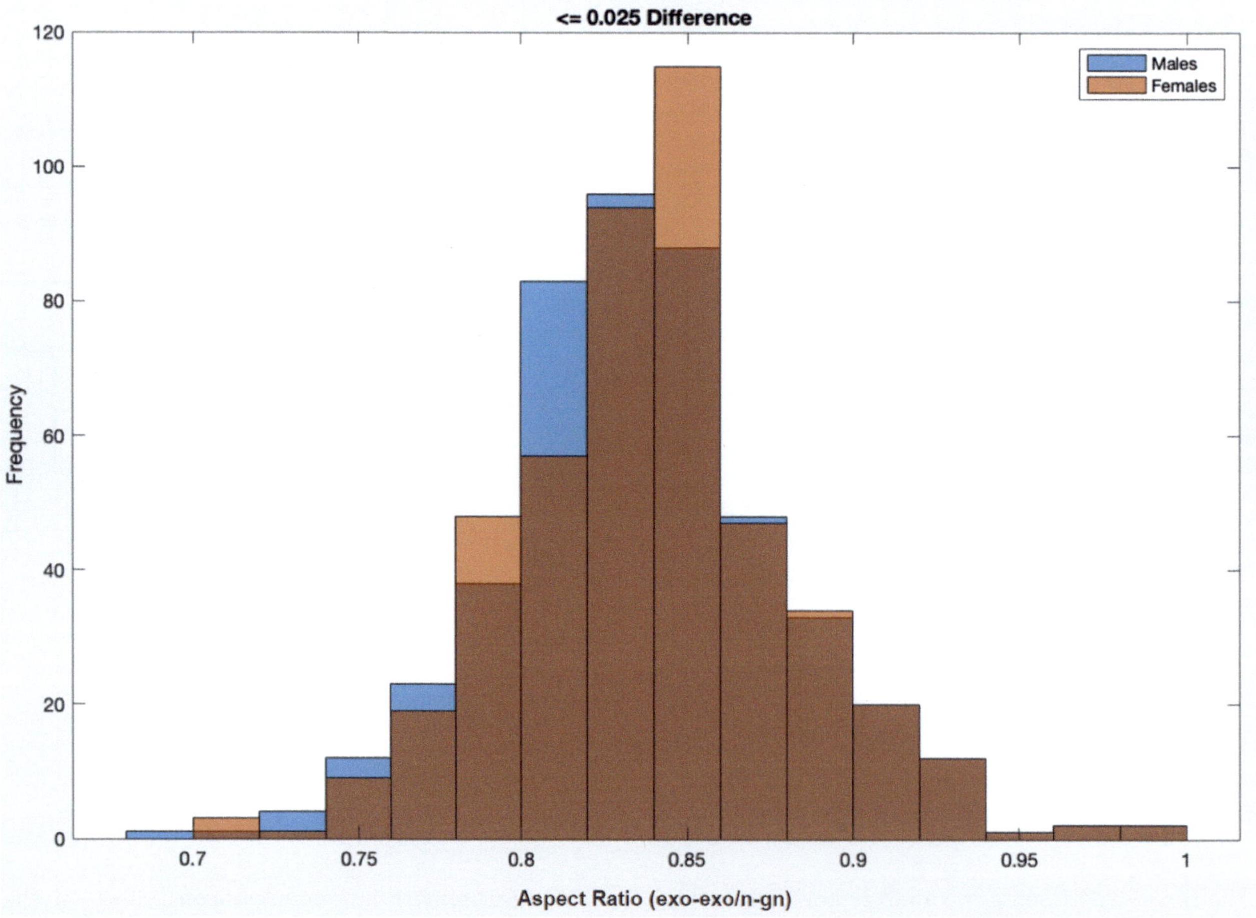

FIGURE 4.18

Balanced female/male aspect ratio distributions.

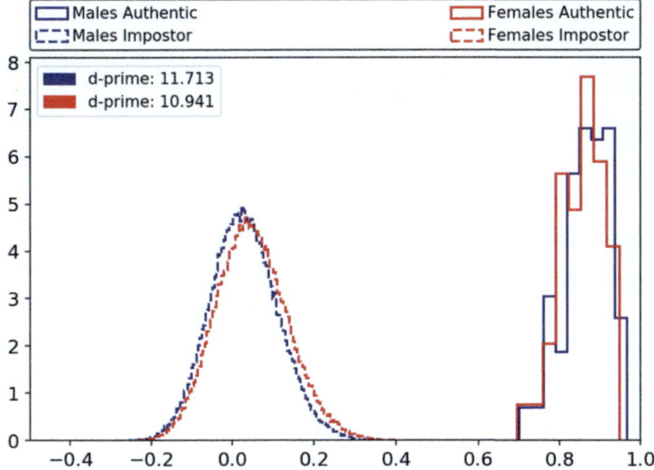

FIGURE 4.19
Impostor and genuine (authentic) distributions for female/male image sets balanced on aspect ratio.

accurate assessment of how balancing training data impacts recognition, they trained a network from scratch. Using the ResNet-50 backbone, they trained two separate networks with combined margin loss on subsets of the VGGFace2 [27] and MS1Mv2 datasets. These subsets were explicitly balanced on a number of male and female persons and images.

Results for each model are given in Figure 4.20, tested on four datasets: MORPH's African American and Caucasian cohorts, the Notre Dame dataset [34], and the Asian Faces Dataset (AFD) [39], curated as described in Section 4.2.

The MS1Mv2-trained model achieves the best distributions across all datasets and both genders. Both models give a decreased d-prime difference between males and females in the Notre Dame dataset versus initial ArcFace results. However, all results in Figure 4.20 still feature female impostor and genuine distributions that are closer together than the male distributions.

4.5.2 Training with Skewed Data

With the previous result in mind, they broadened their analysis to include both balanced and methodically **imbalanced** data in *How Does Gender Balance in Training Data Affect Face Recognition Accuracy?* [40]. For both the VGGFace2 and MS1Mv2 datasets, they constructed seven training subsets with images explicitly selected to vary gender representation. The descriptions of these subsets, and a visualization adapted from [40], are given below.

- **Full:** Full dataset with no modifications

- **Balanced:** Male images removed to equal the number of female images

- *F100*: 100% female images

- *M25F75*: 75% female, 25% male images

- *M50F50*: 50% female, 50% male images

- *M75F25*: 25% female, 75% male images

- *M100*: 100% male images

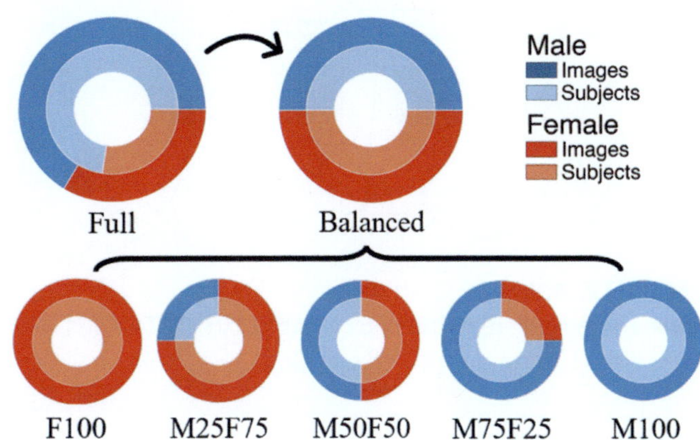

The five latter subsets consist of images randomly selected from the balanced subset. Final counts of subjects and images in each training subset are given in Figure 4.5. They continued to use MORPH's Caucasian and African American cohorts and the Notre Dame dataset for testing.

They again selected the ResNet-50 architecture to train on the subsets, with three different loss functions: the standard softmax loss, the newer combined margin loss, and the non-classification triplet loss. First, they analyzed male, female, and average accuracy on the full dataset and balanced subset. Results for each of the testing datasets are given in Table 4.6.

Interestingly, with a softmax loss function, both balanced subsets give the best male, female, and average accuracy versus full. Combined margin loss gives the opposite result: best cross-gender and average accuracy is generally achieved with the full subsets. Triplet loss gives inconsistent results across the two subsets. In general, Albiero et al. find that the MS1Mv2 training dataset and combined margin loss yield the highest overall accuracy.

Selecting the best training set (MS1Mv2) and loss function (combined margin), they proceeded to analyze the five remaining subsets. They expected the best single-gender accuracy when training with M100 for males and F100 for females. They expected higher average accuracy on the balanced (M50F50) subset. The results are reported in Table 4.7. Accuracy values are colored from green to red, where green indicates the best result is achieved with the expected training and red the opposite.

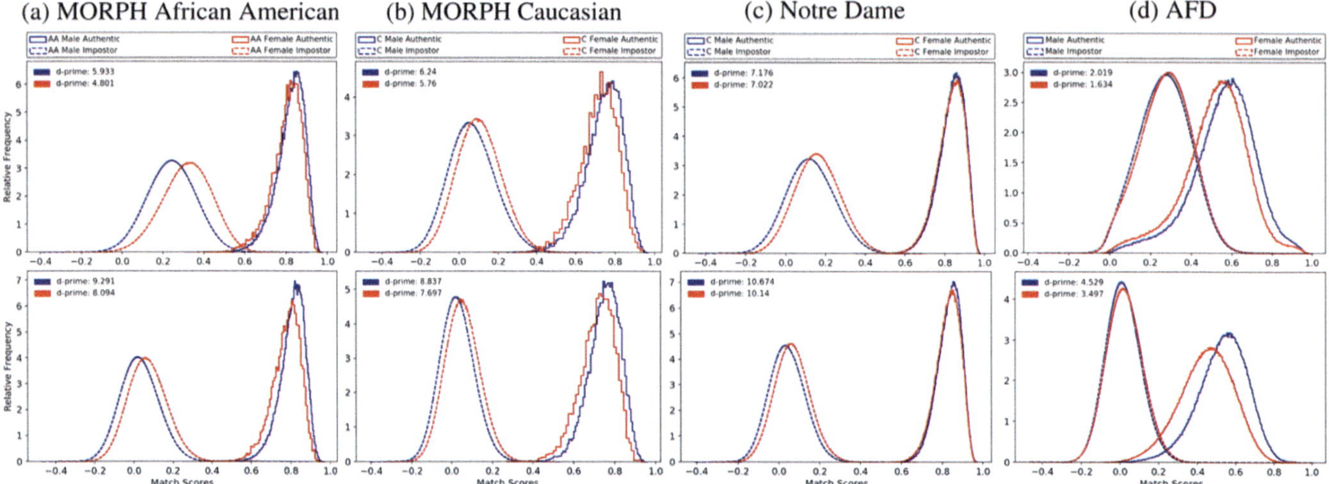

FIGURE 4.20
Male and female distributions using networks trained on the gender-balanced VGGFace2 (top) and MS1Mv2 (bottom) datasets. (Figure adapted from [38].)

TABLE 4.5
VGGFace2 (Top Half) and MS1MV2 (Bottom Half) Training Subsets Created. (*Source:* Table adapted from [40].)

Subset Name	# Subjects		# Images	
	Males	Females	Males	Females
Full	5154	3477	1,828,987	1,291,873
Balanced	3477	3477	1,291,873	1,291,873
F100	0	3477	0	1,291,873
M25F75	870	2607	322,969	968,904
M50F50	1739	1739	645,937	645,937
M75F25	2607	870	968,904	322,969
M100	3477	0	1,291,873	0
Full	59,563	22,499	3,741,274	1,890,773
Balanced	22,499	22,499	1,890,773	1,890,773
F100	0	22,499	0	1,890,773
M25F75	5624	16,875	472,693	1,418,080
M50F50	11,249	11,249	945,386	945,386
M75F25	16,875	5624	1,418,080	472,693
M100	22,499	0	1,890,773	0

For the MORPH Caucasian cohort, each best accuracy is achieved with the expected training. The African American cohort, however, achieves the best female accuracy on the 25% male training subset, and best average accuracy on the 75% male subset. The 75% male subset also gives the best overall accuracy for the Notre Dame dataset.

4.5.3 Summary of Balanced Training Results

In their first study [38] on gender-balancing the training data, Albiero et al. found that an equal number of male and female subjects and images does *not* improve the separation between the female impostor and genuine distributions. Instead, Albiero et al. [40] showed that when training set and loss function are both selected to maximize accuracy, *imbalanced*, male-dominated training data results in higher female, male, and average accuracy. They propose that, with a good combination of training set and loss function, 75%-male training data may give the best average accuracy. The 75%-female training data yields the smallest gender gap in accuracy and implies that a gender ratio may be chosen to aim for approximately equal test accuracy – though this accuracy will not be the highest for males, females, or on average. Ultimately, they find no empirical support for the speculation that gender-balanced training data equalizes test accuracy.

4.6 Does Balancing Pixels in Test Data Equalize Genuine Distributions?

In *Gendered Differences in Face Recognition Accuracy Explained by Hairstyles, Makeup, and Facial Morphology* [14], Albiero et al. honed in on another speculated cause of the gender gap: a disparity in the number of "face pixels" between male and female images. Face images are typically pre-processed with detection, alignment, and cropping. Face pixels compose the resultant segment of the image that is directly evaluated by the matcher. The initial experiment analyzes how the quantity of face pixels varies by gender and relates to gender-specific accuracy.

TABLE 4.6
Gender Accuracy (%) with TAR@FAR=0.001% When Trained Using the Entire Training Dataset (full) and the Gender Balanced Version (Balanced). (*Source:* Table adapted from [40].)

| Loss | Training Dataset | Subset | Testing Dataset | | | | | | | | |
| | | | MORPH C | | | MORPH AA | | | Notre Dame | | |
			Male	Female	Avg.	Male	Female	Avg.	Male	Female	Avg.
Softmax	VGGFace2	Full	89.23	76.7	82.97	91.38	70.08	80.73	98.9	98.04	98.47
		Balanced	90.1	83.71	86.91	94.6	84.77	89.69	99.38	98.93	99.16
	MS1MV2	Full	91.55	83.07	87.31	94.84	83.19	89.02	99.74	99.41	99.58
		Balanced	95.04	88.42	91.73	97.58	84.28	90.93	99.86	99.54	99.7
Combined Margin	VGGFace2	Full	93.69	86.44	90.07	95.93	86.68	91.31	99.71	99.43	99.57
		Balanced	90.38	86.29	88.34	93.31	82.76	88.04	99.33	99.23	99.28
	MS1MV2	Full	99.81	99.39	99.6	99.93	99.7	99.82	100	99.97	99.99
		Balanced	99.65	99.09	99.37	99.88	99.59	99.74	100	99.99	100
Triplet	VGGFace2	Full	78.19	63.67	70.93	82.38	59.27	70.83	93.47	91.26	92.37
		Balanced	77.25	62.79	70.02	81.75	63.01	72.38	93.85	89.43	91.64
	MS1MV2	Full	90.68	79.56	85.12	95.03	84.17	89.6	98.78	97.77	98.28
		Balanced	89.98	85.1	87.54	94.35	85.28	89.82	98.36	99.07	98.72

TABLE 4.7
Gender Accuracy (%) with Different Balancing Proportions. All Trainings Have the Same Number of Subjects and Images. (*Source:* Table adapted from [40].)

		F100	M25F75	M50F50	M75F25	M100
MORPH Caucasian	Male	90.91	98.59	99.24	99.21	99.5
	Female	97.71	97.57	97.45	96.51	94.94
	Average	94.31	98.08	98.35	97.86	97.22
MORPH African American	Male	94.79	99.43	99.72	99.78	99.81
	Female	98.35	98.85	98.44	98.58	97.88
	Average	96.57	99.14	99.08	99.18	98.85
Notre Dame	Male	99.56	99.98	99.99	100	100
	Female	99.92	99.97	99.95	99.98	99.86
	Average	99.74	99.98	99.97	99.99	99.93

A natural follow-up experiment examines recognition accuracy when a dataset is controlled for equal face pixels across genders.

4.6.1 Assessing Impact of Pixels of the Face

In order to discern between face and non-face pixels, Albiero et al. used Bilateral Segmentation Network ("BiSeNet") [41] to segment each face and generate a corresponding binary mask. Segmented regions classified as skin, eyebrows, eyes, ears, nose, and mouth are considered "face pixels" in the mask; regions like neck and hair are excluded. Sample masks are shown in Figure 4.21, adapted from [14].

The distribution of the fraction of images labeled *face* is provided in Figure 4.22. For the majority of female faces,

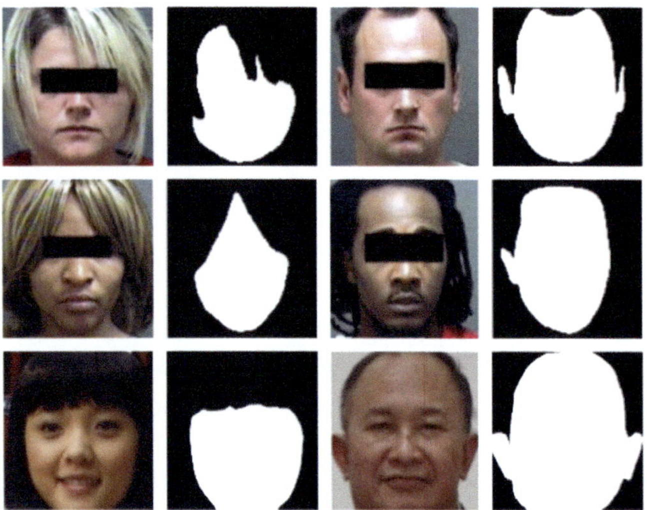

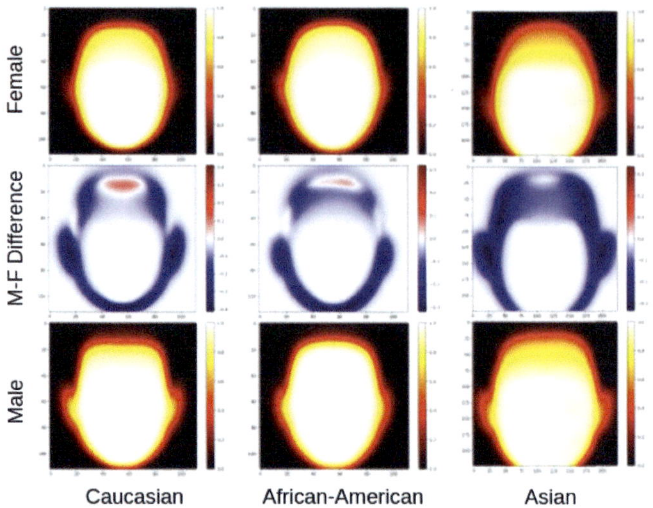

FIGURE 4.21

Example images and their BiSeNet-generated binary masks, where white indicates *face* and black *non-face* regions. Note the significant impact of hairstyle on the size of the *face* region – and subsequently, the number of face pixels.

25%–45% of the image represents face for MORPH, and 20%–55% for Asian-Celeb. For the majority of male images, 45%–70% of the image represents face for MORPH, and 55%–80% for Asian-Celeb. The gender-shifted distributions make it clear that, on average, female images contain less usable information about the face.

With the segmented masks, Albiero et al. computed heatmaps to reflect the fraction of images for which a given pixel is labelled *face*. Figure 4.23 shows the frequency of *face*-labelled pixels for each race and gender. The heatmaps in the middle row of Figure 4.23 give the male–female difference ("M–F Difference") for each group. Blue pixels are those more frequently labeled *face*

FIGURE 4.23

Heatmaps representing frequency of pixels labelled *face*. (Figure adapted from [14].)

for males than females, with red showing the opposite. In the male heatmaps, the chin, ear, and jawline regions are more prominent than in the female heatmaps, as evidenced by the larger regions of blue pixels in the difference heatmaps. Females appear to have slightly higher foreheads, as indicated by the red regions in the Caucasian and African American difference heatmaps.

Overall, gendered differences in facial morphology and choice of hairstyle result in fewer average face pixels for female subjects. Females already tend to have smaller heads and facial regions, and their hairstyles "more frequently occlude the ears and part of the sides of the face." [14] As with the percent-face distributions (Figure 4.22), Albiero et al. observed that the average face image of a biological female contains a smaller fraction of usable biometric information than that of a biological male.

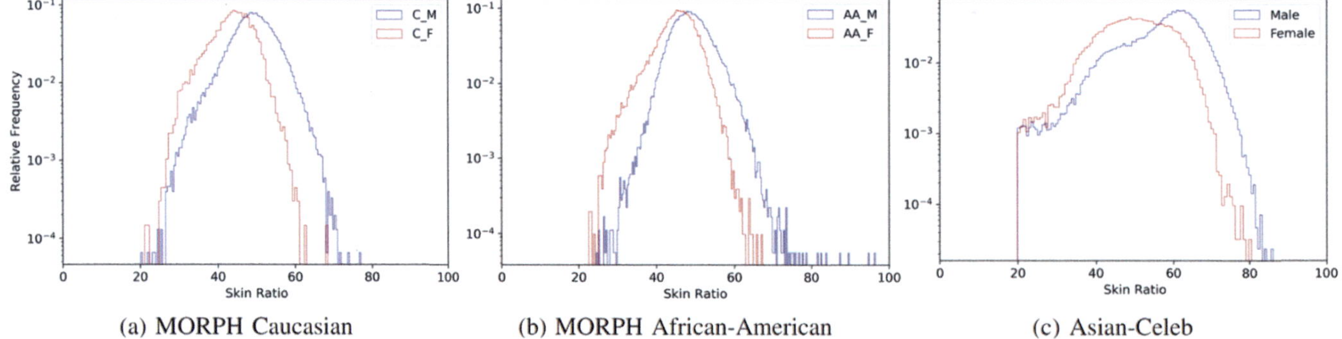

FIGURE 4.22

Comparison of female and male distributions of percent of image labelled *face*: (a) MORPH Caucasian, (b) MORPH African-American, (c) Asian-Celeb. (Figure adapted from [14].)

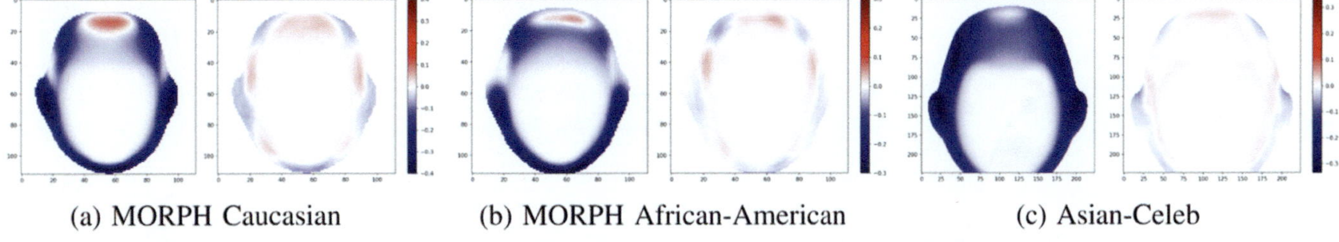

(a) MORPH Caucasian   (b) MORPH African-American   (c) Asian-Celeb

FIGURE 4.24
The original image set is first masked at a 10% level from female heatmap, giving the left difference heatmaps. A subset of the male images is selected using the intersection-over-union (IoU) to balance with the female images, giving the right difference heatmap: (a) MORPH Caucasian, (b) MORPH African-American, (c) Asian-Celeb. (Figure adapted from [14].)

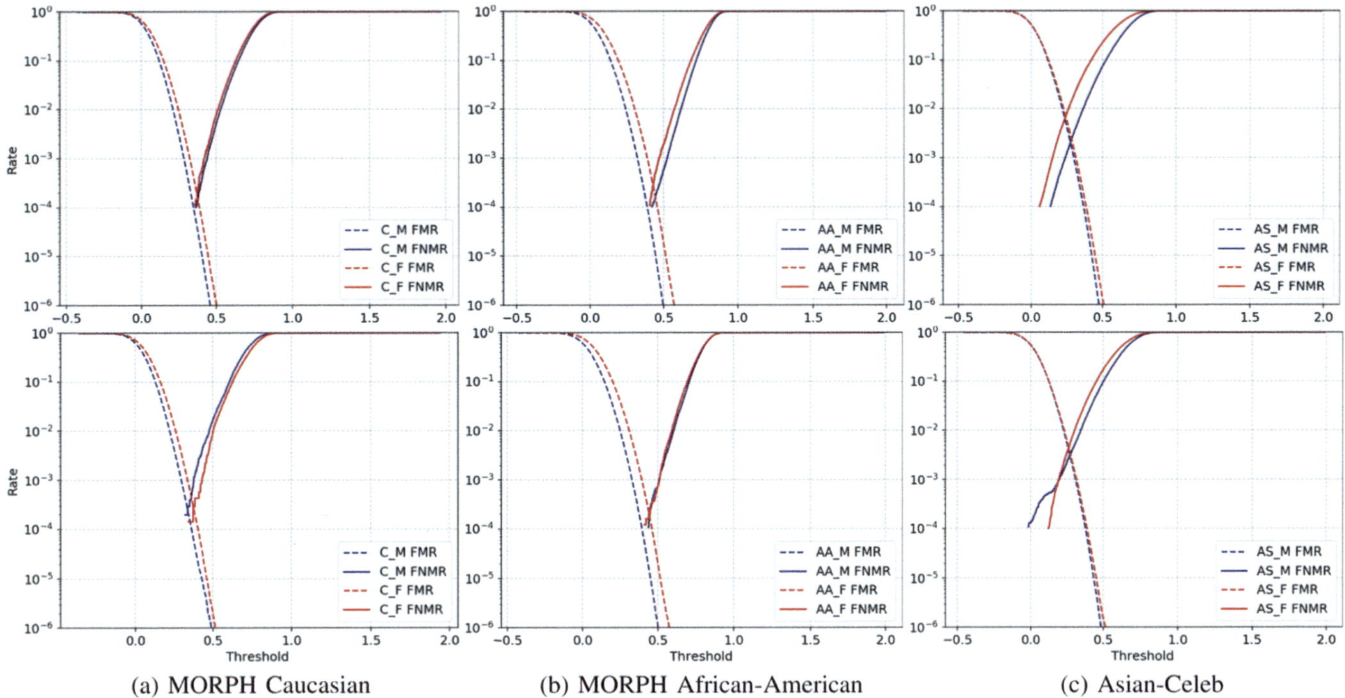

(a) MORPH Caucasian   (b) MORPH African-American   (c) Asian-Celeb

FIGURE 4.25
Impostor and genuine distributions for original (top) and information-equalized (bottom) female and male image sets: (a) MORPH Caucasian, (b) MORPH African-American, (c) Asian-Celeb. (Figure adapted from [14].)

4.6.2 Equalizing Pixels of the Face

Led by the previous finding, Albiero et al. next created a version of the dataset with roughly equal face information. First, to establish the same maximum set of pixels containing face information for males and females, they masked the area outside of the 10% level in the female heatmap in all images. The 10% masks are shown on the left in Figure 4.24. As the female face region is generally a subset of the male, this step preserves the majority of available face information. To decrease the remaining bias toward greater information for males, for every female image, they selected a male image with the maximum intersection-over-union (IoU) of pixels labelled *face*. The final heatmaps are given on the right in Figure 4.24. In the resulting "information equalized" dataset, the male–female difference in information is minimized. At any pixel, remaining differences are generally less than 5% and roughly balanced between the two genders.

The impostor and genuine distributions from the original and information-equalized datasets are given in Figure 4.25. The equalizing step proves highly impactful: though FMR remains worse for females than males, the female FNMR becomes *the same or slightly better* than the male FNMR for African American and Caucasian groups. While Asian-Celeb features the largest

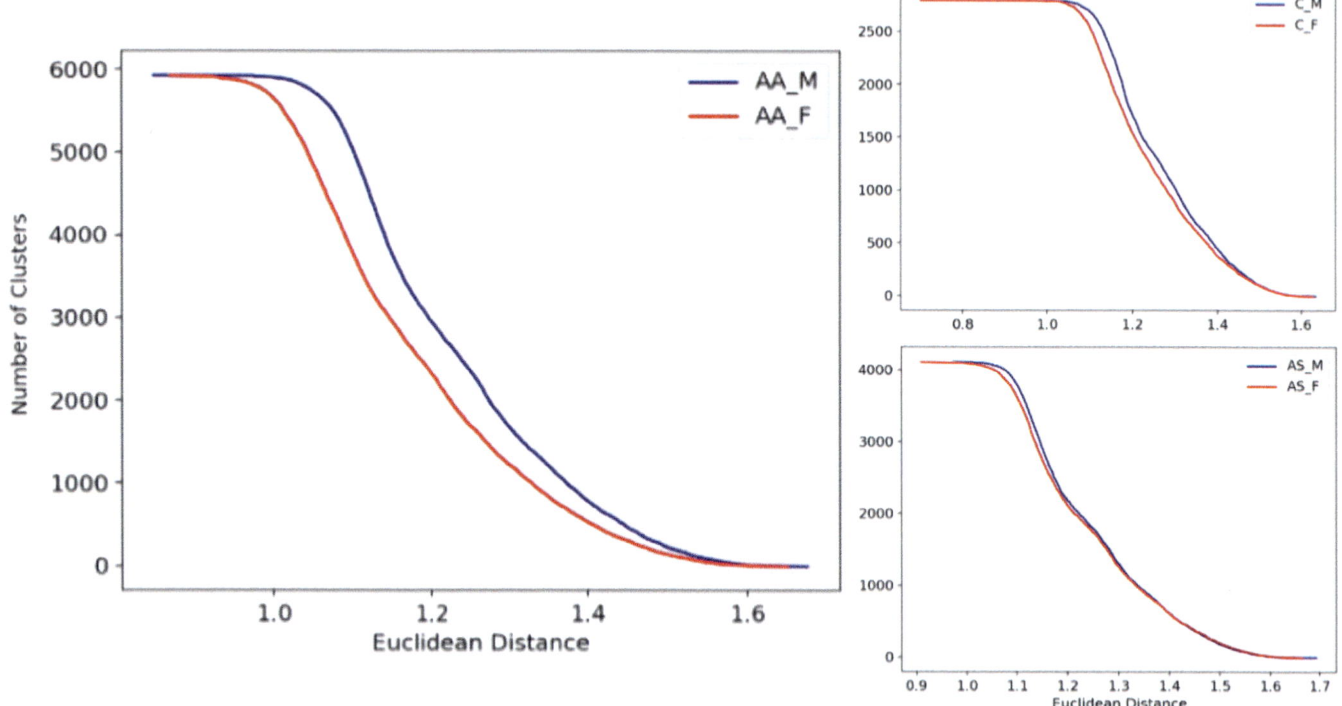

FIGURE 4.26
Number of clusters as distance threshold increases for ArcFace features. Datasets compared have the same number of subjects, and one image per subject, thus clusters are formed by different persons. (Figure adapted from [14].)

gender gap in FNMR, the equalized subset still pushes the male and female genuine distributions significantly closer.

Notably, these results were solely derived from balancing face pixels in the information-equalized dataset. Both experiments used ArcFace trained on a known female-underrepresented dataset. Albiero et al. conclude that the common observation that the female genuine distribution is worse than male is rooted in gendered social conventions for the hairstyles in test images rather than imbalanced training data. When image datasets used to compute test accuracy are controlled for face-pixel information, the female genuine distribution is as good as (or better than!) the male genuine distribution.

Despite promising results for FNMR, Albiero et al. sought to understand why female FMR remains worse. Even in the information-equalized dataset, on average, a non-mated female pair yields a higher similarity score than a non-mated male pair. They analyzed male and female impostors by performing a clustering experiment using each subject's highest face-pixel image. Each image's ArcFace feature vector was generated, and each vector began as its own cluster. Within the same gender, clusters were combined incrementally using Euclidean distance until only one cluster remained.

The male and female clustering curves for each dataset are shown in Figure 4.26. The AAM versus female curves, which feature the greatest separation, are enlarged on the left. The right column features the gender-separated curves for Caucasian (top) and Asian (bottom) groups.

For all three datasets, male clusters form at higher thresholds than female, and the male–female difference in the cluster curves is proportional to the difference in FMR. For MORPH African American, with a large male–female FMR difference, females form clusters at much lower distances than males. For MORPH Caucasian, with a smaller male–female FMR difference, the difference in the cluster curve is smaller. For Asian-Celeb, where the FMR is almost the same for males and females, the number of clusters is also similar. These results imply that images of two different females are more similar than images of two different males, causing females to be clustered together at faster rate, which translates to a worse FMR.

4.6.3   Summary of Balanced Pixel Results

In previous results, females generally have a worse genuine distribution and correspondingly higher FNMR at the same decision threshold as males. However, females also tend to have different hairstyles and face shapes than males, and these factors result in a smaller fraction of an image, on average, containing face information for females. When datasets are controlled to equalize the fraction of the image that represents face, the female genuine distribution is as good as, or better than, the male genuine distribution. The female impostor distribution remains worse and is explained by the finding that non-mated female pairs are more similar than non-mated male pairs.

Thus, the result that females have a higher FNMR than males is found to be caused by gendered differences in facial appearance. Because hairstyle conventions and face morphology may vary between racial groups, this conclusion is consistent with, and may explain, the observation in the NIST report [11] that the gender gap in FNMR is not universal across races.

4.7 Discussion

In recent years, public discourse around race and gender equity has piqued great interest in the subject of face recognition. Research studies have consistently reported that women and darker-skinned individuals are more likely to be falsely accepted or rejected by recognition systems. These demographic-specific accuracy gaps are quantifiable as differences in both false match and non-match rates and impostor and genuine score distributions. However, the *causes* of the observed differences are not so readily apparent.

The previous sections explore four commonly speculated causes. Each experiment is designed to isolate a single factor and remove confounding variables in the analysis of corresponding recognition accuracy. Evidence is not found to support the idea that skin tone or face morphology alone decrease performance or that balancing training data increases performance. Ultimately, controlling for equal cross-gender face pixel information is the only factor found to notably improve the overall female FNMR. However, the gender gap persists in FMR, and the finding that female faces are inherently more similar than male faces suggests that the issue requires further study.

In practice, test data is generally not evaluated for "fairness." That is, research often simply assumes that the test data is free of any factors that would bias the accuracy results. Then any demographic difference in accuracy is attributed to algorithms or training data, rather than inherent demographic-related differences in the test data. As the balanced-pixel experiment reveals, gendered social conventions and aesthetic choices like hairstyle can cause "bias" toward less information available for face recognition for females. These and other image quality issues differ across demographic groups, and, when unaccounted for, manifest as a gender gap in recognition accuracy.

In an increasingly digital world, automated technologies like face recognition have direct, and significant, impact on all people. As of yet, there is no complete or definitive answer to *why* face recognition performance varies across race and gender groups. But we must persist in noting these disparities and assessing them experimentally. Such research is essential to ensuring the very purpose of the technology: to better the lives of its users.

References

[1] T. Kanade. *Picture Processing System by Computer Complex and Recognition of Human Faces.* Nov. 1973.

[2] E. Dans. "The original and the copycats: The case of facial identification". In: *Forbes* (23 December 2018). https://www.forbes.com/sites/enriquedans/2018/12/23/the-original-and-the-copycats-the-case-of-facial-identification

[3] C. Zibreg. "Apple posts a trio of videos highlighting Face ID & Portrait Lighting on iPhone X". In: *iDownload Blog* (17 October 2018). https://www.idownloadblog.com/2017/12/11/apple-iphone-x-ads-portrait-lighting-face-id/

[4] C. Zibreg. "Apple's awesome new iPhone X ad showcases unlocking saved passwords with Face ID". In: *iDownload Blog* (9 July 2018). https://www.idownloadblog.com/2018/07/09/apple-iphone-x-ad-password/

[5] Apple Inc. "Face ID Security". In: *Apple* (November 2017). https://www.apple.com/business-docs/FaceID_Security_Guide.pdf

[6] S. Curtis. "Apple accused of racism after Face ID fails to identify Chinese iPhone X users". In: *The Mirror* (21 December 2017). https://www.mirror.co.uk/tech/apple-accused-racism-after-face-11735152

[7] C. Zhao. "Is the iPhone X racist? Apple refunds device that can't tell Chinese people apart, woman claims". In: *Newsweek* (18 December 2017). https://www.newsweek.com/iphone-x-racist-apple-refunds-device-cant-tell-chinese-people-apart-woman-751263

[8] J. Buolamwini and T. Gebru. "Gender shades: Intersectional accuracy disparities in commercial gender classification". In: *Conference on Fairness, Accountability and Transparency*. PMLR. 2018, pp. 77–91.

[9] S. Lohr. "Facial recognition is accurate, if you're a white guy". In: *The New York Times* (9 February 2018). https://www.nytimes.com/2018/02/09/technology/facial-recognition-race-artificial-intelligence.html

[10] C. M. Cook et al. "Demographic effects in facial recognition and their dependence on image acquisition: An evaluation of eleven commercial systems". In: *IEEE Transactions on Biometrics, Behavior, and Identity Science*, 1.1. 2019, pp. 32–41. DOI: 10.1109/TBIOM.2019.2897801.

[11] P. Grother, M. Ngan, and K. Hanaoka. *Face Recognition Vendor Test Part 3: Demographic Effects*. en. 2019. DOI: https://doi.org/10.6028/NIST.IR.8280.

[12] T. Doshi. "Improving skin tone representation across Google". In: *Google AI Blog* (11 May 2022). https://blog.google/products/search/ monk-skin-tone-scale

[13] M. Hecht. "What are the Fitzpatrick skin types?" In: *Healthline* (7 March 2019). https://www.healthline.com/health/beauty-skin-care/fitzpatrick-skin-types

[14] V. Albiero et al. "Gendered differences in face recognition accuracy explained by hairstyles, makeup, and facial morphology". In: *IEEE Transactions on Information Forensics and Security* 17. 2022, pp. 127–137. DOI: 10.1109/TIFS.2021.3135750.

[15] Jiankang Deng et al. "Arcface: Additive angular margin loss for deep face recognition". In: *Proceedings of the IEEE/CVF Conference on Computer Vision and Pattern Recognition*. 2019, pp. 4690–4699.

[16] Y. Guo et al. "MS-Celeb-1M: A dataset and benchmark for large-scale face recognition". In: *European Conference on Computer Vision*. 2016, pp. 87–102.

[17] Karl Ricanek and Tamirat Tesafaye. "Morph: A longitudinal image database of normal adult age-progression". In: *7th International Conference on Automatic Face and Gesture Recognition*. 2006, pp. 341–345.

[18] X. An et al. "Partial FC: Training 10 million identities on a single machine". In: *2021 IEEE/CVF International Conference on Computer Vision Workshops*. 2021, pp. 1445–1449. DOI: 10.1109/ICCVW54120.2021.00166.

[19] M. Wall. "Biased and wrong? Facial recognition tech in the dock". In: *BBC News* (8 July 2019). https://www.bbc.com/news/business-48842750

[20] P. Grother. "Bias in face recognition: What does that even mean? and is it serious?" In: *Biometrics Congress*. 2017.

[21] M. Wang et al. "Racial faces in the wild: Reducing racial bias by information maximization adaptation network". In: *Proceedings of the IEEE/CVF International Conference on Computer Vision*. 2019, pp. 692–702.

[22] J. J. Howard, Y. B. Sirotin, and A. R. Vemury. "The effect of broad and specific demographic homogeneity on the imposter distributions and false match rates in face recognition algorithm performance". In: *2019 IEEE 10th International Conference on Biometrics Theory, Applications and Systems*. 2019, pp. 1–8. DOI: 10.1109/BTAS46853.2019.9186002.

[23] B. Maze et al. "IARPA Janus Benchmark - C: Face dataset and protocol". In: *2018 International Conference on Biometrics*. 2018, pp. 158–165.

[24] K. S. Krishnapriya et al. "Issues related to face recognition accuracy varying based on race and skin tone". In: *IEEE Transactions on Technology and Society* 1.1. 2020, pp. 8–20. DOI: 10.1109/TTS.2020.2974996.

[25] K. S. Krishnapriya et al. Characterizing the Variability in Face Recognition Accuracy Relative to Race. 2019. doi: 10.48550/ARXIV.1904.07325. url: https://arxiv.org/abs/1904.07325.

[26] R. C. Malli. VGGFace implementation with Keras framework. https://github.com/rcmalli/keras-vggface, last accessed on June 2019.

[27] Q. Cao et al. "VGGFace2: A dataset for recognising faces across pose and age". In: *2018 13th IEEE International Conference on Automatic Face & Gesture Recognition*. 2018, pp. 67–74. DOI: 10.1109/FG.2018.00020.

[28] K. S. Krishnapriya et al. "Analysis of manual and automated skin tone assignments". In: *Proceedings of the IEEE/CVF Winter Conference on Applications of Computer Vision*. 2022, pp. 429–438.

[29] J. J. Howard et al. "Reliability and validity of image-based and self-reported skin phenotype metrics". In: *IEEE Transactions on Biometrics, Behavior, and Identity Science*, 3.4. 2021, pp. 550–560.

[30] S. Del Bino et al. "Relationship between skin response to ultraviolet exposure and skin color type". In: *Pigment Cell Research*, 19.6. 2006, pp. 606–614.

[31] M. Merler et al. "Diversity in faces". In: *arXiv preprint arXiv:1901.10436*, 2019.

[32] L. G. Farkas, M. J. Katic, and C. R. Forrest. "International anthropometric study of facial morphology in various ethnic groups/races". In: *Journal of Craniofacial Surgery*, 16.4. 2005, pp. 615–646.

[33] Z. Zhuang et al. "Facial anthropometric differences among gender, ethnicity, and age groups". In: *Annals of Occupational Hygiene*, 54.4. 2010, pp. 391–402.

[34] P. J. Phillips et al. "Overview of the face recognition grand challenge". In: *2005 IEEE Computer Society Conference on Computer Vision and Pattern Recognition*, Vol. 1. 2005, pp. 947–954.

[35] K. Kärkkäinen and J. Joo. "Fairface: Face attribute dataset for balanced race, gender, and age". In: *arXiv preprint arXiv:1908.04913*, 2019.

[36] R. Vera-Rodriguez, M. Blazquez, and A. Morales. "FaceGenderID: Exploiting Gender information in DCNNs face recognition systems". In: *Conference on Computer Vision and Pattern Recognition Workshops*. 2019.

[37] P. Drozdowski et al. "Demographic bias in biometrics: A survey on an emerging challenge". In: *IEEE Transactions on Technology and Society*, 1.2. 2020.

[38] V. Albiero et al. "Analysis of gender inequality in face recognition accuracy". In: *2020 IEEE Winter Applications of Computer Vision Workshops*. 2020, pp. 81–89. DOI: 10.1109/WACVW50321.2020.9096947.

[39] Z. Xiong et al. "An Asian face dataset and how race influences face recognition". In: *Advances in Multimedia Information Processing*. 2018, pp. 372–383. ISBN: 978-3-030-00767-6.

[40] V. Albiero, K. Zhang, and K. W. Bowyer. "How does gender balance in training data affect face recognition accuracy?" In: *2020 IEEE International Joint Conference on Biometrics*. 2020, pp. 1–10.

[41] C. Yu et al. "Bisenet: Bilateral segmentation network for real-time semantic segmentation". In: *Proceedings of the European Conference on Computer Vision*. 2018, pp. 325–341.

5

AI Knows What You Did Last Summer: Applications in Digital Forensics

Jing Yang, José Nascimento, Gabriel Bertocco, Antonio Theophilo, Rafael Padilha, Aurea Soriano-Vargas, Fernanda A. Andaló, and Anderson Rocha

5.1   Introduction

The worldwide popularization of smartphones and increased access to the Internet made our society more connected than ever. This connectivity bridged geographical distances, allowing every person in the world to reach countless others instantly. As a result, social media became a ubiquitous and popular source of information, surpassing traditional media vehicles such as printed news [1]. Every major event (e.g., political protests, natural disasters, terrorist attempts) is captured by thousands of devices and instantly uploaded to the Internet, rippling through social networks as people discuss and share their thoughts about what happened. Even though the increased accessibility allows anyone to be informed immediately after the fact, each event also creates an immense amount of data — in the form of images, videos, and posts — whose trustworthiness cannot be verified promptly. Consequently, the spread and reach of misinformation have grown significantly in the past decade, impacting many spheres of our society and becoming a concern to Digital Forensics.

Not long ago, forensic experts often relied on witness testaments, news reports, and any information acquired directly in the field to understand what happened. Social media reshaped this setup, offering a data stream comprising an event that complements traditional sources. Unfortunately, the amount of generated data surpasses the human capability of analysis, raising multiple challenges throughout the forensic pipeline (Figure 5.1). First, each massive collection produced in an event needs to be sanitized and filtered to remove irrelevant and redundant data. Relevant samples require proper organization, identifying possible semantic, temporal, and spatial relationships between them that can aid future analyses. Once organized, forensic experts may reason over data, reconstructing how the event unfolded, identifying the main elements and suspects, fact-checking claims and narratives being shared online, and visualizing coherent media or insights distilled during the analyses. Every step throughout this process is affected by the volume and complexity of the collected media. Pictures and videos captured at events present unconstrained lighting conditions, reduced quality and resolution, motion blur, and occlusion that hinder manual inspection and organization. Moreover, they are often accompanied by posts and small text messages that offer insights, adding another layer to forensic analysis.

To cope with the extensive scale of available information and spreading the speed of misinformation, researchers turned their eyes to AI, especially machine learning and computer vision. In the past decade, both research fields gained attention by achieving impressive results in several complex vision problems, such as image captioning [2], human action recognition [3], and medical imaging analysis [4]. Enabled by advances in computing hardware and data availability, deep learning-based approaches also outperformed traditional methods in numerous Digital Forensics tasks—such as, biometrics [5], image provenance analysis [6], and image super-resolution [7]. Similarly, AI-powered methods are essential to deal with the complexity and throughput of social media, aiding forensic experts in their work [8–13].

Inspired by recent breakthroughs, we revisit the forensic event analysis workflow under the lens of AI in this work. We discuss how learning-based methods are helpful throughout the forensic pipeline, enabling experts to filter and organize relevant data effectively, reason about potential suspects, and even verify fake and manipulated narratives surrounding what is shared. We also highlight the importance of information visualization (visual analytics) throughout the workflow, offering insights over input data or supporting AI-based decisions with interpretability and transparency. Finally, we describe the challenges and recent research efforts for each discussed topic while drawing potential future directions.

DOI: 10.1201/9781003328957-5

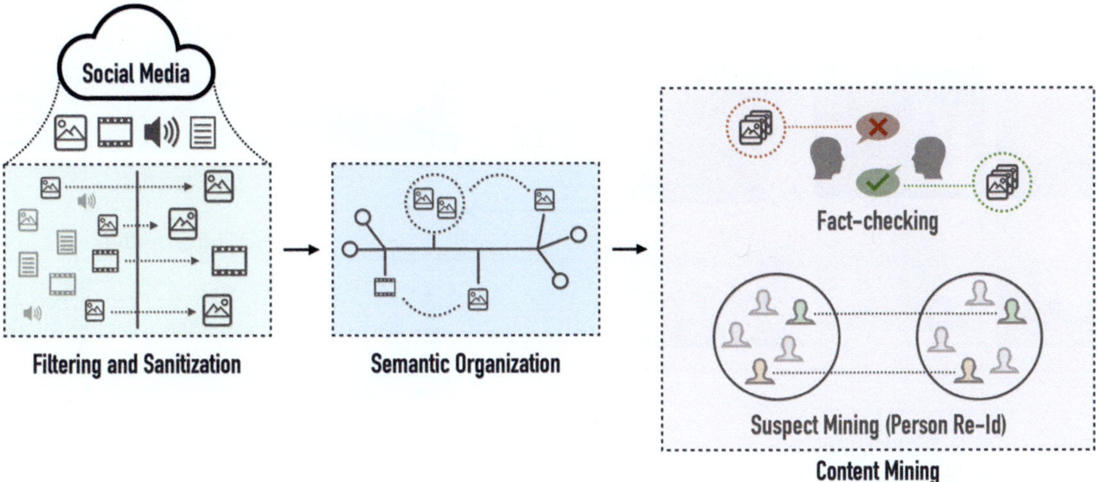

FIGURE 5.1
Forensic event analysis pipeline. Once collected from social media, data is sanitized, and irrelevant or redundant samples are filtered out. The remaining samples are organized, identifying semantic relationships between them before being analyzed. In this work, we discuss AI-based approaches for filtering and mining available data content.

The remainder of this chapter is organized as follows. In Section 5.2, we present the research directions and challenges of filtering and organizing data for forensic events. We discuss how to reason over available data, presenting in Sections 5.3, 5.4, and 5.5 three prominent tasks in this scenario: fact-checking, authorship attribution, and PReID, respectively. In Section 5.6, we present how visualization methods aid the forensic pipeline by tying together different steps throughout the analysis and aiding in interpreting model decisions. Finally, we draw final remarks and conclusions in Section 5.7.

5.2 Event Filtering

A major forensic event is usually an astonishing occurrence that grabs the attention of most people nearby. Let us use as an example the Notre Dame Cathedral Fire, which happened in April 2019. Plenty of imagery was shared across social media in moments ranging from the fire's beginning to the moment when the fire was extinguished. These pictures can assist forensic experts, for instance, in unraveling how the fire spread in the cathedral and in mounting a complete timeline of the event [9, 14]. For events of other nature than fire, such images can aid distinct tasks, such as indicating suspects of a terrorist attack or finding disaster victims.

However, in most cases, we need to input related keywords in a search engine to collect social media imagery (Social Tracker[1] and Twint[2] are some examples of open-source applications in this sense). One of the consequences is the vast amount of irrelevant content present after a collection procedure related to an event, which unavoidably hinders the analysis. For example, these images can be cartoons, memes, or even images from other events, as shown in Figure 5.2. Moreover, since the datasets can be considerably large, manual filtering requires time and much human effort, making the task unfeasible. Because of that, there is an area of research to experiment with machine learning algorithms to assist forensic experts in this task, reducing the overload of information by automatically discarding irrelevant data.

5.2.1 Broad Overview of Event Filtering Research

As most of the images in an event dataset can be irrelevant, some filtering procedures should be applied before further examination. Therefore, we can divide existing methods into two groups: single-modality and multi-modality. As the name suggests, the first group contains methods that consider only features from the images, and the second group includes techniques that take into account additional information that might be available. We describe how the problem has been tackled by different works below.

5.2.1.1 *Single-modality Filtering*

Different studies during the last years tackled this problem by focusing on using only images in a single modality filtering [10, 15, 16]. However, due to the advances in deep learning in the last decade, these works usually employ a pre-trained convolutional network at some step of the

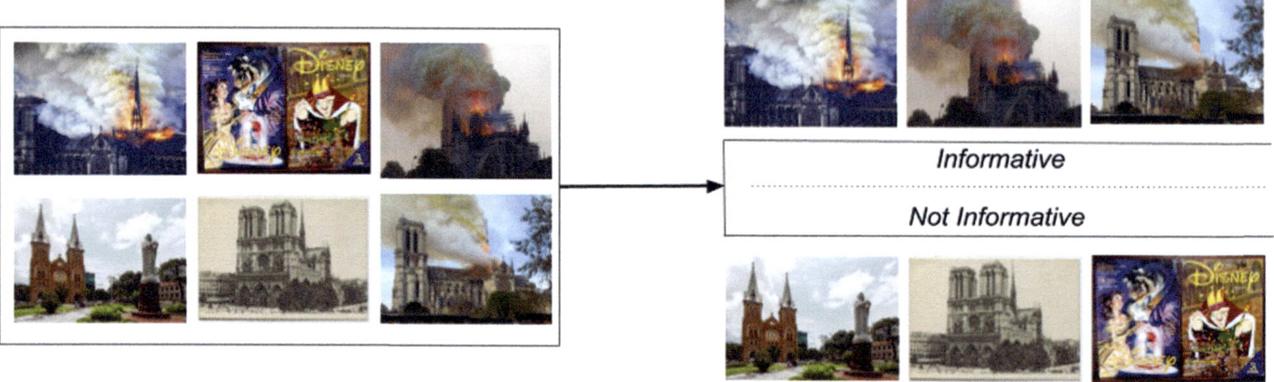

FIGURE 5.2
Sample of images from a collection procedure related to the Notre Dame Cathedral Fire. Noninformative images like pictures from the cathedral before the event, cartoons, pictures from other places, and memes are abundant in the resulting dataset. (Every event image herein is licensed under Creative Commons.)

pipelines to extract representative vectors that describe the information presented on the pictures and, after that, perform classification on them.

However, for most works, the use of pre-trained CNNs is just part of the pipeline. Indeed, some methods are limited to extracting feature vectors and then applying some classification algorithm [17, 18], but the majority of them go beyond and experiment with more complex techniques as well. Some examples of works in this sense use combination with handcrafted features [10], fusion with features from different CNNs [11], active learning [15, 19], federated learning [19], and domain adaptation [20].

An advanced step in this area that is applicable specifically in disaster events (like fires or floods) is not just to automate the removal of noninformative data but also to develop a machine learning model that can classify the damage that a given image portrays. This classification can be either coarse-grained (little, mild, or severe) or fine-grained ("vehicle damage," "injured or dead people") [16]. It is beneficial for disaster-response teams since it can immediately provide an overview of the critical scenes of the event, leading to a faster response.

5.2.1.2 *Multimodality Filtering*

Another way to approach the problem is using side information and images [21–23]. A social media item hardly ever is just a photo but a combination of textual and visual content (when it has an image attached, of course) along with metadata (like spatial or temporal information). This additional information can enhance the filtering by adding context to the picture. In this sense, some studies work on multimodal filtering, where items from

other modalities aid the classification of a specific type of media.

There is almost no work that uses metadata. For example, location tags, which could be decisive for filtering, are attached to only around 2% of items [24]. Despite that, some works use geopositioning information in the analysis of general events [25], but to the best of our knowledge, no work experiments with it specifically to filter forensic event images. Regarding temporal information, a study reports that informative data tends to be higher in the initial hours of the event, but including this data in the model is a challenging task [26]. Other context data like retweets, followers, and quantity of replies did not improve the performance when experimented with for the classification of images from posts [27].

On the other hand, works that use textual data along with images are recurrent. The crux of current methods is usually extracting deep features from both modalities and finding a shared space for classification. This shared space is usually obtained by late-fusion (also called decision-fusion) [21]. Some works in this direction use fully connected layers to perform the fusion [22, 28], but others implement advanced concepts, like cross attention modules [29, 30] and information propagation over graphs [23].

5.2.2 Challenges of Event Filtering

Filtering the immense volume of data from real-world events is a daunting problem. This section discusses some of the challenges associated with this task and the initial efforts to address them.

5.2.2.1 Collection and Annotation of Event Filtering Datasets

One of the main challenges in this area is obtaining labels for classification. Most of the existing methods for this task require many labeled instances from the specific event. It means that each new event requires a new labeling procedure, which takes much time and human labor to annotate such a quantity and might be unfeasible in a real scenario. Therefore, developing models that can reduce this need is a key feature of this task.

Additionally, the definition of informative images might not be straightforward. Even though there are some instances easy to define whether they are informative or not, some pictures can generate disagreement. To illustrate, suppose the Boston Marathon bombing in 2013. In extreme cases, images from the same marathon in other years or from other marathons might be difficult for a human to differentiate [11]. Besides, the datasets have thousands of images, and it is expected annotation errors throughout the process of labeling thousands of images. To work around this, a recent work proposed an annotation procedure where more than one annotator labels the same instance, and then there is a search for an agreement among them [31].

Another issue in this task is that the datasets can be very unbalanced [11, 16], in a situation where most of the items are not informative. This difference between the number of instances of each class jeopardizes the current methods. For example, if the pipeline has an active learning step, the method should be careful about the label distribution since some techniques from this type assume that the dataset is balanced [32]. Therefore, it is not rare for works in this area that experiment with their respective methods only in balanced datasets. Their results are valid, but that should be taken into account when considering their application in a real-world scenario.

5.2.2.2 Fusion of Multimodality Data

Including other types of data brings challenges from the modalities and the fusion of this information into the problem. Due to the issues described in the previous section, hardly ever a work includes metadata in its analyses. Besides, some problems from event filtering on texts that arises when adding textual data to the model are that they may be from different languages and that text from different social media may have different standards [24]. However, the most common problem related to the fusion of text and images is the frequent lack of correlation between the text and the image from the same post. Situations where the image is informative but the text is not (or vice-versa) are common, as shown in Figure 5.3. The authors in [30] try to address this issue by employing an independent classification of global features and correlated classification of local features from the modalities.

5.2.2.3 Lack of Event Filtering Benchmarks

Forensic events happen with some frequency, and there are many tools for data collection available online. Therefore, there are different datasets available for this task, yet we note an absence of a definite benchmark for this problem. We can find works that conduct experiments on data related to a single event [18, 34] or to different events [11, 20]. As a consequence, the comparison among methods might be difficult. The closest dataset we have in this sense is the CrisisMMD dataset [31]. However, it consists of only seven different disaster events, where three

Text: *Not Informative*	Text: *Informative*	Text: *Informative*
Image: *Not Informative*	Image: *Informative*	Image: *Not Informative*

FIGURE 5.3
Examples from the CrisisMMD dataset (with courtesy from [15]) for the 2017 Mexico Earthquake (with courtesy from [33].) We can see both posts where the labels from the text and images agree (left and middle), and posts where the labels disagree (right).

of which are hurricane events. From the perspective of forensic events (including but not limited to disasters), there is a lack of diversity among the events. A recent work released the MEDIC dataset [16], which assembles different available datasets to propose a baseline for the field. There we can find 30 events from a more diverse set of types (terrorist attacks and war conflicts are included, for instance).

5.2.3 Future Directions for Event Filtering

With the ubiquity of smartphones and camera devices, the number of events recorded and, consequently, the volume of data generated surpasses the human ability to analyze it. This section discusses promising future directions in reducing the need for annotated data, as well as the ability to reuse previous knowledge to filter unseen events.

5.2.3.1 Training with Few Labeled Data

One of the main pain points in this problem relies on reducing the need for labeled data so the task can be executed faster and with less human effort. Different approaches might be applied to achieve this goal. One straightforward possibility is few-shot learning, where algorithms that can generalize well using few samples are favored. An example of work in this sense uses semi-supervised learning [10] to reduce the need for annotations since algorithms of this type can learn by both labeled and unlabeled data, using the graph structure generated by the entire dataset [35].

Another approach to reduce the requirements for labels is by adding the human to the loop, using an active learning framework. In this case, the method uses a query strategy to improve the training set by selecting instances from the unlabeled set for a human to label, where these instances are expected to be the most decisive in reducing the error in a posterior classification procedure. Work in this sense experimented with different query strategies and presented promising results [15], while others presented a method that combines active learning with federated learning [19], so the labeling procedure can be made in a decentralized way by multiple annotators.

5.2.3.2 Adapting Existing Models to New Events

Since new forensic events always happen with some frequency, developing a model that can use the knowledge obtained from previous events for the newer one is also a promising field of research in this area. Correct usage of a previously labeled event dataset can reduce or even remove the need for labeled data for a new unlabeled event [36]. In this sense, some works focus on applying domain adaptation in this task, where the goal is to reduce the difference between the feature representations of the source and target data. One work in this direction applies a domain-adversarial neural network to find a transformation function to make the source and target data indistinguishable for the classification [20].

5.3 Fact-checking

With the prevalence of social media, we live in an age when it has never been easier to share information and opinions online. However, often social media has been misused to spread misinformation.[3] In traditional news outlets such as television, reporters are the main spreader of news. Nowadays, everyone can spread a piece of news with a simple click. As a result, small ideas and voices can gain monumental influence because of the way they are amplified online. However, in many cases, misinformation can cause a significant negative impact. For example, in Brazil, propaganda spread by official authorities toward ineffective scientific treatments (such as hydroxychloroquine [37]) puts at risk patients' (already debilitated) health, drowning the public health system into chaos.

As misinformation becomes a growing concern to the public, news fact-checking organizations are proliferating. However, the generation and spreading speed of the former are much faster than the latter. Many automated solutions have been proposed to speed up the fight against misinformation. Most algorithms treat it as a fake news classification problem in the research community. Among these algorithms, some utilize news articles' content [38], and others exploit additional information such as users' comments [39]. However, simply solving it by classification cannot surmount real-world issues, as these algorithms are likely over-relying on different news writing styles and user biases. Besides, the term fake news is ambiguous [40], thus during labeling, annotators may have different definitions of fake in different datasets, potentially bringing biases to algorithms trained on them. Another type of work is automated fact-checking, which adds information retrieval on top of the classification. This task has a more rigorous definition: check if a given piece of information is factually correct based on evidence retrieved from reliable sources. According to a recent survey [41], the gap between human fact-checking and automated solutions is still large. On the one hand, automated fact-checking methods can largely increase checking speed; on the other hand, these solutions are learned from biased and simplistic datasets, failing to meet real-world fact-checking needs. The use of AI for automated fact-checking can have significant negative impact with the case of false positives, that is, assigning false information as genuine.

TABLE 5.1

Fact-checking Datasets. N/A means not applicable.

Dataset	Input Modality	Claim Source	Evidence Source	#input	#classes
FEVER [42]	Text	Annotators written	Wikipedia	185,445	3
SciFact [43]	Text	Annotators written	Scientific Paper Abstracts	1409	3
MultiFC [44]	Text, Metadata	Fact-checking websites	Google Search API	36,534	2-27
FEVEROUS [45]	Text, Table	Annotators written	Wikipedia	97,026	3
VitaminC [46]	Text	Annotators written	Wikipedia	488,904	3
Fool-me-twice [47]	Text	Annotators written	Wikipedia	12,968	2
FAVIQ [48]	Text	Synthetic, from questions	Wikipedia	188,376	2
Fauxtography [49]	Text, Image	Snopes, Reuters	Image Reverse Search	1233	2
COSMOS [50]	Text, Image	Web	N/A	200k	2
NewsCLIPpings [51]	Text, Image	News	N/A	85,360 (balanced)	2
MOCHEG [52]	Text, Image	Web (Text)	Web (Text, Image)	21,184	3

According to The Washington Post,[4] studies have shown that individuals who are exposed to a piece of misinformation may be influenced by it, even if it is later corrected, leading to confusion and potentially harmful effects on their thinking and decision-making. Additionally, when an AI falsely asserts a piece of fake news as being true with a high level of confidence, it can undermine public trust in legitimate news sources.

5.3.1 Broad Overview of Fact-checking Research

According to Wikipedia, fact-checking is verifying factual information to promote the veracity and correctness of reporting.[5] In journalism, fact-checking content was mostly textual claims related to political speeches. However, with the growing number of multimedia content and the extensive influence of disinformation, fact-checkers have extended their work to check various forms of information. This brings more complexity to fact-checking. Not only can the content be false, but pristine images or videos can be put in a different textual context, creating a mixed reality to spread disinformation. For example, during the recent Russian invasion of Ukraine, many old images/videos were misused in the context of war.[6] Therefore, there are mainly two types of fakes: falsified content (text, audio, images, and videos); and real natural content but out of context. For the first type of fake, current automation for fact-checking mainly focuses on checking the veracity of textual claims; another branch of work mainly focuses on DeepFake, which detects AI-generated images, videos, and audio. The second type of fake—detecting visual content out of context—has attracted attention from research communities only recently.

This section will focus on automated fact-checking approaches in text and images. Specifically, we will list some seminal datasets and related work for automated fact-checking on textual claims and out-of-context visual content.

5.3.1.1 Fact-checking Datasets

In 2018, the first seminal fact-checking dataset was proposed [42]. Since then, numerous related datasets have been released; we list some of them in Table 5.1.

In Table 5.1, we compare the datasets from three aspects: input modality, claim source, and evidence source. Most datasets consider the text for input modality, as most initial fact-checking contents were from this modality. With the abundant availability of multimedia, content published on social media usually are images/videos attached with captions. A few datasets have been published to enable automated fact-checking of multimodal content. Comparing claim and evidence sources, in most datasets, textual claims are written by human annotators, and most evidence sources are from Wikipedia. A few works consider more realistic scenarios and use Google Search API as the evidence source, considering that Wikipedia does not include all the evidence needed.

Next, we introduce some works of automated textual and multimedia fact-checking.

5.3.1.2 Textual Fact-checking

Fact-checking on textual claims has been the main focus in automated fact-checking research communities. The benchmark dataset and pipeline were first proposed by Thorne et al. in FEVER [42]. The pipeline consists of three steps: document retrieval, sentence selection, and claim verification. Since then, many researchers have adopted this pipeline and dataset to have a standard evaluation of fact-checking solutions. In addition to the above steps, researchers have also proposed other steps to automate further fact-checking: check-worthiness detection and explanation generation.

Claim check-worthiness detection is usually the first step in automated fact-checking. Depending on fact-checking content, we could define check worthiness in different ways. Most work focuses on checking claims related to political debates. Among the work, the most well-known one is ClaimBuster [53]. They proposed a system to extract and rank sentences from presidential debates, and then identify important factual claims among the selected sentences. Overall, check-worthiness detection has been treated as a sentence ranking problem. More recent methods [54] tend to use state-of-the-art language models for sentence representation instead of hand-crafted features. However, check-worthiness detection as supervised learning needs manual labeling, which is bias prone.

Document retrieval and sentence selection are called evidence retrieval, the process of retrieving relevant information using the claim to be verified as a query. Information retrieval plays a very important role in knowledge-intensive tasks, such as fact-checking and open-domain question answering. With the advance of large pre-trained language models, retrieval methods have also been improved significantly [55]. However, these models are harder to generalize when the database changes.

Claim verification is a natural language inference task that predicts if a claim is supported or refuted by the evidence retrieved or if the evidence is not enough to verify the claim. Although automated fact-checking methods have shown high performances in some datasets, researchers have pointed out some challenges. For example, in [56], Thorne et al. showed that existing fact verification systems are vulnerable to several adversarial attacks, such as paraphrasing and adding noisy words. Another issue with automated fact-checking is dataset bias. This problem was pointed out firstly by Schuster et al. [57], who studied the existence of bias in the FEVER dataset. They learned that a fact-checking verification model could easily predict the label of a claim by some specific words. For example, "did not," "yet to," and "does not" usually appears in factually incorrect claims. This study has alerted researchers that automated solutions must be careful before being deployed in real-world scenarios.

To build more trustworthy systems, Atanasova et al. [58] proposed the first work to generate explanations for the fact-checking result. First, they performed an optimization to learn veracity prediction and its explanation extracted from the selected evidence sentences. Subsequently, Kotonya et al. [59] proposed a joint extractive and abstractive text summarization method for explanation generation. The authors also published a survey about generating fact-checking explanations [60]. However, although generating explanations can provide more precise evidence to understand fact-checking decisions, existing systems lack a way to evaluate the explanations properly. Moreover, especially for explanations based on abstractive summarization, researchers have shown that such models have problems of hallucination [61, 62], generating summaries factually inconsistent with their source.

5.3.1.3 Multimedia Fact-checking

In the multimodal scenario, there are usually two types of disinformation: (1) falsified image with a false claim; and (2) pristine image with a false claim.[7] Image forensic tools can be used to detect if an image has been tampered with, which is a challenging task. Especially with the increasing capability of DeepFake, the forensic communities have been actively studying new methods to detect DeepFake. Nevertheless, in social media, the more common and rarely explored area of disinformation is the second type—using images/videos out of their context to spread disinformation. This task has attracted attention recently as more posts shared on social media usually contain multimedia content.

Unlike fact-checking on textual claims, there is no existing benchmark dataset or pipeline for multimedia fact-checking. A few works have been proposed to deal with multimodal fact-checking, and most proposed each of their datasets.

In Fauxtography [49], Zlatkova et al. proposed a new dataset for verifying claims about images. Their dataset consists of samples of image and text pairs from Snopes and Reuters websites. A sample is considered false if the image or the claim about the image is false. Following the textual fact-checking pipeline, they also perform evidence retrieval. They use Google's Vision API for reverse image search and extract tags and URLs from the returned results. For the evidence of a claim, they use the article titles retrieved from reverse image search. The main problem with the dataset is that most fake samples are from Snopes and all real samples are from Reuters, which may mislead automated methods to learn this bias.

In COSMOS [50], Aneja et al. proposed to detect out-of-context misinformation with self-supervised learning. The dataset requires two input captions for the same image to determine if the image was being used out of context or not. Compared with Fauxtography, this work considers all sample images are pristine. Compared with textual fact-checking, in which we need to retrieve evidence for the claim, the two captions serve as a kind of evidence to support or refute the image. However, it is not always possible to have two captions for the same image in real-world scenarios.

In NewsCLIPpings [51], Luo et al. proposed automatically generating out-of-context image-text pairs. Intended to improve the COSMOS, which needed two captions for an image for detection. To make detection more difficult, that is, extracting features from only images or text information cannot make a good prediction, they ensure the images and text are natural instead of synthetically generated or manipulated. They then proposed a baseline method to detect image-text mismatch using the state-of-the-art multimodal embedding model CLIP [63]. Inspired by the textual fact-checking pipeline, in [64], Abdelnabi et al. try to improve performance on the dataset by adding evidence retrieval. They retrieve text entities from images using Google Vision APIs; and images from captions using Google Custom Search API. The retrieved images and captions can then be used as evidence to support or refute the original image-caption pair.

In MOCHEG [52], Yao et al. proposed multimodal fact-checking datasets, where the claims are text and evidence are from the web, containing text, images, and videos. Different from previous multimodal datasets, they annotated gold evidence for every claim. This work is actually more similar to fact-checking of textual claims as the claims are only text, and they follow the same pipeline of fact-checking of textual claims, except that the evidence contains multimodal content. In addition, the dataset also provides fact-checking explanations.

In addition to the above-mentioned fact-checking datasets, there are a few datasets for Visual Entailment task, in which given a pair of image/video and its caption/description, a model needs to predict if the caption/description matches the content of the image/video [65, 66]. Comparing Image Entailment with Video Entailment, the latter is much harder than former due to the size and complexity of video data. Overall, the Visual Entailment task is comparable to the fact verification part in the fact-checking pipeline; and these datasets can be useful for training and evaluating machine learning models for visual fact-checking; However, they are usually synthetic and simplistic thus may not accurately reflect the complexity and diversity of real-world fact-checking scenarios.

In summary, only a few datasets have been published for multimedia fact-checking. Furthermore, most of them did not consider evidence retrieval, and the ones that do, Google APIs are used for the purpose. Because of this, there is a lack of proper evaluation of the credibility and precision of returned evidence. Therefore, it is important for researchers to carefully evaluate the suitability of these datasets for their specific purposes and to consider using other types of data and evaluation methods to develop more robust and reliable fact-checking systems.

5.3.2   Challenges of Fact-Checking

This section addresses the challenges in adapting existing automated fact-checking methods to tackle complex real-world misinformation problems.

5.3.2.1   *Lack of Multimedia Fact-checking Datasets*

In the literature, most datasets and methods developed for fact-checking focus only on text; however, most real-world misinformation contains images and videos. Only a few datasets have been released for visual fact-checking and mostly annotated for checking mismatches between images and captions; none of the multimodal datasets contain annotations for evidence retrieval. Creating a big dataset for multimodal fact-checking is time-consuming — it requires annotation of not only the labels but also visual and textual evidence that supports the label (supported, refutes, or not enough information), and a big database for evidence retrieval.

5.3.2.2   *Lack of Reliable and Diverse Evidence Sources*

Fact-checking datasets need to consider having an evidence database that is reliable and contains a variety of topics. Most datasets use Wikipedia as the evidence source. Wikipedia contains a large and diverse source of knowledge and is mostly trustworthy. Nevertheless, some pages are constantly being updated, and malicious users could intentionally modify the content to attack fact-checking methods, making the results inaccurate. Also, Wikipedia mostly contains textual information; for verifying images, much content is unavailable in the database. Some work considers using Google Search API, which extends the database much more, but the disadvantage is that the returned results will contain low credible sources.

5.3.2.3   *Lack of Explainability*

Automated fact-checking results are usually just a label. However, real-world fact-checking results always provide explanations and are a key part of human fact-checking. If automated fact-checking wants to gain human trust, it is paramount to have reasonable explainability. So far, only a few works have focused on the explainability of automated fact-checking [12, 58, 59]. One problem is that few fact-checking datasets contain human annotated explanations besides labels, so algorithms have to resort to human evaluation for the generated explanation, which is expensive.

5.3.3   Future Directions for Fact-checking

To deal with the challenges raised above, we discuss some promising future directions for fact-checking.

5.3.3.1   *Benchmark Dataset and Evaluation*

To encourage more work on multimodal fact-checking, it is paramount to provide a database related to visual disinformation and bring benchmark evaluations for evidence retrieval and fact verification to the table. One way to acquire accurate annotation is by resorting to available fact-checking articles. For example, Duke Reporters' Lab provides an excellent website[8] that updates many active fact-checking websites approved by the

International Fact-checking Network (IFCN). Researchers could develop a tool to collect fact-checked content and automatically maintain a multimodal fact-checking database. Based on this database, annotators can further extract useful sources of evidence for the claims.

5.3.3.2   Multimedia Evidence Retrieval

Like textual fact-checking, evidence retrieval can take two steps: retrieval and re-ranking based on relevance and similarity. With the state-of-the-art large multimodal embedding model, given a trustworthy evidence database, we can project image and text embedding in the same semantic space. This will facilitate operations on different modalities. For instance, we can retrieve images using text and vice versa, enriching retrieved evidence.

5.3.3.3   Explainable Fact-checking

Another direction that is important yet under-explored is explainability. Automated methods need to be explainable to be trusted by users, or at least by professional fact-checkers. For example, when checking the inconsistency between image and text, methods need to predict a label and output the suspicious area of the image or claim, so fact-checkers can better examine the results. One way to provide more explainability is to introduce more intermediate steps in the pipeline. With more steps, we bring more transparency by evaluating the intermediate results, allowing better understanding and inspection.

5.4   Authorship Attribution

Authorship attribution is a well-known forensics task that aims to find the author of a textual message by solely inspecting the text without relying on any metadata or additional information. It has been long studied, and probably the most famous and well-investigated work is The Federalist Papers [67], a collection of 85 articles anonymously written by Alexander Hamilton, James Madison, and John Jay to promote the ratification of the constitution of the United States.

Since then, authorship attribution has developed and spread over a wide range of different domains. It evolved from the research over traditional [68–70] and literary [71–74] long texts to more modern types of textual production such as e-mail messages [75, 76], software code [77, 78], and small messages posted on social media platforms [79–88].

Indeed, the current study of authorship attribution over social media platforms can be considered one of the weapons against the modern threat of fake news propagation discussed in Section 5.3. Currently, anonymous messages, fake profiles, and even rogue personifications are being extensively used in social media to commit a myriad of crimes such as racism, misogyny, bullying, phishing, misinformation, and even opening doors to public opinion manipulation at a much grander scale. For example, in an article published by The New York Times [89], an official Russian agency was accused of supporting a team of workers responsible for creating fictitious publications on social networks under false identities to foster propaganda and disinformation campaigns aligned with this country's interests. According to major social media companies, millions of people saw and shared this content. The same newspaper recently showed how these fake identities are created and nurtured to earn trustworthiness over the public [90]. Another article published by convolutional neural network (CNN) [91] explains how social media platforms were manipulated to bias the 2016 American presidential election by creating invented profiles and false news.

There is a vast market involving selling false information through fake identities. Confessore et al. [92] showed how these rogue profiles are created and marketed, often using real data of innocent people, such as names and images of minors. According to Varol et al. [93], nearly 15% of Twitter accounts (more than 30 million) are automated profiles devised to simulate real people, while, for Facebook, this number exceeds 60 million accounts [94].

The most recent and amazing use of authorship attribution over this social media misuse landscape is presented by Kirkpatrick [95]. He showed how machine learning techniques were used to attribute the authorship of QAnon messages (an anonymous far-right political conspiracy theory) to two individuals. This group is associated with serious threats to democracy, such as the U.S. Capitol attack in January 2021. Applying machine learning techniques and features as character 3-grams, two teams of researchers — one Swiss and another French — shared data and concluded that one individual wrote QAnon messages, and then another started to write its messages. The results were supported by two renowned researchers in the authorship attribution field. Despite the success of applying authorship attribution techniques in this critical and relevant scenario for our society, this episode also shows how difficult and challenging the task is. The researchers dealt with a small set of 13 suspects, a large dataset of 100,000 words from QAnon, and at least 12,000 words for each suspect. Parallel investigations outside the textual universe confirmed the results (e.g., messages' timestamps from the suspects claiming to discover the QAnon's existence). However, it is unclear whether the same investigations helped and biased the results.

5.4.1 Broad Overview of Authorship Attribution Research

Authorship attribution is a broad denomination that encompasses the analysis of texts to be attributed. It is usually split into more specific subtasks when we consider the characteristics of the suspects.

5.4.1.1 Authorship Identification

This is probably the most common subtask seen in research articles [81, 86, 88]. It assumes that we have a fixed number of suspects and that the target text to be attributed was written by one of these suspects (closed-set scenario).

5.4.1.2 Authorship Verification

This is one of the most challenging tasks in authorship attribution. It assumes that we have a set of documents from the same author, and we want to answer if this author wrote or not an unknown document [85, 88, 96, 97]. It can be considered a true open-set scenario since it is possible that we do not have any documents from the real author.

5.4.1.3 Authorship Profiling

This refers to the problem of distinguishing between classes of authors by studying how people share the use of a language. It helps identify authors' characteristics, such as age, gender, and language variety [98, 99].

5.4.2 Challenges of Authorship Attribution

Traditional authorship attribution techniques based on stylometry and frequency of linguistic features provide good performance in scenarios where we have long texts and a few suspects. However, in this setting, the big challenge is dealing with the cross-genre problem, that is, having training data from one genre and target data to attribute from a different one [100]. It is a common forensic scenario, for example, when a harassing message needs to be attributed, but we only have texts of different genres from the suspects, like business letters, essays, or blog postings.

However, additional challenges arise when dealing with social media messages. This scenario presents a potentially large suspect set (maybe thousands) with limited samples for each one, what Koppel et al. [101] call the *needle-in-a-haystack* setting. We can add additional obstacles to this description, such as the size of the messages and a very noisy and wide context. The texts in these platforms present a broader range of symbols like non-ASCII Unicode characters, emojis, and emoticons, along-

side a significantly wider parlance span, with the publishing of intentionally misspelled words, onomatopoeia, and extensive use of Internet jargon.

These challenges can be observed in Kirkpatrick's work presented previously. Nevertheless, the researchers achieved acceptable accuracy by dealing with a small set of suspects (13) and a significant number of samples (more than 250,000 words).

Another challenge is dealing with adversarial techniques. Jones et al. [102] proposed an interesting application of language models to mimic users' text and deceive authorship attribution models. The authors identified limitations to imitating tweet authorship, evincing that powerful modern language models are still weak in the social media scenario of small and noisy messages, but this is expected to change in the near future.

5.4.3 Future Directions for Authorship Attribution

Current research in authorship attribution focus on the scenario and challenges mentioned previously regarding small messages posted on social media platforms [81–88]. There is a trend in using modern deep learning techniques based on CNNs over features based on words and sequence of characters (character n-grams) to exploit all the information available in this small samples [82, 83, 86–88]. More recently, there have also appeared some groups trying to exploit developments of language models and apply them to the authorship attribution task [103].

Another path researchers are exploring in authorship attribution is related to the explainability of the models' decisions (eXplainable Artificial Intelligence—XAI) [104]. Like any other forensics task, authorship attribution is usually deployed in sensitive contexts such as courts, and it is imperative to explain the models' decisions in such a setting. It is a challenge as models get larger, and we believe much effort will be put into it.

5.5 Person Re-Identification

Person Re-Identification (PReID) aims to match a person's image seen in one camera to its corresponding images in other surveillance or monitoring system cameras. It has attracted much attention in the last years in the computer vision community due to its myriad applications: deployment in smart cities, traffic control, surveillance, informed investigations, and crowd behavior. Current models must be trained over a bounding box dataset

FIGURE 5.4

Two identities in the Person Re-Identification setup. The four leftmost identities belong to one identity and the rightmost images belong to another. We see different resolutions, viewpoints, occlusions, and scales considering images of the same identity. Faces are covered to preserve the privacy of the recorded person.

of detected people in which many cameras might see the same person in different positions in the environment. For example, from one camera to another, the same person might be under different illumination (moving from inside of a building to the outside), occlusions, vantage points (seen from the back in one camera while upfront in another), resolution (zoom or different pixel density) and background, as Figure 5.4 depicts.

When the same camera records two or more people, they usually have similar viewpoints, backgrounds, occlusions, and illumination conditions. Moreover, people in the same environment (e.g., at an airport) usually dress similarly (most people with backpacks, bags, and luggage), making classes semantically closer to others. Due to the difference in camera environments, feature representations of the same person might be far in the feature space. Thus, PReID usually faces a high intra-class discrepancy and high inter-class similarity requiring fine-grained feature learning. In addition, the models must overcome the camera bias and find invariant-view discriminative features able to co-locate the same individual under different view conditions.

5.5.1   Broad Overview of Person Re-Identification Research

Most prior art relies on supervised training by supposing person images are correctly annotated. However, the annotation is an expensive, time-consuming, error- and bias-prone task, mainly on PReID, where it requires a detailed visualization from the annotator to label the same person from a different point of view correctly. In this context, the PReID research has adopted two strategies: supervised and unsupervised PReID.

5.5.1.1   *Supervised Person Re-identification*

One line of research to tackle this problem is to adopt supervised solutions. They assume all the people in the dataset are mostly correctly labeled in the whole camera system. In this way, they can optimize a model to match two bounding boxes with the same identity annotation while keeping the ones with different annotations apart. An overview of the supervised training is depicted in Figure 5.5.

However, what is the best way to perform supervised training? To regularize and improve the performance, some works leverage a weighted multiscale learning [106] where the authors designed an architecture composed of blocks that generate feature maps in multiple scales and combine them to get a final feature map in the output.

Other works [107] employ a part-based strategy where they split the input image into different patches or horizontal stripes and learn from them. This way, the model is less prone to possible occlusions, person misalignment, and misdetections. Recently He et al. [108] leveraged the Vision Transformer [109] strategy to design a multilayer and multihead attention model to focus on more discriminant and longer-range parts of images achieving state-of-the-art performance.

However, the previous models' success strongly relies on data annotation, which is expensive, time-consuming, error- and bias-prone. Then the following question arises: how to learn from unlabeled data to achieve a competitive performance with the supervised counterpart? In this context, researchers have recently adopted an unsupervised perspective of the problem, which can be divided into *generalizable person re-identification, unsupervised domain adaptation on person re-identification,* and *fully unsupervised models.*

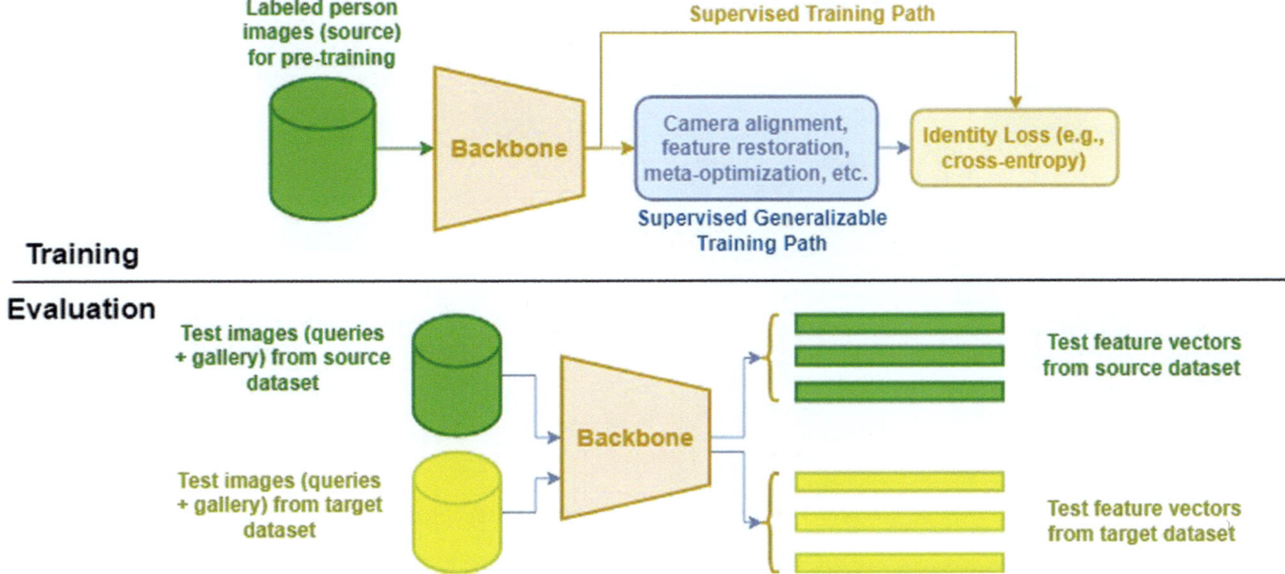

FIGURE 5.5
Pipeline for Supervised and Generalizable person re-identification. In the training stage, both are trained with a source-labeled dataset. For Generalizable person re-identification, further strategies may be applied to achieve robust feature representation across datasets (e.g., camera statistics learning for cross-camera feature alignment (with courtesy from [105])—blue box). On evaluation, the supervised training is evaluated in test images from the same dataset distribution where the source data comes from, as long as in the generalizable re-identification the test images come from other unrelated person re-identification datasets.

5.5.1.2 Generalizable Person Re-identification

Researchers have proposed novel strategies to train the models over the source domains to increase the generalization and apply them directly to unlabeled target domains without fine-tuning. Figure 5.5 shows a general pipeline for this approach. For example, since the PReID problem is highly influenced by camera bias, the authors in [110] redesigned Batch Normalization [105] to employ it only on samples of the same camera in the batch. This way, they learn camera-related statistics and better align the cross-camera features. However, this would require the image's camera label on the target domain.

A well-known way to overcome the style variations due to the different environments is to add the Instance Normalization (IN) [111] on the model's layers, mainly on the early ones, where basic geometric and photometric features are learned related to illumination, shapes, resolutions, contrast, among others. The IN allows the model to be robust again style differences, improving model performance to the unseen domain without training over them. However, the IN removes some identity-relevant features. For this reason, Jin et al. [112] propose a restitution module in which, after applying the instance normalization, they restitute the identity-relevant features improving the identity description and model generalization.

5.5.1.3 Unsupervised Domain Adaptation on Person Re-identification

In this group, the models are pre-trained in a labeled source domain to provide initial knowledge related to matching the same person across the cameras; then, they fine-tune the model over the unlabeled target domain (dataset with person bounding boxes without identity annotation). Since the target domain does not have labels, they usually adopt a cluster-and-finetuning strategy. They obtain the feature representation of the target domain images and cluster them to propose possible groups to fine-tune the model. Person features on the same cluster are treated as from the same class, as long as image features on different clusters are treated as from different classes. In this case, we face three problems: the quality of feature representations, the camera bias, and the choice of clustering and its parameters. Since the model considers samples in the same group as true positives, the feature representation plays a key role in projecting images of the same person as close as possible into the feature space. If there is a significant gap between the source and target images, the representation might be seriously damaged and hinder model performance. A general workflow is presented in Figure 5.6.

As mentioned, images of the same camera tend to be closer; consequently, most of the generated clusters will

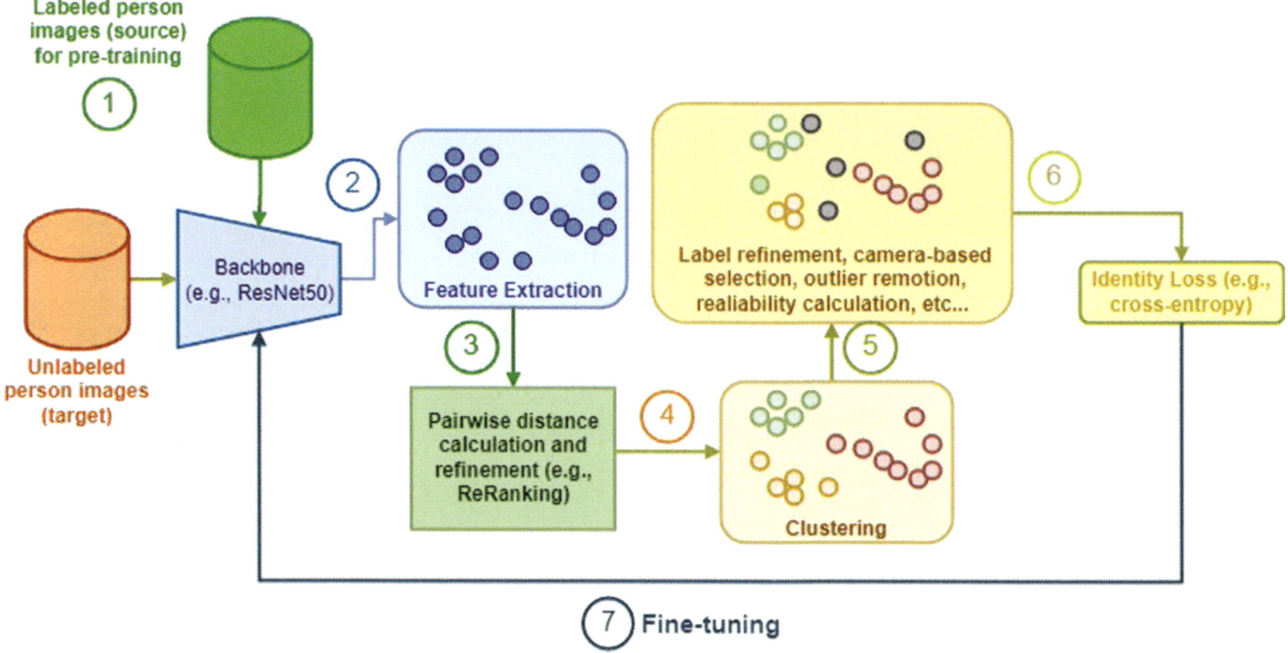

FIGURE 5.6
Pipeline for both unsupervised domain adaptation and fully-unsupervised person re-identification. The common practice is firstly pre-training the model backbone in some labeled person re-identification datasets to initialize its weights and obtain PReID-related features (Step 1). After that, the unlabeled target data is taken and its features are extracted (Step 2). The pair-wise distance is calculated and usually refined based on the samples' neighborhood (e.g., ReRanking (with courtesy from [113])) (Step 3). The clustering is performed (Step 4). Some researchers perform further processing to refine the assignments such as outlier and noisy removing, reliability calculation, camera-based selection, etc. (Step 5). Finally, the loss function is calculated (Step 6) considering the pseudo-labels as a guide, and the backbone is updated (Step 7). The whole pipeline starts again in Step 2. For Fully Unsupervised person re-identification, Step 1 is skipped and the backbone is used directly on the pipeline without initialization.

have feature representations from images captured by the same camera. Furthermore, since the PReID is a cross-camera retrieval task, it is fundamental to have clusters with images from different cameras but from the same identity to achieve a camera-invariant model. Finally, the clustering algorithm must be chosen considering the non-convex nature of the PReID problem, that is, images from the same person are prone to significant changes from one camera to another, then it is expected to have nonconvex clusters with arbitrary shape and density on the feature space. To address these challenges, different methods have been proposed. The most preferred clustering method is the density-based spatial clustering of applications with noise (DBSCAN) [114], since it can find nonconvex and arbitrary shape clusters and detect outliers (noisy samples that are unreliable enough to train the models). DB-SCAN is fed with a context-based distance matrix, in which the feature vectors' neighborhood is considered to calculate the distance among them. The greater the number of neighbors in common between two feature vectors, the lower their distance will be. The most used strategy

is the k-Reciprocal Encoding [113]. It sets hard positive samples (images of the same person seen from different cameras) closer and better groups them reducing the false positive rate in the clusters. For this reason, the combination of DBSCAN and ReRanking is well used in prior art [115, 116]. After clustering, some works propose a further refinement to improve cluster representation. Some of them leverage label refinement [115, 117], camera labels to create triplets or define proxies [13, 118], and reliability based on the output from different pooling layers [116].

In feature representation, some works adopted multiple models with distinct architectures initialized over the source domain to learn different concepts and obtain better feature representation and clustering results [119, 120]. To reduce the camera bias impact, other works assume they have the camera label for each image bounding box, that is, they do not know "who" is on the bounding box (there is no identity annotation), but they know which camera recorded it. In this way, they can guide model training to be camera invariant yielding camera-invariant features [121, 122].

Bertocco et al. [13] proposed an Unsupervised Domain Adaptation method that takes all these strategies simultaneously and achieves state-of-the-art results. The authors propose a novel cross-camera triplet creation strategy, where triplets are created considering the camera labels in each cluster. Generated triplets encourage the model to match images from different cameras with a high probability of having the same identity while keeping apart images that likely present different identities seen from the same camera. A self-ensembling strategy is also proposed to consider all checkpoints generated along training to obtain a final model, and different architectures are also trained in parallel and combined only in the evaluation phase to obtain a better feature representation.

5.5.1.4 Fully Unsupervised Person Re-identification

Despite the recent efforts, generalizable PReID and the unsupervised domain adaptation solutions still require a labeled source domain, which might not have a close gap to the target domain and are prone to the annotation drawbacks exposed at the beginning of this section. To overcome this, researchers have proposed fully unsupervised models where only the image bounding box is available without further annotation, and the models were initialized over ImageNet [123] to have general knowledge instead of PReID-related knowledge. The general workflow is also presented in Figure 5.6, however, without performing Step 1. Unfortunately, camera bias is even stronger since it cannot be alleviated by previous source domain training. Chen et al. [118] proposed a model to optimize a loss function to get samples of a cluster closer to its respective proxy, along with a loss term to align distance distribution between images with and without augmentation. A momentum model is also kept along the training that accumulates the model's weights during optimization to self-ensemble them in a final model. The alignment between different augmentations encourages their model to reduce camera bias.

In cluster-and-finetune solutions, the clustering is usually performed in each iteration of the algorithms; then, a sample might change the cluster it belongs to along the training. To avoid instabilities, Zhang et al. [117] propose to refine clusters by a consensus among iterations. Pseudo-labels on a certain iteration are created by considering the ones generated on a previous iteration, keeping the training stable.

Bertocco et al. [124] proposed a method to tackle the fully unsupervised scenario. Because of the great success of ensemble methods on unsupervised domain adaptation on PReID [13, 119, 120], they adopt model ensembling directly on the pipeline, employing three models to obtain the feature representation for each sample on the unlabeled dataset. Due to the ability of Re-Ranking to reduce the false-positive samples on the clusters, the authors calculate the distances between feature vectors using the k-Reciprocal Encoding [113] for each model separately. After that, the distance matrix is averaged element-wise to ensemble the knowledge from different models.

Additionally, Bertocco et al. designed a novel clustering algorithm to deal with the false-positive samples and camera bias trade-offs. The authors argue that by selecting a DBSCAN hyper-parameter value too restrictive, the generated clusters tend to have a low false-positive rate, but most of them are biased to the camera that records their images, that is, most of the clusters have images from the same identity but captured by only one camera. This might limit training generalization across cameras. In counterpart, if the DBSCAN hyper-parameter is relaxed, the number of false-positives is increased, as clusters tend to have more images from different cameras. To tackle this, the authors propose an ensemble-based clustering where DBSCAN is run with different hyper-parameter values and then their results are combined into a single clustering that alleviates the false positive samples and allows different cameras on the clusters.

Their solution can overcome bias, select discriminant features along the training, and achieve state-of-the-art performance in the fully unsupervised PReID scenario. To test its generalization ability, Bertocco et al. evaluated their solution in the text authorship attribution problem in social media, in which the main goal is to group *tweets* of the same author without any annotation. They also achieved state-of-the-art performance, surpassing prior work that adopts a supervised strategy, demonstrating the model's capacity to reason with complex data.

5.5.2 Challenges of Person Re-Identification

Despite significant progress on PReID, there are still outstanding open challenges. In this section, we describe two of them and the initial efforts addressing them.

5.5.2.1 Self-supervised Pre-training

Due to the difficulties in annotating comprehensive PReID datasets, it is crucial to make the most out of unlabeled data. Despite all the advances shown in previous sections with clustering-based strategies, a promising research direction is to consider the self-supervised solutions [125–128] to help the unsupervised training. Notwithstanding the comparable performance of self-supervised learning to supervised counterparts, those works rely on the ImageNet dataset in which the majority of the classes have significant differences in semantics (e.g., it contains classes representing airplanes, animals, and sports), allowing the models to rely more on coarse details than on fine-grained details to distinguish classes. However, PReID strongly relies on fine-grained details for learning. A recent work [129] proposes a large-scale unlabeled PReID dataset and uses Momentum Contrast (MoCo) [126] to initialize the backbone before applying the pipeline. However, MoCo is used only for pretraining and is not applied directly to their supervised solution. How to leverage the self-supervised solutions to initialize or train along with the current clustering-based PReID solution is still an open question.

5.5.2.2   *Generalization and Scalability to Diverse Re-identification Scenarios*

Beyond PReID, researchers have proposed methods for other Re-Identification scenarios, for instance, vehicles [130–132] and luggage [133]. Similar to PReID, the main goal is to retrieve and group images of the same object captured from different camera views. Therefore, one challenge is to design a common method to deal with these problems with minimal changes to achieve a general re-identification solution.

Despite the performance of the current methods on unsupervised PReID, the prior art requires the full distance matrix between all target samples to be loaded into memory at once to perform clustering. This makes the methods expensive in terms of time and memory footprint if the number of examples on the target domain grows to thousands or millions, introducing scalability issues. One of the first works to address this challenge is [134], but with less than 72 thousand images in the validation. Solutions that can handle more than one hundred thousand or even millions of unlabeled images, alleviate camera bias, and find discriminative features are promising research directions.

5.5.3   Future Directions for Person Re-Identification

Considering a general re-identification solution, the main goal is to reduce the viewpoint bias to enable the model to match the same person or object from a different point of view under different illumination, background, resolution, and occlusion conditions. Researchers have also aimed to design general solutions for different re-identifications scenarios. One way is to assume we do not know if there is a human or an object in the image and adopt a solution with the least annotation possible, aiming to group the most varied images from different cameras without increasing the false-positive samples on the clusters. Some methods on fully unsupervised setups have already reported results in different tasks [117, 135].

With increasing interest in fully unsupervised models in different ReID scenarios, one possible research direction is to take different re-identification solutions (person, vehicles, and objects) and propose a joint analysis by looking at possible relations among them during the analysis of an event or the dynamics of an environment. We could be able to tell what and how the event happened, further investigate it, and contribute to recreating its history.

A large-scale and fully unsupervised solution that could handle an arbitrary amount of data could alleviate camera and other biases present in the data. Thus, learning discriminative features and being generalizable to different modalities is a promising step for event investigations. Furthermore, with solutions that do not require task-related annotation, we could extend them beyond forensics. For instance, the solutions can be tested in the text authorship attribution task [124], medical analysis, general recognition, cross-modality tasks, reasoning, and further application to general AI.

5.6   Visual Analytics in Forensics

The Forensics Sciences, in general, have a primary objective, to find who, in what circumstances, and why an event happened, based on Locard's principle: "Every contact leaves a trace" [136]. With the advent of the digital world, digital forensics became important precisely to propose techniques for the identification, analysis, and validation of digital evidence in order to reconstruct the event and to help answer those questions.

Nonetheless, nowadays, the rapidness, variety, and facility of generating, sharing, and storing data, by different people (specialists or not), from different angles, positions, and devices, have been a tremendous challenge. The ability of a human specialist to understand the enormous volume of data cannot keep pace with the velocity to collect and store the data [137]. In addition, the data obtained has evolved from simple data types, such as documents, pictures, and videos, including simple metadata related to the dates of creation or modification and the device name, to more personalized data covering sensitive information, such as social media, multimedia, with more sophisticated metadata related to geolocation, camera features, and others.

In this way, the cases seem to have multiplied due to the ease of generating digital evidence, they seem to be more complex due to the variety of evidence that can be generated, and it seems that the time required for accurate answers has decreased due to easy access to case tracking [138]. These challenges have been addressed for some time now with AI strategies, which show the capacity for learning and recognizing.

Many researchers have developed several exciting AI solutions to enhance images, videos, and audio, including automatic categorization, knowledge extraction, and indexing [11, 139, 140], but the analysis become a black-box process, which generates difficulty in understanding the reason for many algorithm results and raising concerns over the reliability of decisions made by AI models. Consequently, applying AI solutions to real decision-making in forensics is not yet widely applicable. However, this problem is not specific to this area. Given the importance of explainability, various information visualization techniques are embedded in the different tasks to make the processes transparent and self-explanatory. This inclusion is often not trivial and needs to respond to certain rationale bases.

5.6.1 Broad Overview of Visual Analytics in Forensics

We need to answer four key questions to propose using information visualization techniques in AI Forensics: Why, Who, What, and How.

5.6.1.1 Why: Motivation for Visual Analytics in AI Forensics

The main goal of Information Visualization is to generate visual mappings of abstract data [141] in order to show and convey information easily. In this context, strategies can be proposed to deal with a large amount of digital evidence data, providing an integral part in the analytical process [142]. Moreover, they could provide ways of understanding the methods and results of the AI analysis, requiring less cognitive effort.

5.6.1.2 Who: Users of Visual Analytics in AI Forensics

Information visualization strategies are inherently user-driven, where meeting the user requirements is critical for the visual design, including the needs, tasks, and work they want to perform with the visual representations. Nevertheless, sometimes this step is complicated to accomplish due to the subjective interests of experts. The goal must be to accelerate and improve decision-making for domain experts without harming the reliability of processes and results.

5.6.1.3 What: Forensics Data Used in Visual Analytics

In the early stages of data preprocessing, data visualizations could help to understand the data's nature, exposing possible anomalies (which can direct the investigation toward data that might be of greater importance), creating timelines that help order events (which leads to insight regarding cause and effect, and gaps identification). This could help quickly determine a starting point or a next point to be investigated in a large dataset from a reduced data representation (highlighting important characteristics that can help drive the next step in the examination).

Information Visualization strategies are not only useful in the early data preprocessing stages but also crucial when using complex methods whose processes are opaque to human understanding. Most AI models only show the final results, do not outline the main characteristics of their choices, and do not facilitate the presentation of the findings to other specialists. Consequently, AI models pursuing prediction accuracy and becoming increasingly opaque such as ensemble methods and deep neural networks are problematic. Techniques emerge to clarify black-box models to address the trade-off between interpretability and model performance.

5.6.1.4 How: Visual Analytics Methods

A useful criterion to classify the visualization methods includes some keys [143–145]:

- Local or global [146]: global solutions refer to the interpretability of the entire model behavior, giving insights into how the trained model makes predictions. Local strategies will explain which parts of a single instance were important for the prediction, influencing a user's confidence in the prediction and, consequently, the user's action.

- Model agnostic [146] or model specific [147]: model-agnostic solutions are independent of the model structure and can be coupled to any AI model after this is trained. On the other hand, in model-specific strategies, visualizations are incorporated from the internals, inherent to the structure and learning mechanisms, and consequently limited to specific models.

- Intrinsic or post-hoc based: a visual solution will be intrinsic if the AI model is interpretable and easily visualized because of its simple structures, such as tree models. A post-hoc solution includes methods like feature importance and partial dependency plots to explain the models after a training stage.

- Perturbation or saliency-based [148]: in solutions based on perturbations, simulated data points are generated around an instance through random perturbation. On the other hand, when deep neural networks are used, an efficient way of pointing out what causes a certain outcome, especially when images or texts are treated, consists in using masks visually highlighting the determining aspects of the record analyzed, which is called Saliency Masks.

- Example-based or feature-based [149]: in example-based solutions, the visualizations highlight the points in the training set that resemble the test point we need to explain. In a feature-based context, through reverse engineering, features are scored by contribution to predictive performance.

Many visualization metaphors were proposed to deal with the specific data structure. The most common and useful strategies are Bar Charts, Parallel Coordinates, Trilinear Coordinates, Scatter Plot, Circle Packing, TreeMaps, Voronoi Maps, and Chord Diagram.

5.6.2   Challenges of Visual Analytics in Forensics

One of the major concerns of visual analytics in Forensics is the processing of vast amounts of data to identify and distill the most valuable and relevant information for a certain event. The visual metaphors should reveal patterns and relevant characteristics to support specialists. Nevertheless, there are relevant scientific challenges that need to be faced:

- **Space and visual representations for scalability:** visual analytics solutions need to scale with the size and dimensionality of the input forensic data space, considering data from multiple, heterogeneous sources which need to be integrated and processed jointly. In forensic analysis, the available data can easily surpass the available space in orders of magnitude, which is a characteristic of big data. Many authors [150–152] proposed solutions to deal with this overwhelming data by data transformations, aggregated representations, or coordinated visualizations. However, aggregating or transforming data would represent a loss of information, which could strain our cognitive ability. Finding patterns or relevant properties over several representations is almost impossible as shown by Keim et al., [153]. Even if we include zooming strategies, our perceptual ability could not handle the complete data representation. In addition, the solutions need to be efficient enough to work in interactive analysis. A successful solution could consider intelligently combining visualizations of details and a global overview at different detail levels using appropriate visual abstraction.

- **Quality of data and graphical representation for interpretability:** an important issue in visual analytics for forensic analysis is the avoidance of misinterpretations. In this domain, strategies could be affected by the typical nature of evidence data: unstructured data and the often-varying data quality, which could include manipulated data. Some approaches [154, 155] deal with the identification of data manipulation. However, the visual representations need to take into account the detection of missing, faulty, or inconsistent data, which represents a challenge to visual analytics for forensic analysis. Errors or uncertainty in the input data may produce misleading analysis results.

- **User interfaces for real-time explainable analysis:** Visual analytics solutions need to be easily understood by the specialist, who must be able to focus on the task at hand fully and not on complex user interfaces. In addition, the velocity of forensics data generation implies that the time to extract value from data must be included in the visual analytics workflow. There must be a balance between the representation of data streams and the preservation of interactive capabilities to create sensemaking.

5.6.3   Future Directions for Visual Analytics in Forensics

We summarize the key directions of visual analytics research in AI Forensics.

5.6.3.1   *Scalability*

The increasing volume of digital forensic data presents an overwhelming challenge for effective visualizations. Unfortunately, most visual analytics strategies are not scalable to extreme-scale data.

5.6.3.2   *Report Generation*

Visual analytics tools often allow the creation of reports on the analysis results. The visual metaphors can provide insights into data patterns during the analysis. Nevertheless, report generation from visualization is still heuristic and subjective.

5.6.3.3   *Specialist Knowledge in the Loop*

In many of the strategies, the specialist acts as a simple spectator who observes what is happening with the data scene. However, the inclusion of domain knowledge in order to guide the research process would be important. Also, the need for integrating visual displays of data between different strategies will continue to be highly relevant in the future.

5.6.3.4   *Usability*

To allow efficient decision support, visualization interfaces should be well-designed and esthetically pleasing

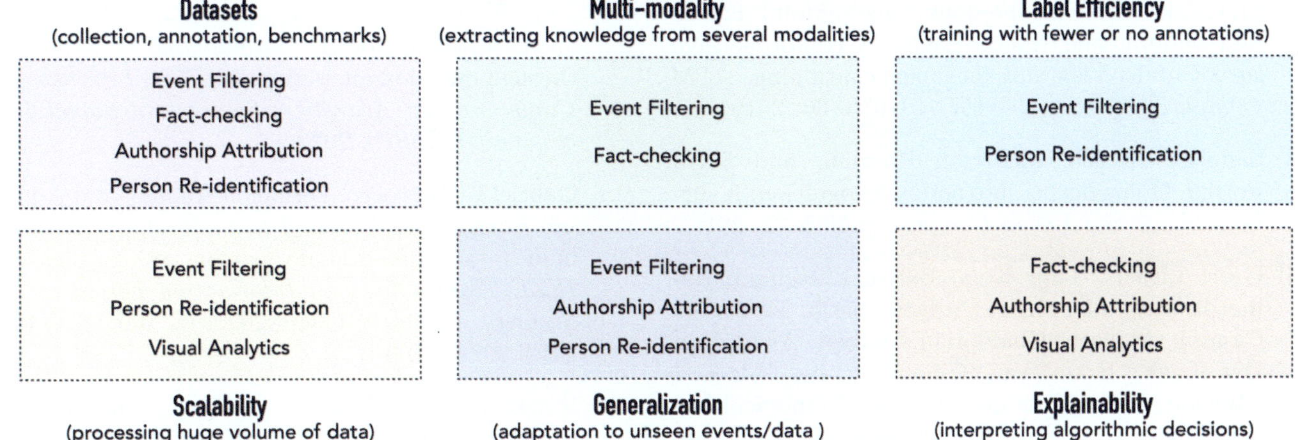

FIGURE 5.7
Summary of the challenges discussed in this chapter for event filtering, fact-checking, authorship attribution, person re-identification, and visual analytics research.

while ensuring the requirements of the target audience of a system are satisfied, using effective color, texture, shape, and motion to capture human attention.

5.7 Conclusion

The popularization of social media in the past decade brought about a new wave of challenges and opportunities for Digital Forensics. The massive volume of information generated daily is a fertile ground for misinformation and fake narratives, negatively impacting many spheres of our society. Nonetheless, such data availability is also a driving force for many AI-based models that aid forensic experts in fact-checking, analyzing, and reconstructing real-world events. In this chapter, we presented modern computer vision and machine learning applications within the forensic event analysis pipeline. In particular, we discuss the research efforts of event filtering, fact-checking, authorship attribution, PReID, and visual analytics tasks.

Digital forensics was forced to evolve in the last 10 years from basic computer vision and simple natural language processing techniques to powerful AI-driven methods to deal with the signs of the new age. Innovation at such an explosive speed is always followed by a series of challenges. In this chapter, we also presented many such problems, as well as future directions to overcome them. In Figure 5.7, we summarize the overall challenges associated with the research problems discussed in this chapter.

As the times go by we are sure AI will keep playing a significant role not only in the detection of misinformation but also in the generation of fake content. After all, the *Era of Synthetic Reality* [156–158] has just begun.

Notes

1. https://github.com/MKLab-ITI/mmdemo-dockerized

2. https://github.com/twintproject/twint

3. In this chapter, we use fake news, misinformation, and disinformation interchangeably, although in some articles they have different definitions.

4. https://www.instagram.com/p/ClybwAjjB9q/

5. https://en.wikipedia.org/wiki/Fact-checking, last updated 6 May 2024.

6. https://ukrainefacts.org

7. Cases with a falsified image with a true claim may happen, but they are not common and usually unintentional, so we do not consider here.

8. https://reporterslab.org/fact-checking/

References

[1] Nic Newman, Richard Fletcher, Anne Schulz, Simge Andi, Craig T Robertson, and Rasmus Kleis Nielsen. Reuters institute digital news report 2021. *Reuters Institute for the Study of Journalism*, 2021.

[2] MD Zakir Hossain, Ferdous Sohel, Mohd Fairuz Shiratuddin, and Hamid Laga. A comprehensive survey of deep learning for image captioning. *ACM Computing Surveys (CsUR)*, 51(6):1–36, 2019.

[3] Samitha Herath, Mehrtash Harandi, and Fatih Porikli. Going deeper into action recognition: A survey. *Image and Vision Computing*, 60:4–21, 2017.

[4] Geert Litjens, Thijs Kooi, Babak Ehteshami Bejnordi, Arnaud Arindra Adiyoso Setio, Francesco Ciompi, Mohsen Ghafoorian, Jeroen Awm Van Der Laak, Bram Van Ginneken, and Clara I Sánchez. A survey on deep learning in medical Image Analysis. *Medical Image Analysis*, 42:60–88, 2017.

[5] Kalaivani Sundararajan and Damon L Woodard. Deep learning for biometrics: A survey. *ACM Computing Surveys (CSUR)*, 51(3):1–34, 2018.

[6] Daniel Moreira, Aparna Bharati, Joel Brogan, Allan Pinto, Michael Parowski, Kevin W Bowyer, Patrick J Flynn, Anderson Rocha, and Walter J Scheirer. Image provenance analysis at scale. *IEEE Transactions on Image Processing*, 27(12):6109–6123, 2018.

[7] Zhihao Wang, Jian Chen, and Steven CH Hoi. Deep learning for image super-resolution: A survey. *IEEE Transactions on Pattern Analysis and Machine Intelligence*, 43(10):3365–3387, 2020.

[8] Rafael Padilha, Tawfiq Salem, Scott Workman, Fernanda A Andaló, Anderson Rocha, and Nathan Jacobs. Content-aware detection of temporal metadata manipulation. *IEEE Transactions on Information Forensics and Security*, 17:1316–1327, 2022.

[9] Rafael Padilha, Fernanda A Andaló, Bahram Lavi, Luís AM Pereira, and Anderson Rocha. Temporally sorting images from real-world events. *Pattern Recognition Letters*, 147:212–219, 2021.

[10] Bahram Lavi, José Nascimento, and Anderson Rocha. Semi-supervised feature embedding for data sanitization in real-world events. In *ICASSP 2021 - 2021 IEEE International Conference on Acoustics, Speech and Signal Processing (ICASSP)*, pages 2495–2499, 2021.

[11] Caroline Mazini Rodrigues, Aurea Soriano-Vargas, Bahram Lavi, Anderson Rocha, and Zanoni Dias. Manifold learning for real-world event understanding. *IEEE Transactions on Information Forensics and Security*, 16:2957–2972, 2021.

[12] Jing Yang, Didier Vega-Oliveros, Tais Seibt, and Anderson Rocha. Explainable fact-checking through question answering. In *IEEE International Conference on Acoustics, Speech and Signal Processing (ICASSP)*, 2022.

[13] Gabriel C. Bertocco, Fernanda Andaló, and Anderson Rocha. Unsupervised and self-adaptative techniques for cross-domain person re-identification. *IEEE Transactions on Information Forensics and Security*, 16:4419–4434, 2021. doi: 10.1109/TIFS.2021.3107157.

[14] Rafael Padilha, Fernanda Andaló, Luís Pereira, and Anderson Rocha. *Unraveling the Notre Dame Cathedral fire in space and time: An X-coherence approach*, pages 3–19. CRC Press, 2021. ISBN 9780429322877.

[15] Kashif Ahmad, Naina Said, Nicola Conci, and Ala Al-Fuqaha. Active learning for event detection in support of disaster analysis applications. *Signal Image and Video Processing*, 15, 2021.

[16] Firoj Alam, Tanvirul Alam, Md. Arid Hasan, Abul Hasnat, Muhammad Imran, and Ferda Ofli. MEDIC: A multi-task learning dataset for disaster image classification. *Neural Computing & Applications*, 2022.

[17] Pakhee Kumar, Ferda Ofli, Muhammad Imran, and Carlos Castillo. Detection of disaster-affected cultural heritage sites from social media images using deep learning techniques. *Journal on Computing and Cultural Heritage*, 13(3), 2020. ISSN 1556-4673.

[18] Matthew Johnson, Dhiraj Murthy, Brett Roberstson, Roth Smith, and Keri Stephens. Disasternet: Evaluating the performance of transfer learning to classify hurricane-related images posted on twitter. In *HICSS*, 2020.

[19] Lulwa Ahmed, Kashif Ahmad, Naina Said, Basheer Qolomany, Junaid Qadir, and Ala Al-Fuqaha. Active learning based federated learning for waste and natural disaster image classification. *IEEE Access*, 2020.

[20] Xukun Li and Cornelia Caragea. Identifying disaster damage images using a domain adaptation approach. In *ISCRAM*, 2019.

[21] Pradeep K. Atrey, M. Anwar Hossain, Abdulmotaleb El Saddik, and M. Kankanhalli. Multimodal fusion for multimedia analysis: A survey. *Multimedia Systems*, 16:345–379, 2010.

[22] Abhinav Kumar, Jyoti Singh, Yogesh Dwivedi, and Nripendra Rana. A deep multi-modal neural network for informative twitter content classification during emergencies. *Annals of Operations Research*, 2020.

[23] Jiankai Li, Yunhong Wang, and Weixin Li. MGMP: Multimodal graph message propagation network for event detection. In *International Conference on Multimedia Modeling (MMM)*, 2022.

[24] Imad Afyouni, Zaher Al Aghbari, and Reshma Abdul Razack. Multi-feature, multi-modal, and multi-source social event detection: A comprehensive survey. *Information Fusion*, 79:279–308, 2022. ISSN 1566-2535.

[25] Y. Yilmaz and Alfred O. Hero. Multimodal event detection in twitter hashtag networks. *Journal of Signal Processing Systems*, 90:185–200, 2018.

[26] Po-Yao Huang, Junwei Liang, Jean-Baptiste Lamare, and Alexander G. Hauptmann. Multimodal filtering of social media for temporal monitoring and event analysis. In *Proceedings of the 2018 ACM on International Conference on Multimedia Retrieval*, ICMR '18, pages 450–457, New York, NY, USA, 2018. Association for Computing Machinery. ISBN 9781450350464.

[27] Tao Chen, Dongyuan Lu, Min-Yen Kan, and Peng Cui. Understanding and classifying image tweets. In *Proceedings of the 21st ACM International Conference on Multimedia*, MM '13, pages 781–784, New York, NY, USA, 2013. Association for Computing Machinery. ISBN 9781450324045.

[28] Ferda Ofli, Firoj Alam, and Muhammad Imran. Analysis of social media data using multimodal deep learning for disaster response. In *17th International Conference on Information Systems for Crisis Response and Management*. ISCRAM, 2020.

[29] Mahdi Abavisani, Liwei Wu, Shengli Hu, Joel Tetreault, and Alejandro Jaimes. Multimodal categorization of crisis events in social media. In *2020 IEEE/CVF Conference on Computer Vision and Pattern Recognition (CVPR)*, pages 14667–14677, 2020.

[30] Xuehua Wu, Jin Mao, Hao Xie, and Gang Li. Identifying humanitarian information for emergency response by modeling the correlation and independence between text and images. *Information Processing & Management*, 59(4):102977, 2022. ISSN 0306-4573.

[31] Firoj Alam, Ferda Ofli, and Muhammad Imran. Crisismmd: Multimodal twitter datasets from natural disasters, 2018.

[32] Umang Aggarwal, Adrian Popescu, and Céline Hudelot. Active learning for imbalanced datasets. In *2020 IEEE Winter Conference on Applications of Computer Vision (WACV)*, pages 1417–1426, 2020.

[33] Nicole Chaves, Steve Almasy, Ray Sanchez, and Darran Simon. Central Mexico earthquake kills more than 200, topples buildings. https://edition.cnn.com/2017/09/19/americas/mexico-earthquake/index.html, 2017. [Online; accessed 30-May-2022].

[34] Ryan Lagerstrom, Yulia Arzhaeva, Piotr Szul, Oliver Obst, Robert Power, Bella Robinson, and Tomasz Bednarz. Image classification to support emergency situation awareness. *Frontiers in Robotics and AI*, 3, 2016. ISSN 2296-9144.

[35] Dengyong Zhou, Olivier Bousquet, Thomas Lal, Jason Weston, and Bernhard Schölkopf. Learning with local and global consistency. In S. Thrun, L. Saul, and B. Schölkopf, editors, *Advances in Neural Information Processing Systems*, volume 16. MIT Press, 2003. URL https://proceedings.neurips.cc/paper/2003/file/87682805257e619d49b8e0dfdc14affa-Paper.pdf.

[36] Wen Xu, Jing He, and Yanfeng Shu. Transfer learning and deep domain adaptation. In Marco Antonio Aceves-Fernandez, editor, *Advances and Applications in Deep Learning*, chapter 3. IntechOpen, Rijeka, 2020. doi: 10.5772/intechopen.94072. URL https://doi.org/10.5772/intechopen.94072.

[37] Paula Adamo Idoeta. A história de Bolsonaro com a hidroxicloroquina em 6 pontos: de tuítes de Trump à CPI da Covid. https://www.bbc.com/portuguese/brasil-57166743, 2021. [Online; accessed 09-September-2021].

[38] Xinyi Zhou, Jindi Wu, and Reza Zafarani. SAFE: Similarity-aware multi-modal fake news detection. In *Pacific-Asia Conference on Knowledge Discovery and Data Mining (PAKDD)*, pages 354–367. Springer, 2020.

[39] Stephane Schwarz, Antônio Theóphilo, and Anderson Rocha. EMET: Embeddings from multilingual-encoder transformer for fake news detection. In *IEEE International Conference on Acoustics, Speech and Signal Processing (ICASSP)*, pages 2777–2781. IEEE, 2020.

[40] Claire Wardle. Understanding Information disorder. https://firstdraftnews.org/long-form-article/understanding-information-disorder/, 2019. [Online; accessed 22-May-2022].

[41] Preslav Nakov, David Corney, Maram Hasanain, Firoj Alam, Tamer Elsayed, Alberto Barrón-Cedeño, Paolo Papotti, Shaden Shaar, and Giovanni Da San Martino. Automated fact-checking for assisting human fact-checkers. In *International Joint Conference on Artificial Intelligence (IJCAI)*, 2021.

[42] James Thorne, Andreas Vlachos, Christos Christodoulopoulos, and Arpit Mittal. FEVER: A large-scale dataset for fact extraction and verification. In *Conference of the North American Chapter of the Association for Computational Linguistics (NAACL)*, 2018.

[43] David Wadden, Shanchuan Lin, Kyle Lo, Lucy Lu Wang, Madeleine van Zuylen, Arman Cohan, and Hannaneh Hajishirzi. Fact or fiction: Verifying scientific claims. In *Conference on Empirical Methods in Natural Language Processing (EMNLP)*, 2020.

[44] Isabelle Augenstein, Christina Lioma, Dongsheng Wang, Lucas Chaves Lima, Casper Hansen, Christian Hansen, and Jakob Grue Simonsen. Multifc: A real-world multi-domain dataset for evidence-based fact checking of claims. In *Conference on Empirical Methods in Natural Language Processing and International Joint Conference on Natural Language Processing (EMNLP-IJCNLP)*, 2019.

[45] Rami Aly, Zhijiang Guo, Michael Sejr Schlichtkrull, James Thorne, Andreas Vlachos, Christos Christodoulopoulos, Oana Cocarascu, and Arpit Mittal. FEVEROUS: Fact extraction and VERification over unstructured and structured information. In *Conference on Neural Information Processing Systems Datasets and Benchmarks Track (NeurIPS)*, 2021.

[46] Tal Schuster, Adam Fisch, and Regina Barzilay. Get your vitamin c! robust fact verification with contrastive evidence. In *Conference of the North American Chapter of the Association for Computational Linguistics: Human Language Technologies (NAACL)*, 2021.

[47] Julian Eisenschlos, Bhuwan Dhingra, Jannis Bulian, Benjamin Börschinger, and Jordan Boyd-Graber. Fool me twice: Entailment from wikipedia gamification. In *Conference of the North American Chapter of the Association for Computational Linguistics: Human Language Technologies (NAACL)*, 2021.

[48] Jungsoo Park, Sewon Min, Jaewoo Kang, Luke Zettlemoyer, and Hannaneh Hajishirzi. Faviq: Fact verification from information-seeking questions. *arXiv preprint arXiv:2107.02153*, 2021.

[49] Dimitrina Zlatkova, Preslav Nakov, and Ivan Koychev. Fact-checking meets fauxtography: Verifying claims about images. In *Conference on Empirical Methods in Natural Language Processing and International Joint Conference on Natural Language Processing (EMNLP-IJCNLP)*, 2019.

[50] Shivangi Aneja, Chris Bregler, and Matthias Nießner. Cosmos: Catching out-of-context misinformation with self-supervised learning. *arXiv preprint arXiv:2101.06278*, 2021.

[51] Grace Luo, Trevor Darrell, and Anna Rohrbach. Newsclippings: Automatic generation of out-of-context multimodal media. In *Conference on Empirical Methods in Natural Language Processing (EMNLP)*, 2021.

[52] Barry Menglong Yao, Aditya Shah, Lichao Sun, Jin-Hee Cho, and Lifu Huang. End-to-end multimodal fact-checking and explanation generation: A challenging dataset and models. *arXiv preprint arXiv:2205.12487*, 2022.

[53] Naeemul Hassan, Fatma Arslan, Chengkai Li, and Mark Tremayne. Toward automated fact-checking: Detecting check-worthy factual claims by claimbuster. In *ACM SIGKDD International Conference on Knowledge Discovery and Data Mining (SIGKDD)*, 2017.

[54] Dustin Wright and Isabelle Augenstein. Claim check-worthiness detection as positive unlabelled learning. In *Findings of the Association for Computational Linguistics: EMNLP 2020*, pages 476–488, 2020.

[55] Vladimir Karpukhin, Barlas Oguz, Sewon Min, Patrick Lewis, Ledell Wu, Sergey Edunov, Danqi Chen, and Wen-tau Yih. Dense passage retrieval for open-domain question answering. In *Proceedings of the 2020 Conference on Empirical Methods in Natural Language Processing (EMNLP)*, pages 6769–6781, 2020.

[56] James Thorne, Andreas Vlachos, Christos Christodoulopoulos, and Arpit Mittal. Evaluating adversarial attacks against multiple fact

verification systems. In *Conference on Empirical Methods in Natural Language Processing and the 9th International Joint Conference on Natural Language Processing (EMNLP-IJCNLP)*, 2019.

[57] Tal Schuster, Darsh Shah, Yun Jie Serene Yeo, Daniel Roberto Filizzola Ortiz, Enrico Santus, and Regina Barzilay. Towards debiasing fact verification models. In *Conference on Empirical Methods in Natural Language Processing and the 9th International Joint Conference on Natural Language Processing (EMNLP-IJCNLP)*, 2019.

[58] Pepa Atanasova, Jakob Grue Simonsen, Christina Lioma, and Isabelle Augenstein. Generating fact checking explanations. In *Annual Meeting of the Association for Computational Linguistics (ACL)*, 2020.

[59] Neema Kotonya and Francesca Toni. Explainable automated fact-checking for public health claims. In *Conference on Empirical Methods in Natural Language Processing (EMNLP)*, 2020.

[60] Neema Kotonya and Francesca Toni. Explainable automated fact-checking: A survey. In *International Conference on Computational Linguistics (ACL)*, 2020.

[61] Ziqiang Cao, Furu Wei, Wenjie Li, and Sujian Li. Faithful to the original: Fact aware neural abstractive summarization. In *AAAI Conference on Artificial Intelligence (AAAI)*, 2018.

[62] Joshua Maynez, Shashi Narayan, Bernd Bohnet, and Ryan McDonald. On faithfulness and factuality in abstractive summarization. In *Annual Meeting of the Association for Computational Linguistics (ACL)*, 2020.

[63] Alec Radford, Jong Wook Kim, Chris Hallacy, Aditya Ramesh, Gabriel Goh, Sandhini Agarwal, Girish Sastry, Amanda Askell, Pamela Mishkin, Jack Clark, et al. Learning transferable visual models from natural language supervision. In *International Conference on Machine Learning (ICML)*. PMLR, 2021.

[64] Sahar Abdelnabi, Rakibul Hasan, and Mario Fritz. Open-domain, content-based, multi-modal fact-checking of out-of-context images via online resources. *arXiv preprint arXiv:2112.00061*, 2021.

[65] Ning Xie, Farley Lai, Derek Doran, and Asim Kadav. Visual entailment: A novel task for fine-grained image understanding. *arXiv preprint arXiv:1901.06706*, 2019.

[66] Junwen Chen and Yu Kong. Explainable video entailment with grounded visual evidence. In *IEEE/CVF International Conference on Computer Vision (CVPR)*, 2021.

[67] Alexander Hamilton, James Madison, and John Jay. *The federalist papers*. Yale University Press, 2009.

[68] Frederick Mosteller and David L. Wallace. Inference in an authorship problem: A comparative study of discrimination methods applied to the authorship of the disputed federalist papers. *Journal of the American Statistical Association*, 58(302):275–309, 1963.

[69] Alvar Ellegård. *A statistical method for determining authorship: The Junius Letters, 1769-1772*, volume 13. Göteborg: Elander, 1962.

[70] John Wade. *The letters of Junius*, volume 1. G. Bell and Son, 1894.

[71] Andrew Queen Morton. *Literary detection: How to prove authorship and fraud in literature and documents*. Bowker, 1978. ISBN 9780859350624.

[72] John F. Burrows. Word-patterns and story-shapes: The statistical analysis of narrative style. *Literary & Linguistic Computing*, 2(2):61–70, 1987.

[73] Dmitry V. Khmelev and William J. Teahan. A repetition based measure for verification of text collections and for text categorization. In *Proceedings of the 26th Conference on Research and Development in Informaion Retrieval (SIGIR)*, pages 104–110. ACM, 2003.

[74] W. Oliveira Jr., Edson Justino, and Luiz S. Oliveira. Comparing compression models for authorship attribution. *Forensic Science International*, 228(1-3):100–104, 2013.

[75] K. A. Apoorva and S. Sangeetha. Deep neural network and model-based clustering technique for forensic electronic mail author attribution. *SN Applied Sciences*, 3(3):1–12, 2021.

[76] K. A. Apoorva and S Sangeetha. Forensic analysis of e-mail for authorship attribution: Research perspective. In *Proceeding of First Doctoral Symposium on Natural Computing Research*, pages 281–292. Springer, 2021.

[77] Georgia Frantzeskou, Efstathios Stamatatos, Stefanos Gritzalis, Carole E. Chaski, and Blake Stephen Howald. Identifying authorship by byte-level n-grams: The source code author

profile (scap) method. *International Journal of Digital Evidence*, 6(1):1–18, 2007.

[78] Egor Bogomolov, Vladimir Kovalenko, Yurii Rebryk, Alberto Bacchelli, and Timofey Bryksin. Authorship attribution of source code: A language-agnostic approach and applicability in software engineering. In *Proceedings of the 29th ACM Joint Meeting on European Software Engineering Conference and Symposium on the Foundations of Software Engineering*, pages 932–944, 2021.

[79] Robert Layton, Paul Watters, and Richard Dazeley. Authorship attribution for twitter in 140 characters or less. In *2010 Second Cybercrime and Trustworthy Computing Workshop*, pages 1–8. IEEE, 2010.

[80] Roy Schwartz, Oren Tsur, Ari Rappoport, and Moshe Koppel. Authorship attribution of micro-messages. In *Proceedings of the 2013 Conference on Empirical Methods in Natural Language Processing (EMNLP)*, pages 1880–1891. ACL, 2013. URL https://aclanthology.org/D13-1193.

[81] Anderson Rocha, Walter J. Scheirer, Christopher W. Forstall, Thiago Cavalcante, Antonio Theophilo, Bingyu Shen, Ariadne R. B. Carvalho, and Efstathios Stamatatos. Authorship attribution for social media forensics. *Transactions on Information Forensics and Security (T-IFS)*, 12(1):5–33, 2016.

[82] Sebastian Ruder, Parsa Ghaffari, and John G. Breslin. Character-level and multi-channel convolutional neural networks for large-scale authorship attribution. *arXiv preprint:1609.06686*, 2016. URL https://arxiv.org/abs/1609.06686.

[83] Prasha Shrestha, Sebastian Sierra, Fabio A. González, Manuel Montes-y Gómez, Paolo Rosso, and Thamar Solorio. Convolutional neural networks for authorship attribution of short texts. In *Proceedings of the 15th Conference of the European Chapter of the Association for Computational Linguistics (EACL)*, pages 669–674. ACL, 2017. URL https://aclanthology.org/E17-2106.

[84] Steven H. H. Ding, Benjamin C.M. Fung, Farkhund Iqbal, and William K. Cheung. Learning stylometric representations for authorship analysis. *IEEE Transactions on Cybernetics*, 49(1):107–121, 2017.

[85] Benedikt Boenninghoff, Robert M Nickel, Steffen Zeiler, and Dorothea Kolossa. Similarity learning for authorship verification in social media. In *Proceedings of the 2019 IEEE International Conference on Acoustics, Speech and Signal Processing (ICASSP)*, pages 2457–2461. IEEE, 2019.

[86] Antonio Theophilo, Luis A. M. Pereira, and Anderson Rocha. A needle in a haystack? harnessing onomatopoeia and user-specific stylometrics for authorship attribution of micro-messages. In *Proceedings of the 2019 IEEE International Conference on Acoustics, Speech and Signal Processing (ICASSP)*, pages 2692–2696. IEEE, 2019.

[87] Zhiqiang Hu, Roy Ka-Wei Lee, Lei Wang, Ee-peng Lim, and Bo Dai. Deepstyle: User style embedding for authorship attribution of short texts. In *Asia-Pacific Web (APWeb) and Web-Age Information Management (WAIM) Joint International Conference on Web and Big Data*, pages 221–229. Springer, 2020.

[88] Antonio Theophilo, Romain Giot, and Anderson Rocha. Authorship attribution of social media messages. *IEEE Transactions on Computational Social Systems (TCSS)*, 2021.

[89] Adrian Chen. The agency. https://www.nytimes.com/2015/06/07/magazine/the-agency.html, 2015. [Online; accessed on 31-March-2022].

[90] Scott Shane. Mystery of russian fake on facebook solved, by a Brazilian. https://www.nytimes.com/2017/09/13/us/politics/russia-facebook-election.html, 2017. [Online; accessed on 31-March-2022].

[91] Florence Davey-Attlee and Isa Soares. The fake news machine. Inside a town gearing up for 2020. http://money.cnn.com/interactive/media/the-macedonia-story/, 2017. [Online; accessed on 31-March-2022].

[92] Nicholas Confessore, Gabriel J. X. Dance, Richard Harris, and Mark Hansen. The follower factory. https://www.nytimes.com/interactive/2018/01/27/technology/social-media-bots.html, 2018. [Online; accessed on 31-March-2022].

[93] Onur Varol, Emilio Ferrara, Clayton Davis, Filippo Menczer, and Alessandro Flammini. Online human-bot interactions: Detection, estimation, and characterization. In *Proceedings of the International AAAI Conference on Web and Social Media*, volume 11, 2017.

[94] Alex Heath. Facebook quietly updated two key numbers about its user base. http://www.businessinsider.com/facebook-raises-duplicate-fake-account-estimates-q3-earnings-2017-11, 2017. [Online; accessed on 31-March-2022].

[95] David D. Kirkpatrick. Who is behind qanon? linguistic detectives find fingerprints. https://www.nytimes.com/2022/02/19/technology/qanon-messages-authors.html, 2022. [Online; accessed on 31-March-2022].

[96] Moshe Koppel, Jonathan Schler, and Droz Mughaz. Text categorization for authorship verification. In *8th International Symposium on Artificial Intelligence and Mathematics*, 2004.

[97] Nektaria Potha and Efstathios Stamatatos. A profile-based method for authorship verification. In *Hellenic Conference on Artificial Intelligence*, pages 313–326. Springer, 2014.

[98] Janek Bevendorff, Berta Chulvi, Gretel Liz De La Peña Sarracén, Mike Kestemont, Enrique Manjavacas, Ilia Markov, Maximilian Mayerl, Martin Potthast, Francisco Rangel, Paolo Rosso, et al. Overview of pan 2021: Authorship verification, profiling hate speech spreaders on twitter, and style change detection. In *International Conference of the Cross-Language Evaluation Forum for European Languages*, pages 419–431. Springer, 2021.

[99] Caio Deutsch and Ivandré Paraboni. Authorship attribution using author profiling classifiers. *Natural Language Engineering*, pages 1–28, 2022.

[100] Efstathios Stamatatos. A survey of modern authorship attribution methods. *Journal of the American Society for information Science and Technology*, 60 (3):538–556, 2009.

[101] Moshe Koppel, Jonathan Schler, and Shlomo Argamon. Computational methods in authorship attribution. *Journal of the American Society for information Science and Technology*, 60(1):9–26, 2009.

[102] Keenan Jones, Jason RC Nurse, and Shujun Li. Are you robert or roberta? Deceiving online authorship attribution models using neural text generators. In *Proceedings of the International AAAI Conference on Web and Social Media*, 2022.

[103] Zhenhao Ge, Yufang Sun, and Mark Smith. Authorship attribution using a neural network language model. In *Proceedings of the AAAI Conference on Artificial Intelligence (AAAI)*, volume 30, 2016.

[104] Antonio Theophilo, Rafael Padilha, Fernanda A. Andaló, and Anderson Rocha. Explainable artificial intelligence for authorship attribution on social media. In *Proceedings of the 2022 IEEE International Conference on Acoustics, Speech and Signal Processing (ICASSP)*. IEEE, 2022.

[105] Sergey Ioffe and Christian Szegedy. Batch normalization: Accelerating deep network training by reducing internal covariate shift. In *International Conference on Machine Learning*, pages 448–456. PMLR, 2015.

[106] Kaiyang Zhou, Yongxin Yang, Andrea Cavallaro, and Tao Xiang. Learning generalisable omni-scale representations for person re-identification. *IEEE Transactions on Pattern Analysis and Machine Intelligence*, 2021.

[107] Changxing Ding, Kan Wang, Pengfei Wang, and Dacheng Tao. Multi-task learning with coarse priors for robust part-aware person re-identification. *IEEE Transactions on Pattern Analysis and Machine Intelligence*, 2020.

[108] Shuting He, Hao Luo, Pichao Wang, Fan Wang, Hao Li, and Wei Jiang. Transreid: Transformer-based object re-identification. In *Proceedings of the IEEE/CVF International Conference on Computer Vision*, pages 15013–15022, 2021.

[109] Alexey Dosovitskiy, Lucas Beyer, Alexander Kolesnikov, Dirk Weissenborn, Xiaohua Zhai, Thomas Unterthiner, Mostafa Dehghani, Matthias Minderer, Georg Heigold, Sylvain Gelly, et al. An image is worth 16x16 words: Transformers for image recognition at scale. *arXiv preprint arXiv:2010.11929*, 2020.

[110] Zijie Zhuang, Longhui Wei, Lingxi Xie, Tianyu Zhang, Hengheng Zhang, Haozhe Wu, Haizhou Ai, and Qi Tian. Rethinking the distribution gap of person re-identification with camera-based batch normalization. In *European Conference on Computer Vision*, pages 140–157. Springer, 2020.

[111] Dmitry Ulyanov, Andrea Vedaldi, and Victor Lempitsky. Improved texture networks: Maximizing quality and diversity in feed-forward stylization and texture synthesis. In *Proceedings of the IEEE Conference on Computer Vision and Pattern Recognition*, pages 6924–6932, 2017.

[112] Xin Jin, Cuiling Lan, Wenjun Zeng, Zhibo Chen, and Li Zhang. Style normalization and restitution for generalizable person re-identification. In *Proceedings of the IEEE/CVF Conference on Computer Vision and Pattern Recognition*, pages 3143–3152, 2020.

[113] Zhun Zhong, Liang Zheng, Donglin Cao, and Shaozi Li. Re-ranking person re-identification with k-reciprocal encoding. In *IEEE Conference on Computer Vision and Pattern Recognition (CVPR)*, pages 1318–1327, 2017.

[114] Martin Ester, Hans-Peter Kriegel, Jörg Sander, Xiaowei Xu, et al. A density-based algorithm for discovering clusters in large spatial databases with noise. In *kdd*, volume 96, pages 226–231, 1996.

[115] Yongxing Dai, Jun Liu, Yan Bai, Zekun Tong, and Ling-Yu Duan. Dual-refinement: Joint label and feature refinement for unsupervised domain adaptive person re-identification. *IEEE Transactions on Image Processing*, 30:7815–7829, 2021.

[116] Hao Chen, Benoit Lagadec, and Francois Bremond. Enhancing diversity in teacher-student networks via asymmetric branches for unsupervised person re-identification. In *Proceedings of the IEEE/CVF Winter Conference on Applications of Computer Vision*, pages 1–10, 2021.

[117] Xiao Zhang, Yixiao Ge, Yu Qiao, and Hongsheng Li. Refining pseudo labels with clustering consensus over generations for unsupervised object re-identification. In *Proceedings of the IEEE/CVF Conference on Computer Vision and Pattern Recognition*, pages 3436–3445, 2021.

[118] Hao Chen, Benoit Lagadec, and Francois Bremond. Ice: Inter-instance contrastive encoding for unsupervised person re-identification. In *Proceedings of the IEEE/CVF International Conference on Computer Vision*, pages 14960–14969, 2021.

[119] Yixiao Ge, Dapeng Chen, and Hongsheng Li. Mutual mean-teaching: Pseudo label refinery for unsupervised domain adaptation on person re-identification. *arXiv preprint:*, arXiv:2001.01526, 2020.

[120] Yunpeng Zhai, Qixiang Ye, Shijian Lu, Mengxi Jia, Rongrong Ji, and Yonghong Tian. Multiple expert brainstorming for domain adaptive person re-identification. *arXiv preprint:*, arXiv:2007.01546, 2020.

[121] Zhun Zhong, Liang Zheng, Zhiming Luo, Shaozi Li, and Yi Yang. Learning to adapt invariance in memory for person re-identification. *IEEE Transactions on Pattern Analysis and Machine Intelligence (PAMI)*, pages 1–1, 2020.

[122] Yunpeng Zhai, Shijian Lu, Qixiang Ye, Xuebo Shan, Jie Chen, Rongrong Ji, and Yonghong Tian. Ad-cluster: Augmented discriminative clustering for domain adaptive person re-identification. In *IEEE Conference on Computer Vision and Pattern Recognition (CVPR)*, pages 9021–9030, 2020.

[123] Jia Deng, Wei Dong, Richard Socher, Li-Jia Li, Kai Li, and Li Fei-Fei. Imagenet: A large-scale hierarchical image database. In *IEEE Conference on Computer Vision and Pattern Recognition (CVPR)*, pages 248–255, 2009.

[124] Gabriel Bertocco, Antônio Theófilo, Fernanda Andaló, and Anderson Rocha. Reasoning for complex data through ensemble-based self-supervised learning. *arXiv preprint arXiv:2202.03126*, 2022.

[125] Mathilde Caron, Ishan Misra, Julien Mairal, Priya Goyal, Piotr Bojanowski, and Armand Joulin. Unsupervised learning of visual features by contrasting cluster assignments. *arXiv preprint arXiv:2006.09882*, 2020.

[126] Kaiming He, Haoqi Fan, Yuxin Wu, Saining Xie, and Ross Girshick. Momentum contrast for unsupervised visual representation learning. In *Proceedings of the IEEE/CVF Conference on Computer Vision and Pattern Recognition*, pages 9729–9738, 2020.

[127] Ting Chen, Simon Kornblith, Mohammad Norouzi, and Geoffrey Hinton. A simple framework for contrastive learning of visual representations. In *International Conference on machine Learning*, pages 1597–1607. PMLR, 2020.

[128] Mathilde Caron, Hugo Touvron, Ishan Misra, Hervé Jégou, Julien Mairal, Piotr Bojanowski, and Armand Joulin. Emerging properties in self-supervised vision transformers. *arXiv preprint arXiv:2104.14294*, 2021.

[129] Dengpan Fu, Dongdong Chen, Jianmin Bao, Hao Yang, Lu Yuan, Lei Zhang, Houqiang Li, and Dong Chen. Unsupervised pre-training for person re-identification. In *Proceedings of the IEEE/CVF Conference on Computer Vision and Pattern Recognition*, pages 14750–14759, 2021.

[130] Zefeng Lu, Ronghao Lin, Xulei Lou, Lifeng Zheng, and Haifeng Hu. Identity-unrelated information decoupling model for vehicle re-identification. *IEEE Transactions on Intelligent Transportation Systems*, 2022.

[131] Hongchao Li, Chenglong Li, Aihua Zheng, Jin Tang, and Bin Luo. Mskat: Multi-scale knowledge-aware transformer for vehicle re-identification. *IEEE Transactions on Intelligent Transportation Systems*, 2022.

[132] Sultan Daud Khan and Habib Ullah. A survey of advances in vision-based vehicle re-identification.

Computer Vision and Image Understanding, 182: 50–63, 2019.

[133] Hao Yang, Xiuxiu Chu, Li Zhang, Yunda Sun, Dong Li, and Stephen J Maybank. Quadnet: Quadruplet loss for multi-view learning in baggage re-identification. *Pattern Recognition*, page 108546, 2022.

[134] Xin Jin, Tianyu He, Xu Shen, Tongliang Liu, Xinchao Wang, Jianqiang Huang, Zhibo Chen, and Xian-Sheng Hua. Meta clustering learning for large-scale unsupervised person re-identification. In *Proceedings of the 30th ACM International Conference on Multimedia*, pages 2163–2172, 2022.

[135] Yixiao Ge, Feng Zhu, Dapeng Chen, Rui Zhao, et al. Self-paced contrastive learning with hybrid memory for domain adaptive object re-ID. *Advances in Neural Information Processing Systems*, 33:11309–11321, 2020.

[136] Richard E Overill and Jan Collie. Quantitative evaluation of the results of digital forensic investigations: A review of progress. *Forensic Sciences Research*, 6(1):13–18, 2021.

[137] Darren Quick and Kim-Kwang Raymond Choo. Impacts of increasing volume of digital forensic data: A survey and future research challenges. *Digital Investigation*, 11(4):273–294, 2014.

[138] Mayank Lovanshi and Pratosh Bansal. Comparative study of digital forensic tools. In *Data, Engineering and Applications*, pages 195–204. Springer, 2019.

[139] Zeno Geradts. Digital, big data and computational forensics, 2018.

[140] Patricio Domingues and Alexandre Frazão Rosário. Deep learning-based facial detection and recognition in still images for digital forensics. In *Proceedings of the 14th International Conference on Availability, Reliability and Security*, pages 1–10, 2019.

[141] Mackinlay Card. *Readings in information visualization: Using vision to think*. Morgan Kaufmann, 1999.

[142] Ben Shneiderman. The eyes have it: A task by data type taxonomy for information visualizations. In *The Craft of Information Visualization*, pages 364–371. Elsevier, 2003.

[143] Christoph Molnar. *Interpretable machine learning*. Lulu. com, 2020.

[144] Gregory Plumb, Denali Molitor, and Ameet S Talwalkar. Model agnostic supervised local explanations. *Advances in Neural Information Processing Systems*, 31, 2018.

[145] Riccardo Guidotti and Salvatore Ruggieri. Assessing the stability of interpretable models. *arXiv preprint arXiv:1810.09352*, 2018.

[146] Scott M Lundberg and Su-In Lee. A unified approach to interpreting model predictions. *Advances in Neural Information Processing Systems*, 30, 2017.

[147] Avanti Shrikumar, Peyton Greenside, and Anshul Kundaje. Learning important features through propagating activation differences. In *International Conference on Machine Learning*, pages 3145–3153. PMLR, 2017.

[148] Sharath M Shankaranarayana and Davor Runje. Alime: Autoencoder based approach for local interpretability. In *International Conference on Intelligent Data Engineering and Automated Learning*, pages 454–463. Springer, 2019.

[149] Amina Adadi and Mohammed Berrada. Peeking inside the black-box: A survey on explainable artificial intelligence (XAI). *IEEE Access*, 6:52138–52160, 2018.

[150] Binu Melit Devassy and Sony George. Dimensionality reduction and visualisation of hyperspectral ink data using t-SNE. *Forensic Science International*, 311:110194, 2020.

[151] Myungjong Kim, Wooyeon Jo, Jaehoon Kim, and Taeshik Shon. Visualization for internet of things: Power system and financial network cases. *Multimedia Tools and Applications*, 78(3):3241–3265, 2019.

[152] J Shropshire and R Benton. Container and V visualization for rapid incident response. In *The 53rd Hawaii International Conference on System Sciences (HICSS-53)*, 2020.

[153] Daniel A Keim, Florian Mansmann, Jörn Schneidewind, Jim Thomas, and Hartmut Ziegler. Visual analytics: Scope and challenges. In *Visual Data Mining*, pages 76–90. Springer, 2008.

[154] Ricardo Moura, Rui Sousa-Silva, and Henrique Lopes Cardoso. Automated fake news detection using computational forensic linguistics. In *EPIA Conference on Artificial Intelligence*, pages 788–800. Springer, 2021.

[155] Vincent Itier, Olivier Strauss, Laurent Morel, and William Puech. Color noise correlation-based splicing detection for image forensics. *Multimedia Tools and Applications*, 80(9):13215–13233, 2021.

[156] Chenqi Kong, Baoliang Chen, Haoliang Li, Shiqi Wang, Anderson Rocha, and Sam Kwong. Detect and locate: Exposing face manipulation by semantic- and noise-level telltales. *Transactions on Information Forensics and Security (T-IFS)*, 17:1741–1756, 2022. doi: 10.1109/TIFS.2022.3169921.

[157] Robert C. Wolcott. Beyond virtual reality: Synthetic reality and our co-created futures. https://www.forbes.com/sites/robertwolcott/2017/08/18/beyond-virtual-reality-synthetic-reality-and-our-co-created-futures/?sh=256f15851320, 2017. [Online; accessed on 31-March-2022].

[158] Walter Pasquarelli. Towards synthetic reality: When deepfakes meet AR/VR. https://www.oxfordinsights.com/insights/2019/8/6/towards-synthetic-reality-when-deepfakes-meet-arvr, 2019. [Online; accessed on 31-March-2022].

6

Advances in Computer Vision for Home-Based Stroke Rehabilitation

Kowshik Thopalli, Niccolo Meniconi, Tamim Ahmed, Sai Krishna Yeshala, Aisling Kelliher, Thanassis Rikakis, and Pavan Turaga

6.1 Introduction

The past decade witnessed explosive growth in the development of state-of-the-art methods for computer vision (CV) tasks such as image and video analysis using solutions based on deep learning (DL). Consequently, these cyber-intelligent techniques are becoming widely adopted in many critical applications such as in physical sciences [1], autonomous vehicles [2], and medicine [3–5]. In healthcare settings, the merits of these methods are being established in tasks such as X-ray and CT analysis [6], protein folding [7], drug discovery [8] and more. With (1) advances in CV/DL and (2) the ever-increasing costs of physical therapy, CV-based rehabilitation systems are increasing in popularity [9]. Consequently, there is both a demand for and adoption of, cyber-human intelligent techniques to assist therapists and patients in realizing reliable and affordable solutions. The impressive performance of DL-based CV in tasks such as human pose estimation [10], human activity recognition [11], and human-object interaction [12] is accelerating the design and development of new rehabilitation systems. In this chapter, we primarily focus on the application of CV within the context of rehabilitation for stroke survivors. We begin by reviewing solutions that rely on motion-capture-based data and highlight their inherent challenges and limitations. Building on this, we discuss requirements for the next-generation home-based rehabilitation for stroke survivors using simple RGB camera sensors. We focus on two main challenges, namely (1) the lack of large amounts of high-quality data in healthcare settings and (2), the limited explainability and usability of CV techniques due to the black-box nature of these methods. We additionally draw attention to an important yet under-discussed component in the interaction between the solutions and healthcare providers and not just rehabilitation systems and patients. We show through a series of experiments in collaboration with multiple hospitals that through the means of a cyber-human intelligent system design, these challenges can be effectively addressed. Specifically, we will outline a number of design principles required to build critical rehabilitation systems and demonstrate how those principles can make promising progress in developing low-cost minimally intrusive rehabilitation systems that can be deployed in the homes of patients. Lastly, we conclude by discussing proposed techniques and reviewing opportunities for utilizing advances in 3D CV to design the next generation of rehabilitation systems.

6.2 Systems for 3D Depth-Capture

6.2.1 Marker-based motion tracking system

Figure 6.1 shows a typical example of a marker-based motion tracking system, where a series of calibrated cameras capture the location and trajectory of the required object with the help of active or passive markers placed on the desired object. These systems typically use 8–16 cameras, with infrared emitters, and the cameras capture the light reflected by the markers. Moreover, since the markers only reflect light rather than generate it, these systems can be classified as passive motion tracking systems [13]. The 2-D information captured by multiple cameras along with the known intrinsic and extrinsic parameters of cameras is merged to compute the 3-D position of the desired target. This rich 3-D information is then used to build analytic pipelines to understand and evaluate parameters of human activity for various applications such as for rehabilitation [14], Parkinson's analysis [15], gait analysis [16]. Consequently, commercial solutions have started becoming available such as Qualisys,[1] OptiTrack,[2] etc.

While the aforementioned systems have capabilities to produce rich data, enabling superior quality in the analysis, they are, however, bulky, expensive, and cumbersome to both setup and use. Consequently, markerless systems such as CODA and Microsoft Kinect [17] have

DOI: 10.1201/9781003328957-6

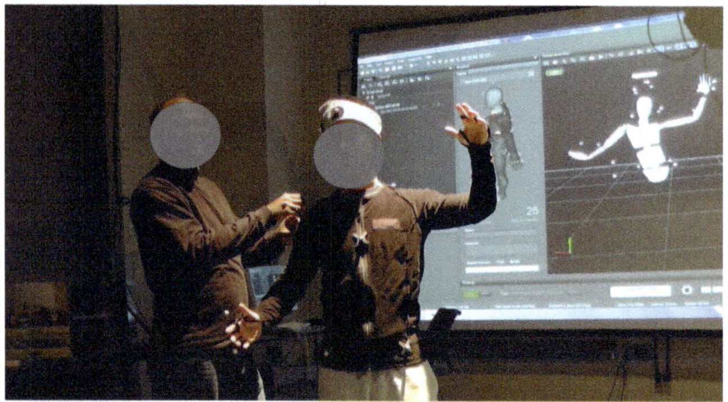

FIGURE 6.1
Demonstration of a typical marker-based motion capture system.

been widely used to develop affordable solutions with applications in multiple clinical contexts. In the past decade, there was a plethora of Kinect-based setups and their analysis [18]. A typical Kinect consists of RGB sensors and 3D depth sensors that use structured light to infer the depth of objects. In particular, these sensors project infrared light with a known pattern and then estimate the depth according to the deformation of the pattern. This information is combined with two classical CV depth estimation methods—depth from focus and depth from the stereo. Subsequently, Kinect uses a pre-trained decision forest to estimate the body positions in real time [17]. The portable real-time 3D skeletal tracking enabled by Kinect has been a game-changer for many rehabilitation programs and studies. A number of articles reviewing Kinect-based research have been published in the literature. Webster *et al.* focused on the formulation of exercises for rehabilitation and monitoring [19], while Da Gama *et al.* [20] focused on the analysis of different comparison methods for rehabilitation and introduced a new taxonomy from a clinical perspective. Hondori *et al.* [21], on the other hand, reviewed and highlighted the impact of Kinect in clinical settings, and Sathyanarayana *et al.* [22] reviewed articles through the lens of CV as opposed to a clinical perspective for different applications such as activities of daily living (ADLs) or fall detection and not just rehabilitation. Most recently, Ahad *et al.* [23] provided a review of CV-based action understanding for assistive healthcare. We refer our readers to these articles for in-depth reading and understanding of these approaches in rehabilitation.

6.2.2 Limitations

Movement quality-based rehab systems often rely on real-time feedback, via visual, sonic, or haptic feedback. To do this effectively, one needs a high frame-rate and low tracking error. While there are multiple benefits in using cost-intensive sensing modalities such as marker-

based motion-capture systems, including, high frame rate (100fps or more) readily accessible depth information, accurate 3D reconstruction of joints, ability to deliver immersive experiences, among others, the limitations are much strongly pronounced. The foremost limitation is that these are expensive. This makes it really challenging to develop solutions that can be implemented at home. Moreover, these systems are often limited to lab-based establishments with expert supervision. On the other hand, solutions, such as the Kinect, are inexpensive but suffer from low tracking quality, and low frame rate (less than 30fps), which is not typically good enough for detailed movement quality assessment. Most applications in movement-based rehab require us to compute detailed movement information from very short segments of actions (such as no more than two seconds long actions), where tracking errors and low frame rate make it extremely hard to derive good features. Due to these issues, many studies have found unclear evidence on the efficacy of these systems [19, 21].

6.3 Moving Towards RGB-based Systems

The most commonly cited reasons for adopting solutions that rely on more than simple RGB cameras in the past such as motion capture are primarily due to the low fidelity of CV approaches. Before the success of DL, tasks such as dense pose estimation from single RGB cameras were very difficult. Often, the earlier algorithms failed to generalize to test samples that are drawn from a different distribution than on which they were trained. For example, a common failure mode is the model's inability to distinguish hands and wooden tables [21] owing to similarities in skin and table color. As a result, the systems did not recognize hands reliably. To overcome these, the community started adopting more intensive sensing modalities such as Microsoft Kinect or motion capture systems that are generally occlusion-free and provide reliable data. However, with the phenomenal success of modern deep CV where the models are trained on large quantities of data, there is a renewed interest in using simple RGB camera-based systems. Such systems are adapted in stroke rehabilitation to assist clinicians with movement quality assessment. For example, real-time robust solutions for pose recognition such as OpenPose [24] have improved the fidelity of body joint estimation in these applications greatly. There are a number of advantages of rehabilitation setup that rely only on RGB systems. First of all, they are cheap and with minimal supervision these systems can be easily set up within a limited space at patients' homes. However, the variable behavior of stroke survivors and space constraints make it challenging to implement such systems.

Design principles

Constraint sensing: (1) space limitation; (2) two-camera requirements/optimum setup (sensing, modeling, analysis, and feedback issues)—interactive computing system—cyber-human system; (3) Who will set up the system? The person setting up the system is the therapist; (4) cost of the system; (5) lighting condition.

Constraint modeling and analyzing data: (1) very difficult to get stroke data; (2) two types of noise—due to patient variability and noise system; (3) unsupervised machine learning is not possible with the limited dataset. So, labeling from expert clinicians is essential. (4) Decision complexity of human movement analysis: humans give you a high level of labels. We need to establish a mapping to low-level features. Talk about hierarchy here.

Constraint feedback: (1) interaction incentive for the clinicians: increase the impact; (2) need for increased explainability and interpretability: segmentation, decision tree-based rating annotation tool; (3) strategy during the exercises

Activity space principle (1) common anchor points; (2) common objects; (3) how to get the composite features.

In the following paragraphs, we espouse core design principles required for building a system that only uses RGB camera setup. Our principles are necessitated for the following reasons:

1. **Limited and noisy data:** Healthcare data is most often limited and consequently modern DL systems tend to poorly generalize. This is because the success of these systems depends on the availability of large amounts of training data, which is not practical in rehabilitation. In addition, low camera resolution, variable focusing, and change in lighting conditions make DL models with RGB systems prone to errors in estimating body joints [25]. Activity space invariance can also enhance the noise.

2. **Need for increased explainability and interpretability:** One of the cornerstone requirements for the adoption of rehabilitation systems by both clinicians and patients is that the decisions from computational systems need to be explainable and interpretable by humans. Especially in stroke rehab, where multiple movement quality components are associated with one outcome, and clinicians need to understand the contribution of those factors. In the absence of this, the model is tantamount to a black box and such systems fail to enhance the clinician's ability to assess stroke survivors' performance. In addition, the black-box systems are inherently untrustworthy, especially for critical applications. In order to appreciate this better, let us consider a straightforward example. Let's say a patient is trying to utilize an automated system at their home for physical therapy, and an expert therapist will review the patient's progress after a certain time. The information and feedback available to both the stakeholders will be severely crippled if the output of the system is just one measurement, such as a movement quality score even more so when we do not understand why the system gave such a score.

3. **Decision complexity of human movement analysis:** Human activity is a highly complex system and so is the complexity of analyzing it. Therapists require years of training to effectively understand and propose new exercise regimes for their patients. Thus, their decision-making relies heavily on their subjective experience, which cannot be easily represented in a form that can be leveraged by computational systems thus shunting its performance in both accuracy and interpretability measures.

In this section, we will talk about the development of a pure RGB camera-based interactive rehabilitation system called SARAH (*Semi-Automated Rehabilitation At the Home*). SARAH is currently used for upper extremity stroke rehabilitation and satisfies all the above constraints having three core design principles: (1) co-design of the activity space with Algorithms, (2) construction of an ensemble of different machine-learning models, (3) integration of therapists' knowledge in the computational system to realize a true cyber-human system. We show that through these core design principles (DP) one could realize an effective and interpretable system in the form of SARAH that can empower clinicians.

6.3.1 DP1: Co-design of Activity Space with Algorithms

Achieving activity space invariance is key to reduce variability in the captured data and improving the performance of machine learning models. Generally, camera calibration and activity space standardization can make a system invariant to noise discussed in [25]. In this section, we will demonstrate the efficacy of the activity space invariance in the SARAH setup.

6.3.1.1 Description of SARAH Setup

It consists of two RGB video cameras, a tablet computer, a flexible activity mat, and 15 custom-designed 3D printed objects, as shown in Figure 6.2.

The objects are designed to support a broad range of perceived affordances [26], that is, they can be moved, grasped, and manipulated in a wide variety of ways [27]. Each object is designed to be distinct in terms of both size and color to assist identification of objects by patients and

FIGURE 6.2
SARAH system and objects setup (left); SARAH activity mat (right).

to make identification and tracking using CV methods easier. The four primary activity regions (near and distal ipsilateral and contralateral) are delineated through increasingly colored high-contrast guidance lines, which we screen-printed on the flexible activity mat. Moreover, we also added four rows of circles on the mat to aid the CV systems for detecting boundaries between different activity spaces, which in turn helps maintain consistency in analyzing patient activities. As we only rely on commercially available inexpensive RGB cameras, the system can be easily installed on a regular kitchen or living room table. In addition, we designed a custom application for the patients to initiate, activate, and control the cameras through their tablet computer.

6.3.1.2   *Activity Space Calibration*

To achieve invariance in the captured data, we first adopted denoising techniques and activity space normal-ization methods [25]. We can mitigate the noise to some extent; however, the necessity of a calibrated activity space was prominent. The calibration pipeline of the SARAH system is shown in Figure 6.3. The first step to calibrate the activity space is to set up the table and the mat according to the instructions from an intuitive interface. Then, the cameras are placed via a computational approach to ensure that the requirements to achieve invariance across captures are met for a smoother analysis of the captured video data. This is a fiducial marker-based approach that utilizes two fiducial markers per camera. The first one is used to calculate the distance between the edge of the mat and the camera. The second one is used to align the camera frame correctly. The clinicians are guided thoroughly during this process with a both visual and textual display of instructions to ensure that the camera is being moved as required. The next steps involve detecting the patient bounding box (E, F, and G) and automatically segmenting the activity space into five

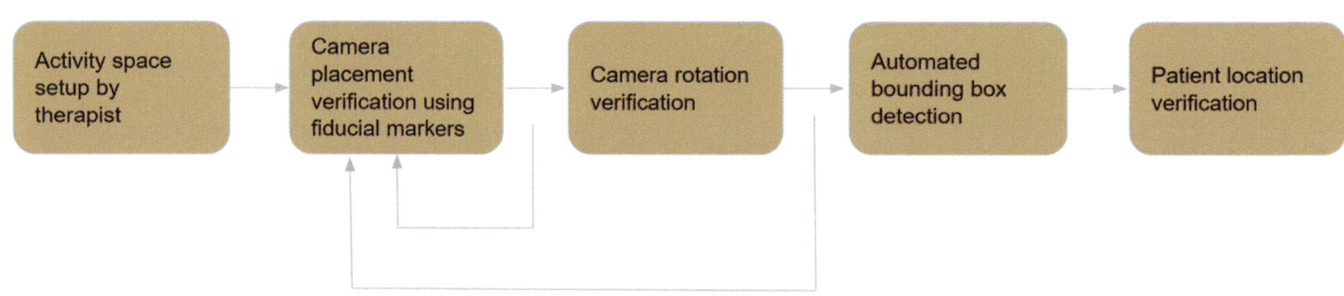

FIGURE 6.3
The activity space calibration pipeline of the SARAH system.

TABLE 6.1

Experiments with Unimpaired Invariant Data

	Train	Test
Experiment 2a	Trained model using impaired	Unimpaired data
Experiment 2b	75% impaired 25% unimpaired	Test set using impaired
Experiment 2c	75% impaired 25% unimpaired	60% impaired 40% unimpaired

TABLE 6.2

Frame-wise Segmentation Results for Experiments 2a, 2b and 2c in Table 6.1

	Experiment 2a	Experiment 2b		Experiment 2c	
	Mean	Mean	STD	Mean	STD
ACC	80.52	86.01	1.28	87.85	0.58

regions (A, B, C, D, and H) demonstrated in Figure 6.4. The details of these steps will be presented in a separate article.

In [28], the efficacy of activity space calibration on the accuracy of segmentation models is demonstrated. We have captured unimpaired movement using the calibrated setup. We performed three experiments using mixes of impaired patient data and unimpaired subject data and compared the results to the experiments from [25]. The data mixes of the three new experiments are shown in Table 6.1. In Table 6.2, we show the comparative analysis for frame-wise label prediction for Experiments 2a, 2b, and 2c from Table 6.1. As evident from the results, Experiment C has the highest accuracy for per-frame label prediction. In Experiment 2a, the models were trained using noisy patient data captured in a variant activity space. However, we evaluated using unimpaired subject data captured in an invariant space. We observed that the models achieved 88.5% accuracy in segment detection of unimpaired subject data even when the model is trained using only noisy patient data.

The focus of our system is to integrate expert knowledge with purely RGB data-driven CV algorithms to produce a coarse real-time automated assessment of the movement of stroke survivors during therapy at the home [29]. Upon the execution of each task, this assessment can provide the patients with high-level feedback on their performance and results. This feedback can enable patients to self-evaluate and help them plan their next attempts at the task. Remote therapists will be provided with daily summaries of the assessment reports so that they can assess overall progress, make adjustments to therapy structure and also provide valuable feedback and directions to the patient directly through the tablet. We incorporated continuous monitoring accompanied by feedback along with avenues for therapists' customization of therapy to suit patients' learning styles and needs so as to increase the likelihood of patients adopting and utilizing home-based rehabilitation systems [30].

It has been shown that stroke survivors who score above 30 on the Fugl-Meyer test and can initiate even minimal movement into an extension of the elbow, wrist, and digits when discharged from the clinic post-stroke can show significant improvement in function through repetitive training [31–33]. Our SARAH system is optimized for at-home therapy of stroke survivors with

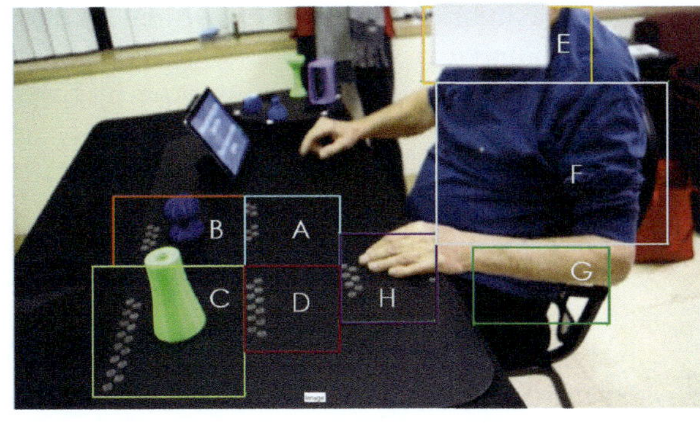

FIGURE 6.4

Drawn bounding boxes on the activity space and patient upper body; there are five bounding boxes on the mat and three on the patient's body.

moderate and moderate-to-mild impairment. The system promotes active learning by the patient and importantly the generalizable design of the objects and tasks of SARAH encourages the patient to actively map the task to multiple ADLs. These active learning characteristics of the SARAH system combined with the variety of included tasks in the system make the system feasible for in-home rehabilitation lasting two to eight weeks [27, 29].

To this end, inspired by clinical measures for rating rehabilitation movement, the SARAH system utilizes a standardized activity space with eight well-defined subspaces that are drawn as bounding boxes on the video capture of therapy (see Figure 6.4).

To generate high-quality labeled training data, we worked with the therapists and custom-designed interactive video rating tools to aid expert therapists reveal and reflect on their rating process and internalized (tacit) rating schemas [29].

In order to better understand and rate the complex human activity, especially of stroke victims, therapists typically segment the task into multiple segments intuitively, and depending on the performance of all those segments individually, they provide a task-level evaluation. As most therapists use intuition/experience for segmentation, the segmentation vocabulary in itself is not standardized. We worked with therapists to produce such a standardized vocabulary for the segmentation of a therapy task and they consist of

1. Initiation + Progression + Termination (IPT)

2. Manipulate and Transport (MTR), and

3. Release and Return (RR).

As an example, a drinking-related task can be described by the following codification: subject reaches out and grasps a cone object (IPT) and brings it to their mouth (MTR), then returns the object to the original position (MTR), and releases the object and returns the hand to the rest position (RR). To achieve real-time segmentation and thus evaluation, most therapists usually focus on a limited number of kinematic/composite features (e.g., velocity of the action, etc.) using their experience. This process too is not standardized and thus different therapists could look at different features for evaluation. Note that this nonstandardization of segment vocabulary and features used for segmentation and evaluation poses interesting challenges in designing a computational system. To alleviate this, we worked with therapists to define a consensus-limited set of composite movement features that are important when assessing the performance of each segment in our model [29].

6.3.2 DP2: Implicit Integration of Expert Knowledge Through a Hybrid ML Ensemble

While one building block of SARAH is the acquisition of movement data through home-based low-cost cameras, for the system to be successful and used by many patients the system has to generate accurate and robust assessments automatically. Generation of high-fidelity automated assessment is highly challenging because the processes used by therapists are largely implicit and not well standardized [34], and this makes the design of SARAH fundamentally different than the design of a CV/DL techniques for standard vision tasks. It is pertinent to note that clinicians are often trained to use validated clinical measures (e.g., Fugl-Meyer and Wolf motor functional test), utilizing a small range of quantitative scales (0-2, 0-3, and 0-5, respectively) for assessing the performance of functional tasks that map to activities of daily life [35, 36]. However, this high-level feedback can not be easily translated into feedback for specific aspects of movement or for kinematic features extracted through computational algorithms. More importantly, it is also well-known that different therapists focus on different aspects of movement [37–40] thus making the development of approaches for structuring and customizing therapy partly based on subjective experience rather than a standard quantifiable framework [34, 38]. Consequently, this creates challenges in the collection and generation of large-scale data on the structuring, customization, and net effect of therapy in everyday life [41]. The aforementioned SARAH system is designed to overcome these limitations through its interactive design, and custom objects that map to multiple ADLs, providing avenues for real-time automated feedback along with customization and structuring of therapy through experts. Crucially, to overcome the limitation of subjective experience, we collect and label data through standardized procedures from therapists with our custom interface.

In contrast to a standard data-driven approach, we focus on a hybrid knowledge-based data-driven approach for automated assessment of human movement in the home. Note that, this hybrid approach is well-suited for applications such as rehabilitation as the data is typically limited, noisy, and highly variable as the patients have different degrees of impairment. Another requirement in critical healthcare applications is that the assessment process and outcomes need to be explainable and compatible with the assessment approaches of therapists. This enables remote therapists to better design, structure, and customize therapy for individual patients according to their learning styles and progress. Our approach at its heart attempts to leverage the expert rating

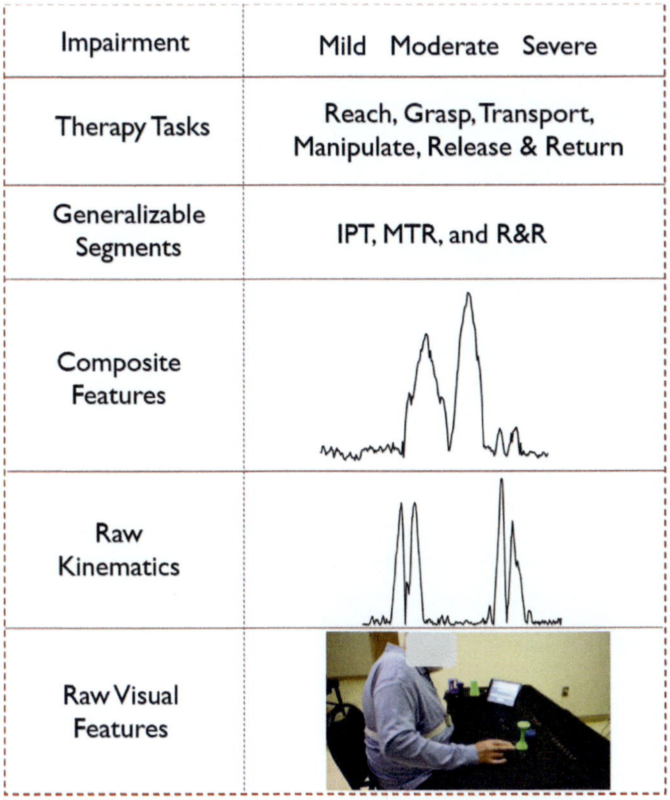

Impairment	Mild Moderate Severe
Therapy Tasks	Reach, Grasp, Transport, Manipulate, Release & Return
Generalizable Segments	IPT, MTR, and R&R
Composite Features	
Raw Kinematics	
Raw Visual Features	

FIGURE 6.5
Hierarchical representation of the cyber–human system; (from the bottom) a video frame as an example of raw visual features, velocity profile as an example of raw kinematic features, composite feature for task 5 generated using PCA, segment labels, different types of tasks and impairment levels.

process to inform the structuring and improved performance of computational algorithms.

Before we proceed with the construction of different ML models, we will outline why we need them through the help of a hierarchical framework for cyber–human automated movement assessment.

1. **Framework:** In the proposed hierarchy (see also Figure 6.5), we have five layers. Listing them from top to bottom-

   a Overall task rating
   b Segment rating
   c Composite feature assessment
   d Raw kinematics and
   e Raw RGB images.

Through Figure 6.5, we remark that the expert knowledge becomes less observable and less standardized as we move down. Pure data-driven computational knowledge on the other hand works in reverse with much higher confidence in lower levels while being very uncertain at the top level as the assessment relies on an expert therapist's knowledge and experience, which are not easily quantifiable [42, 43]. While we present the hierarchy suited for upper extremity stroke rehabilitation, with minor modifications, it can potentially be used in developing automated assessments for many other rehabilitation contexts such as lower-extremity training.

Achieving accurate and reliable movement assessment automatically requires the development of solutions for several subtasks and solutions for the subtasks of those subtasks. For example, one of the cornerstone tasks for movement assessment is temporally segmenting a video into segments defined above (IPT, MTR, and RR). Even though there are a number of DL-based temporal video segmentation algorithms [44] proposed in the CV literature, almost all of them are fully data-driven and were trained on large amounts of labeled data. However, as explained, most healthcare data is limited and noisy. This limits the efficacy of the popular approaches to rehabilitation data straight out of the box or even after fine-tuning. In the following sections, we instead show that we can leverage the observable expert knowledge at the top levels of the hierarchy to better inform and improve the performance of computational algorithms that use raw visual RGB frames or raw kinematic features. Furthermore, we also leverage the prior knowledge of the task design, its expected sequence, and objects present in the scene to post-process the outputs of ML solutions to achieve reliable results.

2. **Methods:** Since a detailed exposition of the methods is beyond the scope of this chapter, we instead focus on explaining the core principles of building an ensemble of different kinds of ML systems and exploring the corresponding choices. We consider the crucial problem of temporally segmenting a given RGB video and show how our design principles help achieve more reliable, robust, and accurate performance.

Our ensemble system consists of three different machine learning models each incorporating different levels of expert or prior knowledge into them and they are (1) Hidden Markov Models (HMMs), (2) Transformers, and (3) Temporal convolutional neural networks (CNN). The state-time characteristics of our hierarchy (i.e., segment state machine) motivate our use of automated segmentation through a HMM. Similarly, since transformers

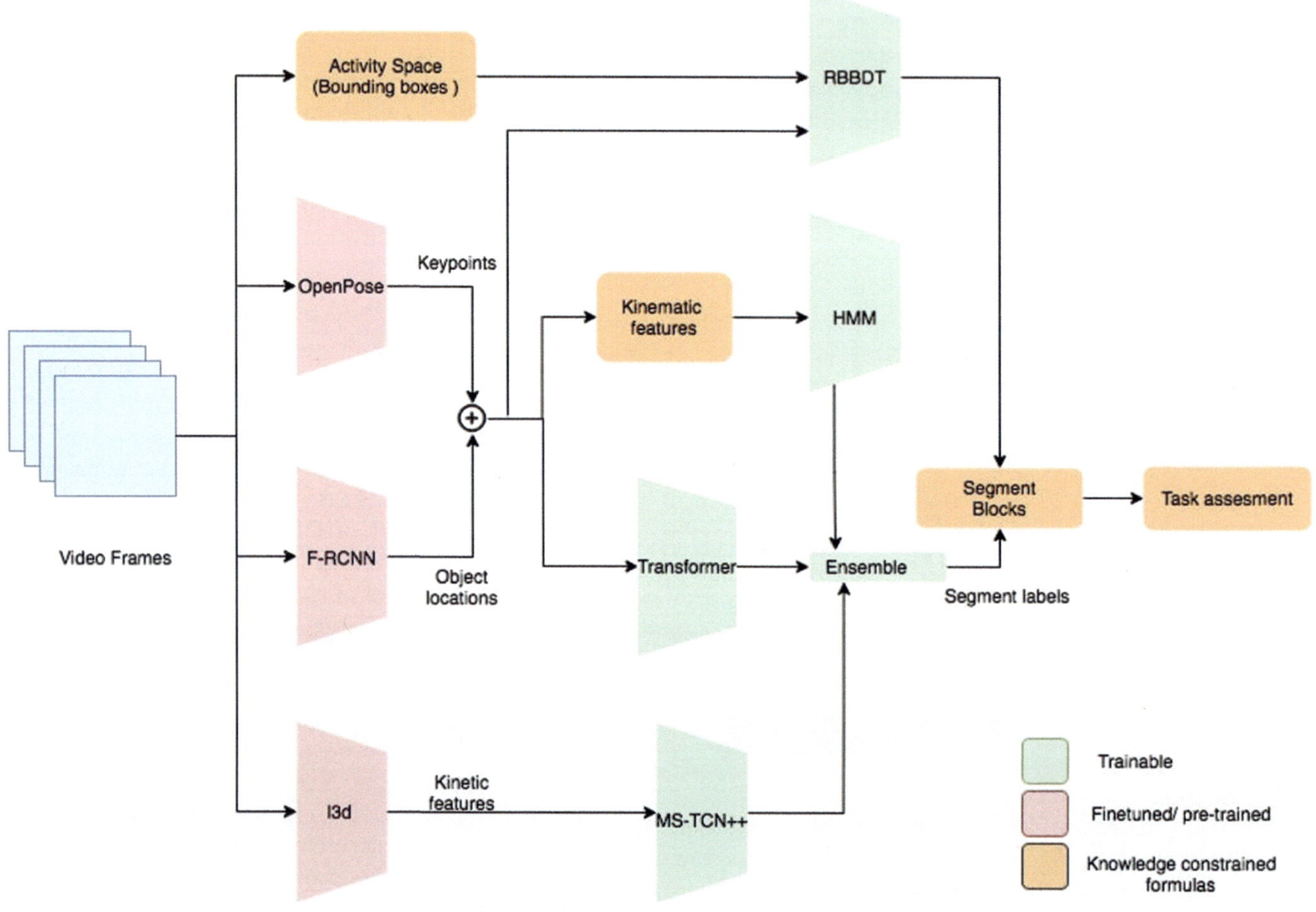

FIGURE 6.6
Block diagram of the proposed analysis framework. Keypoints and object locations are extracted from OpenPose and Faster-RCNN followed by Kalman filtering, respectively. RBBDT along with the model ensemble determines the segment blocks from the per-frame segment labels, which are then used to assess the performance of the subject for the given task.

[45] have been shown to achieve remarkable performance in processing sequential data, we utilized them to operate on the low-level action features (tracking of individual body joints, object tracking, etc.). Note that both these algorithms do not directly rely on RGB pixel values but on features computed on top of them. We utilize a temporal transformer architecture to better capture the hand–object interaction, thus the input to this algorithm was a set of kinematic features such as the velocity of the keypoint of hand and the location of the object in the frame, etc. To compute the features of the object, we utilized a fine-tuned faster-RCNN [46] pre-trained on the COCO [47] on our manually annotated dataset as our object detection model. To compute the kinematic features of the hand, we leveraged OpenPose [24] to detect the keypoints and then used analytic expres-

sions to obtain features. We also considered MS-TCN [48, 49] a state-of-the-art video segmentation algorithm that outputs the start and end times of different segments with a sequence of RGB frames as input. MS-TCN relies on a novel temporal CNN architecture to handle the temporal continuity that exists between different frames of the video. MS-TCN has been shown to outperform other video segmentation networks for human activity and this makes it very appealing in our scenario too. MS-TCN operates on features obtained from an Inception3D [50] model which is trained on large-scale kinetics dataset [51]. Since the I3D model has been trained on this large-scale dataset, features extracted through the model can capture representations of complex human activity, including human–object interaction, thus making the model an appropriate choice for composite feature extraction for our analysis. Each of these three different approaches showed different strengths and weaknesses, thus making an ensemble approach a natural choice to finalize the automated segmentation as well as segment classification decisions. We now expand on the design of our ensemble.

3. **Ensemble:** Through a weighted approach, we fuse the outputs of data-driven models with outputs from expert knowledge-constrained models. The HMM algorithm uses kinematic features and design priors that are constrained by expert knowledge by leveraging the state transition probabilities guided by expert knowledge. The predictions from the HMM thus encode information about the segments and task level of the movement analysis hierarchy shown in Figure 6.5. While methods such as Transformer and MSTCN++ use kinematic and composite features and perform better at the frame level segment label, they however fail to encode information about the segments and the tasks. To combine the strengths of each approach, we employ a standard weighing scheme to fuse these outputs, that is, $\hat{y}_j = \sum_i W^i P_j^i$ where W_i is the weight assigned to each model's prediction P^i for the jth sample. To determine the weights W_i, while we could do a sophisticated tuning or learn them, we instead chose a much simpler approach and conducted a grid search for each of them in the range of $[0.1, 0.2, 0.3, \ldots, 1.0]$. Thus, for a three-model ensemble, we run the evaluation $10 \times 10 \times 10 = 1000$ times for each combination of weights. We experimented with three ensemble schemes to find the combination that provides the most robust results on validation data.

4. **Denoising Ensemble Predictions:** While the ensemble produces improved performance on per-frame segment labels, however, due to miss-classifications in certain frames, the transitions and thus the accuracy of a segment's start and end times decrease. Apart from this, we also filter out false predictions through a Gaussian filter for each transition model that implements these priors. In Figure 6.7, we illustrate the per-segment filters for a 3×3 transition matrix model. This can be mathematically expressed as $N_s(\mu_s, \sigma_s)$, where μ_s and σ_s denote the mean duration and standard deviation of each segment, s, respectively, calculated from the training data. For a test data of length, T, we calculate the posterior distribution and use a 75% confidence cutoff to get the upper and lower bounds for each segment. We only accept a predicted transition, if it falls within the bounds. Moreover, since all task segments in our system are at least greater than 1 second, we pose

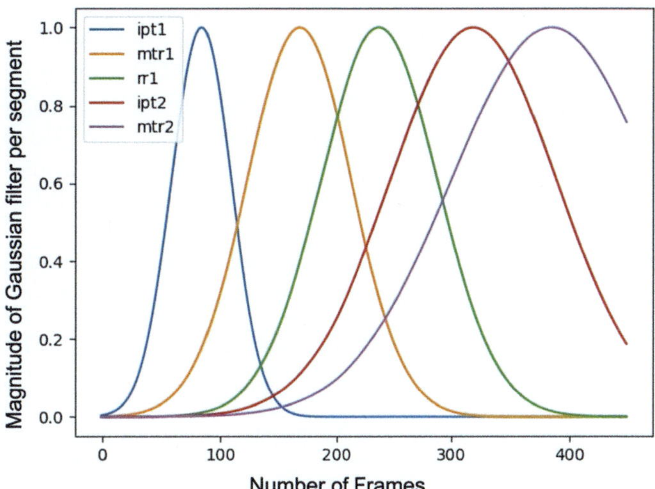

FIGURE 6.7
Gaussian windows per segment to filter out false transitions outside the first standard deviation mark.

a minimum 20 frame window constraint on the predictions to filter out miss-transitions, that is, any transitions within 20 frames are discarded. Through these steps, we achieve a significant reduction in false transitions.

6.3.2.1 *Evaluation of SARAH*

1. **Data:** We recorded videos of nine (seven men and two women) stroke survivors using our setup performing 12 different tasks. Each patient was asked to repeat the tasks four times. These stroke survivors had differing amounts of impairment- from mild, to moderate to severe impairment as measured by standard clinical scores such as Fugl-Meyer tests. Thus, the overall dataset can be viewed as a representative dataset for different movement challenges. In total, we collected 610 videos, out of which we filtered out 206 videos where due to severe limb impairment, for certain tasks, subjects dropped multiple objects or videos had multiple segments implying that the exercises were incomplete. Including these in the training, data would have severely skewed the dataset and thus make the models suffer a strong performance gap. We have, however, included some of them in the test set. All videos are manually annotated to generate ground truth data.

2. **Results:** We review some of the important findings from our experimental analysis. More results and in-depth ablations can be found in [25]. All experiments are repeated five times with five different random seeds. For each seed, available 404 videos are split into 370 train and 34 test

TABLE 6.3
Per-frame Segmentation Results Using the Proposed Different Machine Learning Models Along with Results from Ensembles. Ensemble 3 Performs Signifying the Complementary Nature of Knowledge Learnt by the Three Models. (Ensemble 1- HMM and Transformer, Ensemble 2- HMM and MSTCN++, and Ensemble 3- HMM, Transformer, and MSTCN++).

	Accuracy	Precision	Recall
HMM	77.82 ± 2.88	78.60 ± 2.47	78.26 ± 2.76
Transformer	76.34 ± 2.34	70.11± 2.59	68.36 ± 4.10
MS-TCN++	81.91 ± 2.38	78.71 ± 4.25	78.82 ± 3.7
1-4 Ensemble 1	81.01 ± 1.99	77.28 ± 4.38	77.88 ± 4.46
Ensemble 2	83.76 ± 2.77	81.40 ± 3.92	81.57 ± 3.66
Ensemble 3	85.08 ± 2.14	84.34 ± 2.61	84.6 ± 2.68

videos, and the split is performed such that for any task, the same patient ID and session ID appear either in train or test but not in both. This was needed because our subjects did each task multiple times in the same session. Including an attempt from the same session in the train set and another attempt in the test set would lead to an unfair evaluation. We report three metrics—frame-wise accuracy, precision, and recall for all the methods including ensembles to evaluate their performance for the task of temporal segmentation.

From Table 6.3, it can be seen that among the individual models, MS-TCN++ achieves the best performance in terms of *per-frame classification error*. This is because MSTCN++ [49]—a data-driven technique—primarily relies on the RGB data layer of our hierarchy and on kinetic features obtained from I3D [50] model, which is trained through generic videos of human activity. These generic composite features are not fully observable and thus not directly related to upper extremity functional tasks. Importantly, this algorithm cannot incorporate the constraints imposed by the design of the system easily resulting in large errors in the order of segments. For example, it can classify the beginning frames of a task as belonging to an MTR segment when all the SARAH tasks start with IPT. Due to these ordering errors, MSTCN++, even with its superior performance, cannot be solely relied upon for assessing the completion of either the segment or the task as a whole. The Transformer operates on only expert-selected features derived from select keypoints on the hand and wrist rather than on all the keypoints. It can thus be said that some form of prior expert knowledge is used to constrain the Transformer. However, due to the large number of parameters in Transformer architectures coupled with very limited and highly noisy data, this model achieves the lowest segmentation performance. HMMs, on the other hand, use kinematic features that are prominent in characterizing upper limb functional movement [52] and integrate segment and task layer information through the customized transition matrices. The prior transition matrices that signify the probabilities of transition from one segment to another are constructed through expert knowledge and the size of the transition matrix depends on the state machine of the exercise. While this model slightly underperforms compared to MS-TCN++, it does not suffer from the segment ordering issues because of the carefully constructed priors from the therapist's knowledge. Strikingly, the ensemble with its constituents as all three algorithms significantly outperforms individual models showing the importance of the fusion of algorithms that have different sensitivities to different layers of the hierarchy. This result also highlights the complementary nature of our algorithms. In summary, by utilizing different machine learning models each with various levels of design and prior constraints, we achieve more robust and reliable performance. These observations will hold beyond the case of upper extremity rehabilitation and will be generally applicable for automating movement tasks across different applications.

6.3.3 DP3: Explicit Incorporation of Therapists Knowledge

After discussing our approach to implicitly incorporate expert knowledge, we now make a case for additional regularization by explicitly encoding therapist knowledge. We propose to do so by developing a decision tree that would operate at the highest level of hierarchy as shown in Figure 6.6. Before we proceed to outline the construction mechanism of such a rule-based system, we quickly introduce another important task in stroke rehabilitation—automatically assessing if and to what degree an exercise has been completed by a patient. To achieve this, one needs to organize the per-frame segment labels into segment blocks. To do so, we leverage therapist expertise in the form of a decision tree for grouping segment samples into segment blocks (as embedded in the SARAH system design) and then assess the order, duration, and continuity of the blocks.

Rule-Based Binary Decision Tree (RBBDT) is developed to integrate the process that therapists use to segment the patient movement that we observed while developing the SARAH system with therapists and subsequent labeling of data into multiple segments to produce ground truth data. We have noticed that therapists, instead of classifying each frame of movement, utilize a few

TABLE 6.4

List of RBBDT Features Calculated Using Patient Keypoints, Object Co-ordinates, and Center of Eight bounding Boxes

Index	Feature Notation	Definition
1	d_{Fl}	Distance between upper torso, F, and the limb
2	d_{Gl}	Distance between the lower torso, G, and the limb
3	d_{El}	Distance between head, E, and the limb
4	d_{hl}	Distance between limb and hand bounding box, h
5	$d_{Al}, d_{Bl}, d_{Cl}, d_{Dl}$	Distance between limb and bounding box A, B, C, and D
6	d_{Fo}	Distance between upper torso, F, and the object
7	d_{Go}	Distance between lower torso, G, and the object
8	d_{Eo}	Distance between head, E, and the object
9	d_{ho}	Distance between object and hand bounding box, h
10	$d_{Ao}, d_{Bo}, d_{Co}, d_{Do}$	Distance between the object and bounding box A, B, C, and D
11	d_{lo}	Distance between limb and object
12	d_{oo}	Distance between two objects
13	$\frac{d}{dt}()$	First derivative of the above features

key events to find transition point candidates between segments. These events are primarily focused on three relationships: (1) patient–object interaction, (2) patient–activity space relationship, and (3) object–activity space relation. As there is a lot of variability, especially in the case of stroke victims on how they perform functional tasks, we observed that therapists organize and focus on generalizable regions that are robust to variation. To capture those regions and to model relationships between the object(s), patient, and generalizable activity regions, we recreate the activity regions as eight bounding boxes on each frame of the video. These bounding boxes are shown in Figure 6.4. We use these eight bounding boxes to calculate the space–patient, space–object, and patient–object relationships that appear in Table 6.4. To compute these features, we re-utilize the keypoints obtained from Open-Pose and object bounding boxes achieved through our fine-tuned faster-RCNN. However, these features that encode distances do not exhibit a simple statistical pattern (e.g., Gaussian) across different patients for each type of segment and hence dropping their utility to train the HMM algorithm, but they become useful under a decision tree framework to detect segment blocks since they explicitly encode features used by experts. As a result through RBBDT, we observe a significant boost in segment block detection accuracy. Note that, through RBBDT we are not interested in improving the per-frame accuracy (this is improved through ensemble); rather, we use it to identify the correct transition points. In order to obtain the thresholds so as to split branches (to estimate if a given frame is a transition point or not) in the Tree, we could employ a CART (Classification and Regression Tree) [53]. However, as we are operating with low and noisy data, we chose to manually tune these thresholds supported through training data. This tuning is required to re-order

TABLE 6.5

Segmentation Block Results

Split	Segment-wise Accuracy	Precision	Recall
1	99.33	99	98.5
2	98.48	98.18	97.04
3	98.66	98	97
4	100	100	100
5	98.66	98.5	97.5
1-4 Mean	99.03	98.73	98.01
STD	0.56	0.71	1.13

the features in descending order of observability and error for each type of transition prediction with the most observable and accurate feature coming first in the decision sequence. If a candidate sample of a feature stream meets this threshold condition, the confidence value of that candidate as a transition point increases. More details are presented in [25].

In summary, the RBBDT aims to only find transition points organizing the data stream into segments blocks that are feasible (could have been performed by a stroke patient), while minimizing missed transitions and false transitions between blocks. This module completes our system which has multiple models with the incorporation of decreasing amount of therapist's knowledge (from top to bottom in Figure 6.6). It can be seen from Table 6.5 that through the incorporation of the proposed denoising and RBBDT, around 99% of the segment blocks can be labeled correctly. Note that, automatically assessing the completion and order of segment blocks cannot be done solely by ensemble models as they have many false-positive blocks and transitions. We introduced RBBDT

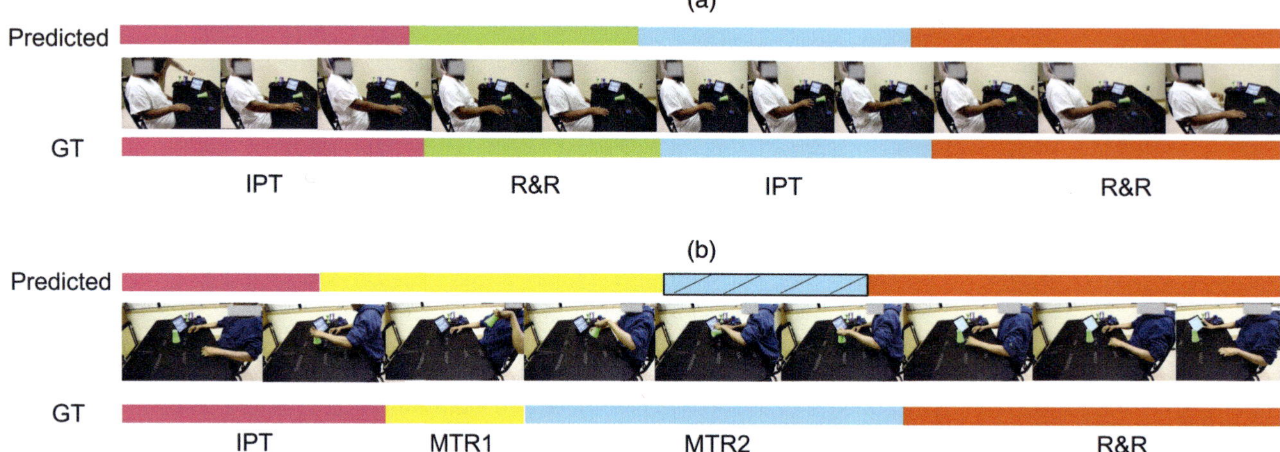

FIGURE 6.8

Qualitative representation of the task segmentation; (a) almost 97% accurate segmentation using the proposed ensemble approach; (b) A case where the algorithm misses the segment MTR2 using the ensemble model and mistakenly identifies it as MTR1. The striped blue zone is the correction in the prediction after incorporating the block-based segmentation using RBBDT.

an algorithm that is heavily biased by expert knowledge to connect the segmentation layer to the task assessment layer. The proposed RBBDT algorithm successfully incorporates important unimpaired and impaired movement priors for segment block transition points in one integrated decision sequence, thus greatly improving the performance of task completion assessment.

Through Figure 6.8, we also show the qualitative performance of the proposed ensemble approach. In Figure 6.8(a), we show the segmentation performance for exercise 1, which is one of the easiest tasks among the 15 tasks. As evident from the illustration, the proposed ensemble model is almost 97% accurate in this example. In Figure 6.8(b), we show the performance of task 5, which is a more complex task. In this case, the model fails to predict one of the transportation segments as the patient drops the object at the end of the "MTR2" segment resulting in distortion of features. However, after incorporating the block-based segmentation using the design constraint denoising and RBBDT, we were able to recover more than 50% of the "MTR2" segment as indicated by the striped blue region in Figure 6.8(b).

Summing up, unobtrusive, low cost and portable rehabilitation systems are increasingly in demand, especially since remote healthcare continues to become popular. However, based on available studies and our own experience, we see that portable systems often produce noisy and highly variable kinematic data. Through the design principles that we have described and by careful construction of a hierarchical model which fuses expert knowledge-based approaches with data-driven techniques, we can aim for robust segmentation performance and high-fidelity task completion assessment. The outputs from the proposed system can be used to monitor whether rehabilitation exercises have been completed by the patient correctly and with ease. This system can also drive quick feedback to the patient to incentivize good performance, encourage them to repeat the exercise, or move to a more difficult task. Toward this end, the feedback should be generated in real-time.

Note that the proposed knowledge-driven assessment realizes a hierarchical analysis in a bottom-up manner (features conditioning segment frames, conditioning segment blocks, and conditioning task completion). Therapists utilize a similar hierarchical analysis in a top-down manner (task conditioning segments, conditioning composite features, conditioning raw features) so as to leverage their heuristics about functional task performance. As a result, our analysis can be used to send periodic reports (e.g., daily) of training results to a remote therapist. As we make the incorporation of interpretability an integral part of our design, the computational analysis will be highly compatible with the therapist assessment heuristics thus enabling a seamless utilization by therapists so as to structure the next day's therapy. The feasibility of the low-cost semiautomated rehabilitation at the home using the SARAH system will allow the collection of many more videos used to further train machine learning algorithms for several component problems.

While throughout this section, we used SARAH as a case study to expand and demonstrate the benefits of the proposed design principles, we mention that these principles are widely applicable, albeit with minor alterations, for assisting automation in many other movement rehabilitation contexts (such as lower extremity training) as well.

6.4 Monocular 3D Reconstruction of Hands and Objects to Aid Rehabilitation Systems

Having outlined the principles behind building a low-cost, minimally intrusive home-based systems for rehabilitation that only uses RGB cameras, we now turn our attention to understanding their limitations and studying possible solutions harping on the advances in 3D CV. A foremost limitation of using pure RGB cameras is the lack of depth information. As a result, analysis in the ambient 3D space in which the exercises are performed is not feasible. The second important limitation is low-fidelity analysis due to occlusions. They can happen for various practical reasons, such as, patients' hands may obstruct the object being grasped, resulting in inaccurate object detection; other people may be present and block a particular view; or patients may drop the objects. Although these limitations are not new, past research attempts in this approach relied on exploiting depth information and consequently 3D representations using either high-quality and extremely expensive marker-based motion capture systems or low-quality systems like Microsoft Kinect. (See Section 6.2 for a review of rehabilitation approaches that rely on these systems and their corresponding limitations). In this context, it becomes imperative to develop novel solutions that can both capture and utilize 3D information while still operating under the low-cost, minimally intrusive home-based setting.

A tempting solution could be to utilize two or more RGB cameras and use stereo reconstruction to get 3D information [54]. However, in our experiments, when we positioned one camera to the side of the patient and another directly above providing a top-view, we observed that standard algorithms failed to yield satisfactory results. It is partly unsurprising because both cameras have an orthogonal field of view. Attempts to alleviate this by adding more cameras would make the system bulky and expensive. Secondly, understanding patients' movement quality through a fine-grained 3D analysis will be challenging with standard reconstruction techniques as they require known corresponding points from both camera views. Thirdly, since they do not take human anatomy into account the resulting 3D keypoints could be erroneous and anatomically impossible. Lastly, even though these methods are shown to be successful in some applications, joint positions are generally not sufficient to describe complex human behavior that involves fine motor skills, especially for those who suffer from extremities caused due to stroke.

Recently, DL-based 3D vision has emerged as a strong alternative in enabling high-quality dense reconstructions of human hands from a single view, albeit for unimpaired people. These approaches rely on having high-fidelity ground truth data, which is made possible by expensive range scanners [55]. Obtaining the ground truth data of 3D motion of hands is, however, nontrivial and a number of research efforts have been made in this direction since decades [56–60]. One of the successful methods for obtaining dense 3D annotations of human bodies is proposed by Loper *et al.* in their seminal work SMPL [55]. Importantly, apart from collecting a large amount of aligned scans of different people in different poses, they learn a statistical skeletally driven human body model that can capture large variations in body shapes and poses. Romero *et al.* extend this to build MANO [61], a low-dimensional parametric model of the human hand, that is learned from examples and is easy to pose and fit to data. It has also been shown that MANO outperforms previously proposed hand models, which include models that relied on shape primitives [62–65], sum of gaussians model [66, 67] an the sphere-mesh models [68]. Ever since MANO, there has been an explosion of DL-based methods for 3D reconstruction of hands from a single RGB image. This is mainly because, in contrast to the extremely difficult challenge of creating a mesh directly, it is now sufficient to regress to the MANO parameters given an RGB image, which can be addressed relatively easily by fine-tuning pre-trained CNNs. The CNN parameters are optimized with the objective of producing the optimal MANO parameters such that the distance between predicted and ground truth meshes is minimized. In this framework, a number of methods and novel loss functions have been proposed [69–73]. There has also been an increase in the RGB to 3D datasets for hand reconstruction such as [74–76]. Note that, almost all of the publicly available datasets consider only people who have no impairments. Therefore, a direct application of the machine learning models developed on these datasets to impaired stroke subjects will produce poor results. To address this issue, we propose a novel approach to develop a database of 3D motion capturing of such individuals who suffer from extremities.

However, similar approaches to that of SMPL or MANO cannot be easily developed as they require extensive technical setup, including the use of multiple cameras to obtain the scans. In light of this, we suggest adopting the mean hand template of the MANO model, to develop, in collaboration with a 3D animation artist, an annotation tool. The 3D animation artist helped us with rigging the MANO hand model with bones and armature in Blender allowing for manipulation of each finger position and rotation based on RGB images. An example of such a process is shown in Figure 6.9. Instead of labeling every single frame, only keyframes will be manually

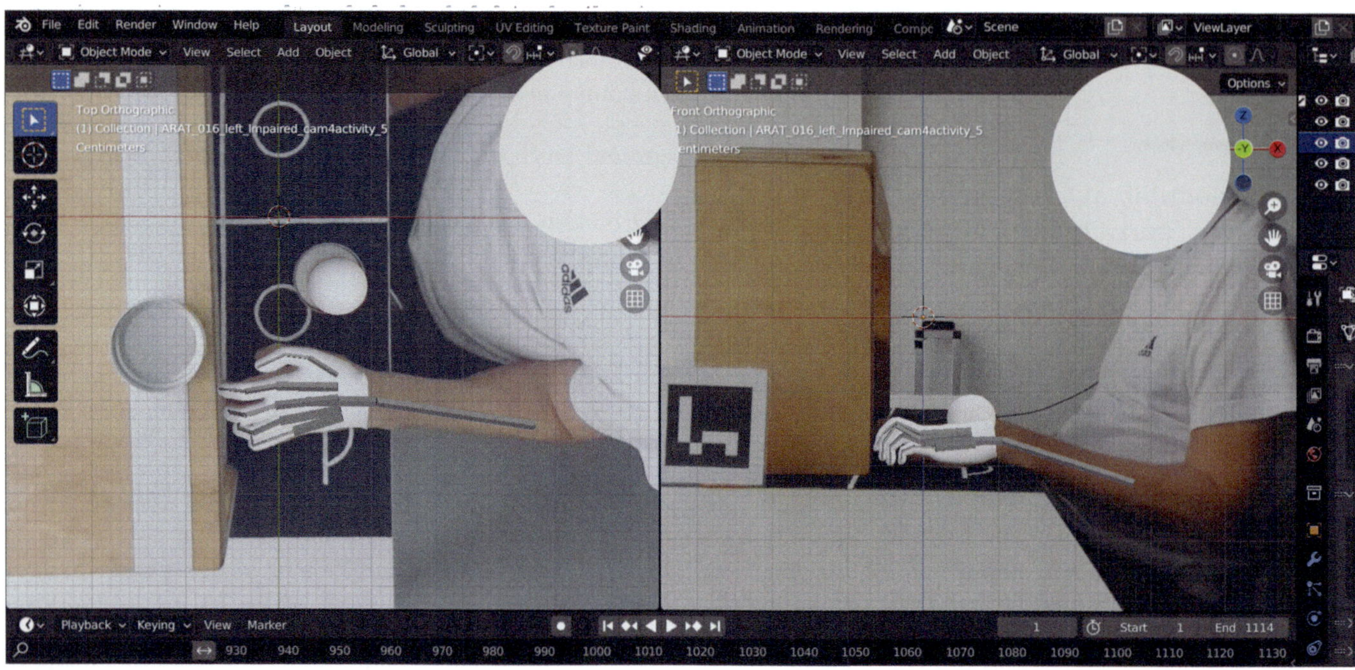

FIGURE 6.9
Illustration of our annotation approach. We worked with a 3D artist to rig the MANO mean hand template with an armature in Blender so that we can manually modify the position of the hand to match the RGB view. This is done for certain keyframes and we use Blender's inbuilt hand tracking to interpolate between the frames.

annotated, with Blender's inbuilt hand-tracking being utilized to interpolate the intermediate frames. This process will be verified for quality and intervention will be performed if necessary. Our current efforts are focused on annotating a total of 400 available videos. We are currently in the process of labeling these videos. To reconstruct objects, however, is a slightly easier process as the objects are rigid in nature. We have obtained high-quality 3D scans of objects using photogrammetry and scaled the meshes to have 600 vertices as required by Atlasnet [77]. Our future work involves training a DL model, similar to ObMAN, for hand and object reconstruction along with developing novel loss functions to incorporate therapists' knowledge into this framework.

We believe that our proposed approach, due to its unique annotation process, carefully considered design principles, and incorporation of novel loss functions, will outperform existing hand and object reconstruction systems. We posit that our approach will enable superior 3D analysis for minimally invasive, low-cost upper extremity stroke rehabilitation.

systems. The primary emphasis was on addressing the obstacles involved in developing a cost-effective, home-based setup, and we presented the necessary design principles to achieve such a system. Through these design principles, we demonstrated an effective and reliable rehabilitation system that operates solely on noisy and highly variable data captured from RGB cameras. We also discussed the limitations of this setup and reviewed recent advances in enabling 3D analysis from single-view images. In conclusion, the careful design of CV systems has the potential to greatly improve rehabilitation for stroke patients, especially in the development of home-based therapy programs.

Acknowledgments

This work was supported in part by NSF grant 2230762 and NIH-NIDILRR award number 90REGE0010.

6.5 Conclusion

In summary, this chapter provided an overview of CV systems for stroke rehabilitation, specifically discussing the use of marker-based and marker-less motion capture

Notes

1. https://www.qualisys.com/

2. https://optitrack.com/

References

[1] Rushil Anirudh, Jayaraman J Thiagarajan, Peer-Timo Bremer, and Brian K Spears. Improved surrogates in inertial confinement fusion with manifold and cycle consistencies. *Proceedings of the National Academy of Sciences*, 117(18):9741–9746, 2020.

[2] Badr Ben Elallid, Nabil Benamar, Abdelhakim Senhaji Hafid, Tajjeeddine Rachidi, and Nabil Mrani. A comprehensive survey on the application of deep and reinforcement learning approaches in autonomous driving. *Journal of King Saud University - Computer and Information Sciences*, 34(9):7366–7390, 2022.

[3] Andre Esteva, Katherine Chou, Serena Yeung, Nikhil Naik, Ali Madani, Ali Mottaghi, Yun Liu, Eric Topol, Jeff Dean, and Richard Socher. Deep learning-enabled medical computer vision. *NPJ Digital Medicine*, 4(1):1–9, 2021.

[4] Visar Berisha, Chelsea Krantsevich, P Richard Hahn, Shira Hahn, Gautam Dasarathy, Pavan Turaga, and Julie Liss. Digital medicine and the curse of dimensionality. *NPJ Digital Medicine*, 4(1):1–8, 2021.

[5] Jayaraman J Thiagarajan, Kowshik Thopalli, Deepta Rajan, and Pavan Turaga. Training calibration-based counterfactual explainers for deep learning models in medical image analysis. *Scientific Reports*, 12(1):1–15, 2022.

[6] Erdi Çallı, Ecem Sogancioglu, Bram van Ginneken, Kicky G van Leeuwen, and Keelin Murphy. Deep learning for chest x-ray analysis: A survey. *Medical Image Analysis*, 72:102125, 2021.

[7] John Jumper, Richard Evans, Alexander Pritzel, Tim Green, Michael Figurnov, Olaf Ronneberger, Kathryn Tunyasuvunakool, Russ Bates, Augustin Žídek, Anna Potapenko, et al. Highly accurate protein structure prediction with alphafold. *Nature*, 596(7873):583–589, 2021.

[8] Debleena Paul, Gaurav Sanap, Snehal Shenoy, Dnyaneshwar Kalyane, Kiran Kalia, and Rakesh K Tekade. Artificial intelligence in drug discovery and development. *Drug Discovery Today*, 26(1):80, 2021.

[9] Bappaditya Debnath, Mary O'brien, Motonori Yamaguchi, and Ardhendu Behera. A review of computer vision-based approaches for physical rehabilitation and assessment. *Multimedia Systems*, pages 1–31, 2021.

[10] C Zheng, W Wu, T Yang, S Zhu, C Chen, R Liu, J Shen, N Kehtarnavaz, and M Shah. Deep learning-based human pose estimation: A survey. arxiv 2020. *arXiv preprint arXiv:2012.13392.*

[11] Jing Li, Aixin Sun, Jianglei Han, and Chenliang Li. A survey on deep learning for named entity recognition. *IEEE Transactions on Knowledge and Data Engineering*, 34(1):50–70, 2020.

[12] Tiancai Wang, Tong Yang, Martin Danelljan, Fahad Shahbaz Khan, Xiangyu Zhang, and Jian Sun. Learning human-object interaction detection using interaction points. In *Proceedings of the IEEE/CVF Conference on Computer Vision and Pattern Recognition*, pages 4116–4125, 2020.

[13] Alana Elza Fontes Da Gama, Thiago de Menezes Chaves, Pascal Fallavollita, Lucas Silva Figueiredo, and Veronica Teichrieb. Rehabilitation motion recognition based on the international biomechanical standards. *Expert Systems with Applications*, 116:396–409, 2019.

[14] Andrea Catherine Alarcón-Aldana, Mauro Callejas-Cuervo, and Antonio Padilha Lanari Bo. Upper limb physical rehabilitation using serious videogames and motion capture systems: A systematic review. *Sensors*, 20(21):5989, 2020.

[15] Samarjit Das, Laura Trutoiu, Akihiko Murai, Dunbar Alcindor, Michael Oh, Fernando De la Torre, and Jessica Hodgins. Quantitative measurement of motor symptoms in Parkinson's disease: A study with full-body motion capture data. In *2011 Annual International Conference of the IEEE Engineering in Medicine and Biology Society*, pages 6789–6792. IEEE, 2011.

[16] S. Mihradi, Ferryanto, T. Dirgantara, and A.I. Mahyuddin. Development of an optical motion-capture system for 3D gait analysis. In *2011 2nd International Conference on Instrumentation, Communications, Information Technology, and Biomedical Engineering*, pages 391–394, 2011.

[17] Jamie Shotton, Andrew Fitzgibbon, Mat Cook, Toby Sharp, Mark Finocchio, Richard Moore, Alex Kipman, and Andrew Blake. Real-time human pose recognition in parts from single depth images. In *CVPR 2011*, pages 1297–1304, 2011.

[18] Sohrab Almasi, Hossein Ahmadi, Farkhondeh Asadi, Leila Shahmoradi, Goli Arji, Mojtaba Alizadeh, and Hoshang Kolivand. Kinect-based rehabilitation systems for stroke patients: A scoping review. *BioMed Research International*, 2022, 2022.

[19] David Webster and Ozkan Celik. Systematic review of kinect applications in elderly care and stroke rehabilitation. *Journal of Neuroengineering and Rehabilitation*, 11(1):1–24, 2014.

[20] Alana Da Gama, Pascal Fallavollita, Veronica Teichrieb, and Nassir Navab. Motor rehabilitation using kinect: A systematic review. *Games for Health Journal*, 4(2):123–135, 2015.

[21] Hossein Mousavi Hondori and Maryam Khademi. A review on technical and clinical impact of microsoft kinect on physical therapy and rehabilitation. *Journal of Medical Engineering*, 2014, 2014.

[22] Supriya Sathyanarayana, Ravi Kumar Satzoda, Suchitra Sathyanarayana, and Srikanthan Thambipillai. Vision-based patient monitoring: A comprehensive review of algorithms and technologies. *Journal of Ambient Intelligence and Humanized Computing*, 9(2):225–251, 2018.

[23] Md Atiqur Rahman Ahad, Anindya Das Antar, and Omar Shahid. Vision-based action understanding for assistive healthcare: A short review. In *CVPR Workshops*, pages 1–11, 2019.

[24] Z. Cao, G. Hidalgo Martinez, T. Simon, S. Wei, and Y. A. Sheikh. Openpose: Realtime multi-person 2D pose estimation using part affinity fields. *IEEE Transactions on Pattern Analysis and Machine Intelligence*, 2019.

[25] Tamim Ahmed, Kowshik Thopalli, Thanassis Rikakis, Pavan Turaga, Aisling Kelliher, Jia-Bin Huang, and Steven L Wolf. Automated movement assessment in stroke rehabilitation. *Frontiers in Neurology*, page 1396, 2021.

[26] Donald A. Norman. *The design of everyday things.* Basic Books, [New York], 2002.

[27] Aisling Kelliher, Andrew Gibson, Eric Bottelsen, and Edward Coe. Designing modular rehabilitation objects for interactive therapy in the home. In *Proceedings of the Thirteenth International Conference on Tangible, Embedded, and Embodied Interaction*, TEI '19, pages 251–257, New York, NY, USA, 2019. Association for Computing Machinery.

[28] Tamim Ahmed, Thanassis Rikakis, Setor Zilevu, Aisling Kelliher, Kowshik Thopalli, Pavan Turaga, and Steven L. Wolf. A hierarchical Bayesian model for cyber-human assessment of rehabilitation movement. *medRxiv*, 2022.

[29] Aisling Kelliher, Setor Zilevu, Thanassis Rikakis, Tamim Ahmed, Yen Truong, and Steven L. Wolf. Towards standardized processes for physical therapists to quantify patient rehabilitation. In *Proceedings of the 2020 CHI Conference on Human Factors in Computing Systems*, CHI '20, pages 1–13, New York, NY, USA, 2020. Association for Computing Machinery.

[30] Kelsey Picha and Dana Howell. A model to increase rehabilitation adherence to home exercise programmes in patients with varying levels of self-efficacy. *Musculoskeletal Care*, 16, 2017.

[31] Robert Teasell, Swati Mehta, Shelialah Pereira, Amanda McIntyre, Shannon Janzen, Laura Allen, Liane Lobo, and Ricardo Viana. Time to rethink long-term rehabilitation management of stroke patients. *Topics in Stroke Rehabilitation*, 19(6):457–462, 2012. PMID: 23192711.

[32] Elizabeth J Woytowicz, Jeremy C Rietschel, Ronald N Goodman, Susan S Conroy, John D Sorkin, Jill Whitall, and Sandy McCombe Waller. Determining levels of upper extremity movement impairment by applying a cluster analysis to the Fugl-Meyer assessment of the upper extremity in chronic stroke. *Archives of Physical Medicine and Rehabilitation*, 98(3):456–462, 2017.

[33] Steven L Wolf, Paul A Thompson, David M Morris, Dorian K Rose, Carolee J Winstein, Edward Taub, Carol Giuliani, and Sonya L Pearson. The excite trial: Attributes of the wolf motor function test in patients with subacute stroke. *Neurorehabilitation and Neural Repair*, 19(3):194–205, 2005.

[34] M. F. Levin, J. Kleim, and SL. Wolf. What do motor recovery and compensation mean in patients following stroke? *Neurorehabilitation and Neural Repair*, 23(4):313–319.

[35] M. H. Rabadi and F. M. Rabadi. Comparison of the action research arm test and the Fugl-Meyer assessment as measures of upper-extremity motor weakness after stroke. *Archives of Physical Medicine and Rehabilitation*, 87(7):962–966, 2006.

[36] Steven Wolf, Pamela Catlin, Michael Ellis, Audrey Archer, Bryn Morgan, and Aimee Piacentino. Assessing wolf motor function test as outcome measure for research in patients after stroke. *Stroke*, 32(7):1635–1639, 2001.

[37] Yinpeng Chen, Michael Baran, Hari Sundaram, and Thanassis Rikakis. A low cost, adaptive mixed reality system for home-based stroke rehabilitation. In 2011 *Annual International Conference of the IEEE Engineering in Medicine and Biology Society.*

IEEE Engineering in Medicine and Biology Society. 2011:1827–1830, 2011.

[38] M. C. Cirstea and M. F. Levin. Improvement of arm movement patterns and endpoint control depends on type of feedback during practice in stroke survivors. *Neurorehabilitation and Neural Repair*, 21(5):398–411, 2007.

[39] Asa Nordin, Margit Alt Murphy, and Anna Danielsson. Intra-rater and inter-rater reliability at the item level of the action research arm test for patients with stroke. *Journal of Rehabilitation Medicine*, 46, 2014.

[40] SL Wolf, CJ Winstein, JP Miller, E. Taub, G. Uswatte, D. Morris, C Giuliani, KE Light, and D. Nichols-Larsen. Effect of constraint-induced movement therapy on upper extremity function 3 to 9 months after stroke: The excite randomized clinical trial. *JAMA*, 296(17):2095–2104, 2006.

[41] David J. Reinkensmeyer, Etienne Burdet, Maura Casadio, John W. Krakauer, Gert Kwakkel, Catherine E. Lang, Stephan P. Swinnen, Nick S. Ward, and Nicolas Schweighofer. Computational neurorehabilitation: Modeling plasticity and learning to predict recovery. *Journal of Neuroengineering and Rehabilitation*, 13(1):42, 2016.

[42] Herbert A. Simon. *The sciences of the artificial.* MIT Press, Cambridge, Mass., 2d edition, 1981.

[43] Vinay Venkataraman, Pavan K. Turaga, Michael Baran, Nicole Lehrer, Tingfang Du, L. Cheng, Thanassis Rikakis, and Steven Wolf. Component-level tuning of kinematic features from composite therapist impressions of movement quality. *IEEE Journal of Biomedical and Health Informatics*, 20(1):143–152, 2016.

[44] Elahe Vahdani and Yingli Tian. Deep learning-based action detection in untrimmed videos: A survey. *IEEE Transactions on Pattern Analysis and Machine Intelligence*, 2022.

[45] Ashish Vaswani, Noam Shazeer, Niki Parmar, Jakob Uszkoreit, Llion Jones, Aidan N Gomez, Ł ukasz Kaiser, and Illia Polosukhin. Attention is all you need. In I. Guyon, U. V. Luxburg, S. Bengio, H. Wallach, R. Fergus, S. Vishwanathan, and R. Garnett, editors, *Advances in Neural Information Processing Systems*, volume 30. Curran Associates, Inc., 2017.

[46] Shaoqing Ren, Kaiming He, Ross Girshick, and Jian Sun. Faster r-cnn: Towards real-time object detection with region proposal networks. In C. Cortes, N. Lawrence, D. Lee, M. Sugiyama, and R. Garnett,

editors, *Advances in Neural Information Processing Systems*, volume 28. Curran Associates, Inc., 2015.

[47] Tsung-Yi Lin, Michael Maire, Serge Belongie, Lubomir Bourdev, Ross Girshick, James Hays, Pietro Perona, Deva Ramanan, C. Lawrence Zitnick, and Piotr Dollár. Microsoft coco: Common objects in context, 2014.

[48] Yazan Abu Farha and Jürgen Gall. MS-TCN: Multi-stage temporal convolutional network for action segmentation. In *CVPR*, pages 3575–3584. Computer Vision Foundation/IEEE, 2019.

[49] Shi-Jie Li, Yazan AbuFarha, Yun Liu, Ming-Ming Cheng, and Juergen Gall. MS-TCN++: Multi-stage temporal convolutional network for action segmentation. *IEEE Transactions on Pattern Analysis and Machine Intelligence*, 2020.

[50] João Carreira and Andrew Zisserman. Quo vadis, action recognition? A new model and the kinetics dataset. In *2017 IEEE Conference on Computer Vision and Pattern Recognition (CVPR)*, pages 4724–4733, 2017.

[51] Will Kay, Joao Carreira, Karen Simonyan, Brian Zhang, Chloe Hillier, Sudheendra Vijayanarasimhan, Fabio Viola, Tim Green, Trevor Back, Paul Natsev, et al. The kinetics human action video dataset. *arXiv preprint arXiv:1705.06950*, 2017.

[52] Yinpeng Chen, Margaret Duff, Nicole Lehrer, Hari Sundaram, Jiping He, Steven Wolf, and Thanassis Rikakis. A computational framework for quantitative evaluation of movement during rehabilitation. *AIP Conference Proceedings*, 1371, 2011.

[53] Vinay Venkataraman, Pavan Turaga, Nicole Lehrer, Michael Baran, Thanassis Rikakis, and Steven Wolf. Decision support for stroke rehabilitation therapy via describable attribute-based decision trees. volume 2014, pages 3154–9, 2014.

[54] Yasutaka Furukawa, Carlos Hernández, et al. Multiview stereo: A tutorial. *Foundations and Trends® in Computer Graphics and Vision*, 9(1-2):1–148, 2015.

[55] Matthew Loper, Naureen Mahmood, Javier Romero, Gerard Pons-Moll, and Michael J. Black. SMPL: A skinned multi-person linear model. *ACM Transactions on Graphics, (Proc. SIGGRAPH Asia)*, 34(6):248:1–248:16, October 2015.

[56] Tony Heap and David Hogg. Towards 3D hand tracking using a deformable model. In *IEEE International Conference on Automatic Face and Gesture Recognition (FG)*, pages 140–145, 1996.

[57] Ali Erol, George Bebis, Mircea Nicolescu, Richard D. Boyle, and Xander Twombly. Vision-based hand pose estimation: A review. *Computer Vision and Image Understanding (CVIU)*, 108(1-2):52–73, 2007.

[58] Mao Ye, Qing Zhang, Liang Wang, Jiejie Zhu, Ruigang Yang, and Juergen Gall. A survey on human motion analysis from depth data. In *Time-of-Flight and Depth Imaging. Sensors, Algorithms, and Applications*, pages 149–187. 2013.

[59] James S. Supančič III, Grégory Rogez, Yi Yang, Jamie Shotton, and Deva Ramanan. Depth-based hand pose estimation: Data, methods, and challenges. In *International Conference on Computer Vision (ICCV)*, pages 1868–1876, 2015.

[60] Lin Huang, Boshen Zhang, Zhilin Guo, Yang Xiao, Zhiguo Cao, and Junsong Yuan. Survey on depth and RGB image-based 3D hand shape and pose estimation. *Virtual Reality & Intelligent Hardware*, 3(3):207–234, 2021.

[61] Javier Romero, Dimitrios Tzionas, and Michael J. Black. Embodied hands: Modeling and capturing hands and bodies together. *ACM Transactions on Graphics*, 36(6):245:1–245:17, 2017.

[62] Iason Oikonomidis, Nikolaos Kyriazis, and Antonis A. Argyros. Efficient model-based 3D tracking of hand articulations using Kinect. In *British Machine Vision Conference (BMVC)*, pages 101.1–101.11, 2011.

[63] Iasonas Oikonomidis, Nikolaos Kyriazis, and Antonis A. Argyros. Tracking the articulated motion of two strongly interacting hands. In *IEEE Conference on Computer Vision and Pattern Recognition (CVPR)*, pages 1862–1869, 2012.

[64] Chen Qian, Xiao Sun, Yichen Wei, Xiaoou Tang, and Jian Sun. Realtime and robust hand tracking from depth. In *IEEE Conference on Computer Vision and Pattern Recognition (CVPR)*, pages 1106–1113, 2014.

[65] James M. Rehg and Takeo Kanade. Visual tracking of high dof articulated structures: An application to human hand tracking. In *European Conference on Computer Vision (ECCV)*, pages 35–46, 1994.

[66] Srinath Sridhar, Antti Oulasvirta, and Christian Theobalt. Interactive markerless articulated hand motion tracking using RGB and depth data. In *International Conference on Computer Vision (ICCV)*, pages 2456–2463, 2013.

[67] Srinath Sridhar, Franziska Mueller, Antti Oulasvirta, and Christian Theobalt. Fast and robust hand tracking using detection-guided optimization. In *IEEE Conference on Computer Vision and Pattern Recognition (CVPR)*, pages 3213–3221, 2015.

[68] Anastasia Tkach, Mark Pauly, and Andrea Tagliasacchi. Sphere-meshes for real-time hand modeling and tracking. *ACM Transactions on Graphics (TOG)*, 35(6), 2016.

[69] Yufei Ye, Abhinav Gupta, and Shubham Tulsiani. What's in your hands? 3d reconstruction of generic objects in hands. In *Proceedings of the IEEE/CVF Conference on Computer Vision and Pattern Recognition*, pages 3895–3905, 2022.

[70] Yana Hasson, Gül Varol, Dimitrios Tzionas, Igor Kalevatykh, Michael J. Black, Ivan Laptev, and Cordelia Schmid. Learning joint reconstruction of hands and manipulated objects. In *CVPR*, 2019.

[71] Lim Guan Ming, Jatesiktat Prayook, and Ang Wei Tech. Mobilehand: Real-time 3D hand shape and pose estimation from color image. In *27th International Conference on Neural Information Processing (ICONIP)*, 2020.

[72] Dominik Kulon, Riza Alp Guler, Iasonas Kokkinos, Michael M Bronstein, and Stefanos Zafeiriou. Weakly-supervised mesh-convolutional hand reconstruction in the wild. In *Proceedings of the IEEE/CVF Conference on Computer Vision and Pattern Recognition*, pages 4990–5000, 2020.

[73] Adnane Boukhayma, Rodrigo de Bem, and Philip HS Torr. 3D hand shape and pose from images in the wild. In *Proceedings of the IEEE/CVF Conference on Computer Vision and Pattern Recognition*, pages 10843–10852, 2019.

[74] Christian Zimmermann, Duygu Ceylan, Jimei Yang, Bryan Russell, Max Argus, and Thomas Brox. Freihand: A dataset for markerless capture of hand pose and shape from single RGB images. In *Proceedings of the IEEE/CVF International Conference on Computer Vision*, pages 813–822, 2019.

[75] Shreyas Hampali, Mahdi Rad, Markus Oberweger, and Vincent Lepetit. Honnotate: A method for 3D annotation of hand and object poses. In *Proceedings of the IEEE/CVF Conference on Computer Vision and Pattern Recognition*, pages 3196–3206, 2020.

[76] Zhe Cao, Ilija Radosavovic, Angjoo Kanazawa, and Jitendra Malik. Reconstructing hand-object interactions in the wild. In *Proceedings of the IEEE/CVF International Conference on Computer Vision*, pages 12417–12426, 2021.

[77] Thibault Groueix, Matthew Fisher, Vladimir G Kim, Bryan C Russell, and Mathieu Aubry. A Papier-Mâché approach to learning 3D surface generation. In *Proceedings of the IEEE Conference on Computer Vision and Pattern Recognition*, pages 216–224, 2018.

7

U-Net-based Medical Image Segmentation: A Comparative Analysis and Future Trends

Sidike Paheding, Abel A. Reyes-Angulo, and Mohammad S. Alam

Recent advances in deep learning, a subfield of machine learning, have been contributing to diverse modalities of health care, encompassing the identification of large-artery occlusion in the brain, prediction of cardiovascular risk and diabetic retinopathy, and detection of breast lesions in mammograms. In healthcare-related problems, biomedical image segmentation is an important pivot in computational analysis and clinical diagnosis. U-Net, one of the popular deep neural network architectures designed for biomedical image segmentation, has become the most referred segmentation method for biomedical imaging applications. U-Net has been effectively utilized for many image modalities, such as CT scans, MRIs, X-rays, and microscopy. The unique structural design of U-Net allows it to be easily integrated with various novel techniques to produce state-of-the-art results. This chapter explores recent developments in U-Net-based deep learning architectures and empirically analyzes the performance of these models using standard evaluation metrics on popular benchmark datasets. This chapter ends with a discussion of the limitations, challenges, and future trends of U-Net architectures.

7.1 Introduction

Deep Learning (DL) [50] has demonstrated great promises and advances in a variety of computer vision tasks, including image segmentation [59], object detection and tracking [37], and image captioning. In biomedical image segmentation, manually labeling each pixel of an image with a corresponding category can be performed by radiologists. However, it is labor-intensive and time-consuming, and even there could be conflicting opinions among clinicians. In addition, the segmentation task for biomedical images is particularly different from segmentation in regular images. Biomedical image segmentation involves identifying and segmenting regions of interest within a variety of image types such as MRI or CT scans

rather than regular RGB images. Those image protocols can have significant variability due to differences in patient anatomy, imaging modality, and the presence of unstructured objects [27, 57]. Therefore, an automated image segmentation tool specialized for biomedical images is essential, especially in the current era of digital transformation. Although many segmentation algorithms have been developed and are facilitating clinicians to perform more efficient work, they are mostly based on pixel intensity values or handcrafted engineering features, which suffer from low generalization and precision. With the rapid advancement of DL, more sophisticated algorithms, such as U-Net [69], which is one of the most popular Fully Convolutional Networks (FCN) [53] approaches for biomedical image segmentation, have been introduced. Convolutional layers with elegant encoder and decoder design in the U-Net architecture ensure better feature representation of input images and produce precise segmentation maps. U-Net has become the de-facto architecture in medical image segmentation.

Before the development of the U-Net or any of its improvements, there were a few constraints of medical image segmentation, including but not limited to data shortage, class irregularity, and model training time [30, 80]. U-Net alleviated these problems by effectively utilizing a data augmentation strategy with a more elegant network architecture that consists of a contracting path to capture context and a symmetric expanding path to output precise localization. However, the performance of vanilla U-Net is constrained by its network design from certain aspects [41]. For instance, the disparity between the encoder–decoder features leads to feature discrepancy during the network training and thereby adversely affecting the prediction accuracy. Besides, a sequence of two 3×3 convolutional layers in a U-Net limits its ability to segment objects of interest of irregular and different scales. Moreover, the number of layers in U-Net architecture is relatively less compared to the state-of-the-art DL methods, which hinders its capability to gain improved performance [96].

In recent years, there have been rapid improvements in U-Net architecture. Popular models such as 3D U-Net

DOI: 10.1201/9781003328957-7

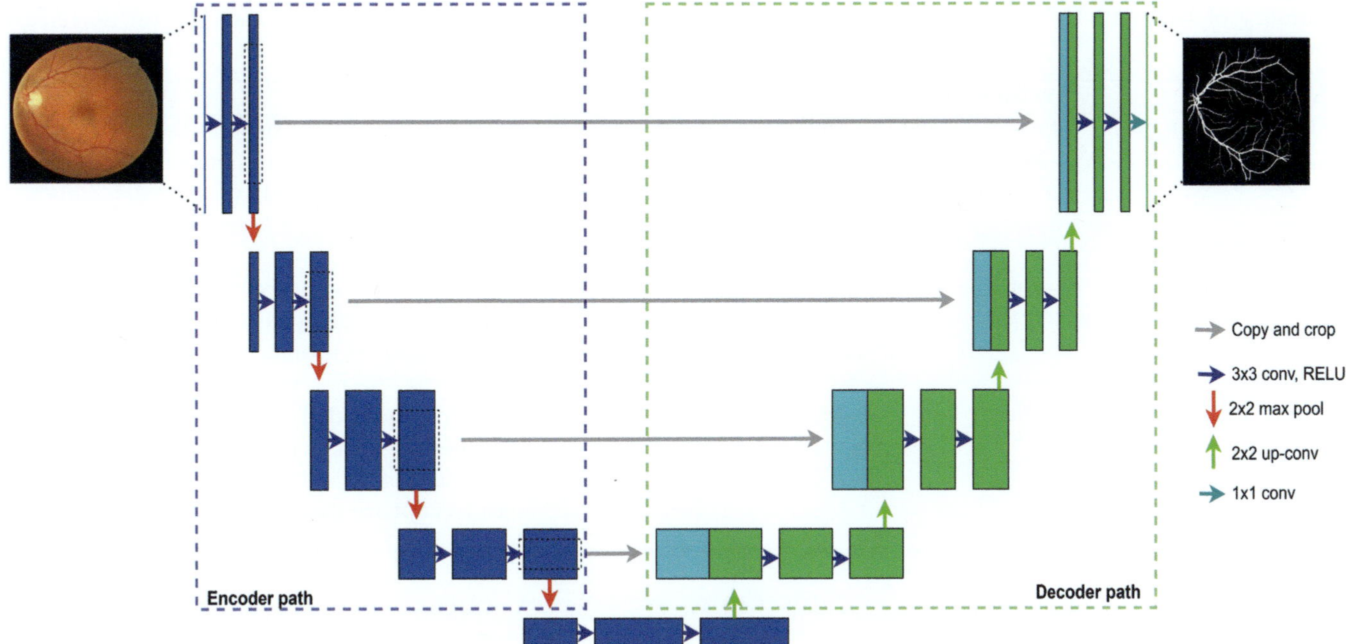

FIGURE 7.1
The U-Net architecture. The U-shape is formed by convolutional encoding (blue blocks) and decoding (green blocks) paths. A concatenation block (light blue blocks) acts over every correspondent level in the decoder from the encoder path. U-Net takes an image as an input and produces the segmentation output with the respective pixel class.

[18], Residual U-Net (ResU-Net) [95], Recurrent Residual U-Net (R2U-Net) [3], Efficient Dense U-Net[76], and Multi-scale Densely Connected U-Net (MDU-Net) [93] that are developed for various medical image segmentation tasks. This study provides a review and evaluate the performance of some popular U-Net-based architectures, including Attention U-Net [61], ResU-Net [95], R2U-Net [3], V-net [58], U-Net 3+ [40], U2-net [65], Efficient Dense U-Net, and Efficient R2U-Net [76], using three standard datasets. The remainder of this chapter is structured as follows: Section 7.2 describes vanilla U-Net architecture and its variants that will be evaluated in this work. Section 7.3 introduces three standard databases in our experiments. Sections 7.4 and 7.5 detail the experimental setup and results. Section 7.6 discusses limitations, challenges, and future trends of U-Net-based models. Finally, Section 7.7 concludes our findings from this study.

7.2 U-Net Architectures

7.2.1 U-Net

U-Net architecture consists of a contracting (encoder) path and an expansive (decoder) path. The upsampling layers in the encoder path contain feature maps to propagate context information to higher-resolution layers. The symmetry of the contracting and the expansive path yields a U-shaped architecture. The details of the encoder and decoder subnetwork are given as follows:

- *Encoder path* involves the repeated operation of two 3×3 convolutions, where each convolution is followed by a ReLU activation, and a batch normalization layer [42]. A 2×2 max-pooling operation is then applied to downsample feature maps. At each downsampling step, the model doubles the number of feature maps, while reducing spatial dimension by half.

- *Decoder path* consists of transposed convolution or upsampled convolution, which upsamples the input feature map to match spatial resolution in the encoder path. At the same scale/resolution of feature maps in the encoder–decoder subnetwork, there is a skip connection to concatenate the corresponding feature maps between encoder and decoder paths, which allows for retrieving the spatial information lost by pooling operations. At the final layer, a 1×1 convolution is computed to map the feature channels to the desired number of classes in the dataset.

The U-Net architecture is depicted in Figure 7.1, in which the encoder path is represented with blue color blocks, while the decoder path is represented with green

TABLE 7.1

Summary of Different U-Net Architectures for Medical Image Segmentation Presented in This Chapter Sorted by Year ($\downarrow$)

Author	Architecture	Year	#Param (M)	Key Features
Siddique et al. [76]	Efficient Dense U-Net and R2U-Net	2022	(-)	Proposed the use of Efficient-Net as backbones with the use of either dense convolutional blocks or residual recurrent convolutional blocks.
Huang et al. [40]	U-Net 3+	2020	26.97	A U-Net-based architecture with the introduction of full-scale skip connections and deep supervisions.
Qin et al. [65]	U2-Net	2020	4.70	Two-level nested U-structure, designed for salient object detection and proposed a mixture of receptive fields of different sizes to capture more contextual information.
Alom et al. [3]	R2U-Net	2019	13.34	A variant of U-Net architecture that leverages the strength of the recurrent residual neural network.
Oktay et al. [61]	Attention U-Net	2018	6.40	A U-Net-based architecture with the addition of an attention gate (AF) unit to enhance the feature representation of target structures is proposed.
Zhang et al. [95]	ResU-Net	2018	7.80	A U-Net alike architecture with the replacement of the basic neural units by residual units.
Milletari et al. [58]	V-Net	2016	(-)	A V-shape network architecture that performs volumetric convolutions to process MRI volumes for image segmentation tasks.

color blocks. The modularity property of U-Net architecture allows the introduction of different configurations into the architecture to accomplish various segmentation tasks. This work explores U-Net and its popular variants for biomedical image segmentation tasks.

7.2.2 U-Net Variants

Since the introduction of U-Net architecture for medical image segmentation, there have been many attempts to improve the performance of this vanilla architecture and the adaptation for different modalities (i.e., volumetric and multimodal images). In Table 7.1, a summary of U-Net and its popular variants, which are reviewed in this chapter, is provided. An extended description of each U-Net variant is provided in the following.

- **Attention U-Net:** Oktay et al. [61] proposed an approach based on an attention gate (AG) unit for medical imaging. AG automatically learns to focus on target structures of varying shapes and sizes and suppresses irrelevant regions in an input image. This takes the input feature map from the encoder and a gating signal (which is a feature map from the encoder at a lower resolution) to selectively weight the feature map in the decoder to focus on the most relevant parts of the encoder feature map while ignoring the noisy information. This approach addressed the general problem of using explicit external tissue/organ localization modules from cascaded convolutional neural networks (CNNs), in which there is excessive and redundant use of computational resources and model parameters. Their work presented the integration of AGs into the U-Net model, and they named it Attention U-Net architecture. Attention U-Net architecture along with the attention gate component is illustrated in Figure 7.2.

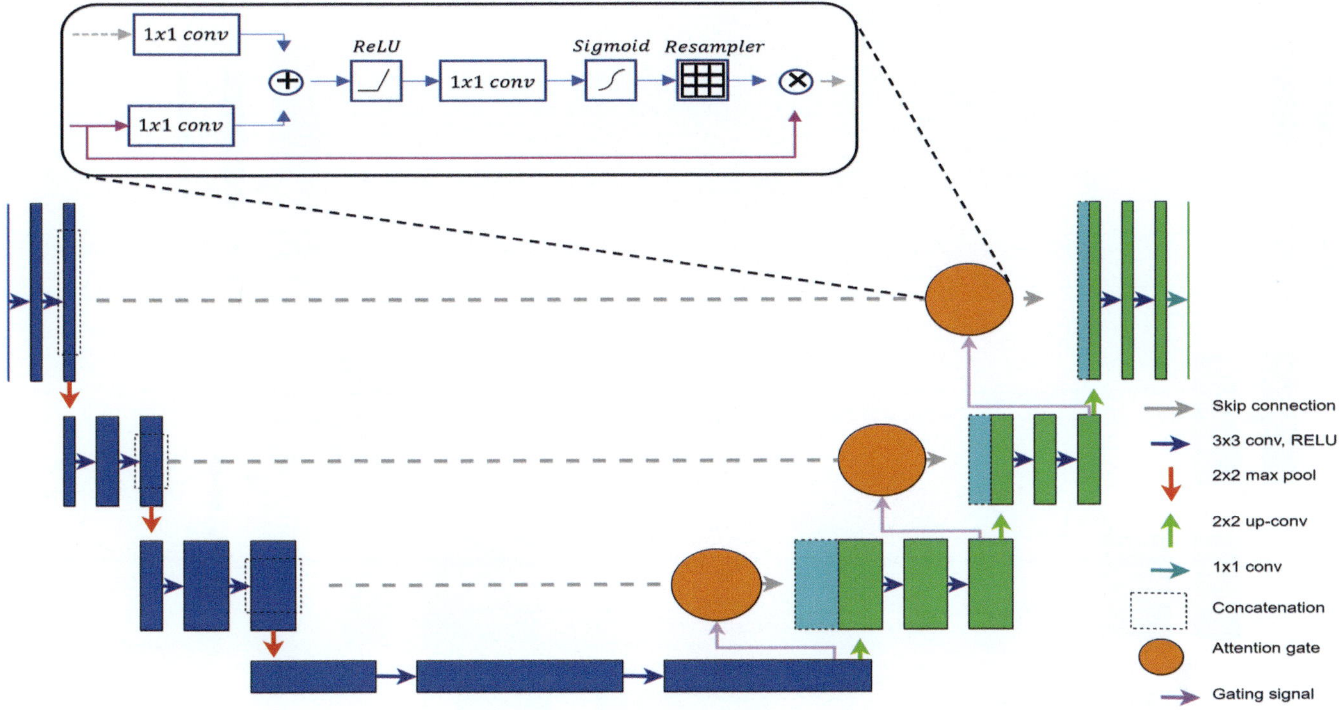

FIGURE 7.2
Illustration of Attention U-Net with attention module.

- **R2U-Net:** R2U-Net was introduced by Alom et al. [3] where they proposed two U-Net variants: one based on Recurrent Convolutional Neural Network (RCNN) [52] named RU-Net, and the second variant based on Recurrent Residual Convolutional Neural Network (RRCNN) named as R2U-Net. The proposed models leverage the power of U-Net, RCNN, and RRCNN. Advantages of R2U-Net [3] are described as follows: (1) the residual unit helps with deeper architectures design, (2) incorporation of recurrent residual convolutional layers ensures a better feature representation, and (3) the combination of residual units and recurrent residual convolutional layers allows better performance for medical image segmentation task. The R2U-Net model showed better results compared to existing models, such as U-Net and ResU-Net, not only during the training process but also during the testing phase. R2U-Net architecture is illustrated in Figure 7.3.

- **ResU-Net:** Residual U-Net or ResU-Net [95] was proposed as a U-Net-like model with the use of residual units instead of plain neural units on the encoder and decoder paths. By adding the residual units, the model was claimed to be capable to learn more discriminative components of the input data. This assumption was examined over a road dataset [88] achieved promising performance in this segmentation

task. The ResU-Net architecture is illustrated in Figure 7.4.

- **V-Net:** V-Net was presented by Milletari et al. [58] as an approach to leverage the power of fully CNNs to process MRI volumes for medical image segmentation tasks in a fast and accurate manner. They introduced a new objective function that is optimized during training based on the Dice coefficient. The V-net architecture is inspired by U-Net; however, V-net divided the architecture into stages that learn a residual function by adding the input of each stage to the last output from the convolutional layer of that stage, this procedure demonstrated to improve both: results and time of convergence. Different from the original V-net proposal that uses volumetric convolution. In this work, we present a modified version for 2-D inputs instead, as shown in Figure 7.5.

- **U-Net 3+:** Huang et al. [40] proposed a U-Net 3+ model. In comparison to U-Net (plain skip connections) and U-Net++ (nested and dense skip connections), U-Net 3+ (U-Net+++) takes advantage of the full-scale skip connections (incorporating low-level details with high-level semantics) and deep supervision. This proposed method specializes in the image segmentation task for organs that appears at varying scales. This method was tested in two organs (the liver and spleen). Experimental results showed that

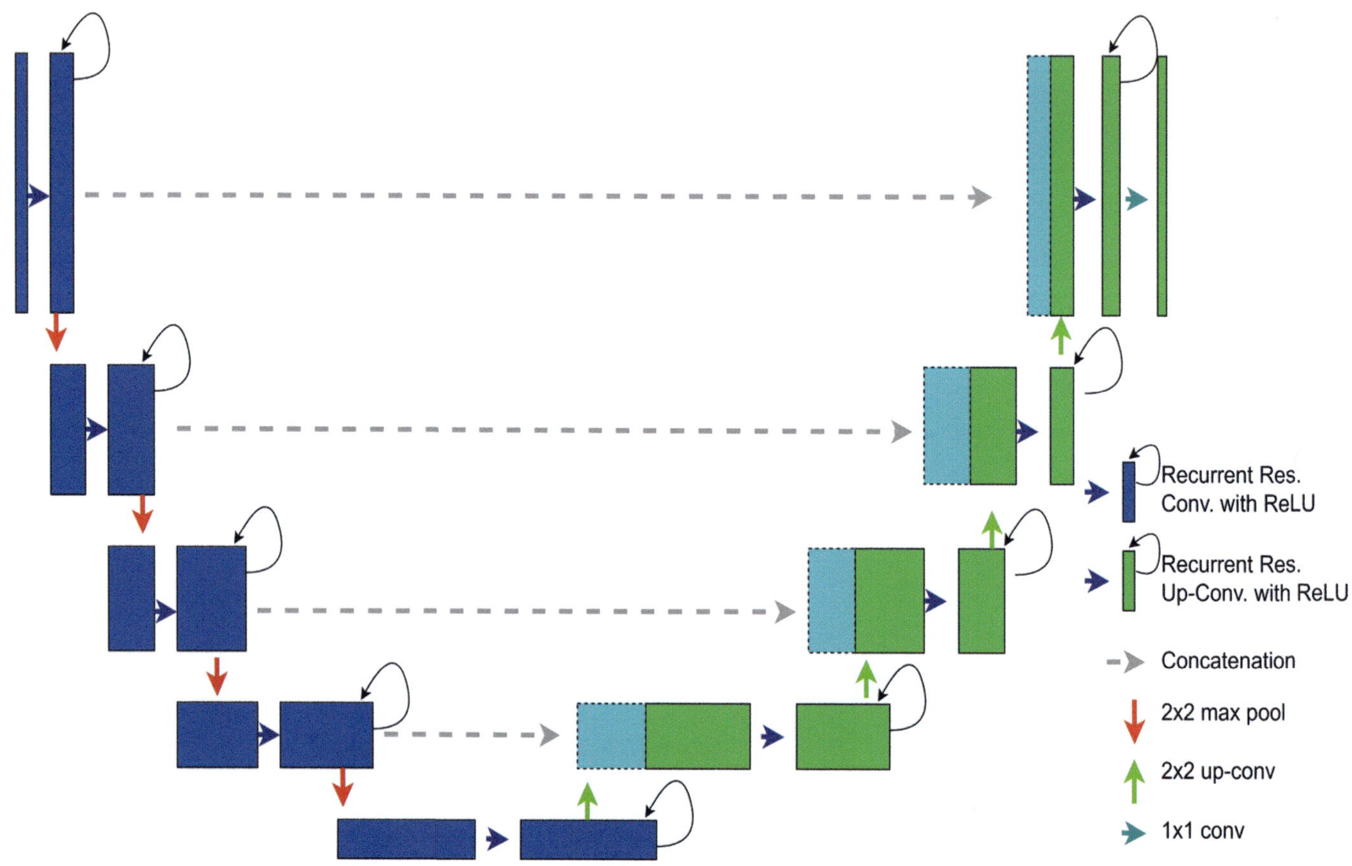

FIGURE 7.3
Illustration of R2U-Net with recurrent residual convolutional neural network module.

U-Net 3+ surpassed the state-of-the-art approaches in terms of boundary-aware segmentation maps. U-Net 3+ architecture is illustrated in Figure 7.6.

- **U2-Net:** U2-Net [65] was designed for salient object detection (SOD), which consists of a two-level nested U-Structure (U2), and it is claimed to have the following advantages: firstly, Residual U-blocks (RSU) enables capturing more contextual information from various scales; and secondly, the computational cost is not affected due to the depth increment of the whole architecture because of the pooling operations used in the RSU blocks. The experimental results on different datasets demonstrated that the U2-net model can achieve very competitive performance compared to other state-of-the-art U-Net-based architectures in terms of both qualitative and quantitative measures. U2-net architecture is illustrated in Figure 7.7.

- **Efficient Dense U-Net and R2 U-Net:** Efficient Dense U-Net and R2 U-Net were proposed by Siddique et al. [76]. Efficient Dense U-Net introduces a convolutional block based on densely connected convolutional networks namely DenseNet.

DenseNet is a DL architecture based on ResNet with two major characteristics: first, every layer in a block receives the features map from all the preceding layers, then these feature maps are passed onto the next layers; and second, the feature maps are combined into tensors, allowing DenseNet to preserve identity maps from prior layers. This proposed convolutional block (DenseNet blocks) replaced all standard convolutional blocks in the expansive path in U-Net. This architecture is illustrated in Figure 7.8(a).

Additionally, Efficient R2U-Net [76] architecture was developed by combining two major neural network architectures into one. The purpose of this architecture is to take advantage of the residual learning characteristics for the development of deeper neural networks, where the residual skip connections can transfer feature maps from previous layers. This approach is helping to improve gradient propagation. Another purpose of using RCNN with the residual skip connections is that the recurrent connections can be input to the preceding layer at discrete time steps. In this way, the model is capable of

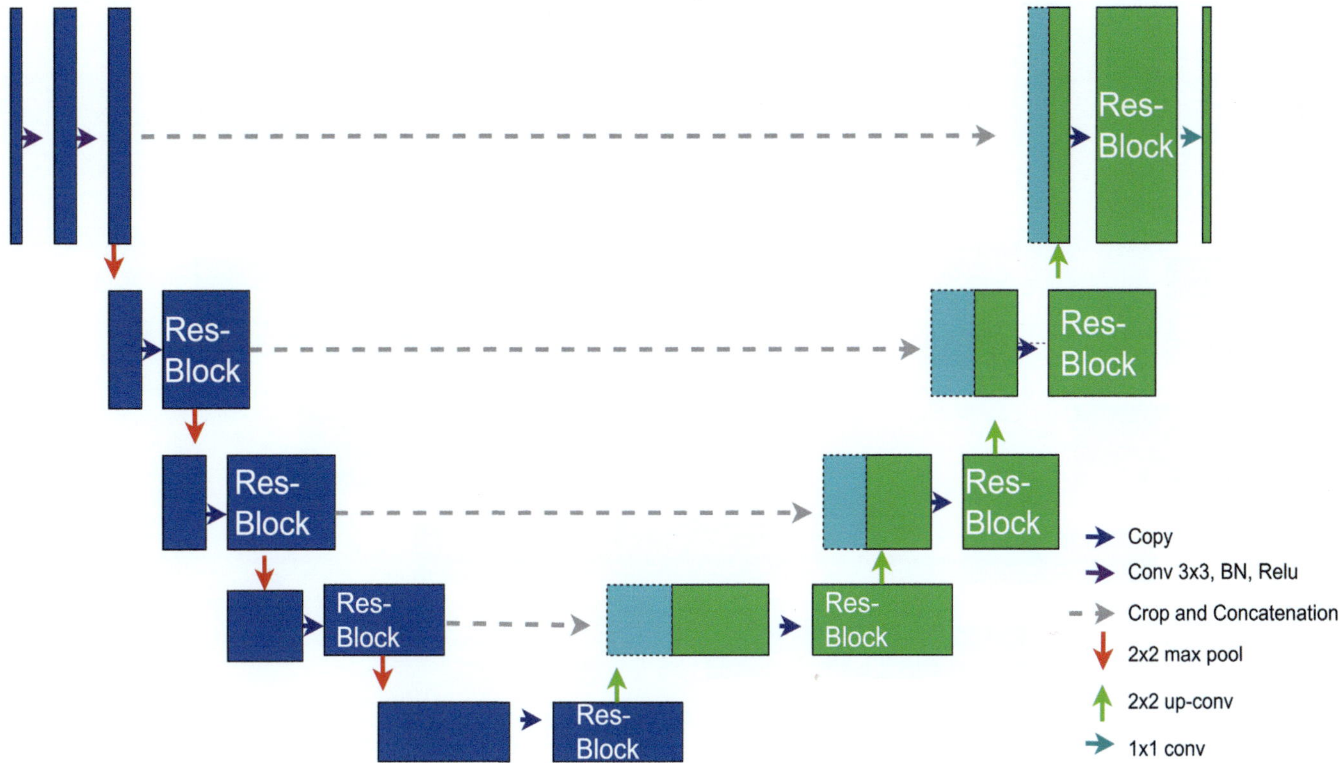

FIGURE 7.4
Illustration of Residual U-Net with residual convolutional network blocks.

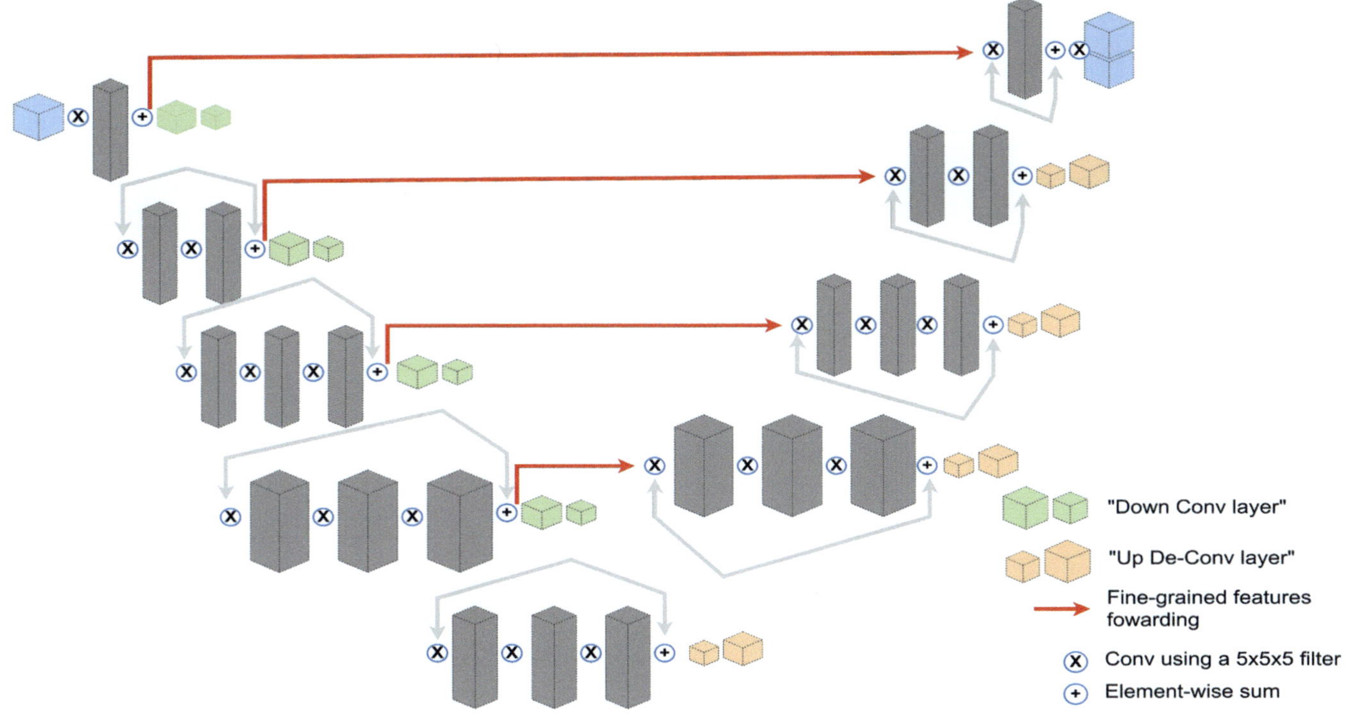

FIGURE 7.5
Illustration of V-Net architecture with volumetric kernels.

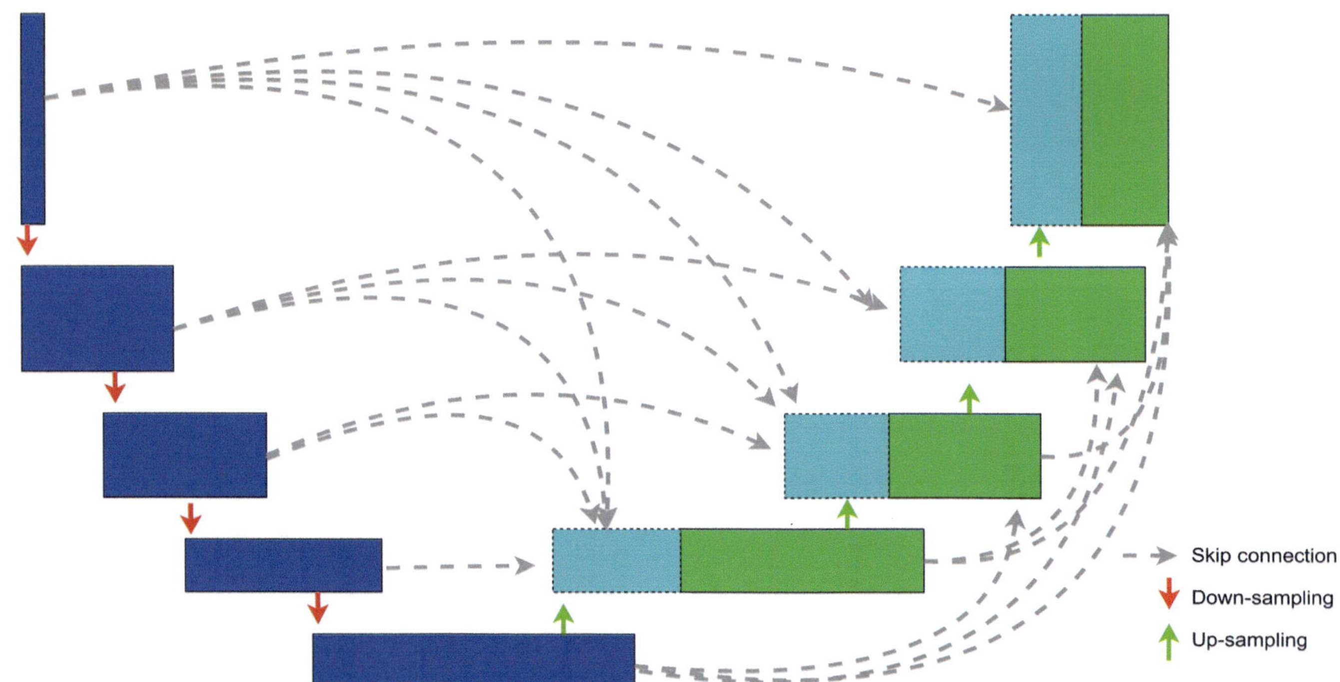

FIGURE 7.6
Illustration of U-Net 3+ with full scale skip connection.

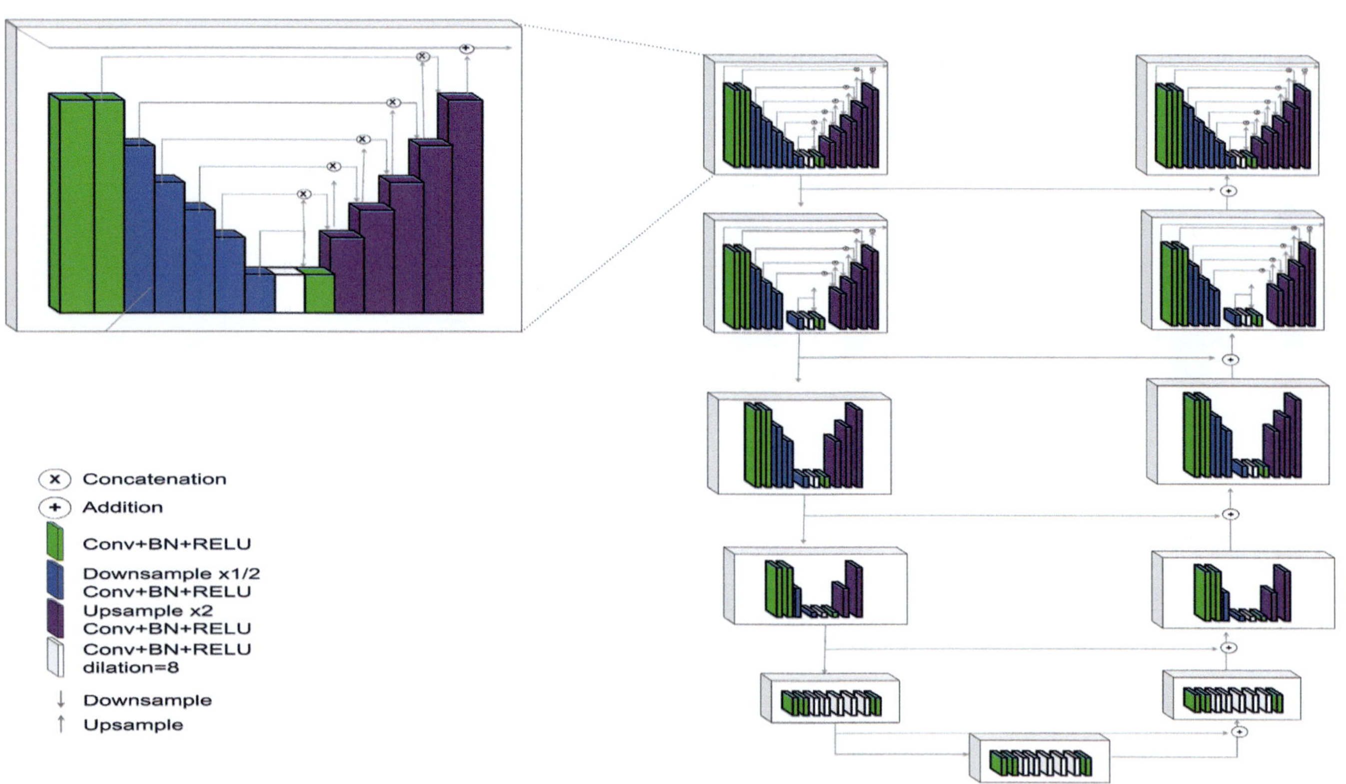

FIGURE 7.7
Illustration of U2-Net with nested U-structure.

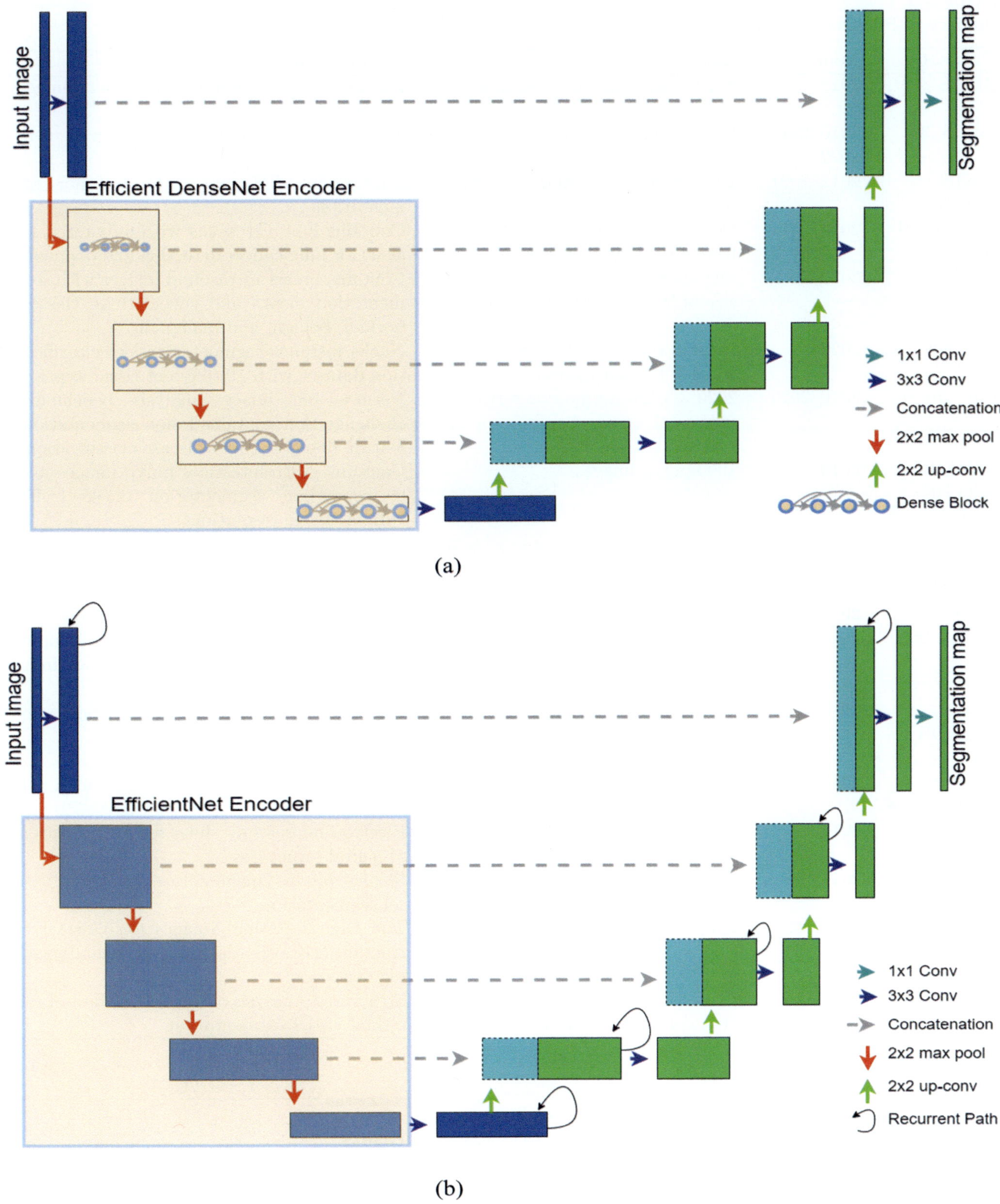

FIGURE 7.8
Illustration of U-Net variants' architecture: (a) Efficient Dense U-Net consisting in EfficientNet encoder integrated with dense blocks; (b) Efficient R2U-Net consisting in EfficientNet encoder integrated with R2U-Net.

TABLE 7.2
Summary of Popular Biomedical Image Segmentation Datasets Sorted by Year of Publication ($\downarrow$).

Dataset	Year	Size	Description
BUSIS [94]	2022	562 breast ultrasound images.	Dataset for B-mode breast ultrasound image segmentation
ISLES [38]	2016–2018, 2022	2022: 400 multi-vendor MRI.	MRI scans for ischemic stroke lesion segmentation
BraTS [5, 6, 56]	2012–2021	2021: 2000 cases (8000 mpMRI scans)	Multimodal MRI scans for brain tumor segmentation
LIDC-IDRI [4]	2020	Contains 1018 low-dose lung CTs from 1010 lung patients	Contains lung CT scans with annotated lesions for lung cancer detection and diagnosis
BUSI [1]	2020	780 images with an average image size of 500 × 500 pixels.	Contains breast ultrasound images with segmentation masks and covers three classes: normal, benign, and malignant.
PanNuke [28]	2020	In total the dataset contains 205,343 labeled nuclei	Nuclei instance segmentation and classification dataset with 19 different tissue types
ISIC [70]	2016–2020	2020: 33,126 dermoscopic training images of unique benign and malignant skin lesions.	Dermoscopic image analysis benchmark challenge that contains lesion segmentation, feature detection, and disease classification.
KVASIR-SEG [46]	2019	1000 polyp images.	Contains gastrointestinal polyp images and corresponding segmentation masks, with various image resolutions.
KDSB 2018 [10]	2018	This dataset contains a large number of segmented nuclei images.	A large volume of segmented nuclei images with varied cell type, magnification, and imaging modality
LiTS [8]	2017	200 CT scans.	CT scans for liver tumor segmentation
LUNA 2016 [72]	2016	This dataset contains 888 CT scans	Uses LIDC/IDRI database and it is open challenges for localization of nodules from volumetric CT images
CVC-ClinicDB [7]	2015	612 images with a resolution of 384 × 288 from 31 colonoscopy sequences.	Endoscopic colonoscopy frames for polyp detection
xVertSeg [49]	2015	Contains 25 CT lumbar spine images	Contains CT lumbar spine images with references for vertebra segmentation
ISBI2012 [12]	2012	30 ssTEM (serial section Transmission Electron Microscopy) images	Electron microscopy slides for segmentation of neural structures
CHASEDB1 [25]	2011	28 color retina images with the size of 999 × 960 pixels	Retinal fundus image dataset for blood vessel segmentation
DRIVE [78]	2004	It consists of a total of JPEG 40 color fundus images.	The Digital Retinal Images for Vessel Extraction (DRIVE) dataset for retinal vessel segmentation.
STARE [39]	2000	It contains 20 equal-sized (700 × 605) color fundus images.	Structured Analysis of the Retina dataset for retinal vessel segmentation.

extracting pertinent information from the data. This type of design replaces the standard convolutional blocks in the expansive path with the modified residual RCNN blocks (R2) and incorporates EfficientNet encoders in the contracting path of the original U-Net architecture. Efficient Residual Recurrent U-Net architecture (i.e., Efficient R2U-Net) is illustrated in Figure 7.8(b).

7.3 Datasets

In order to provide a widely accepted approach, most of the architectures reviewed in this work utilize publicly available biomedical image segmentation (BIS) datasets, a summary of some popular used BIS datasets is provided in Table 7.2. It is worth mentioning the capabilities of the original U-Net and the U-Net-based architectures to discriminate between more than two classes for pixel-wise classification [21, 35, 43]. However, this work focuses on U-Net-based architectures for binary segmentation tasks. For a fair comparison and analysis of the performance of the aforementioned U-Net-based architectures, this work provides experimental results from the implementation of the U-Net-based architectures over three specific biomedical segmentation datasets: CVC-ClinicDB, KVASIR-SEG, and ISIC2016. The selection of the datasets was based on the computational resources availability as well as the diversity of the application domain. Details of those datasets are provided as follows.

1. **CVC-ClinicDB:** The CVC-ClinicDB [7] is a publicly available dataset of frames extracted from colonoscopy videos. The frames contain examples of polyps and each frame has a correspondent ground truth mask of the region covered by the polyp in the image. A total of 29 different video sequences produced 612 images associated with a correspondent ground truth. Each image and the correspondent mask have a size of 384 × 288 pixels and those are encoded using PNG compression.

2. **KVASIR-SEG:** This is a variant of the original Kvasir dataset, meant to be used for semantic segmentation modeling. The Kvasir dataset comprises 8000 gastrointestinal tract images, divided into eight different classes of 1000 images each. The Kvasir-SEG dataset [46] contains 1000 RGB images from the gastrointestinal tract and its corresponding ground-truth mask for the polyp detection task. The images and corresponding masks vary in size from 332 × 487 to 1920 × 1072 pixels and they are encoded using JPEG compression. This dataset is publicly accessible for educational purposes.

3. **ISIC2016:** The ISIC Archive contains over 13,000 dermoscopic images of skin lesions, this accomplishment is given by the effort of the International Skin Imaging Collaboration (ISIC), sponsored by the International Society for Digital Imaging of the Skin (ISDIS). The images were collected from leading clinical centers internationally. The ISIC2016 Dataset [19] contains 900 dermoscopic lesion RGB images in JPEG format and their corresponding 900 binary mask images in PNG format.

7.4　Experimental Setup

In this research, to show the robustness of the U-Net architecture for biomedical image segmentation tasks, the three different datasets were used to train DL models with the different U-Net architecture variants mentioned in the previous sections. The three different datasets were resized to have a similar input size of 128×128 pixels and the three RGB channels. The ground truth was also resized to 128×128 pixels and kept just one channel. In all the experiments, neither data augmentation nor pre-training was performed, and the models were trained to segment regions of interest. Each DL model was set up to run for 200 epochs, and the dataset is randomly split into 90% for training and 10% for testing. During the training, we save the history of the loss and evaluation metrics such as prediction accuracy and dice coefficient. Once the model was built, it was used to test the performance by computing the mean IoU over the predictions using the test set.

The code was implemented using Python 3.7 and TensorFlow—Keras 2.7, with a hardware setup: i9 CPU at 3.7GHz Intel processor, and a GPU NVIDIA GeForce RTX 3070 with 8GB dedicated GPU.

7.5　Results

7.5.1　Evaluation Metrics

In order to measure and compare the performance of different U-Net-based models, the use of the appropriate performance metrics became an important factor. According to Fenster and Chiu [24], the lack of standardized metrics, and the intrinsic differences in systems, forced the use of specific metrics to evaluate the overall performance of a model. Because most of the BIS datasets are imbalanced, thus evaluation metrics such as accuracy are not sufficient for BIS assessment. Intersection over Union (IoU) and Dice coefficient are two of the most widely used performance metrics for image segmentation tasks [36, 59]. Those evaluation metrics are described below.

- **Dice coefficient:** The Dice coefficient is a measure of the similarity between two samples [20]: ground truth (GT) for the input sample and the segmentation output (SO) generated by the U-Net model. Both, GT and SO are binary segmentation masks,

TABLE 7.3

Distribution of the Samples on Each Dataset for Training and Testing the Different DL Models.

Dataset	Total Samples	Training Samples	Testing Samples
CVC-ClinicDB	612	551	61
KVASIR	1000	900	100
ISIC2016	900	810	90

TABLE 7.4

Performance Comparison of U-Net and Its Variants on Three Different Datasets. The Best Performance on Each Dataset is Highlighted in Bold Font.

U-Net Models	CVC ClinicDB		KVASIR-SEGDB		ISIC2016DB	
	Val Loss	*Test Mean IoU*	*Val Loss*	*Test Mean IoU*	*Val Loss*	*Test Mean IoU*
U-Net	0.0936	0.8759	0.3219	0.8454	0.2985	0.8611
Attention U-Net	0.1565	0.8618	0.2293	0.8233	0.4191	0.8637
R2 U-Net	0.1596	0.8673	0.3854	0.8403	0.2855	0.8446
Res U-Net-a	0.1471	0.8434	0.3948	0.7743	0.3598	0.8349
V-Net	0.2048	0.8504	0.5521	0.7822	0.3305	0.8500
U-Net 3+	0.2428	0.8598	0.5574	0.8106	0.4541	0.8489
U2-Net	0.1353	0.8594	0.2647	0.8254	0.2768	0.8594
Eff. Den. U-Net	0.0757	0.8972	0.2049	0.8566	**0.1359**	**0.8852**
Eff. R2U-Net	**0.0716**	**0.9068**	**0.1754**	**0.8592**	0.1809	0.8788

and we set a commonly used threshold value of 0.5 for differentiating between background and region of interest, keeping consistency with the related forehead mentioned works. The Dice coefficient is expressed as:

$$Dice-coeff = \frac{2|GT \cap SO|}{|GT| + |SO|} \tag{7.1}$$

- **Mean IoU:** IoU, also known as Jaccard index, is the measurement of the overlap between two samples [45]. The Mean IoU is computed using the IoU on each class (denoted as $c \in [1, \ldots, C]$) between GT and SO, and then the values are averaged for each class. The IoU and the Mean IoU can be computed as in Eq. (7.2) and Eq. (7.3), respectively. TP indicates true positives, FP is false positives, and FN represents false negatives.

$$IoU = \frac{|GT \cap SO|}{|GT \cup SO|} = \frac{TP}{(TP + FP + FN)} \tag{7.2}$$

$$MeanIoU = \frac{1}{C} \sum_c \frac{TP_c}{(TP_c + FP_c + FN_c)} \tag{7.3}$$

7.5.2 Model Evaluation

In this research work, three different datasets were used, and their distribution of the training and testing samples are detailed in Table 7.3. U-Net architecture is well known to work well with small datasets since it was designed to accomplish segmentation tasks for biomedical imaging, where the data tends to be scarce. However, some techniques, such as patching or data augmentation, are implemented to alleviate the lack of enough data during the training process of the DL models. The use of unannotated data is being extensively discussed in the computer vision community. Weak supervision learning works with limited data, and there is a different approach to address this scenario. For instance, Jee et al. [51] proposed the use of DL model to capture attention maps and create segmentation masks from them, avoiding the process of manual annotation, and leveraging the same information as if the dataset were annotated. A total of nine DL models based on the U-Net architecture were evaluated in this research. The quantitative results of each DL model are detailed in Table 7.4.

Results for CVC ClinicDB: Figure 7.9(a) shows the Dice-coefficient on the validation set when using CVC ClinicDB dataset. Efficient Dense U-Net and Efficient R2-Unet models provide better performances during the validation phase, outperforming the other seven U-Net-based models. The quantitative results shown in Table 7.4 also support the superior accuracy of the Efficient Dense R2 U-Net models. The results obtained by Efficient R2U-Net and Efficient Dense U-Net (in that order) are by far higher than those obtained when using the other state-of-the-art models, achieving a mean IoU of 0.9068 (Efficient R2U-Net) and 0.8972 (Efficient Dense U-Net) on the test samples.

Results for KVASIR-SEG: The quantitative comparison of nine U-Net-based models on the KVASIR-SEG

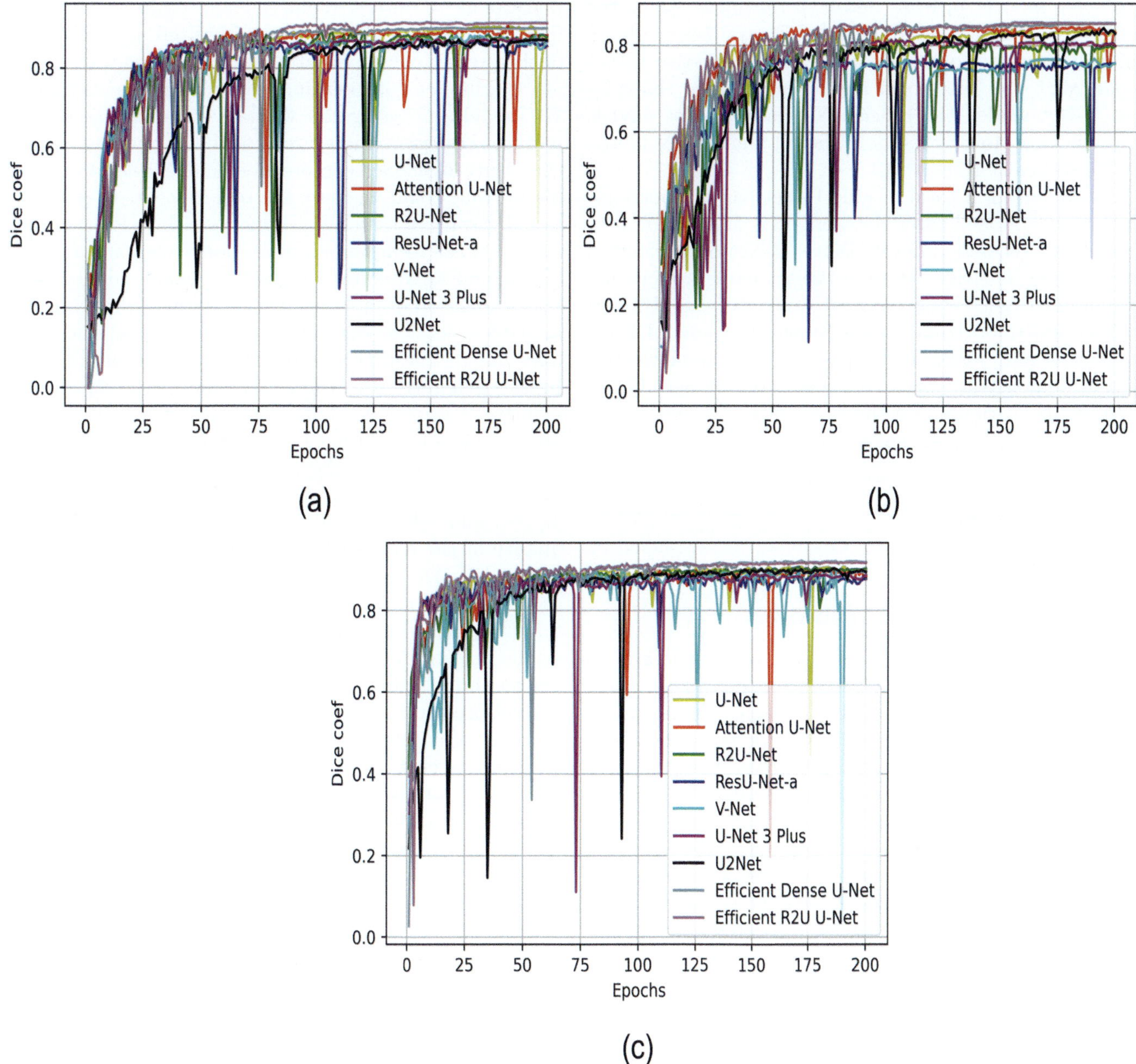

FIGURE 7.9
Dice coefficient of the different DL models on the three different validation datasets: (a) CVC ClinicDB, (b) KVASIR-SEG, and (c) ISIC2016.

dataset is shown in Table 7.4. A mean IoU of 0.8520 is achieved with an Efficient R2U-Net for the KVASIR-SEG, which is 1.82% better than the result obtained by the original U-Net model architecture. In addition, the results indicated the superior performance of the Efficient Dense U-Net and the Efficient R2U-Net models compared with the other six state-of-the-art U-Net variant models.

It is worth noting that these above-mentioned two datasets (CVC ClinicDB and KVASIR-SEG) are similar in terms of images contained that are from the gastrointestinal tract for polyp detection tasks. And it can be seen that the Efficient R2U-Net yielded the best accuracy in both dataset consistency, which proves that this U-Net variant generalizes well for this specific task.

Results for ISIC2016 dataset: Quantitative results of nine U-Net-based models on the ISIC2016 dataset are summarized in Table 7.4. From the table, it is noticeable that the Efficient Dense U-Net and Efficient

R2U-Net models provide better performance than the other seven U-Net-based models. Figure 7.9(c) shows the validation performance of the Efficient Dense U-Net model with the Dice-coefficient. Compared to the original U-Net model, the EfficientNet-based U-Net models improve mean IoU by 2.41% (Efficient Dense U-Net) and 1.77% (Efficient R2U-Net).

By evaluating a total of nine U-Net-based models for the aforementioned three datasets, it is found that Efficient R2U-Net and Efficient Dense U-Net outperformed the other U-Net-based state-of-the-art models, including Attention U-Net, U-Net 3+, ResU-Net, R2U-Net, U2-net, V-Net, and the original U-Net model, in terms of Dice-coefficient and Mean IoU. Therefore, the quantitative results demonstrate the effectiveness of the new U-Net architecture variants and the versatility of U-Net architecture for medical segmentation tasks.

7.6 Discussion

7.6.1 Limitations and Challenges

Although U-Net architecture and its respective variants have shown unprecedented results for BIS tasks, there are still various limitations of U-net and challenges to be addressed. Shahedi et al. [73] conducted experiments with the use of synthetic data and showed the limitation of U-Net architecture for extracting complex features in localizing objects of interest. The lack of capability for location-related feature extraction implies difficulty to localize irregular lesions (by shape) or irregular tumors (by size)[34] in biomedical images. Zhou et al. [97] argued that the performance of U-Net can be restricted by only fusing feature maps at the same level of the encoder–decoder path via skip connection. In addition, it was found that there is a need to find a sophisticated solution to define the depth of the network for a specific segmentation task since most of U-Net models use some typical pre-fixed depth parameter. Furthermore, Su et al. [79] suggested the use of kernel size with a fixed receptive field limits the capability of feature extraction and feature restoration of the architecture, since the use of small kernels could lead to the extraction of redundant features, and the use of larger kernels could avoid the presence of smalls targets. To sum up, the limitations of U-Net and its variants could be tackled from the following aspects: effective layer connection, robust feature extraction scheme, efficient parameter setting, and generalization for different BIS tasks.

According to Punn et al. [64], general challenges that DL architectures such as U-Net are facing include: computational resources requirements, data scarcity, low quality of the available data, and limited interpretability of the predictions. Computational power tends to threshold the implementation and design of new cutting-edge DL models, including new variants of U-Net-based architecture. However, there are a series of progress toward the study of model compression and acceleration techniques that could alleviate the requirements of computational resources to train a model [16]. In addition, there are attempts to develop automation techniques for optimal model architecture design [67], which could help in the production of more efficient and less complex models for BIS tasks. The data availability is crucial due to the data-hungry nature of DL models. However, it is affected by the cost related to the acquisition of the data or the security implications. This causes a bottleneck in developing real-world systems based on deep learning models. This challenge is alleviated by the exploration of techniques such as data augmentation [75], transfer learning [9, 81], and self-supervised learning (SSL) [47]. The low quality of the data available could also limit the performance of a DL model for BIS. Images with low quality, as a result of noise and artifacts, may end up with altered features, that is, obscure features due to the noise or the presence of irrelevant features due artifact. In order to address this challenge, there have been proposed several pre-processing strategies to eliminate or hide the presence of noise and artifacts from the data. Techniques such as wavelets thresholding, partial differential equations, and anisotropic diffusion are used to denoise the data [62, 66], meanwhile, newer reconstruction [82] or metal artifact reduction [14] are techniques used to reduce the presence of artifacts. Another challenge is the black-box effect caused by the lack of transparency of the DL models, which makes them hard to interpret and understand. Few visualization tools are proposed for interpretability of the model predictions (i.e., LIME [60], SHAP [54], PDP [26], and anchor [68]). However, it still requires a concrete benchmark scheme to fulfill the necessity of understanding the reasons behind a prediction.

In terms of feasibility of implementation in clinical setup, since the level of performance achieved by a DL model in BIS depends on the specific task and the quality of the training data, there is not a consensual threshold that would warrant a satisfactory performance for medical cases [44, 55]. This statement is aside from several factors to be considered such as the level of accuracy, robustness, and interpretability of the DL model. In addition, it is important to consider the ethical and legal implications, which involve data privacy and patient consents to ensure that the DL model has been implemented and trained following the standards dictated by the scientific community [17, 86].

7.6.2 Future Trends

There is no doubt that the use of DL to mine clinical information is one of the major trends in the medical industry, and the research community is actively integrating new technologies that align with this current trend. Although U-Net is still considered the baseline for BIS, there are several promising approaches that are being considered to eventually lead the new trends in the field. For instance, the use of spatiotemporal information is being considered to improve the performance of deep neural networks for segmentation tasks [15, 31, 63]. Klymenko et al. [48] proposed the use of information from longitudinal CT scans with the implementation of a spatiotemporal memory network, named Butterfly-Net, in which the prior scan information is stored and used for segmentation at a target time, obtaining a better performance comparing to V-Net. In addition, they explored the use of clinical factors along with the temporal information boosting the performance of the initial study.

On the other hand, graph convolutional neural networks (GCN) [87] is another popular methodology since this provides a powerful and intuitive modeling approach [92], and it has been incorporated in the encoder–decoder networks for segmentation tasks [29], obtaining better performance than the popular U-Net, showing an improved capability to encode features with a small number of parameters. This methodology has been widely used for different BIS [74, 90, 91].

The success of Transformers [84] in NLP has called the attention of researchers to leverage the advantage of self-attention modules to learn global and long-range semantic information. Vision Transformers (ViT) [22] accomplished the task of image classification using Transformers, and this trend has been followed by several works where Transformers are applied for BIS and demonstrated outstanding performance. However, researchers agree that just replacing CNN operations with Transformers may yield high computational costs. Thus, a suitable direction for research could be the combination of Transformers with CNN [11, 13, 32, 83, 89] in an effective manner.

Another important trend is the development of techniques that helps in the ability to understand the reasoning behind the DL models' predictions, and thus learn what the feature and the decision boundaries learned by those models. Explainability can benefit the deployment of DL models in clinical setups since this can provide the robustness required to be trusted as a solution by professionals in medicine and patients in general [77]. The use of saliency and feature activation maps are approach trending to improve the explainability of DL models for BIS [23], as well as the use of interpretable ML models [71], such as decision trees or rule-based models, which tend to be easier to explain in comparison with complex neural networks. Model distillation [2, 33], in which a DL model is trained to "mimic" the predictions of a simpler and interpretable model, is a topic of study to enhance the explainability of decision-making for DL models in BIS tasks.

Lastly, the design of novel architecture manually requires experience and strong exposure to the field where the model is targeting. Therefore, the use of neural architecture search (NAS) is expected to improve the design process of architecture. However, the computational cost involved in a direct search limits the implementation, thus the combination of manual design, for backbones, and NAS, for small network modules, seems to be the future direction in this field [85].

7.7 Conclusions

The U-Net architecture has proven to be a powerful DL framework used as a tool for biomedical image segmentation tasks. The performance of U-Net architecture and its different variants have shown high reliability in terms of precise segmentation, which is a desired aspect for a proper diagnosis of diseases. The results from the experiments performed in this work have proven that U-Net architecture can effectively identify objects of interest in biomedical images, and the integration of various DL techniques can help to improve its performance. We empirically showed that (1) the integration of recurrent feedback connections into U-Net can enhance segmentation accuracy; (2) EfficientNet encoded U-Net obtained richer feature representation by leveraging a compound coefficient to uniformly scale network width, depth, and resolution. However, there are some limitations of U-Net-based architectures. For instance, parameter efficiency and generalizability of U-Net models are the two challenges and their innovative solutions that could be the future trend in architecture design. In addition, as U-Net uses CNN structure to extract local features, there is a need to extract both local and global features simultaneously to improve prediction accuracy. To address this problem, ViT [22] based U-net architectures are gradually emerging. However, most ViT-based U-Nets are simply an alternate that combines transformer blocks with plain U-Net, which may not be an effective solution to achieve desired performance. Furthermore, the future U-Net-based architectures are anticipated to be more explainable and clinically acceptable, so that they can be deployed to the clinical systems with confidence to aid in medical diagnostics.

References

[1] Walid Al-Dhabyani, Mohammed Gomaa, Hussien Khaled, and Aly Fahmy. Dataset of breast ultrasound images. *Data in Brief*, 28:104863, 2020.

[2] Abdolmaged Alkhulaifi, Fahad Alsahli, and Irfan Ahmad. Knowledge distillation in deep learning and its applications. *PeerJ Computer Science*, 7:e474, 2021.

[3] Md Zahangir Alom, Chris Yakopcic, Mahmudul Hasan, Tarek M Taha, and Vijayan K Asari. Recurrent residual U-Net for medical image segmentation. *Journal of Medical Imaging*, 6(1):014006, 2019.

[4] Samuel G Armato III, Geoffrey McLennan, Luc Bidaut, Michael F McNitt-Gray, Charles R Meyer, Anthony P Reeves, Binsheng Zhao, Denise R Aberle, Claudia I Henschke, Eric A Hoffman, et al. The lung image database consortium (LIDC) and image database resource initiative (IDRI): A completed reference database of lung nodules on CT scans. *Medical Physics*, 38(2):915–931, 2011.

[5] Ujjwal Baid, Satyam Ghodasara, Suyash Mohan, Michel Bilello, Evan Calabrese, Errol Colak, Keyvan Farahani, Jayashree Kalpathy-Cramer, Felipe C Kitamura, Sarthak Pati, et al. The RSNS-ASNR-MICCAI BraTS 2021 benchmark on brain tumor segmentation and radiogenomic classification. *arXiv preprint arXiv:2107.02314*, 2021.

[6] Spyridon Bakas, Hamed Akbari, Aristeidis Sotiras, Michel Bilello, Martin Rozycki, Justin S Kirby, John B Freymann, Keyvan Farahani, and Christos Davatzikos. Advancing the Cancer Genome Atlas Glioma MRI collections with expert segmentation labels and radiomic features. *Scientific Data*, 4(1):1–13, 2017.

[7] Jorge Bernal, F Javier Sánchez, Gloria Fernández-Esparrach, Debora Gil, Cristina Rodríguez, and Fernando Vilariño. WM-DOMA maps for accurate polyp highlighting in colonoscopy: Validation vs. saliency maps from physicians. *Computerized Medical Imaging and Graphics*, 43:99–111, 2015.

[8] Patrick Bilic, Patrick Ferdinand Christ, Eugene Vorontsov, Grzegorz Chlebus, Hao Chen, Qi Dou, Chi-Wing Fu, Xiao Han, Pheng-Ann Heng, Jürgen Hesser, et al. The liver tumor segmentation benchmark (lits). *arXiv preprint arXiv:1901.04056*, 2019.

[9] Michal Byra, Mei Wu, Xiaodong Zhang, Hyungseok Jang, Ya-Jun Ma, Eric Y Chang, Sameer Shah, and Jiang Du. Knee menisci segmentation and relaxometry of 3D ultrashort echo time cones mr imaging using attention U-Net with transfer learning. *Magnetic Resonance in Medicine*, 83(3):1109–1122, 2020.

[10] Juan C Caicedo, Allen Goodman, Kyle W Karhohs, Beth A Cimini, Jeanelle Ackerman, Marzieh Haghighi, CherKeng Heng, Tim Becker, Minh Doan, Claire McQuin, et al. Nucleus segmentation across imaging experiments: The 2018 data science bowl. *Nature Methods*, 16(12):1247–1253, 2019.

[11] Hu Cao, Yueyue Wang, Joy Chen, Dongsheng Jiang, Xiaopeng Zhang, Qi Tian, and Manning Wang. Swin-unet: UNet-like pure transformer for medical image segmentation. *arXiv preprint arXiv:2105.05537*, 2021.

[12] Albert Cardona, Stephan Saalfeld, Stephan Preibisch, Benjamin Schmid, Anchi Cheng, Jim Pulokas, Pavel Tomancak, and Volker Hartenstein. An integrated micro-and macroarchitectural analysis of the drosophila brain by computer-assisted serial section electron microscopy. *PLoS Biology*, 8(10):e1000502, 2010.

[13] Jieneng Chen, Yongyi Lu, Qihang Yu, Xiangde Luo, Ehsan Adeli, Yan Wang, Le Lu, Alan L Yuille, and Yuyin Zhou. Transunet: Transformers make strong encoders for medical image segmentation. *arXiv preprint arXiv:2102.04306*, 2021.

[14] Ming Chen, Dimeng Xia, Dan Wang, Jingqi Han, and Zhu Liu. An analytical method for reducing metal artifacts in x-ray CT images. *Mathematical Problems in Engineering*, 2019, 2019.

[15] Mingjian Chen, Hao Zheng, Changsheng Lu, Enmei Tu, Jie Yang, and Nikola Kasabov. Accurate breast lesion segmentation by exploiting spatio-temporal information with deep recurrent and convolutional network. *Journal of Ambient Intelligence and Humanized Computing*, pages 1–9, 2019.

[16] Yu Cheng, Duo Wang, Pan Zhou, and Tao Zhang. A survey of model compression and acceleration for deep neural networks. *arXiv preprint arXiv:1710.09282*, 2017.

[17] Emna Chikhaoui, Alanoud Alajmi, and Souad Larabi-Marie-Sainte. Artificial intelligence applications in healthcare sector: Ethical and legal challenges. *Emerging Science Journal*, 6(4):717–738, 2022.

[18] Özgün Çiçek, Ahmed Abdulkadir, Soeren S Lienkamp, Thomas Brox, and Olaf Ronneberger. 3D U-Net: Learning dense volumetric segmentation from sparse annotation. In *International Conference on Medical Image Computing and Computer-Assisted Intervention*, pages 424–432. Springer, 2016.

[19] Noel CF Codella, David Gutman, M Emre Celebi, Brian Helba, Michael A Marchetti, Stephen W Dusza, Aadi Kalloo, Konstantinos Liopyris, Nabin Mishra, Harald Kittler, et al. Skin lesion analysis toward melanoma detection: A challenge at the 2017 international symposium on biomedical imaging (ISBI), hosted by the international skin imaging collaboration (ISIC). In *2018 IEEE 15th International Symposium on Biomedical Imaging (ISBI 2018)*, pages 168–172. IEEE, 2018.

[20] Lee R Dice. Measures of the amount of ecologic association between species. *Ecology*, 26(3):297–302, 1945.

[21] Konstantin Dmitriev and Arie E Kaufman. Learning multi-class segmentations from single-class datasets. In *Proceedings of the IEEE/CVF Conference on Computer Vision and Pattern Recognition*, pages 9501–9511, 2019.

[22] Alexey Dosovitskiy, Lucas Beyer, Alexander Kolesnikov, Dirk Weissenborn, Xiaohua Zhai, Thomas Unterthiner, Mostafa Dehghani, Matthias Minderer, Georg Heigold, Sylvain Gelly, et al. An image is worth 16x16 words: Transformers for image recognition at scale. *arXiv preprint arXiv:2010.11929*, 2020.

[23] Martin Dyrba, Arjun H Pallath, and Eman N Marzban. Comparison of CNN visualization methods to aid model interpretability for detecting Alzheimer's disease. In *Bildverarbeitung für die Medizin 2020: Algorithmen–Systeme–Anwendungen. Proceedings des Workshops vom 15. bis 17. März 2020 in Berlin*, pages 307–312. Springer, 2020.

[24] Araon Fenster and Bernard Chiu. Evaluation of segmentation algorithms for medical imaging. In *2005 IEEE Engineering in Medicine and Biology 27th Annual Conference*, pages 7186–7189. IEEE, 2006.

[25] M. M. Fraz, P. Remagnino, A. Hoppe, B. Uyyanonvara, A. R. Rudnicka, C. G. Owen, and S. A. Barman. An ensemble classification-based approach applied to retinal blood vessel segmentation. *IEEE Transactions on Biomedical Engineering*, 59(9):2538–2548, 2012.

[26] Jerome H Friedman. Greedy function approximation: A gradient boosting machine. *Annals of Statistics*, pages 1189–1232, 2001.

[27] King-Sun Fu and JK Mui. A survey on image segmentation. *Pattern Recognition*, 13(1):3–16, 1981.

[28] Jevgenij Gamper, Navid Alemi Koohbanani, Ksenija Benet, Ali Khuram, and Nasir Rajpoot. Pannuke: An open pan-cancer histology dataset for nuclei instance segmentation and classification. In *European Congress on Digital Pathology*, pages 11–19. Springer, 2019.

[29] Hongyang Gao and Shuiwang Ji. Graph U-Nets. In *International Conference on Machine Learning*, pages 2083–2092. PMLR, 2019.

[30] Long Gao, Lei Zhang, Chang Liu, and Shandong Wu. Handling imbalanced medical image data: A deep-learning-based one-class classification approach. *Artificial Intelligence in Medicine*, 108:101935, 2020.

[31] Yang Gao, Jeff M Phillips, Yan Zheng, Renqiang Min, P Thomas Fletcher, and Guido Gerig. Fully convolutional structured LSTM networks for joint 4D medical image segmentation. In *2018 IEEE 15th International Symposium on Biomedical Imaging (ISBI 2018)*, pages 1104–1108. IEEE, 2018.

[32] Yunhe Gao, Mu Zhou, and Dimitris N Metaxas. Utnet: a hybrid transformer architecture for medical image segmentation. In *International Conference on Medical Image Computing and Computer-Assisted Intervention*, pages 61–71. Springer, 2021.

[33] Jianping Gou, Baosheng Yu, Stephen J Maybank, and Dacheng Tao. Knowledge distillation: A survey. *International Journal of Computer Vision*, 129:1789–1819, 2021.

[34] Naga Raju Gudhe, Hamid Behravan, Mazen Sudah, Hidemi Okuma, Ritva Vanninen, Veli-Matti Kosma, and Arto Mannermaa. Multi-level dilated residual network for biomedical image segmentation. *Scientific Reports*, 11(1):1–18, 2021.

[35] Saumya Gupta, Xiaoling Hu, James Kaan, Michael Jin, Mutshipay Mpoy, Katherine Chung, Gagandeep Singh, Mary Saltz, Tahsin Kurc, Joel Saltz, et al. Learning topological interactions for multi-class medical image segmentation. In *European Conference on Computer Vision*, pages 701–718. Springer, 2022.

[36] Intisar Rizwan I Haque and Jeremiah Neubert. Deep learning approaches to biomedical image segmentation. *Informatics in Medicine Unlocked*, 18:100297, 2020.

[37] Mahmoud Hassaballah, Mourad A Kenk, Khan Muhammad, and Shervin Minaee. Vehicle detection and tracking in adverse weather using a deep learning framework. *IEEE Transactions on Intelligent Transportation Systems*, 22(7):4230–4242, 2020.

[38] Moritz Roman Hernandez Petzsche, Ezequiel de la Rosa, Uta Hanning, Roland Wiest, Waldo Enrique Valenzuela Pinilla, Mauricio Reyes, Maria Ines Meyer, Sook-Lei Liew, Florian Kofler, Ivan Ezhov, et al. Isles 2022: A multi-center magnetic resonance imaging stroke lesion segmentation dataset. *arXiv e-prints*, pages arXiv–2206, 2022.

[39] AD Hoover, Valentina Kouznetsova, and Michael Goldbaum. Locating blood vessels in retinal images by piecewise threshold probing of a matched filter response. *IEEE Transactions on Medical imaging*, 19(3):203–210, 2000.

[40] Huimin Huang, Lanfen Lin, Ruofeng Tong, Hongjie Hu, Qiaowei Zhang, Yutaro Iwamoto, Xianhua Han, Yen-Wei Chen, and Jian Wu. Unet 3+: A full-scale connected unet for medical image segmentation. In *ICASSP 2020-2020 IEEE International Conference on Acoustics, Speech and Signal Processing (ICASSP)*, pages 1055–1059. IEEE, 2020.

[41] Nabil Ibtehaz and M Sohel Rahman. Multiresunet: Rethinking the U-Net architecture for multimodal biomedical image segmentation. *Neural Networks*, 121:74–87, 2020.

[42] Sergey Ioffe and Christian Szegedy. Batch normalization: Accelerating deep network training by reducing internal covariate shift. In *International Conference on Machine Learning*, pages 448–456. PMLR, 2015.

[43] Shu Isaka, Hiroharu Kawanaka, Bruce J Aronow, and VB Surya Prasath. Multi-class segmentation of lung immunofluorescence confocal images using deep learning. In *2019 IEEE International Conference on Bioinformatics and Biomedicine (BIBM)*, pages 2362–2368. IEEE, 2019.

[44] Fabian Isensee, Paul F Jaeger, Simon AA Kohl, Jens Petersen, and Klaus H Maier-Hein. NNU-Net: a self-configuring method for deep learning-based biomedical image segmentation. *Nature Methods*, 18(2):203–211, 2021.

[45] Paul Jaccard. The distribution of the flora in the alpine zone. 1. *New Phytologist*, 11(2):37–50, 1912.

[46] Debesh Jha, Pia H Smedsrud, Michael A Riegler, Pål Halvorsen, Thomas de Lange, Dag Johansen, and Håvard D Johansen. Kvasir-SEG: A segmented polyp dataset. In *International Conference on Multimedia Modeling*, pages 451–462. Springer, 2020.

[47] Longlong Jing and Yingli Tian. Self-supervised visual feature learning with deep neural networks: A survey. *IEEE Transactions on Pattern Analysis and Machine Intelligence*, 43(11):4037–4058, 2020.

[48] Tetiana Klymenko, Seong Tae Kim, Kirsten Lauber, Christopher Kurz, Guillaume Landry, Nassir Navab, and Shadi Albarqouni. Butterfly-net: Spatial-temporal architecture for medical image segmentation. In *2021 IEEE 18th International Symposium on Biomedical Imaging (ISBI)*, pages 616–620, 2021.

[49] Robert Korez, Bulat Ibragimov, Boštjan Likar, Franjo Pernuš, and Tomaž Vrtovec. A framework for automated spine and vertebrae interpolation-based detection and model-based segmentation. *IEEE Transactions on Medical Imaging*, 34(8):1649–1662, 2015.

[50] Yann LeCun, Yoshua Bengio, and Geoffrey Hinton. Deep learning. *Nature*, 521(7553):436–444, 2015.

[51] Jungbeom Lee, Seong Joon Oh, Sangdoo Yun, Junsuk Choe, Eunji Kim, and Sungroh Yoon. Weakly supervised semantic segmentation using out-of-distribution data. In *Proceedings of the IEEE/CVF Conference on Computer Vision and Pattern Recognition*, pages 16897–16906, 2022.

[52] Ming Liang and Xiaolin Hu. Recurrent convolutional neural network for object recognition. In *Proceedings of the IEEE Conference on Computer Vision and Pattern Recognition*, pages 3367–3375, 2015.

[53] Jonathan Long, Evan Shelhamer, and Trevor Darrell. Fully convolutional networks for semantic segmentation. In *Proceedings of the IEEE Conference on Computer Vision and Pattern Recognition*, pages 3431–3440, 2015.

[54] Scott M Lundberg and Su-In Lee. A unified approach to interpreting model predictions. *Advances in Neural Information Processing Systems*, 30, 2017.

[55] Huan Minh Luu and Sung-Hong Park. Extending NN-Unet for brain tumor segmentation. In *Brainlesion: Glioma, Multiple Sclerosis, Stroke and Traumatic Brain Injuries: 7th International Workshop, BrainLes 2021, Held in Conjunction with MICCAI 2021, Virtual Event, September 27, 2021, Revised Selected Papers, Part II*, pages 173–186. Springer, 2022.

[56] Bjoern H Menze, Andras Jakab, Stefan Bauer, Jayashree Kalpathy-Cramer, Keyvan Farahani,

Justin Kirby, Yuliya Burren, Nicole Porz, Johannes Slotboom, Roland Wiest, et al. The multimodal brain tumor image segmentation benchmark (brats). *IEEE Transactions on Medical Imaging*, 34(10):1993–2024, 2014.

[57] Pablo Mesejo, Andrea Valsecchi, Linda Marrakchi-Kacem, Stefano Cagnoni, and Sergio Damas. Biomedical image segmentation using geometric deformable models and metaheuristics. *Computerized Medical Imaging and Graphics*, 43:167–178, 2015.

[58] Fausto Milletari, Nassir Navab, and Seyed-Ahmad Ahmadi. V-Net: Fully convolutional neural networks for volumetric medical image segmentation. In *2016 Fourth International Conference on 3D Vision (3DV)*, pages 565–571. IEEE, 2016.

[59] Shervin Minaee, Yuri Y Boykov, Fatih Porikli, Antonio J Plaza, Nasser Kehtarnavaz, and Demetri Terzopoulos. Image segmentation using deep learning: A survey. *IEEE Transactions on Pattern Analysis and Machine Intelligence*, 2021.

[60] Saumitra Mishra, Bob L Sturm, and Simon Dixon. Local interpretable model-agnostic explanations for music content analysis. In *ISMIR*, volume 53, pages 537–543, 2017.

[61] Ozan Oktay, Jo Schlemper, Loic Le Folgoc, Matthew Lee, Mattias Heinrich, Kazunari Misawa, Kensaku Mori, Steven McDonagh, Nils Y Hammerla, Bernhard Kainz, et al. Attention U-Net: Learning where to look for the pancreas. *arXiv preprint arXiv:1804.03999*, 2018.

[62] Hind Oulhaj, Aouatif Amine, Mohammed Rziza, and Driss Aboutajdine. Noise reduction in medical images-comparison of noise removal algorithms. In *2012 International Conference on Multimedia Computing and Systems*, pages 344–349. IEEE, 2012.

[63] Constance A Owens, Christine B Peterson, Chad Tang, Eugene J Koay, Wen Yu, Dennis S Mackin, Jing Li, Mohammad R Salehpour, David T Fuentes, Laurence E Court, et al. Lung tumor segmentation methods: Impact on the uncertainty of radiomics features for non-small cell lung cancer. *PLoS One*, 13(10):e0205003, 2018.

[64] Narinder Singh Punn and Sonali Agarwal. Modality specific u-net variants for biomedical image segmentation: a survey. *Artificial Intelligence Review*, pages 1–45, 2022.

[65] Xuebin Qin, Zichen Zhang, Chenyang Huang, Masood Dehghan, Osmar R Zaiane, and Martin Jagersand. U2-Net: Going deeper with nested U-structure for salient object detection. *Pattern Recognition*, 106:107404, 2020.

[66] Aditya Ravishankar, S Anusha, HK Akshatha, Anjali Raj, S Jahnavi, and J Madhura. A survey on noise reduction techniques in medical images. In *2017 International Conference of Electronics, Communication and Aerospace Technology (ICECA)*, volume 1, pages 385–389. IEEE, 2017.

[67] Shaoqing Ren, Kaiming He, Ross Girshick, and Jian Sun. Faster R-CNN: Towards real-time object detection with region proposal networks. *Advances in Neural Information Processing Systems*, 28, 2015.

[68] Marco Tulio Ribeiro, Sameer Singh, and Carlos Guestrin. Anchors: High-precision model-agnostic explanations. In *Proceedings of the AAAI Conference on Artificial Intelligence*, volume 32, 2018.

[69] Olaf Ronneberger, Philipp Fischer, and Thomas Brox. U-Net: Convolutional networks for biomedical image segmentation. In *International Conference on Medical Image Computing and Computer-Assisted Intervention*, pages 234–241. Springer, 2015.

[70] Veronica Rotemberg, Nicholas Kurtansky, Brigid Betz-Stablein, Liam Caffery, Emmanouil Chousakos, Noel Codella, Marc Combalia, Stephen Dusza, Pascale Guitera, David Gutman, et al. A patient-centric dataset of images and metadata for identifying melanomas using clinical context. *Scientific Data*, 8(1):34, 2021.

[71] Cynthia Rudin. Stop explaining black box machine learning models for high stakes decisions and use interpretable models instead. *Nature Machine Intelligence*, 1(5):206–215, 2019.

[72] Arnaud Arindra Adiyoso Setio, Alberto Traverso, Thomas De Bel, Moira SN Berens, Cas Van Den Bogaard, Piergiorgio Cerello, Hao Chen, Qi Dou, Maria Evelina Fantacci, Bram Geurts, et al. Validation, comparison, and combination of algorithms for automatic detection of pulmonary nodules in computed tomography images: The LUNA16 challenge. *Medical Image Analysis*, 42:1–13, 2017.

[73] Maysam Shahedi et al. A study on U-Net limitations in object localization and image segmentation. In *Society for Imaging Informatics in Medicine (SIIM), Virtual Meeting*, 2020.

[74] Seung Yeon Shin, Soochahn Lee, Il Dong Yun, and Kyoung Mu Lee. Deep vessel segmentation by learning graphical connectivity. *Medical Image Analysis*, 58:101556, 2019.

[75] Connor Shorten and Taghi M Khoshgoftaar. A survey on image data augmentation for deep learning. *Journal of Big Data*, 6(1):1–48, 2019.

[76] Nahian Siddique, Sidike Paheding, Abel A Reyes Angulo, Md Zahangir Alom, and Vijay K Devabhaktuni. Fractal, recurrent, and dense U-Net architectures with efficientnet encoder for medical image segmentation. *Journal of Medical Imaging*, 9(6):064004, 2022.

[77] Amitojdeep Singh, Sourya Sengupta, and Vasudevan Lakshminarayanan. Explainable deep learning models in medical image analysis. *Journal of Imaging*, 6(6):52, 2020.

[78] J. Staal, M.D. Abramoff, M. Niemeijer, M.A. Viergever, and B. van Ginneken. Ridge-based vessel segmentation in color images of the retina. *IEEE Transactions on Medical Imaging*, 23(4):501–509, 2004.

[79] Run Su, Deyun Zhang, Jinhuai Liu, and Chuandong Cheng. MSU-Net: multi-scale U-Net for 2D medical image segmentation. *Frontiers in Genetics*, 12:639930, 2021.

[80] Nima Tajbakhsh, Laura Jeyaseelan, Qian Li, Jeffrey N Chiang, Zhihao Wu, and Xiaowei Ding. Embracing imperfect datasets: A review of deep learning solutions for medical image segmentation. *Medical Image Analysis*, 63:101693, 2020.

[81] Chuanqi Tan, Fuchun Sun, Tao Kong, Wenchang Zhang, Chao Yang, and Chunfang Liu. A survey on deep transfer learning. In *International Conference on Artificial Neural Networks*, pages 270–279. Springer, 2018.

[82] Benjamin L Triche, John T Nelson Jr, Noah S McGill, Kristin K Porter, Rupan Sanyal, Franklin N Tessler, Jonathan E McConathy, David M Gauntt, Michael V Yester, and Satinder P Singh. Recognizing and minimizing artifacts at CT, MRI, US, and molecular imaging. *RadioGraphics*, 39(4):1017–1018, 2019.

[83] Jeya Maria Jose Valanarasu, Poojan Oza, Ilker Hacihaliloglu, and Vishal M Patel. Medical transformer: Gated axial-attention for medical image segmentation. In *International Conference on Medical Image Computing and Computer-Assisted Intervention*, pages 36–46. Springer, 2021.

[84] Ashish Vaswani, Noam Shazeer, Niki Parmar, Jakob Uszkoreit, Llion Jones, Aidan N Gomez, Lukasz Kaiser, and Illia Polosukhin. Attention is all you need. *Advances in Neural Information Processing Systems*, 30, 2017.

[85] Yu Weng, Tianbao Zhou, Yujie Li, and Xiaoyu Qiu. NAS-Unet: Neural architecture search for medical image segmentation. *IEEE Access*, 7:44247–44257, 2019.

[86] Julian L Wichmann, Martin J Willemink, and Carlo N De Cecco. Artificial intelligence and machine learning in radiology: Current state and considerations for routine clinical implementation. *Investigative Radiology*, 55(9):619–627, 2020.

[87] Zonghan Wu, Shirui Pan, Fengwen Chen, Guodong Long, Chengqi Zhang, and S Yu Philip. A comprehensive survey on graph neural networks. *IEEE Transactions on Neural Networks and Learning Systems*, 32(1):4–24, 2020.

[88] Xiao Xiao, Shen Lian, Zhiming Luo, and Shaozi Li. Weighted RES-Unet for high-quality retina vessel segmentation. In *2018 9th International Conference on Information Technology in Medicine and Education (ITME)*, pages 327–331. IEEE, 2018.

[89] Enze Xie, Wenhai Wang, Zhiding Yu, Anima Anandkumar, Jose M Alvarez, and Ping Luo. Segformer: Simple and efficient design for semantic segmentation with transformers. *Advances in Neural Information Processing Systems*, 34:12077–12090, 2021.

[90] Bin Yang, Haiwei Pan, Jieyao Yu, Kun Han, and Yanan Wang. Classification of medical images with synergic graph convolutional networks. In *2019 IEEE 35th International Conference on Data Engineering Workshops (ICDEW)*, pages 253–258. IEEE, 2019.

[91] Han Yang, Xingjian Zhen, Ying Chi, Lei Zhang, and Xian-Sheng Hua. CPR-GCN: Conditional partial-residual graph convolutional network in automated anatomical labeling of coronary arteries. In *Proceedings of the IEEE/CVF Conference on Computer Vision and Pattern Recognition*, pages 3803–3811, 2020.

[92] Bingfeng Zhang, Jimin Xiao, Jianbo Jiao, Yunchao Wei, and Yao Zhao. Affinity attention graph neural network for weakly supervised semantic segmentation. *IEEE Transactions on Pattern Analysis and Machine Intelligence*, 2021.

[93] Jiawei Zhang, Yuzhen Jin, Jilan Xu, Xiaowei Xu, and Yanchun Zhang. MDU-NET: Multi-scale densely connected U-Net for biomedical image segmentation. *arXiv preprint arXiv:1812.00352*, 2018.

[94] Yingtao Zhang, Min Xian, Heng-Da Cheng, Bryar Shareef, Jianrui Ding, Fei Xu, Kuan Huang, Boyu Zhang, Chunping Ning, and Ying Wang. Busis: A benchmark for breast ultrasound image segmentation. In *Healthcare*, volume 10, page 729. MDPI, 2022.

[95] Zhengxin Zhang, Qingjie Liu, and Yunhong Wang. Road extraction by deep residual U-Net. *IEEE Geoscience and Remote Sensing Letters*, 15(5):749–753, 2018.

[96] Ziang Zhang, Chengdong Wu, Sonya Coleman, and Dermot Kerr. Dense-inception U-Net for medical image segmentation. *Computer Methods and Programs in Biomedicine*, 192:105395, 2020.

[97] Zongwei Zhou, Md Mahfuzur Rahman Siddiquee, Nima Tajbakhsh, and Jianming Liang. Unet++: Redesigning skip connections to exploit multiscale features in image segmentation. *IEEE Transactions on Medical Imaging*, 39(6):1856–1867, 2019.

8

Robot-Mediated Assistance: Opportunities and Challenges in Computer Vision and Human–Robot Interaction

Md Alimoor Reza and Syed Masum Billah

Robot-mediated assistance is poised for wide acceptance in sectors such as households, healthcare, distance learning, and assisted living in the near future. In this chapter, we analyze a curated collection of 29 existing robots. This analysis leads us to extend the current human–robot interaction (HRI) taxonomy by incorporating three additional lenses: robot appearance (including human-like, animal-like, and machine-like), collaborators (involving local users, bystanders, tele-users, tele-operators, other robots, and AI developers), and abilities (such as locomotion, perception, grasping, manipulation, and communication with humans or robots). We examine how these factors influence the quality, utility, and nature of collaboration in robot-mediated interaction. Further, we scrutinize the challenges and opportunities in robotics, computer vision, and HRI, aiming to stimulate and inform future research in these areas.

8.1 Introduction

Robot-mediated assistance, where robots increasingly assume pivotal roles as collaborative partners with humans, stands as a technological breakthrough on the brink of reshaping society. Robots, proving indispensable across diverse contexts [60, 73, 101], range from aiding in assisted care to handling routine tasks. For instance, in the *general home service* sector, robots undertake tasks like clearing the dinner table, doing laundry, or engaging in crafting activities. In *assisted living service*, they assist individuals by grasping canes, navigating environments, and performing various support tasks. Additionally, in the realm of *tele-medicine service*, robots can interpret medicine labels and retrieve specified medications. Depending on the nature of their tasks, these robots operate from fixed locations—like workstations or bedside setups—or navigate within a defined environment. Their autonomy ranges from independent navigation and semi-autonomous maneuvering to tele-operation under human supervision.

Two primary categories typify robots based on their roles and modes of human interaction: *social* and *service* robots. Social robots are engineered to engage in socially meaningful interactions with humans, while service robots concentrate on offering functional assistance to simplify and expedite the execution of physical tasks. Service robots are rapidly evolving to serve citizens more effectively in the future [144]. Cases in point include the tele-operation of a robotic arm via virtual reality (VR) to set a dining table from halfway across the globe [161], and the automatic synthesis of virtual wheelchair training environments [89]. Such advancements can not only blur geographical boundaries but also extend the reach of service robots beyond physical constraints, enhancing human skills and quality of life [28]. Concurrently, social robots are beginning to find applications in sectors such as mental healthcare and education [120, 121].

In traditional tele-assistance settings, a human provider remotely aids another human receiver through audio and video communication. However, introducing a robot into this framework ushers in a new paradigm of robot-mediated tele-assistance. In this arrangement, a remote human provider can instruct a robot to perform various tasks such as picking out the correct medicine bottle from a cabinet for an older adult, locating dropped objects like keys or wallets for a visually impaired user, or serving food to patrons at a restaurant table.

Four critical dimensions characterize robot-mediated interactions. Firstly, while humans and robots form a team and work jointly toward a shared goal, the actions available to each team member may vary [30]. For example, robots are well-suited to manage dangerous or unpleasant tasks under remote human operations. Secondly, the responsibilities of each team member can shift as a task progresses, fostering dynamic and adaptive human–robot collaboration [27]. Thirdly, such collaborations may span considerable durations, with repeated interactions fostering shared knowledge and conventions such as role

DOI: 10.1201/9781003328957-8

specialization based on skill recognition [30, 132]. Lastly, collaborations can also be brief, with the human partner frequently rotating.

In the exploration of robot-mediated interaction, we can scrutinize the roles of humans and robots from multiple perspectives. One foundational classification, as proposed by Yanco et al., provides broad categories of interaction [157].

- **Autonomy and Intervention Levels:** These measure the percentage of time the robot performs tasks independently and the percentage of time it requires human intervention, respectively. The two sums to 100%, denoting the division of task completion between robotic autonomy and human involvement.

- **Space-Time Taxonomy:** This approach classifies human-robot interaction (HRI) based on its synchronicity (simultaneous or time-separated) and location (same or different places).

- **Composition of Robot Teams:** Multiple robots can form either homogeneous teams (e.g., multiple Roombas cleaning a house) or heterogeneous teams (e.g., an Astro playing with a pet before a Roomba cleans up).

- **Human-to-Robot Ratio:** This simply counts the number of humans and robots involved (e.g., 1:2 represents one human operating two robots).

- **Shared Interaction:** The level of shared interaction can vary from one human interacting with a single robot or a team of robots to multiple humans interacting with a single robot.

- **Decision Support:** Factors such as available sensor information, sensor fusion type, and preprocessing can facilitate HRI.

- **Task Criticality:** This assesses the importance of accurately completing a task and the potential negative effects of failure, categorized as high, medium, or low.

While this taxonomy offers a comprehensive framework, it leaves out certain aspects of robot characteristics and functionalities, such as appearance, capabilities, and collaborative partners. For instance, a robot's appearance can significantly influence its reception and effectiveness [23]. Similarly, factors contributing to a robot's autonomy, such as locomotion, perception, communication, and manipulation abilities, are not distinctly highlighted. To address this gap, we examine a sample of 29 existing robots (Table 8.1). Our selection of these robots is informed by prevalent trends in AI and HRI literature, industry movements, and relevant newsletters (e.g., IEEE Spectrum Robotics). Based on our analysis, we propose the following three additional lenses:

- **Robots' Appearance:** We consider the varying appearances of robots, including human-like, machine-like, animal-like, and combinations of these appearances (Section 8.2).

- **Robots' Collaborators:** We look at the diverse roles humans can assume in collaboration with robots, such as users, tele-operators, bystanders, and robot developers (Section 8.3).

- **Robots' Abilities:** We examine the broad range of robotic capabilities, including locomotion, perception, grasping and manipulation, and communication with humans and other robots (Section 8.4).

This chapter is organized as follows: Section 8.2 describes the crucial role that the physical appearance of a robot plays in shaping HRI. The challenges presented in this section revolve around designing a robot's appearance to accurately reflect its behavioral capacities, operational domain, and intended functionality. The importance of striking a balance between the function and design of a robot's appearance is emphasized. The section also reveals a promising path toward creating robots with hybrid and shape-changing appearances. Such adaptive robots, capable of altering their form based on situational needs, could cater to specific user requirements, marking an exciting direction for future research in robotics and HRI.

Section 8.3 details several distinct roles within the robot-mediated assistance framework: the robot, user, bystanders, tele-operator/tele-user, AI developer, environments, and providers. The majority of robots are designed for single-user tasks, limiting their potential. Introducing a tele-operation interface could enhance their utility, but it poses challenges in observing social norms and privacy. Coordinating multiple robots or users presents difficulties due to limited interaction possibilities. Robots designed for interaction with other robots lack a uniform communication protocol. Allowing end users to reprogram robots is crucial for improvement. Integrating bystanders' input can enhance a robot's task performance, offering a significant opportunity.

Section 8.4 explores the opportunities and challenges associated with the robots' abilities, such as locomotion, perception, grasping and manipulation, and communication with humans and other robots. Robots' locomotion (Section 8.4.1) capability addresses their movement from one location to another. While some robots are anchored to a specific platform, many assistive applications require autonomous or semiautonomous movement. Robots can employ various locomotion mechanisms, including crawling, sliding, or walking. They typically use

TABLE 8.1

A Sample of Robots Classified According to the Proposed Robot Appearance Dimension. "H" stands for "Human-like", "A" stands for "Animal-like", "M" stands for "Machine-like", and "HM" stands for "Human- and Machine-like".

ID	Robot Name	Appearance	Locomotion	Perception	Grasping & Manipulation	Communication with Human	Communication with Robots	Local User	Tele-operator /tele-user	Bystander	AI Developer	Purpose
1	Astro	HM	1	1	0	1	0	1	1	1	0	Household, service, social
2	Atlas	H	1	1	1	1	0	1	1	1	1	Industrial
3	Ava	HM	1	1	0	1	0	0	1	1	0	Telepresence, household, service
4	Baxter/Sawyer	HM	1	1	1	1	0	1	1	1	1	Industrial/Research and education
5	Bluerov	M	1	1	0	0	0	0	0	1	1	Underwater navigation
6	Da Vinci	M	0	1	1	1	0	0	1	1	0	Surgical robot
7	DJI drone	M	1	1	0	0	0	0	0	1	1	Aerial navigation
8	EMO	HM	1	1	0	1	0	1	0	1	0	Small pet, social
9	Everyday robot	HM	1	1	1	1	0	1	1	1	0	Service (discontinued)
10	Fetch	HM	1	1	1	1	0	1	1	1	1	Research and education
11	Franka	M	0	1	1	1	0	1	1	1	1	Research and education
12	Kilobot	A	1	1	0	1	1	0	0	0	1	Swarms of robots
13	LoCoBot	M	1	1	1	1	0	1	1	1	1	Research and education
14	Moley	HM	0	1	1	1	0	1	0	0	0	Household, service, food prep
15	NAO	H	1	1	1	1	0	1	1	1	1	Social
16	Ohmni	HM	1	1	0	1	0	0	1	1	0	Telepresence
17	PARO	A	0	1	0	1	0	1	0	0	0	Social, therapeutic pet
18	Pepper	H	1	1	1	1	0	1	1	1	1	Social
19	Phoenix	H	1	1	1	1	0	0	1	1	1	Service, tele-operated
20	Plato	HM	1	1	0	1	0	1	1	1	0	Service, food delivery, social
21	PR2	HM	1	1	1	1	0	1	1	1	1	Household, service, industrial
22	Reachy	HM	1	1	1	1	0	0	1	1	1	Telepresence, service, social
23	Roomba	M	1	1	1	1	0	1	1	0	0	Household, service
24	Relay	HM	1	1	0	1	0	1	0	1	0	Service, hospitality
25	Spot	A	1	1	1	1	0	1	1	1	1	Service, industrial, dog-like
26	Stretch	M	1	1	1	1	0	1	1	1	1	Industrial
27	TurtleBot	M	1	1	1	1	0	1	1	1	1	Research and education
28	Unitree Go 1	A	1	1	1	1	0	1	1	1	1	Research and education
29	Vgo	HM	1	0	0	1	0	0	1	1	0	Telepresence

wheeled, chain/track, or legged systems, each with hardware design and software challenges to address. Robotic perception (Section 8.4.2.1) relies on various computer vision tasks, such as object recognition, object tracking, and segmentation. Opportunities involve the use of active learning, human-in-the-loop techniques, and other robust approaches. Challenges include handling lighting variations, occlusions, and dynamic environments. Further, robotic grasping and manipulation (Section 8.4.3) in unstructured environments pose challenges in selecting suitable configurations and performing varied tasks. Opportunities include developing universal object-picking robots, improving algorithms and hardware, and facilitating adaptive tool use for enhanced real-world assistance. Finally, the communication between humans and robots (Section 8.4.4) poses key challenges in HRI. These include fostering long-term engagement, managing dependence on robots, and ensuring cultural sensitivity. Creating enduring human–robot relationships, and designing culturally sensitive robots are critical. Deciphering human nonverbal cues and gestures is challenging but crucial for nuanced interactions. Security, privacy, and trust are also significant concerns in HRI. Evaluating HRI is complex due to the scarcity of open robot platforms, the lack of real-world datasets, and limited access to evaluation settings. Innovative interaction modalities, such as virtual reality/augmented reality (VR/AR) technologies and brain–control interfaces (BCIs), have the potential to enable shared and embodied interactions in HRI.

8.2 Robots' Appearance

A robot's appearance, encompassing its form, structure, and visual characteristics, greatly influences HRIs. Robots designated for social roles often showcase anthropomorphic features, which users tend to prefer [83, 93, 154]. However, overly intricate designs may inflate user expectations, resulting in diminished satisfaction when these expectations remain unfulfilled [15, 39, 113]. We categorize robot appearances into three main groups:

- **Human-like:** Robots bearing human-like features such as a face, arms, legs, or even a complete exoskeleton. These robots operate in diverse environments and possess the ability to navigate various terrains.

- **Animal-like:** Robots resembling animals, equipped with multiple legs or arms. Like their human-like counterparts, these robots can navigate an array of terrains, including flat surfaces, stairs, and uneven ground.

- **Machine-like:** Robots that emulate man-made machinery often incorporate wheels or tracks for move-

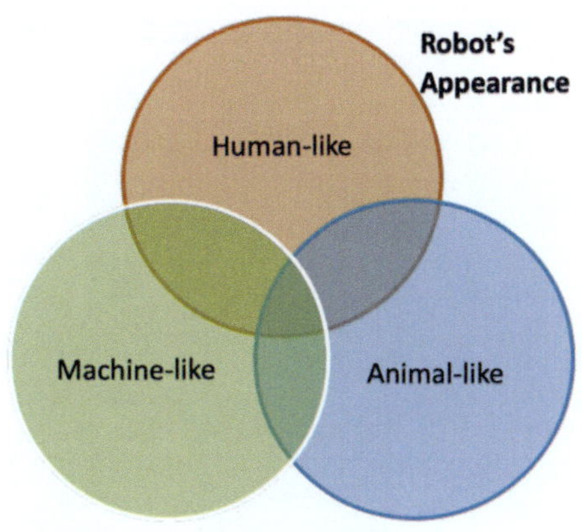

FIGURE 8.1
Robots' appearance lens.

FIGURE 8.2
NAO: a human-like robot. (With courtesy from [147].)

ment, thus allowing for efficient traversal on flat surfaces.

8.2.1 Challenges and Opportunities in Robot Appearance

Mori's uncanny valley theory [106] offers a valuable guideline regarding robot appearance. The theory proposes a critical point of human-likeness in robots, where positive

FIGURE 8.3
AVA: a human- and machine-like robot. (With courtesy from [163].)

FIGURE 8.4
LoCoBot: a machine-like robot.

emotional responses suddenly give way to aversion. However, if the human-likeness continues to escalate beyond this point, positive responses resume, approaching levels of empathy typically associated with human–human interactions. Given that the current state of robot autonomy and appearance falls short of human-like, reducing a robot's human-like appearance and behavior might prove beneficial in managing user expectations [56]. Therefore, it becomes a challenge to align a robot's appearance with its social and behavioral capacities, operational domain, and intended function.

Earlier generations of manufacturing robots adopted machine-like forms and were employed for tasks such as moving items, assembling components, and inspecting products. Today's advancements have led to a hybrid design approach, combining human and machine-like features, as seen in models like Baxter (refer to Table 8.1). In the service sector, robots designed to assist humans often incorporate human-like features. Examples include Pepper and NAO, which serve in classrooms and hospitality roles; Relay for transporting objects in restaurants; Moley for kitchen assistance; and Astro for home monitoring (see Table 8.1). The human-like faces of these robots create high expectations for their dialogue capabilities,

FIGURE 8.5
Paro: an animal-like robot. It looks like a pet seal. (With courtesy from [13].)

leading to reduced interaction when these expectations remain unmet. Telepresence robots, designed to represent the user remotely, also commonly feature a human-like appearance. Although anthropomorphism is not necessary for effective teleoperation [34], the degree of

FIGURE 8.6
Baxter: an industrial human- and machine-like robot. (With courtesy from [62].)

FIGURE 8.7
Kilobots: a group of animal (insect)-like robots. (With courtesy from [119].)

anthropomorphism can influence user perceptions of presence in a virtual environment [113]. In healthcare, animal-inspired robots such as PARO (shown in Table 8.1) have proven effective due to their lower expectation levels. These instances underscore the importance of achieving a balance between the function and design of a robot's appearance.

Most robots exhibit either human- and machine-like attributes (e.g., 13 out of 29 robots in Table 8.1), followed by those purely machine-like (8), human-like (4), or animal-like (4). However, an intriguing opportunity resides in exploring hybrid designs that combine human-

and animal-like features (e.g., a snake with a human face), or in investigating the potential of shape-changing robots that adapt their form based on context (e.g., adopting human-, animal-, or machine-like appearances as required).

Such adaptive, shape-changing robots could comprise reprogrammable, Lego-like modules, and assume task-specific configurations, such as robotic limbs to assist individuals with disabilities. Additionally, the form of the robot could be tailored to specific personal needs, enabling it to perform various tasks like opening and closing doors and windows, forming a staircase-like shape to aid users with motor impairments, or locating dropped objects for users with vision impairments. The integration of swarm robotics, akin to Kilobot in Table 8.1, could facilitate this on-demand adaptability of robot appearance and functionality. This advancement in robot design presents a compelling direction for future research in robotics and HRI.

8.3 Robot's Collaborators

The robot-mediated assistance paradigm encompasses several distinct roles for individuals who can interact with the robot, as shown in Figure 8.8. These roles are inspired by the design challenges and guidelines for social interacting using mobile telepresence robots [145].

- **Robot:** This term straightforwardly refers to the robot itself.

- **User:** The user, who is co-located with the robot, actively interacts with and benefits from the robot's services.

- **Bystanders:** These are individuals physically present in the same environment as the user and the robot, who may interact with the robot directly or indirectly.

- **Tele-operator or Tele-user:** A tele-operator remotely controls the robot to serve the user through a specific human–robot interface. In cases where the tele-operator uses the robot for personal needs, such as remote conference attendance, we refer to this role as the tele-user.

- **AI Developers:** These individuals can build, install software and firmware, and test and debug the performance of the AI modules in the robot, either frequently or infrequently, locally or remotely.

- **Providers**: For the sake of simplicity, we employ this term to denote nonuser human entities, such as bystanders, tele-operators, and AI developers.

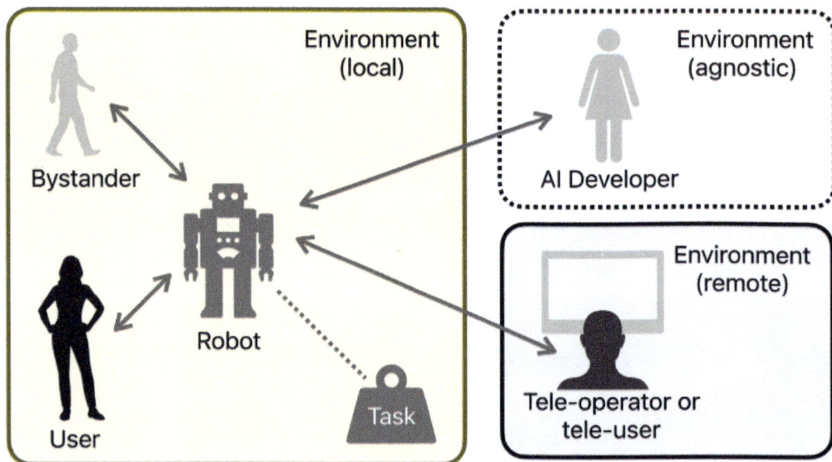

FIGURE 8.8
Robots' collaborators' lens.

We distinguish between two types of environments: the local environment where the robot, user, and bystanders are co-located; and the remote environment from where the tele-operator or tele-user controls the robot.

8.3.1 Challenges and Opportunities in Robots' Collaborators

A majority of robots are designed to serve a single user, either local or remote, for a specific task, as illustrated in Table 8.1. This design philosophy can limit the robot's potential because introducing a tele-operation interface could allow remote human operators to assist local users through the robot, particularly in complex, unstructured tasks. Implementing such an interface would thus enhance the robot's utility.

However, incorporating a tele-operation interface presents its own set of difficulties. For instance, observing social norms regarding personal space (proxemics) is critical for creating comfortable HRIs. Robots should also follow social etiquette, such as preserving private communication and avoiding unauthorized audio or video recording. Moreover, robots could be designed to disclose user status and identification to promote transparency in interactions. Anthropomorphic robots, in particular, could maintain eye contact, facilitate joint attention, and adjust their height to sit or stand in alignment with the context, thus emulating human behavior [145]. We provide a more detailed discussion on this challenge in Section 8.4.4.

Another difficulty lies in coordinating various collaborators. Consider the scenarios where a single user owns multiple robots (e.g., a Roomba and an Astro), or a single robot serves multiple users (as when a service robot delivers a sandwich to one person and a salad to another

in a care facility). At present, such coordination relies primarily on audio and video connections, with insufficient support for natural language, gestures, facial expressions, speech, and various body and posture cues [20, 67, 145]. This limitation results in unequal interaction possibilities for different collaborators, reducing the overall utility of robot-mediated assistance.

Robots intended for interaction with other robots face an added challenge in achieving uniform communication protocol, given that many robots do not provide a Software Development Kit (SDK). Current inter-robot communication is ad hoc, and for those robots with SDK, only AI developers can update firmware. To improve this situation, it is crucial to enable end users to reprogram their robots, an action currently restricted to programming.

Lastly, it is noteworthy that most robots regard bystanders merely as obstacles to be avoided for safety. However, bystanders can offer valuable input to robots to enhance their task performance. Therefore, integrating bystanders into a robot's decision-making process presents another significant opportunity.

8.4 Robots' Abilities

The abilities of robots are most accurately characterized by Moravec's paradox [105], which posits that tasks humans find simple, such as sensorimotor activities and perception, tend to be challenging for robots. On the other hand, tasks that are often demanding for humans, like computations, are more readily performed by robots. This paradox is deeply rooted in human evolution [105].

Through the process of natural selection, the sensory and motor regions of the human brain have undergone significant evolution, enabling humans to effectively engage with the natural world and survive within it. Therefore, to reverse-engineer these biologically refined and seemingly effortless human skills into robots is expected to be challenging [102, 105].

In the current landscape of robotics, human-like skills are grouped into three fundamental categories: **sensing**: the processing of sensor-derived information; **planning**: the computation of suitable directives; and **actuation**: the generation of commands for mechanical actions. Depending on the tasks, robots may also require **communication** skills for interaction with humans or other robots and might need to exhibit **locomotive** abilities. Arranging these skills differently yields various robot abilities. In this section, we identify four such abilities, each discussed further below, along with their respective challenges and opportunities.

- **Locomotion:** This pertains to the robot's navigational ability, specifically moving from point A to point B. Key sensors and hardware for locomotion include wheels, legs, and chains.

- **Sensing or Perception:** This domain involves interpreting visual scenes, decoding natural language and speech, and formulating plans. Necessary sensors and hardware for perception encompass RGB cameras, depth sensors (like LiDAR or IR sensors), bump sensors, tactile sensors, temperature sensors, and text and audio input devices.

- **Grasping and Manipulation:** This domain refers to a robot's capacity to handle physical objects and perform actions that modify the environment. Essential sensors and hardware for grasping and manipulation include suction cups, grippers, tails, arms, motors, and tactile sensors.

- **Communication with Humans and Other Robots:** Robots can possess the ability to interact with humans and coordinate with other homogeneous or heterogeneous robots.

8.4.1 Challenges and Opportunities in Locomotion

While not all robots require locomotion capabilities, many do (as shown in Table 8.1). For instance, robots like *PARO*, *Da Vinci*, and *Moley*, which provide therapeutic, surgical, or culinary assistance, respectively, are anchored to a specific platform or surface. Their motion is limited to the range of their manipulators, denying them independent mobility. However, a plethora of assistive applications necessitates the ability of a robot to move autonomously or semiautonomously from one location to another, a capability known as locomotion [135]. Robotic systems often take cues from biological locomotion mechanisms, adopting strategies such as crawling, sliding, running, jumping, or walking. The majority of locomotive robots employ three types of hardware: (1) wheeled robots (*Reachy, Everyday robot, Relay, Pepper, NAO, Pluto, Astro, Turtlebot, LoCoBot, Roomba, Ohmni*); (2) chain or track systems; and (3) legged robots (*Atlas, Phoenix, Spot, Unitree Go 1*). The design of the mechanics, materials, and degrees of freedom of this hardware pose significant challenges. Software design challenges for enabling locomotion include addressing high-level robotic tasks such as simultaneous localization and mapping (SLAM) and visual navigation, along with the associated issues these areas present.

8.4.1.1 SLAM

In the SLAM task, a robot must concurrently reconstruct its environment and determine its position (rotation and translation) relative to a global coordinate frame. Various types of SLAM tasks, including Visual SLAM and Semantic SLAM, face the common challenges of accurate feature extraction, loop closure, and real-time computation [110, 143]. These tasks require substantial computational resources and rich sensor data, which can be prohibitive in terms of cost for many robots.

In our robot-mediated assistance context, an *opportunity* arises: the robot could request the bystanders or tele-operators to supply a coarse map or landmarks, thereby simplifying the map-building process and reducing computing resource demands. For instance, users could sketch a map on their smartphone app for the cleaning robot, *Roomba*. An additional *opportunity* arises in the realm of tele-operation, where the objective is to localize the robot under low-bandwidth communication channels. In such circumstances, rather than streaming the full video feed from the robot's camera, which requires a higher bandwidth, the robot could opt to transmit only a select few anchor points, specifically, image feature descriptors, grounded in visual-inertial odometry [41].

8.4.1.2 Visual Navigation

Traditional visual navigation encompasses three fundamental stages:

1. **Mapping:** the process of generating a representation of the environment, capturing relevant features and landmarks;

2. **Localization:** the robot identifies its position within the mapped environment, typically through sensor data comparison with mapped features;

3. **Path planning:** given the mapped environment and the robot's current location, an optimal path is determined to reach the desired destination [114]. More recently, datasets like Habitat 2.0 [140] offer simulation platforms that enable the exploration of unseen 3D environments for benchmarking two versions of visual navigation tasks: image-goal navigation and object-goal navigation. In the context of image-goal navigation, the objective for a robot is to autonomously reach a specific goal indicated by an image representing the target location. This task requires the robot to interpret the visual information provided by the image and make decisions on how to navigate the environment to successfully reach the desired destination [44, 109, 166]. Visual navigation presents numerous challenges, such as addressing variations in lighting conditions, adapting to dynamic environments, and maintaining real-time processing capabilities.

An opportunity exists to design robots that can navigate within an environment semiautonomously, using input or cues from providers (i.e., nonuser humans) [161]. For example, the robot could follow a human traveling to the same destination. This approach could also assist the robot in recovering from navigational failures.

8.4.2 Challenges and Opportunities in Perception

Perception in robotics hinges on the extraction of meaningful data from the robot's array of sensors. Modern robots come equipped with a variety of sensors, such as visual cameras, thermal sensors, infrared sensors, and LiDAR (see Section 8.4). Of these, the visual camera emerges as the most impactful sensor due to its prevalent use and cost-effectiveness on a large scale. Notably, the evolution of machine learning, specifically deep learning techniques, has pushed computer vision methods to a stage where they can reliably provide solutions for robot-mediated assistance. Despite these positive developments, challenges persist. The system must handle visual data to tackle a range of computer vision tasks. These tasks span basic, mid-level, and high-level functions, such as object recognition, object detection, object tracking, segmentation, scene comprehension, and 3D reconstruction. In the following sections, we explore these challenges and opportunities in the context of robot-mediated assistance, considering the computer vision aspect, the robot assisting the human user and the provider.

8.4.2.1 Object Recognition

Object recognition involves classifying an image into specific categories using distinct feature descriptors. In the past decade, significant strides have been made in the field, predominantly through the application of convolutional neural networks (CNNs) [137] and transformers [33]. Groundbreaking networks like AlexNet [75], VGG [137], and ResNet [54] have been pivotal in driving these advancements. Despite progress, challenges persist in object recognition, such as accommodating objects with differing appearances, dealing with occlusions, navigating background clutter, and adjusting to changes in orientation and perspective. Robust recognition, capable of withstanding variations in lighting conditions [61], is essential for a robot to interact effectively with its environment. Domain adaptation strategies provide potential solutions to these challenges, particularly when dealing with variations in appearance, lighting conditions, and object distribution [43, 124, 138].

There lies an *opportunity* to apply active learning [65] in object recognition. Users can identify and collect instances where they observe the robot's failures in recognition scenarios, and they can provide the correct annotations. This user-provided information can subsequently improve the model. Additionally, there is a potential *opportunity* for the development of novel techniques specifically designed to recognize content on digital displays [81]. This difficulty arises from the factors such as varying screen brightness, the mingling of different light sources (e.g., LCD backlight, sunlight, lamplight), and mismatches in frame rates between the camera and the screen—all can influence the robot's ability to accurately interpret digital content. Further complications arise from inconsistencies between the pixel grid dimensions of the camera and the digital screen, leading to moiré patterns, which manifest as strobe or striping optical effects[115].

8.4.2.2 Object Detection

Object detection involves the classification of object categories within an image and determining their positions, typically depicted by rectangular bounding boxes. Challenges in object detection span various factors such as diverse object appearances, occlusions, background clutter, variations in object pose, and detection of object parts [38]. For robot-mediated assistance, it is vital for the robot to accurately detect the objects with which it interacts [18, 53]. This requirement becomes especially significant when the robot needs to grasp or manipulate objects.

This arises an opportunity for applying a human-in-the-loop technique [17]: the human provider can assist the robot in detecting objects in cases where it encounters difficulties; then they can proactively identify challenging scenarios and capture images of those situations. These challenging images can then be utilized to train a robust supervised model through a combination of online learning and interactive labeling, even in the presence of weak supervision.

8.4.2.3   Object Tracking.

Object tracking refers to the continuous detection of an object within a spatio-temporal setting, such as a video. Object tracking faces numerous challenges, such as illumination variations, abrupt camera motion, motion blur, fast motion [55], rotation of the target object (within or outside the image plane), object deformation, occlusion (partial or complete), the target object going out of view, changes in viewpoint, scale variations, low-resolution images, background clutter, and distinguishing between similar objects [74]. The robot might be required to track the human user's movements. Moreover, there may be scenarios where the robot needs to track a moving target [45].

An opportunity is to develop a single object tracker [36] to continuously track a unique individual based on their non-facial features (e.g., heights, body shapes, and gaits). An extension of this tracker is to track multiple objects based on natural language commands (e.g., "keep an eye on Tom, the cat, and Jerry, the mouse") [42, 152]. The second opportunity is to improve the trackers by robust feature learning [156].

8.4.2.4   Segmentation

Segmentation involves partitioning an image into distinct regions based on particular grouping criteria. The main challenge in segmentation stems from the diverse and intricate nature of environments, including indoor [136], outdoor [3], aerial [100], and underwater settings [59]. Various types of segmentation, such as semantic segmentation [8, 24, 94], instance segmentation [32, 123], and panoptic segmentation [70], are employed depending on the specific application needs. A significant challenge in semantic segmentation is the inconsistencies found in labeling across different benchmarks [79].

Several opportunities present themselves in the realm of robot-mediated assistance. First, there is the potential to create a unified model for segmentation [26]. Second, it is possible to design segmentation models with robust zero-shot generalization capability, which can effectively segment any object found in diverse environments [71]. The third opportunity involves creating segmentation models that can respond to natural language prompts (e.g., *segment the brown dog in the image*). This concept of *promptable segmentation* aligns with challenges in visual question answering (VQA) [48], but with a focus on segmentation. Finally, there is a promising opportunity to develop optical character recognition (OCR) techniques that can adeptly segment text on curved or warped surfaces, like reading text from a round medicine bottle or a crumpled piece of paper.

8.4.2.5   Scene Understanding.

Scene understanding aims to build a system with the capacity to fully perceive and interpret a scene, taking into account its layout, components, and environmental characteristics. This task is particularly challenging due to the complexities and diversity inherent in different environments; the robot should be capable of answering a multitude of queries, such as identifying the labels and locations of various regions, discerning the spatial extent of objects, understanding the layout of the scene, and recognizing different scene types, like a kitchen, office, or bedroom [91, 128, 151, 159].

A promising avenue to explore is the development of the robot's ability to detect changes within dynamic environments alongside the human user and other active or passive bystanders [1]. These could include changes in the number of objects, either added or removed, as well as shifts in the lighting ambiance.

8.4.2.6   Visual Question Answering

VQA presents the challenge of predicting an open-ended answer to a question given an image [7]. Swift advancements in computer vision and natural language processing (NLP) methodologies, along with the accessibility of large-scale datasets, have sparked significant interest in VQA [64]. There is a wide range of deep neural architectural choices used in the design of existing VQA models, including attention mechanisms [97], stacks of attention networks [158], combinations of LSTM and CNN modules [96], compositional models [58], and transformer-based attention networks [25, 51, 87, 95, 141]. Successful VQA models hold great promise for robot-mediated assistance as they allow users to pose natural language questions to robots on a variety of topics, such as visual recognition, reasoning, object counting, and common sense. A notable challenge in VQA models centers on their robustness, that is, their capacity to reliably respond to slightly varied forms of the same question [88] such as *linguistic shifts, logical reasoning adjustments, visual content manipulation,* and *changes in answer distribution between training and testing datasets.* Current strategies to boost robustness revolve around crafting purpose-designed benchmarks and robustness metrics [5, 49, 66, 130].

A crucial opportunity within robot-mediated assistance is the generation of a more sophisticated context for each question, leveraging other computer vision tasks, particularly scene understanding and real-time segmentation. Such enriched contextual understanding can contribute substantially to the accuracy and applicability of the VQA models in complex real-world scenarios.

8.4.2.7   3D Scene Reconstruction.

Several approaches can be taken for 3D scene reconstruction, depending on the availability of cameras and their relative calibration. For instance, *monocular depth estimation* is a technique that aims to predict the depth of each pixel from a single image [12, 35, 125]. *Stereo matching* estimates the disparity between corresponding pixels in a pair of images taken by a binocular camera setup. This process determines the displacement value for each pixel, providing vital depth information about the scene [90, 127, 155]. Finally, *multiview stereo* attempts to 3D reconstruct a scene using multiple images. These images can be taken by a single camera from various viewpoints or multiple calibrated cameras [40, 129, 160]. There are several challenges associated with 3D reconstruction, including the handling of textureless regions, self-occlusion, object occlusion, deformable objects, dynamic scenes with moving objects, low-light or night-time scenarios, and the requirement for real-time processing [50, 68, 77, 165]. These challenges pose obstacles in accurately reconstructing the three-dimensional structure of the scene and require innovative solutions to address them.

Considerable improvements to these challenges would create several opportunities for robot-mediated assistance. Firstly, effective 3D scene reconstruction can significantly improve the robot's spatial awareness, leading to a more comprehensive understanding of its surroundings. Secondly, 3D object recognition can enhance the robot's ability to interact with its environment, particularly in terms of object manipulation and grasping.

8.4.3   Challenges and Opportunities in Grasping and Manipulation

Robotic grasping and manipulation are essential in unstructured environments such as homes, offices, warehouses, and logistics centers where there is a physical necessity to retrieve objects. Two key representations for grasp configurations exist: (1) planar grasping, which utilizes a simpler representation (a 2D location) but is limited in its applicability [84, 99], and (2) 6-DOF (Degrees of Freedom) grasping, which is more complex but provides a greater range of flexibility [108].

- **Grasping challenges:** Grasping is challenging because the task of grasp synthesis, that is, finding a grasp configuration for a given object, involves determining the most suitable configuration from a wide range of plausible candidates that meet specific criteria [16]. The grasping problem can be approached as an object detection problem that involves predicting the rectangular location on the object that can be grasped [84]. Large-scale grasp datasets are typ-

ically built by sampling from a simulated environment, using an analytical method [99, 108] or self-supervision [85, 118]. The challenge of grasping becomes even more pronounced when the target object is occluded by clutter, a scenario framed as *mechanical search* [31, 76]. Common mechanical search strategies involve a series of actions, including parallel jaw grasping, suction grasping, and pushing, until the target object is successfully retrieved [31].

- **Manipulation challenges:** Dexterous manipulation tasks also remain challenging, including actions such as *throwing, sliding, poking, pushing* in unclutter environments, *stacking boxes according to their sizes/weights*, to more complex tasks such as *inserting pegs in holes, hammering, sorting objects, packaging objects, folding clothes, pouring fluid into a glass*. Additionally, challenges arise in manipulating deformable objects (e.g., clothes), performing actions that cause changes to the object (such as cutting or crushing), and executing in-hand manipulations where an object is moved while being held (e.g., twirling a pen between fingers). While robots can learn manipulation tasks through human demonstrations with fewer trials, these methods often struggle with generalization [14].

- **Opportunities:** Opportunities exist in developing robots capable of universally picking objects of different sizes and shapes. Improvements in these areas could involve developing more efficient algorithms or enhancing the motion capabilities of robotic hardware. Dexterous manipulation presents open opportunities, specifically in developing flexible gripping mechanisms that require research in new hardware and materials for effective grasping. Indeed, in robot-mediated assistance, the ability of a robot to adapt and use tools or utensils appropriately presents a significant opportunity. This requires the development of adaptive algorithms that can teach robots how and when to use these tools effectively. For instance, consider a scenario where a robot needs to pick up a small object like an M&M chocolate. In this case, it may be more effective for the robot to use kitchen clamps or tongs instead of trying to grip the small piece with its own mechanical fingers or suction apparatus. This adaptive tool use can make robotic assistance more practical and efficient in various real-world situations. It also presents an opportunity for robots to engage in more complex tasks that would traditionally be challenging for them. Adapting to use tools also means that the robot needs to understand the properties and functionality of different tools, which might involve a combination of techniques from computer vision, machine learning, and reinforcement learning. It can also entail the robot learning from past experiences, much like how humans learn to use tools effectively over time. Further discussions on challenges and opportunities can be found in works such as [14, 28].

8.4.4 Challenges and Opportunities in Human–Robot Interaction

Recent developments in robotics have seen a transition from stationary, solitary robotic systems to emerging categories of mobile manipulators. These new robot types are intended to operate within human-centric environments and work in direct cooperation with people. They are also accommodating a wide spectrum of user profiles that span diverse backgrounds, physical and cognitive capabilities, training, and receptivity to technological adaptation. As such, the significance of HRI and specifically, a robot's capability to communicate with humans, has been amplified. Despite the considerable progress made in HRI over the past decade, this field continues to face several significant challenges and opportunities. A few key ones are discussed below:

8.4.4.1 *Promoting Sustained Interaction, Mitigating Overreliance, and Ensuring Cultural Sensitivity*

- **Sustained interaction:** Ensuring substantive, long-term interaction between robots and users represents a crucial challenge. Within HRI literature, "long-term" is commonly understood to mean a period of approximately two months [67, 139], which is typically sufficient for a user to become familiar with a robot, thus neutralizing novelty bias [83].

  Various social robots, such as PARO [149] and Jibo [116], provide immediate comfort or diversion but struggle to foster lasting connections or adapt their behaviors based on past user interactions. For example, an ideal in-home robot caregiver should adjust its care strategies over time according to the resident's preferences, habits, and health transitions. However, most social robots presently serve merely as conversation partners, limited to basic dialogues [83]. Similarly, while assistive robotic systems like exoskeletons and rehabilitation robots can provide physical support, their adaptability to individual user needs and preferences remains insufficient [98].

- **Mitigating overreliance:** Conversely, an essential challenge is the potential for users to become overly dependent on robots. As humans tend to treat anthropomorphic technologies as social enti-

ties [113, 122, 131], there's a risk that strong social bonds with robots might replace meaningful human relationships, for instance, in the context of elderly care [146]. This dynamic also raises concerns about possible exploitation via anthropomorphic robots.

- **Cultural sensitivity:** The necessity of cross-cultural interaction and cultural sensitivity is equally significant for successful HRI. Robots such as Pepper can recognize and respond to a variety of human emotions; however, cultural differences in expressing emotions could lead to misunderstandings. For instance, excessive direct eye contact, often seen as disrespectful in Japan, is typically associated with honesty and engagement in Western cultures. Thus, a robot designed based on Western norms might inadvertently offend users from non-Western cultures. The development of culturally adaptable robots, therefore, constitutes an essential component of HRI research [11].

Consequently, fostering personalization, maintaining long-term adaptability, striking a balance between human and robot interaction, and ensuring cultural appropriateness, continue to be active areas of investigation within HRI [83].

8.4.4.2 *Robots' Understanding of Humans*

Interpreting and understanding human behavior remains a major challenge in the areas of AI and HRI [10, 19]. This problem encompasses the ability to perceive and interpret human nonverbal cues, gestures, and actions, as well as learn to emulate their behaviors, strategies, and actions.

- **Interpreting nonverbal cues:** Nonverbal communication is intricate, often representing a significant stumbling block for robots. For example, the Kismet robot, designed for social interaction [20], can recognize and respond to a certain range of predefined facial expressions and vocal cues. However, it generally falls short in dealing with the subtleties and diversity intrinsic to human nonverbal communication. Humans use body language, facial expressions, voice modulation, and even periods of silence to convey complex emotions and intentions. The task of equipping robots with the ability to decipher these nonverbal cues and respond appropriately remains a central challenge in AI and HRI [10].

- **Gestural recognition:** Recognition of gestures, including hand movements and body language, can enable more intuitive, nonverbal interaction between humans and robots. The inherent challenges associated with gestural recognition involve distinguishing ambiguous gestures, managing occlusions, and handling the complexity of human body movements [9, 92, 153, 162]. It also includes real-time gestural classification with minimal latency [103].

- **Learning from human interactions:** Facilitating the acquisition of complex, long-term domestic tasks by robots also remains an active area of research. Presently, most assistive robots function under open-loop paradigms [111]—performing tasks based on explicit human commands—either locally or remotely [29], or adhering to pre-programmed routines. Some research attempts to enable humans to guide robots through demonstrations or performance feedback. However, providing exhaustive feedback can prove taxing for users, particularly when it requires comprehensive task demonstrations or intricate instruction. Certain feedback mechanisms may not be feasible, especially when preferences only emerge in response to specific stimuli. Furthermore, swiftly adapting robots to new partners is a challenging learning problem, given that partners may have varying preferences, capabilities, knowledge, intentions, or strategies for accomplishing shared objectives [46, 107, 164].

Various strategies are proposed to address these challenges. For example, one approach involves modeling humans as partially observable Markov decision processes (POMDPs) [63, 78, 104] in an adaptation parameter space, which a robot can learn and apply to direct its actions. By learning the human's adaptation parameter [164], the robot can modify its behavior to foster more effective collaboration [72]. This modification can be assessed using metrics such as task performance, collaborative fluency [57], team trust [52, 86], or engagement [134]. Recent advances also utilize deep neural networks, including large language models [6], to help robots incorporate feedback from natural user behavior, framing the learning problem as an online inverse reinforcement learning (IRL) process [2, 111].

8.4.4.3 *Security, Privacy, Trust, and Ethical Considerations*

With the increasing integration of robots into human society, the urgency of addressing security, privacy, and ethical concerns is magnified. Consider, for instance, a caregiving robot designed to alleviate the physical strain on human caregivers by assisting with lifting and moving patients. This scenario prompts ethical questions surrounding patient consent.

- **Security concerns:** Three primary categories emerge in relation to robots and security and privacy issues:

(1) facilitation of surveillance, as demonstrated by police drones overseeing public protests; (2) infringement upon traditionally private domains, such as home robots that fall victim to hackers; and (3) creation of a perception of being watched, a prevalent concern associated with anthropomorphic (human-like) robots [21, 28]. Moreover, the possibility of security vulnerabilities can increase the potential for physical harm that a compromised robot could cause.

- **Privacy challenges:** Ensuring visual privacy presents another hurdle, as it necessitates the anonymization or obscuration of identifiable characteristics without impairing the accuracy of other vision tasks [150]. Determining which information qualifies as "identifiable" or "private" contributes to this complexity. Critically, the algorithms propelling these vision tasks should not direct users toward corporate or other objectives conflicting with their own interests [112]. Hence, a balance between the beneficial and detrimental impacts of robots on humans mandates thorough research and mindful consideration within the realm of HRI.

- **Trust in HRI:** Trust forms a vital component of successful HRI. Users must understand a robot's abilities and intentions and feel a sufficient level of control over the robot's actions [52]. Trust extends beyond merely the user's faith in the robot, encompassing the robot's estimation of the user's trust as well. While there is prior work on the factors influencing trust in robots, constructing robust models to manage trust proves to be a challenge [86, 126]. For instance, while consistency and reliability may be desired characteristics, unexpected behavior might be appreciated in certain contexts due to its novelty, such as a robot cheating during gameplay to maximize entertainment [133]. Striking a balance between trust, reliability, and novelty, therefore, presents an area for exploration.

- **Rights of robots:** One of the significant ethical issues in robot-mediated assistance concerns the rights of robots against abusive behavior, such as wanton kicking of a robot for amusement. This concern also prompts questions about the ramifications of assigning rights to robots. A contrasting viewpoint proposes considering robots as bearers of "rites" instead of "rights" [69]. "Rites" denote a sequence of actions, often involving multiple actors, such as robots, users, and providers, collectively bearing symbolic significance. Through these rites, actors acknowledge the value of the interaction and delineate their reciprocal roles. For instance, in a basketball game, each player has specific roles or duties, thus, their rites. Should player 1 neglect to pass the ball to a better-positioned teammate, player 2, player 1 is not infringing upon player 2's rights but failing in their own rites. Enhancing our comprehension of the nuanced roles, rites, and rights affiliated with robots offers another promising area for investigation within HRI.

8.4.4.4 *Lack of Open Robot Platforms, Datasets, and Real-World Evaluation Environments*

The spectrum of HRIs is extensive and poses considerable challenges in objective evaluation [28]. This spectrum spans from scenarios where humans have complete control over robots to those where robots operate with minimal human oversight and even scenarios where robots are engaged in social interactions.

The first challenge lies in the scarcity of open robot platforms suitable for a wide range of HRI research. The majority of research platforms are primarily designed for mobility in 2D or 3D space (e.g., ground, air), but effective interaction with humans demands careful consideration of a robot's appearance (form) and its perceptual and operational capabilities (function). We note that middleware such as the Robot Operating System (ROS) has spurred considerable adoption in the field. Affordably priced humanoid robots, expressive enough for social HRI research, are not available at present. For instance, *Pepper* robot, despite lacking facial features, is still priced well above a thousand dollars. Some startups occasionally introduce potentially promising platforms (e.g., Jibo, Kuri), but these offerings tend to be short-lived. Further, there is a dearth of economically priced tabletop robot platforms designed for social and socially assistive HRI. These platforms would aid in the creation, deployment, and testing of large-scale user studies.

The second challenge is the lack of large-scale, real-world datasets for HRI. Advancements in HRI for specific user groups, such as the elderly, children, stroke patients, or individuals on the autism spectrum, are contingent on the availability of interaction data with these populations. However, the collection and dissemination of such data are hindered by substantial privacy constraints; consequently, only a limited number of researchers are able to conduct extensive studies with real-world datasets. A few community datasets and testbeds are available at present, such as the Minedojo [37], a Minecraft-based simulated environment for developing and testing reinforcement learning algorithms. Other platforms include Overcooked multi-agent environment [22], IKEA furniture assembly environment [82], and scalable data collection platform via teleoperation [117]. However, these simulations, while useful, are still different from real-world interactions.

Lastly, there is a lack of access to real-world evaluation settings, such as nursing homes, schools, and hospitals. This restriction often compels HRI researchers to rely on university students as study participants, potentially skewing the results due to this biased sample. Mirroring the impact of large-scale datasets in computer vision research, creating shared resources, datasets, and testbeds presents an open opportunity in HRI that can allow for methodical comparisons, fostering collaboration and cumulative progress across the field.

8.4.4.5 *Novel Interaction Modalities in HRI*

HRI brings fresh opportunities to interaction design, extending beyond the traditional scope of human–computer interaction. Traditionally, the human–computer interaction paradigm is rooted in the windows-icons-menu-pointer (WIMP) model [148]. In this paradigm, users interact with application windows, menus, and visual metaphors (icons) using pointing devices like a mouse. However, HRI offers the potential to revolutionize this model with innovative interaction modalities. This includes the integration of VR/AR technologies, enabling shared viewpoints, contexts, and embodied interactions between humans and robots. These advancements promise a wealth of design possibilities for various input devices, the development of new interaction models, and the production of reference implementations for hardware and software layers.

- **Immersive interfaces in VR:** VR-based immersive interfaces are emerging as pivotal tools for teleoperated robots such as *Reachy*, enabling human users to remotely supervise and control multiple robots [47]. This interface can facilitate collaboration and mimic the spontaneity found in face-to-face interactions. For instance, robots possessing movable necks can mimic human gestures like nodding or adjusting their gaze direction, which amplifies the naturalness of interactions. Furthermore, the user's avatar in the VR environment can perform actions such as shaking hands with bystander avatars or even engaging in tasks like cleaning a kitchen countertop located in a remote environment. These dynamic, embodied interactions can substantially boost user engagement, cultivating a sensation of true presence in the virtual scene.

This novel interface also presents opportunities to minimize interaction latency by improving rendering pipelines and utilizing progressive compression in the communication channels. Additional opportunities include developing technologies that can accurately create realistic 3D environments with less manual effort, such as generative AI models that produce images from textual prompts. This development aims to enhance not just visual realism but also interaction realism, pertaining to how the robot responds to user commands or behaves within the VR environment.

- **Brain-control interfaces:** BCIs have driven substantial advancements in HRI, notably allowing individuals with quadriplegia to operate robotic arms via thought processes alone [80, 142]. These interfaces can translate brain activity into control signals, establishing a new communication pathway between humans and robots. Despite significant progress, research efforts are focused on enhancing the reliability, responsiveness, and versatility of BCIs. Reliability ensures consistent operation, a critical requirement in healthcare or assistive technologies. Response time improvement facilitates smoother and more timely interactions, which is crucial in emergency or surgical applications. Lastly, improving adaptability encompasses making the interface user-friendly and capable of catering to various users and conditions. Such advances are poised to revolutionize HRI.

- **Collaborative robot interfaces:** Collaborative robots, often referred to as "cobots," represent a significant development in HRI [4]. These machines are engineered with a host of advanced sensors, actuators, and control algorithms to ensure both safe and efficient interactions with human colleagues in shared work environments.

Despite the potential benefits, several challenges lie ahead for the successful integration of cobots into a wide variety of work settings. Firstly, refining their perception capabilities and decision-making processes is vital to enhance their awareness and responsiveness toward human co-workers. Secondly, managing the change in workflow that comes with introducing cobots into traditional work environments is a significant hurdle. Lastly, there is a need for rigorous, context-specific evaluations of their performance, especially in scenarios requiring close collaboration between humans and cobots. Addressing these challenges is critical to ensure seamless and intuitive human-cobot communication, thereby enhancing the efficacy of tasks that require shared effort and understanding.

8.5 Conclusion

In this work, we present a nuanced analysis of robot-mediated assistance by introducing three new lenses to extend the current framework [157]. We delve into robot appearances, explore diverse human and entity roles in HRI, and probe the spectrum of necessary robotic abilities for effective human assistance. Our in-depth analysis is derived from a comprehensive review of 29 robots, identifying pivotal challenges and untapped research avenues in computer vision and HRI within the context of robot-mediated assistance. This enriched understanding illuminates precise directions for future exploration. We anticipate our contributions will inspire multidisciplinary research communities, propelling advancements in robot-mediated assistance.

8.A   Appendix

A Sample of Robots and Their Weblinks

ID	Robot Name	Link
1	Astro	https://www.aboutamazon.com/news/devices/meet-astro-a-home-robot-unlike-any-other
2	Atlas	https://www.bostondynamics.com/atlas
3	Ava	https://www.avarobotics.com/ava-webex
4	Baxter	https://robots.ieee.org/robots/baxter
5	Bluerov	https://bluerobotics.com
6	Da Vinci	https://www.intuitive.com/en-us/products-and-services/da-vinci/systems
7	DJI drone	https://www.dji.com/
9	Everyday robot	https://everydayrobots.com/
10	Fetch	https://fetchrobotics.com/freight-base-research/
11	Franka	https://www.franka.de/research
12	Kilobot	https://robotsguide.com/robots/kilobot/
13	LoCoBot	https://www.trossenrobotics.com/locobot-overview.aspx
14	Moley	https://www.moley.com/moley-kitchen
15	NAO	https://www.aldebaran.com/en/industries/government
16	Ohmni	https://ohmnilabs.com/products/ohmni-telepresence-robot/
17	PARO	http://www.parorobots.com/
18	Pepper	https://www.aldebaran.com/en/pepper
19	Phoenix	https://www.sanctuary.ai
20	Plato	https://cobiotx.unitedrobotics.group/en/plato
21	PR2	https://robots.ieee.org/robots/pr2/
22	Reachy	https://www.pollen-robotics.com
23	Roomba	https://www.irobot.com/en_US/why-irobot.html
24	Relay	https://www.relayrobotics.com/robots-in-action
25	Spot	https://www.bostondynamics.com/products/spot
26	Stretch	https://www.bostondynamics.com/products/stretch
27	TurtleBot	https://www.turtlebot.com
28	Unitree Go 1	https://shop.unitree.com/products/unitreeyushutechnologydog-artificial-intelligence-companion-bionic-companion-intelligent-robot-go1-quadruped-robot-dog
29	Vgo	http://www.vgocom.com

References

[1] https://nikosuenderhauf.github.io/roboticvision challenges/cvpr2023. In *Workshop in IEEE/CVF Conference on Computer Vision and Pattern Recognition*, 2023.

[2] Pieter Abbeel and Andrew Y Ng. Apprenticeship learning via inverse reinforcement learning. In *Proceedings of the Twenty-First International Conference on Machine Learning*. 1 (2004).

[3] Hassan Abu Alhaija, Siva Karthik Mustikovela, Lars Mescheder, Andreas Geiger, and Carsten Rother. Augmented reality meets computer vision: Efficient data generation for urban driving scenes. *International Journal of Computer Vision* 126, 9 (2018), 961–972.

[4] A Adriaensen, F Costantino, G Di Gravio, and R Patriarca. Teaming with industrial cobots: A socio-technical perspective on safety analysis. *Human Factors and Ergonomics in Manufacturing & Service Industries* 32, 2 (2022), 173–198.

[5] Aishwarya Agrawal, Dhruv Batra, Devi Parikh, and Aniruddha Kembhavi. Don't just assume; look and answer: Overcoming priors for visual question answering. In *Proceedings of the IEEE Conference on Computer Vision and Pattern Recognition* (2018), 4971–4980.

[6] Michael Ahn, Anthony Brohan, Noah Brown, Yevgen Chebotar, Omar Cortes, Byron David, Chelsea Finn, Keerthana Gopalakrishnan, Karol Hausman, Alex Herzog, et al. 2022. Do as i can, not as i say: Grounding language in robotic affordances. *arXiv preprint arXiv:2204.01691* (2022).

[7] Stanislaw Antol, Aishwarya Agrawal, Jiasen Lu, Margaret Mitchell, Dhruv Batra, C Lawrence Zitnick, and Devi Parikh. VQA: Visual question answering. In *Proceedings of the IEEE International Conference on Computer Vision* (2015).

[8] V. Badrinarayanan, A. Kendall, and R. Cipolla. SegNet: A deep convolutional encloder-decoder architecture for scene segmentation. In *IEEE Transactions on Pattern Analysis and Machine Intelligence* (2017).

[9] Gilles Bailly, Jörg Müller, Michael Rohs, Daniel Wigdor, and Sven Kratz. Shoesense: A new perspective on gestural interaction and wearable applications. In *Proceedings of the SIGCHI Conference on Human Factors in Computing Systems* (2012). 1239–1248.

[10] Samuel Barrett, Avi Rosenfeld, Sarit Kraus, and Peter Stone. Making friends on the fly: Cooperating with new teammates. *Artificial Intelligence* 242 (2017), 132–171.

[11] Tony Belpaeme, James Kennedy, Aditi Ramachandran, Brian Scassellati, and Fumihide Tanaka. Social robots for education: A review. *Science Robotics* 3, 21 (2018), eaat5954.

[12] Shariq Farooq Bhat, Ibraheem Alhashim, and Peter Wonka. Adabins: Depth estimation using adaptive bins. In *Proceedings of the IEEE/CVF Conference on Computer Vision and Pattern Recognition* (2021). 4009–4018.

[13] Aaron Biggs. Paro robot. https://commons. wikimedia.org/wiki/File:Paro_robot.jpg. Licensed under CC BY-SA 2.0 via Wikimedia Commons, 2015.

[14] Aude Billard and Danica Kragic. Trends and challenges in robot manipulation. *Science* 364, 6446 (2019), eaat8414.

[15] Mike Blow, Kerstin Dautenhahn, Andrew Appleby, Chrystopher L Nehaniv, and David Lee. The art of designing robot faces: Dimensions for human-robot interaction. In *Proceedings of the 1st ACM SIGCHI/SIGART Conference on Human-Robot Interaction* (2006). 331–332.

[16] Jeannette Bohg, Antonio Morales, Tamim Asfour, and Danica Kragic. Data-driven grasp synthesis—A survey. *IEEE Transactions on Robotics* 30, 2 (2013), 289–309.

[17] Steve Branson, Pietro Perona, and Serge Belongie. Strong supervision from weak annotation: Interactive training of deformable part models. In *2011 International Conference on Computer Vision*. IEEE (2011) 1832–1839.

[18] Garrick Brazil, Julian Straub, Nikhila Ravi, Justin Johnson, and Georgia Gkioxari. Omni3D: A large benchmark and model for 3D object detection in the wild. *Conference on Computer Vision and Pattern Recognition (CVPR)* (2023).

[19] Cynthia Breazeal. Emotion and sociable humanoid robots. *International Journal of Human-Computer Studies* 59, 1-2 (2003), 119–155.

[20] Cynthia Breazeal. *Designing sociable robots*. MIT press (2004).

[21] Ryan Calo. Open robotics. *Md. L. Rev.* 70 (2010), 571.

[22] Micah Carroll, Rohin Shah, Mark K Ho, Tom Griffiths, Sanjit Seshia, Pieter Abbeel, and Anca Dragan. On the utility of learning about humans for human-ai coordination. *Advances in Neural Information Processing Systems* 32 (2019).

[23] Wan-Ling Chang and Selma Šabanović. Interaction expands function: Social shaping of the therapeutic robot PARO in a nursing home. In *Proceedings of the Tenth Annual ACM/IEEE International Conference on Human-Robot Interaction* (2015). 343–350.

[24] L. Chen, G. Papandreou, I. Kokkinos, K. Murphy, and A. Yuille. DeepLab: Semantic image segmentation with deep convolutional nets, atrous convolution, and fully connected CRFs. *IEEE Transactions on Pattern Analysis and Machine Intelligence* (2018).

[25] Yen-Chun Chen, Linjie Li, Licheng Yu, Ahmed El Kholy, Faisal Ahmed, Zhe Gan, Yu Cheng, and Jingjing Liu. Uniter: Universal image-text representation learning. In *European Conference on Computer Vision (ECCV)* (2020).

[26] Bowen Cheng, Ishan Misra, Alexander G. Schwing, Alexander Kirillov, and Rohit Girdhar. Masked-attention mask transformer for universal image segmentation. *Conference on Computer Vision and Pattern Recognition (CVPR)* (2022).

[27] Michael A Chilton, Bill C Hardgrave, and Deborah J Armstrong. Performance and strain levels of it workers engaged in rapidly changing environments: A person-job fit perspective. *ACM SIGMIS Database: The DATABASE for Advances in Information Systems* 41, 1 (2010), 8–35.

[28] Henrik Christensen, Nancy Amato, Holly Yanco, Maja Mataric, Howie Choset, Ann Drobnis, Ken Goldberg, Jessy Grizzle, Gregory Hager, John Hollerbach, Seth Hutchinson, Venkat Krovi, Daniel Lee, Bill Smart, Jeff Trinkle, and Gaurav Sukhatme. A roadmap for US robotics - from internet to robotics 2020 Edition. *Foundations and Trends in Robotics* 8, 4 (2021), 307–424. https://doi.org/10.1561/2300000066

[29] Matei Ciocarlie, Kaijen Hsiao, Adam Leeper, and David Gossow. Mobile manipulation through an assistive home robot. In *2012 IEEE/RSJ International Conference on Intelligent Robots and Systems*. IEEE (2012) 5313–5320.

[30] H.H. Clark, H.H. Clark, American Council of Learned Societies, H.C. Clark, and H.H. Clark. *Using Language.* Cambridge University Press. https://books.google.ie/books?id=DiWBGOP-YnoC

[31] Michael Danielczuk, Andrey Kurenkov, Ashwin Balakrishna, Matthew Matl, David Wang, Roberto Martín-Martín, Animesh Garg, Silvio Savarese, and Ken Goldberg. Mechanical search: Multi-step retrieval of a target object occluded by clutter. In *2019 International Conference on Robotics and Automation (ICRA)*. IEEE, (2019) 1614–1621.

[32] Bert De Brabandere, Davy Neven, and Luc Van Gool. Semantic instance segmentation with a discriminative loss function. *arXiv preprint arXiv:1708.02551* (2017).

[33] Alexey Dosovitskiy, Lucas Beyer, Alexander Kolesnikov, Dirk Weissenborn, Xiaohua Zhai, Thomas Unterthiner, Mostafa Dehghani, Matthias Minderer, Georg Heigold, Sylvain Gelly, et al. An image is worth 16x16 words: Transformers for image recognition at scale. *arXiv preprint arXiv:2010.11929* (2020).

[34] John V Draper, David B Kaber, and John M Usher. Telepresence. *Human Factors* 40, 3 (1998), 354–375.

[35] David Eigen, Christian Puhrsch, and Rob Fergus. Depth map prediction from a single image using a multi-scale deep network. *Advances in Neural Information Processing Systems* 27 (2014).

[36] Heng Fan, Liting Lin, Fan Yang, Peng Chu, Ge Deng, Sijia Yu, Hexin Bai, Yong Xu, Chunyuan Liao, and Haibin Ling. Lasot: A high-quality benchmark for large-scale single object tracking. In *Proceedings of the IEEE/CVF Conference on Computer Vision and Pattern Recognition.* (2019), 5374–5383.

[37] Linxi Fan, Guanzhi Wang, Yunfan Jiang, Ajay Mandlekar, Yuncong Yang, Haoyi Zhu, Andrew Tang, De-An Huang, Yuke Zhu, and Anima Anandkumar. MineDojo: Building open-ended embodied agents with internet-scale knowledge. In *Thirty-sixth Conference on Neural Information Processing Systems Datasets and Benchmarks Track* (2022). https://openreview.net/forum?id=rc8o_j8I8PX

[38] Pedro Felzenszwalb, David McAllester, and Deva Ramanan. A discriminatively trained, multiscale, deformable part model. In *2008 IEEE Conference on Computer Vision and Pattern Recognition.* IEEE (2008), 1–8.

[39] Terrence Fong, Illah Nourbakhsh, and Kerstin Dautenhahn. A survey of socially interactive robots. *Robotics and Autonomous Systems* 42, 3-4 (2003), 143–166.

[40] Yasutaka Furukawa and Jean Ponce. Accurate, dense, and robust multiview stereopsis. *IEEE Transactions on Pattern Analysis and Machine Intelligence* 32, 8 (2009), 1362–1376.

[41] Giovanni Fusco and James M Coughlan. Indoor localization using computer vision and visual-inertial odometry. In *International Conference on Computers Helping People with Special Needs*. Springer, (2008), 86–93.

[42] Adrien Gaidon, Qiao Wang, Yohann Cabon, and Eleonora Vig. Virtual worlds as proxy for multi-object tracking analysis. In *Proceedings of the IEEE conference on computer vision and pattern recognition*. (2016), 4340–4349.

[43] Yaroslav Ganin and Victor Lempitsky. Unsupervised domain adaptation by backpropagation. In *International Conference on Machine Learning*. PMLR, (2015), 1180–1189.

[44] Georgios Georgakis, Karl Schmeckpeper, Karan Wanchoo, Soham Dan, Eleni Miltsakaki, Dan Roth, and Kostas Daniilidis. Cross-modal map learning for vision and language navigation. In *Proceedings of the IEEE/CVF Conference on Computer Vision and Pattern Recognition*. (2022), 15460–15470.

[45] Samuel Gibbs. Amazon launches home robot Astro and giant Alexa display. *The Guardian* (2021).

[46] Matthew Craig Gombolay, Cindy Huang, and Julie Shah. Coordination of human-robot teaming with human task preferences. In *2015 AAAI Fall Symposium Series*, (2015).

[47] Michael A Goodrich, Jacob W Crandall, and Emilia Barakova. Teleoperation and beyond for assistive humanoid robots. *Reviews of Human Factors and Ergonomics* 9, 1 (2013), 175–226.

[48] Yash Goyal, Tejas Khot, Douglas Summers-Stay, Dhruv Batra, and Devi Parikh. Making the V in VQA Matter: Elevating the role of image understanding in visual question answering. In *Conference on Computer Vision and Pattern Recognition (CVPR)* (2017).

[49] Yash Goyal, Tejas Khot, Douglas Summers-Stay, Dhruv Batra, and Devi Parikh. Making the V in VQA matter: Elevating the role of image understanding in visual question answering. In *Proceedings of the IEEE Conference on Computer Vision and Pattern Recognition*, (2017). 6904–6913.

[50] Yulan Guo, Hanyun Wang, Qingyong Hu, Hao Liu, Li Liu, and Mohammed Bennamoun. Deep learning for 3D point clouds: A survey. *IEEE Transactions on Pattern Analysis and Machine Intelligence* 43, 12 (2020), 4338–4364.

[51] Tanmay Gupta, A. Kamath, Aniruddha Kembhavi, and Derek Hoiem. Towards General Purpose Vision Systems. *ArXiv* abs/2104.00743 (2021).

[52] Peter A Hancock, Deborah R Billings, Kristin E Schaefer, Jessie YC Chen, Ewart J De Visser, and Raja Parasuraman. A meta-analysis of factors affecting trust in human-robot interaction. *Human Factors* 53, 5 (2011), 517–527.

[53] Kaiming He, Georgia Gkioxari, Piotr Dollár, and Ross Girshick. Mask R-CNN. In *Proceedings of the IEEE International Conference on Computer Vision (ICCV)* (2017).

[54] Kaiming He, Xiangyu Zhang, Shaoqing Ren, and Jian Sun. Deep residual learning for image recognition. In *Proceedings of the IEEE Conference on Computer Vision and Pattern Recognition* (2016). 770–778.

[55] Jo ao F Henriques, Rui Caseiro, Pedro Martins, and Jorge Batista. High-speed tracking with kernelized correlation filters. In *Proceedings of the IEEE Conference on Computer Vision and Pattern Recognition* (2015). 583–591.

[56] Wan Ching Ho, Kerstin Dautenhahn, Mei Yii Lim, Patricia A Vargas, Ruth Aylett, and Sibylle Enz. An initial memory model for virtual and robot companions supporting migration and long-term interaction. In *RO-MAN 2009-The 18th IEEE International Symposium on Robot and Human Interactive Communication*. IEEE (2009) 277–284.

[57] Guy Hoffman. Evaluating fluency in human–robot collaboration. *IEEE Transactions on Human-Machine Systems* 49, 3 (2019), 209–218.

[58] Ronghang Hu, Jacob Andreas, Trevor Darrell, and Kate Saenko. Explainable neural computation via stack neural module networks. In *Proceedings of the European Conference on Computer Vision (ECCV)* (2018). 53–69.

[59] Md Jahidul Islam, Chelsey Edge, Yuyang Xiao, Peigen Luo, Muntaqim Mehtaz, Christopher Morse, Sadman Sakib Enan, and Junaed Sattar. Semantic

segmentation of underwater imagery: Dataset and benchmark. In *2020 IEEE/RSJ International Conference on Intelligent Robots and Systems (IROS)*. IEEE (2020) 1769–1776.

[60] David Jaffe, John Thiemer, and Drew Nelson. Perspectives in Assistive Technology (2012). https://web.stanford.edu/class/engr110/2012/04b-Jaffe.pdf

[61] Anne Jorstad, David Jacobs, and Alain Trouvé. A deformation and lighting insensitive metric for face recognition based on dense correspondences. In *CVPR 2011*. IEEE (2011) 2353–2360.

[62] Steve Jurvetson. Baxter Robot Caught Coding (2013). https://commons.wikimedia.org/wiki/File:Caught_Coding_(9690512888).jpg. Licensed under CC BY 2.0 via Wikimedia Commons.

[63] Leslie Pack Kaelbling, Michael L Littman, and Anthony R Cassandra. Planning and acting in partially observable stochastic domains. *Artificial Intelligence* 101, 1-2 (1998), 99–134.

[64] Kushal Kafle and Christopher Kanan. Visual question answering: Datasets, algorithms, and future challenges. *Computer Vision and Image Understanding* 163 (2017), 3–20.

[65] Ashish Kapoor, Kristen Grauman, Raquel Urtasun, and Trevor Darrell. Active learning with gaussian processes for object categorization. In *2007 IEEE 11th International Conference on Computer Vision*. IEEE (2007), 1–8.

[66] Corentin Kervadec, Grigory Antipov, Moez Baccouche, and Christian Wolf. Roses are red, violets are blue... but should vqa expect them to?. In *Proceedings of the IEEE/CVF Conference on Computer Vision and Pattern Recognition* (2021). 2776–2785.

[67] Cory D Kidd and Cynthia Breazeal. Robots at home: Understanding long-term human-robot interaction. In *2008 IEEE/RSJ International Conference on Intelligent Robots and Systems*. IEEE, (2008) 3230–3235.

[68] Namil Kim, Yukyung Choi, Soonmin Hwang, and In So Kweon. Multispectral transfer network: Unsupervised depth estimation for all-day vision. In *Proceedings of the AAAI Conference on Artificial Intelligence*, Vol. 32 (2018).

[69] Tae Wan Kim and Alan Strudler. Should Robots Have Rights or Rites? *Communications of the ACM* 66, 6 (2023), 78–85.

[70] Alexander Kirillov, Kaiming He, Ross Girshick, Carsten Rother, and Piotr Dollár. Panoptic segmentation. In *Proceedings of the IEEE/CVF Conference on Computer Vision and Pattern Recognition* (2019). 9404–9413.

[71] Alexander Kirillov, Eric Mintun, Nikhila Ravi, Hanzi Mao, Chloe Rolland, Laura Gustafson, Tete Xiao, Spencer Whitehead, Alexander C Berg, Wan-Yen Lo, et al. Segment anything. *arXiv preprint arXiv:2304.02643* (2023).

[72] Kheng Lee Koay, Dag Sverre Syrdal, Michael L Walters, and Kerstin Dautenhahn. Living with robots: Investigating the habituation effect in participants' preferences during a longitudinal human-robot interaction study. In *RO-MAN 2007-The 16th IEEE International Symposium on Robot and Human Interactive Communication* IEEE (2007), 564–569.

[73] Hema S Koppula, Ashesh Jain, and Ashutosh Saxena. Anticipatory planning for human-robot teams. In *Experimental Robotics: The 14th International Symposium on Experimental Robotics*. Springer, (2016) 453–470.

[74] Matej Kristan, Ales Leonardis, Jiri Matas, Michael Felsberg, Roman Pflugfelder, Joni-Kristian Kamarainen, Hyung Jin Chang, Martin Danelljan, Luka Čehovin Zajc, Alan Lukežič, Ondrej Drbohlav, Johanna Bjorklund, Yushan Zhang, Zhongqun Zhang, Song Yan, Wenyan Yang, Dingding Cai, Christoph Mayer, and Gustavo Fernandez. The Tenth Visual Object Tracking VOT2022 Challenge Results (2022).

[75] Alex Krizhevsky, Ilya Sutskever, and Geoffrey E Hinton. Imagenet classification with deep convolutional neural networks. *Communications of the ACM* 60, 6 (2017), 84–90.

[76] Andrey Kurenkov, Joseph Taglic, Rohun Kulkarni, Marcus Dominguez-Kuhne, Animesh Garg, Roberto Martín-Martín, and Silvio Savarese. Visuomotor mechanical search: Learning to retrieve target objects in clutter. In *2020 IEEE/RSJ International Conference on Intelligent Robots and Systems (IROS)*. IEEE, (2020) 8408–8414.

[77] Hamid Laga, Laurent Valentin Jospin, Farid Boussaid, and Mohammed Bennamoun. A survey on deep learning techniques for stereo-based depth estimation. *IEEE Transactions on Pattern Analysis and Machine Intelligence* 44, 4 (2020), 1738–1764.

[78] Chi-Pang Lam and S Shankar Sastry. A POMDP framework for human-in-the-loop system. In *53rd IEEE Conference on Decision and Control*. IEEE, (2014) 6031–6036.

[79] John Lambert, Zhuang Liu, Ozan Sener, James Hays, and Vladlen Koltun. MSeg: A composite dataset for multi-domain semantic segmentation. In *Proceedings of the IEEE/CVF Conference on Computer Vision and Pattern Recognition*. (2020) 2879–2888.

[80] Mikhail A Lebedev and Miguel AL Nicolelis. Brain–machine interfaces: Past, present and future. *TRENDS in Neurosciences* 29, 9 (2006), 536–546.

[81] Sooyeon Lee, Rui Yu, Jingyi Xie, Syed Masum Billah, and John M Carroll. Opportunities for human-AI collaboration in remote sighted assistance. In *27th International Conference on Intelligent User Interfaces*. (2022) 63–78.

[82] Youngwoon Lee, Edward S Hu, and Joseph J Lim. IKEA furniture assembly environment for long-horizon complex manipulation tasks. In *2021 IEEE International Conference on Robotics and Automation (ICRA)*. IEEE, (2021) 6343–6349.

[83] Iolanda Leite, Carlos Martinho, and Ana Paiva. Social robots for long-term interaction: A survey. *International Journal of Social Robotics* 5 (2013), 291–308.

[84] Ian Lenz, Honglak Lee, and Ashutosh Saxena. Deep learning for detecting robotic grasps. *The International Journal of Robotics Research* 34, 4-5 (2015), 705–724.

[85] Sergey Levine, Peter Pastor, Alex Krizhevsky, Julian Ibarz, and Deirdre Quillen. Learning hand-eye coordination for robotic grasping with deep learning and large-scale data collection. *The International Journal of Robotics Research* 37, 4-5 (2018), 421–436.

[86] Michael Lewis, Katia Sycara, and Phillip Walker. The role of trust in human-robot interaction. *Foundations of Trusted Autonomy* (2018), 135–159.

[87] Junnan Li, Dongxu Li, Caiming Xiong, and Steven Hoi. Blip: Bootstrapping language-image pre-training for unified vision-language understanding and generation. In *International Conference on Machine Learning*. PMLR, (2022) 12888–12900.

[88] Linjie Li, Zhe Gan, and Jingjing Liu. A closer look at the robustness of vision-and-language pre-trained models. *arXiv preprint arXiv:2012.08673* (2020).

[89] Wanwan Li, Javier Talavera, Amilcar Gomez Samayoa, Jyh-Ming Lien, and Lap-Fai Yu. Automatic synthesis of virtual wheelchair training scenarios. In *IEEE Virtual Reality* (Atlanta, Georgia) (2020).

[90] Zhaoshuo Li, Xingtong Liu, Nathan Drenkow, Andy Ding, Francis X Creighton, Russell H Taylor, and Mathias Unberath. Revisiting stereo depth estimation from a sequence-to-sequence perspective with transformers. In *Proceedings of the IEEE/CVF International Conference on Computer Vision*. (2021) 6197–6206.

[91] Dahua Lin, Sanja Fidler, and Raquel Urtasun. Holistic scene understanding for 3D object detection with RGBD cameras. In *Proceedings of the IEEE International Conference on Computer Vision*. (2013) 1417–1424.

[92] Mingyu Liu, Mathieu Nancel, and Daniel Vogel. Gunslinger: Subtle arms-down mid-air interaction. In *Proceedings of the 28th Annual ACM Symposium on User Interface Software & Technology*. (2015) 63–71.

[93] Manja Lohse, Frank Hegel, and Britta Wrede. Domestic applications for social robots: An online survey on the influence of appearance and capabilities. (2008).

[94] Jonathan Long, Evan Shelhamer, and Trevor Darrell. Fully convolutional networks for semantic segmentation. In *Proceedings of the IEEE Conference on Computer Vision and Pattern Recognition*. (2015) 3431–3440.

[95] Jiasen Lu, Dhruv Batra, Devi Parikh, and Stefan Lee. Vilbert: Pretraining task-agnostic visiolinguistic representations for vision-and-language tasks. *Advances in Neural Information Processing Systems* 32 (2019).

[96] Jiasen Lu, Xiao Lin, Dhruv Batra, and Devi Parikh. Deeper LSTM and normalized CNN visual question answering model. *GitHub Repository* 6 (2015).

[97] Jiasen Lu, Jianwei Yang, Dhruv Batra, and Devi Parikh. Hierarchical question-image co-attention for visual question answering. *Advances in Neural Information Processing Systems* 29 (2016).

[98] Paweł Maciejasz, Jörg Eschweiler, Kurt Gerlach-Hahn, Arne Jansen-Troy, and Steffen Leonhardt. A survey on robotic devices for upper limb rehabilitation. *Journal of Neuroengineering and Rehabilitation* 11, 1 (2014), 1–29.

[99] Jeffrey Mahler, Jacky Liang, Sherdil Niyaz, Michael Laskey, Richard Doan, Xinyu Liu, Juan Aparicio Ojea, and Ken Goldberg. Dex-net 2.0: Deep learning to plan robust grasps with synthetic point clouds and analytic grasp metrics. *arXiv preprint arXiv:1703.09312* (2017).

[100] Gellert Mattyus, Shenlong Wang, Sanja Fidler, and Raquel Urtasun. Enhancing road maps by parsing aerial images around the world. In *Proceedings of the IEEE International Conference on Computer Vision.* (2015) 1689–1697.

[101] Alexander Mertens, Ulrich Reiser, Benedikt Brenken, Mathias Lüdtke, Martin Hägele, Alexander Verl, Christopher Brandl, and Christopher Schlick. Assistive robots in eldercare and daily living: Automation of individual services for senior citizens. In *Intelligent Robotics and Applications: 4th International Conference, ICIRA 2011, Aachen, Germany, December 6-8, 2011, Proceedings, Part I 4.* Springer, (2011) 542–552.

[102] Marvin Minsky. *Society of mind.* Simon and Schuster (1988).

[103] Pavlo Molchanov, Xiaodong Yang, Shalini Gupta, Kihwan Kim, Stephen Tyree, and Jan Kautz. Online detection and classification of dynamic hand gestures with recurrent 3D convolutional neural network. In *Proceedings of the IEEE Conference on Computer Vision and Pattern Recognition.* (2016) 4207–4215.

[104] George E Monahan. State of the art—a survey of partially observable Markov decision processes: Theory, models, and algorithms. *Management Science* 28, 1 (1982), 1–16.

[105] Hans Moravec. *Mind children: The future of robot and human intelligence.* Harvard University Press (1988).

[106] Masahiro Mori, Karl F MacDorman, and Norri Kageki. The uncanny valley [from the field]. *IEEE Robotics & automation magazine* 19, 2 (2012), 98–100.

[107] Stephan J Motowildo, Walter C Borman, and Mark J Schmit. A theory of individual differences in task and contextual performance. *Human Performance* 10, 2 (1997), 71–83.

[108] Arsalan Mousavian, Clemens Eppner, and Dieter Fox. 6-DOF GraspNet: Variational grasp generation for object manipulation. In *Proceedings of the IEEE/CVF International Conference on Computer Vision.* (2019) 2901–2910.

[109] Arsalan Mousavian, Alexander Toshev, Marek Fišer, Jana Košecká, Ayzaan Wahid, and James Davidson. Visual representations for semantic target driven navigation. In *2019 International Conference on Robotics and Automation (ICRA).* IEEE, (2019) 8846–8852.

[110] Ra'ul Mur-Artal, JMM Montiel, and Juan D Tard'os. ORB-SLAM: A versatile and accurate monocular SLAM system. In *IEEE Transactions on Robotics*, Vol. 31. IEEE, (2015) 1147–1163.

[111] Benjamin A Newman, Christopher Jason Paxton, Kris Kitani, and Henny Admoni. Towards online adaptation for autonomous household assistants. In *Companion of the 2023 ACM/IEEE International Conference on Human-Robot Interaction.* (2023) 506–510.

[112] Illah Reza Nourbakhsh. *Robot futures.* MIT Press (2015).

[113] Kristine L Nowak and Frank Biocca. The effect of the agency and anthropomorphism on users' sense of telepresence, copresence, and social presence in virtual environments. *Presence: Teleoperators & Virtual Environments* 12, 5 (2003), 481–494.

[114] Giuseppe Oriolo, Marilena Vendittelli, and Giovanni Ulivi. On-line map building and navigation for autonomous mobile robots. In *Proceedings of 1995 IEEE International Conference on Robotics and Automation*, Vol. 3. IEEE, (1995) 2900–2906.

[115] Gerald Oster and Yasunori Nishijima. Moiré patterns. *Scientific American* 208, 5 (1963), 54–63.

[116] Anastasia K Ostrowski, Cynthia Breazeal, and Hae Won Park. Mixed-method long-term robot usage: Older adults' lived experience of social robots. In *2022 17th ACM/IEEE International Conference on Human-Robot Interaction (HRI).* IEEE, (2022) 33–42.

[117] Karl Pertsch, Hejia Zhang, Youngwoon Lee, Joseph J Lim, Stefanos Nikolaidis, et al. [n. d.]. Assisted teleoperation for scalable robot data collection. In *CoRL 2022 Workshop on Pre-training Robot Learning.*

[118] Lerrel Pinto and Abhinav Gupta. Supersizing self-supervision: Learning to grasp from 50k tries and 700 robot hours. In *2016 IEEE International Conference on Robotics and Automation (ICRA).* IEEE, (2016) 3406–3413.

[119] QuarkyTale. Playing with Kilobots (2018). https://commons.wikimedia.org/wiki/File:Playing_with_

Kilobots.jpg. Licensed under CC BY-SA 4.0 via Wikimedia Commons.

[120] Sarah M Rabbitt, Alan E Kazdin, and Brian Scassellati. Integrating socially assistive robotics into mental healthcare interventions: Applications and recommendations for expanded use. *Clinical Psychology Review* (2015).

[121] Aditi Ramachandran, Chien-Ming Huang, and Brian Scassellati. Toward effective robot–child tutoring: Internal motivation, behavioral intervention, and learning outcomes. *ACM Transactions on Interactive Intelligent Systems (TiiS)* 9, 1 (2019), 1–23.

[122] Byron Reeves and Clifford Nass. The media equation: How people treat computers, television, and new media like real people. *Cambridge, UK* 10 (1996), 236605.

[123] Bernardino Romera-Paredes and Philip Hilaire Sean Torr. Recurrent instance segmentation. In *Computer Vision–ECCV 2016: 14th European Conference, Amsterdam, The Netherlands, October 11-14, 2016, Proceedings, Part VI 14*. Springer, 312–329.

[124] Kate Saenko, Brian Kulis, Mario Fritz, and Trevor Darrell. Adapting visual category models to new domains. In *Computer Vision–ECCV 2010: 11th European Conference on Computer Vision, Heraklion, Crete, Greece, September 5-11, 2010, Proceedings, Part IV 11*. Springer, 213–226.

[125] Ashutosh Saxena, Min Sun, and Andrew Y Ng. Make3D: Learning 3D scene structure from a single still image. *IEEE Transactions on Pattern Analysis and Machine Intelligence* 31, 5 (2008), 824–840.

[126] Kristin E Schaefer, Tracy L Sanders, Ryan E Yordon, Deborah R Billings, and Peter A Hancock. Classification of robot form: Factors predicting perceived trustworthiness. In *Proceedings of the Human Factors and Ergonomics Society Annual Meeting*, Vol. 56. SAGE Publications Sage CA: Los Angeles, CA, (2012) 1548–1552.

[127] Daniel Scharstein and Richard Szeliski. A taxonomy and evaluation of dense two-frame stereo correspondence algorithms. *International Journal of Computer Vision* 47 (2002), 7–42.

[128] Alexander G Schwing, Sanja Fidler, Marc Pollefeys, and Raquel Urtasun. Box in the box: Joint 3d layout and object reasoning from single images. In *Proceedings of the IEEE International Conference on Computer Vision.* (2013) 353–360.

[129] Steven M Seitz, Brian Curless, James Diebel, Daniel Scharstein, and Richard Szeliski. A comparison and evaluation of multi-view stereo reconstruction algorithms. In *2006 IEEE Computer Society Conference on Computer Vision and Pattern Recognition (CVPR'06)*, Vol. 1. IEEE, (2006) 519–528.

[130] Meet Shah, Xinlei Chen, Marcus Rohrbach, and Devi Parikh. Cycle-consistency for robust visual question answering. In *Proceedings of the IEEE/CVF Conference on Computer Vision and Pattern Recognition.* (2019) 6649–6658.

[131] Amanda Sharkey and Noel Sharkey. Granny and the robots: ethical issues in robot care for the elderly. *Ethics and Information Technology* 14 (2012), 27–40.

[132] Andy Shih, Arjun Sawhney, Jovana Kondic, Stefano Ermon, and Dorsa Sadigh. On the critical role of conventions in adaptive human-AI collaboration. *arXiv preprint arXiv:2104.02871* (2021).

[133] Elaine Short, Justin Hart, Michelle Vu, and Brian Scassellati. No fair!! an interaction with a cheating robot. In *2010 5th ACM/IEEE International Conference on Human-Robot Interaction (HRI)*. IEEE, (2010) 219–226.

[134] Candace L Sidner, Christopher Lee, Cory D Kidd, Neal Lesh, and Charles Rich. Explorations in engagement for humans and robots. *Artificial Intelligence* 166, 1-2 (2005), 140–164.

[135] Roland Siegwart, Illah Reza Nourbakhsh, and Davide Scaramuzza. *Introduction to autonomous mobile robots.* MIT press (2011).

[136] N. Silberman, D. Hoiem, P. Kohli, and R. Fergus. Indoor segmentation and support inference from RGBD images. In *European Conference on Computer Vision* (2012).

[137] Karen Simonyan and Andrew Zisserman. Very deep convolutional networks for large-scale image recognition. *arXiv preprint arXiv:1409.1556* (2014).

[138] Baochen Sun, Jiashi Feng, and Kate Saenko. Return of frustratingly easy domain adaptation. In *Proceedings of the AAAI Conference on Artificial Intelligence*, Vol. 30 (2016).

[139] JaYoung Sung, Henrik I Christensen, and Rebecca E Grinter. Robots in the wild: Understanding long-term use. In *Proceedings of the 4th ACM/IEEE International Conference on Human Robot Interaction.* (2009) 45–52.

[140] Andrew Szot, Alexander Clegg, Eric Undersander, Erik Wijmans, Yili Zhao, John Turner, Noah Maestre, Mustafa Mukadam, Devendra Singh Chaplot, Oleksandr Maksymets, et al. Habitat 2.0: Training home assistants to rearrange their habitat. *Advances in Neural Information Processing Systems* 34 (2021), 251–266.

[141] Hao Tan and Mohit Bansal. LXMERT: Learning cross-modality encoder representations from transformers. *arXiv preprint arXiv:1908.07490* (2019).

[142] Jerry Tang, Amanda LeBel, Shailee Jain, and Alexander G Huth. Semantic reconstruction of continuous language from non-invasive brain recordings. *Nature Neuroscience* (2023), 1–9.

[143] Sebastian Thrun. Simultaneous localization and mapping. *Robotics and Cognitive Approaches to Spatial Mapping* (2008), 13–41.

[144] Carme Torras. Service robots for citizens of the future. *European Review* 24, 1 (2016), 17–30.

[145] Katherine M Tsui and Holly A Yanco. Design challenges and guidelines for social interaction using mobile telepresence robots. *Reviews of Human Factors and Ergonomics* 9, 1 (2013), 227–301.

[146] Sherry Turkle. In Constant Digital Contact, We Feel'Alone Together'. *Alone Together* (2012).

[147] ubahnverleih. Nao Robot. (2016) https://commons.wikimedia.org/wiki/File:Nao_Robot_(Robocup_2016).jpg. Online; accessed 22-July-2023.

[148] Andries Van Dam. Post-WIMP user interfaces. *Commun. ACM* 40, 2 (1997), 63–67.

[149] Kazuyoshi Wada and Takanori Shibata. Robot therapy in a care house-its sociopsychological and physiological effects on the residents. In *Proceedings 2006 IEEE International Conference on Robotics and Automation, 2006. ICRA 2006.* IEEE, (2006) 3966–3971.

[150] Sameer Wagh, Divya Gupta, and Nishanth Chandran. SecureNN: 3-party secure computation for neural network training. *Proceedings on Privacy Enhancing Technologies* 2019, 3 (2019), 26–49.

[151] Shenlong Wang, Sanja Fidler, and Raquel Urtasun. Holistic 3D scene understanding from a single geotagged image. In *Proceedings of the IEEE Conference on Computer Vision and Pattern Recognition.* (2015) 3964–3972.

[152] Xinshuo Weng, Jianren Wang, David Held, and Kris Kitani. 2020. 3D multi-object tracking: A baseline and new evaluation metrics. In *Proceedings of (IROS) IEEE/RSJ International Conference on Intelligent Robots and Systems.* (2020) 10359 – 10366.

[153] Jacob O Wobbrock, Andrew D Wilson, and Yang Li. Gestures without libraries, toolkits or training: A \$1 recognizer for user interface prototypes. In *Proceedings of the 20th Annual ACM Symposium on User Interface Software and Technology.* (2007) 159–168.

[154] Sarah Woods, Kerstin Dautenhahn, and Joerg Schulz. The design space of robots: Investigating children's views. In *RO-MAN 2004. 13th IEEE International Workshop on Robot and Human Interactive Communication (IEEE Catalog No. 04TH8759).* IEEE, (2004) 47–52.

[155] Gangwei Xu, Junda Cheng, Peng Guo, and Xin Yang. Attention concatenation volume for accurate and efficient stereo matching. In *Proceedings of the IEEE/CVF Conference on Computer Vision and Pattern Recognition.* (2022) 12981–12990.

[156] Kuan Xu, Chen Wang, Chao Chen, Wei Wu, and Sebastian Scherer. Aircode: A robust object encoding method. *IEEE Robotics and Automation Letters* (2022).

[157] Holly A Yanco and Jill L Drury. A taxonomy for human-robot interaction. In *Proceedings of the AAAI Fall Symposium on Human-Robot Interaction.* (2002) 111–119.

[158] Zichao Yang, Xiaodong He, Jianfeng Gao, Li Deng, and Alex Smola. Stacked attention networks for image question answering. In *Proceedings of the IEEE Conference on Computer Vision and Pattern Recognition.* (2016) 21–29.

[159] Jian Yao, Sanja Fidler, and Raquel Urtasun. Describing the scene as a whole: Joint object detection, scene classification and semantic segmentation. In *2012 IEEE Conference on Computer Vision and Pattern Recognition.* IEEE, (2012) 702–709.

[160] Yao Yao, Zixin Luo, Shiwei Li, Tianwei Shen, Tian Fang, and Long Quan. Recurrent MVSNet for high-resolution multi-view stereo depth inference. In *Proceedings of the IEEE/CVF Conference on Computer Vision and Pattern Recognition.* (2019) 5525–5534.

[161] Lai Sum Yim, Quang TN Vo, Ching-I Huang, Chi-Ruei Wang, Wren McQueary, Hsueh-Cheng Wang, Haikun Huang, and Lap-Fai Yu. WFH-VR: Teleoperating a robot arm to set a dining table across the globe via virtual reality. In *2022 IEEE/RSJ International Conference on Intelligent Robots and Systems (IROS)* (2022).

[162] Gareth Young, Hamish Milne, Daniel Griffiths, Elliot Padfield, Robert Blenkinsopp, and Orestis Georgiou. Designing mid-air haptic gesture controlled user interfaces for cars. *Proceedings of the ACM on Human-Computer Interaction* 4, EICS (2020), 1–23.

[163] Z22. iRobot Ava 500. (2014) https://commons.\penalty-\@Mwikimedia.org/wiki/File:IRobot_Ava_500.jpg. Licensed under CC BY-SA 3.0 via Wikimedia Commons.

[164] Michelle Zhao, Reid Simmons, and Henny Admoni. The role of adaptation in collective human–AI teaming. *Topics in Cognitive Science* (2022).

[165] Tinghui Zhou, Matthew Brown, Noah Snavely, and David G Lowe. 2017. Unsupervised learning of depth and ego-motion from video. *Proceedings of the IEEE Conference on Computer Vision and Pattern Recognition* (2017), 1851–1858.

[166] Yuke Zhu, Roozbeh Mottaghi, Eric Kolve, Joseph J Lim, Abhinav Gupta, Li Fei-Fei, and Ali Farhadi. Target-driven visual navigation in indoor scenes using deep reinforcement learning. In *2017 IEEE International Conference on Robotics and Automation (ICRA)*. IEEE, (2017) 3357–3364.

9

Computer Vision Applications in Underwater Robotics and Oceanography

Md Jahidul Islam, Alberto Quattrini Li, Yogesh A Girdhar, and Ioannis Rekleitis

9.1 Introduction

Computer Vision has surpassed many impressive milestones in recent years in the field of robotics and Artificial Intelligence (AI). Cameras are passive, inexpensive sensors that produce big volumes of data; as visual sensing is one of the primary perception modalities for humans, it is also human interpretable. However, the underwater realm is a challenging domain with severe light attenuation where visual perception techniques that work well above water often fail. Nevertheless, several application areas are looking into vision to obtain semantic as well as metric information. Subsea exploration, environmental monitoring, oceanography, aquaculture, marine archaeology, search and rescue, submarine structure inspection, and resource utilization are some of the major areas where computer vision has been applied with great success. Autonomous Underwater Vehicles (AUVs) and Remotely Operated Vehicles (ROVs) are generally deployed in these applications for autonomous or semi-autonomous missions, often alongside human divers. Unlike terrestrial robots, ensuring robust visual guidance for underwater robots faces unique challenges due to the domain-specific operational complexities associated with underwater sensing and estimation. In this chapter, we present the vibrant scientific literature that attempts to address these challenges for robust underwater visual perception, and then elaborately discuss the emerging vision technologies of subsea robotics applications.

We start the discussion by depicting the inherent challenges and practicalities of underwater vision from the perspective of light propagation and image formation literature. We discuss how the waterbody-specific optical characteristics (*e.g.*,, light absorption and scattering) [1] cause a range of nonlinear chromatic distortions and image degradation, which limit the applicability of many classical vision-based algorithms that otherwise work on atmospheric imagery for the analogous problems. As seen in Figure 9.1, AUVs operate in low-light conditions in

FIGURE 9.1
Aqua2 AUV collecting data over a coral reef.

a scattering medium, hence *visual filtering* is necessary to restore/enhance the perceptual and statistical properties (*e.g.*,, color, contrast, sharpness) of raw camera images. We present the existing systems and state-of-the-art (SOTA) methodologies that tackle this problem by both physics-based formulations and data-driven learning pipelines; see Section 9.2 for the full discussion. In particular, we focus on fast image enhancement and color restoration schemes that can be used in underwater robots' autonomy loop for improved visual perception [2, 3]. We also highlight the best scientific and engineering practices for their efficient implementation and single-board use in standard visual perception tasks of underwater robots.

Along this line, we discuss the prominent SOTA methods and emerging solutions for robust object detection, recognition, and tracking in underwater robotics applications (see Section 9.3). We particularly focus on two important use cases: (1) coral-reef classification, 3D mapping, and coverage estimation [4, 5]; and (2) fish detection, tracking, and segmentation [6, 7]. We also elaborately discuss how contemporary deep visual learning-based models are transforming the AI capabilities of vision-based monitoring and mapping systems. Additionally, we highlight the recent success in data acquisition,

DOI: 10.1201/9781003328957-9

labeling, and deep learning literature for both class-agnostic saliency estimation [8, 9] and semantic scene segmentation [10] in underwater imagery. We then review the utility and effectiveness of several visual saliency estimation and attention modeling approaches for underwater scene analysis [11], subsea inspection [12], and seabed mapping [13].

In Section 9.4, we present the vision-based state estimation algorithms and localization systems, which are critical to ensure safe and efficient underwater robot navigation for autonomous missions [14–16]. Autonomous monitoring, surveying/mapping, or inspection operations in underwater caves, oil rigs, data center pipelines, and shipwrecks are extremely important for the economy, conservation, archaeological, and scientific discoveries. We elaborately discuss the sensory design choices and algorithmic formulations of SOTA Visual-Inertial Odometry (VIO) and Simultaneous Localization and Mapping (SLAM) approaches for AUVs [17–19]. We then provide qualitative and quantitative analyses of the prominent systems with practical considerations for fusing visual, inertial, sonar, and water-pressure information for robust state estimation and localization [17, 20]. We also highlight some emerging techniques for 3D reconstruction [21–23] and depth estimation [24, 25] of natural underwater scenes. To this end, the self-supervised and semi-supervised approaches for depth estimation, depth-guided image enhancement, vision-filtered state estimation, scene parsing, and attention modeling have great potential to transform the underwater robot vision literature in the coming years.

Furthermore, we outline a comprehensive overview of the vision-based underwater Human–Robot Interaction (HRI) frameworks that are widely used for cooperative inspection, exploration, and data collection, especially in shallow-water and coastal-water applications. We first depict the general scenarios where human divers provide continuous or incremental guidance during cooperative task execution, while the *companion robot* maintains interaction throughout the mission [26]. Then, we elaborately discuss the SOTA systems and methodologies for two major components of underwater HRI: autonomous diver following [27, 28] and diver-robot communication [29, 30]. For both topics, we present the key developments in the spatial, temporal, and frequency domain estimation, and then highlight the recent advancements in the deep learning literature of detection, tracking, following, and interaction for ensuring robust and efficient underwater HRI [26, 31, 32]. We also exemplify how these improved AI capabilities can significantly simplify the operational overhead of AUV deployments in challenging real-world applications; see Section 9.5 for the relevant discussions.

Finally, we conclude this chapter by delineating a set of high-impact applications of computer vision for marine life monitoring and conservation, subsea structure inspection, cave mapping, aquaculture, and marine ecology. While existing systems and solutions offer promising results, we anticipate significant advancements in computer vision, embedded deep learning, and underwater robotics technologies over the next few decades. Enabling fully autonomous low-cost subsea operations, coordinated multirobot mission execution, long-term remote monitoring support, smart aquaculture sensing and feedback systems, efficient domain-aware 3D reconstruction and mapping capabilities, to name a few, have tremendous scientific and economic worth. We identify several open problems and potential research directions in these application areas, which can lead us to foster the next AI revolution in the subsea robotics and machine vision technologies.

9.2 Underwater Imaging

The digital image formation process differs underwater from in-air (atmospheric) process due to the inherent optical properties of water. Light attenuates exponentially underwater with the propagation distance [1], resulting in degraded and color distorted images. Various nonlinear artifacts in underwater images occur due to two physical characteristics of light: scattering (forward scatter and backscatter) and absorption. Forward scattering is responsible for blurry images, whereas backscatter causes contrast reduction in underwater images [33–35]. Besides, light absorption by water has a prominent spectral dependency; red light loses most of its energy beyond five meters of depth due to its longer wavelength, causing bluish or greenish tint and other irregular color artifacts in underwater images [36, 37]; see Figure 9.2a. Quite often, at great depths, inside shipwrecks or underwater caves, there is complete absence of ambient light and the only light source is the one provided by the imaging system, as in the case of Figure 9.2b. It is worth noting this scenario produces a continuous movement of shadows with the motion of the camera/light system. Moreover, as shown in Figure 9.2c, extreme intensity variations can also occur in adverse visual conditions.

Some of these aspects can be modeled and well estimated by physics-based solutions, particularly for dehazing and color correction [40, 41]. Classical computer vision approaches are driven by the atmospheric scattering model [42, 43], which considers two components for image formation: direct attenuation and air-light (or atmospheric light). According to this model [44], hazy images are represented as:

$$\mathbf{U}(\mathbf{x}) = \mathbf{I}(\mathbf{x}) \cdot t(\mathbf{x}) + \mathbf{B} \cdot \left(1 - t(\mathbf{x})\right), \qquad (9.1)$$

where $\mathbf{U}(\cdot)$ is the observed image captured by the camera, $\mathbf{B}$ represents atmospheric light (also known as

(a) Sample from the *cemetery* dataset

(b) Sample from the *cave* dataset

(c) Sample from a *sunken bus* dataset

FIGURE 9.2
Sample images from three publicly available datasets [38] are shown; the images are collected by an underwater sensor suite (with courtesy from [39].) (a) a fake cemetery (SC, USA); (b) inside a cave (FL, USA); and (c) outside a sunken bus (NC, USA). Notice the challenges in visual sensing conditions and differences in each water-body.

background light), and $\mathbf{I}(\cdot)$ is the clear latent image which we want to reconstruct. Additionally, $t(\cdot) \in [0,1]$ is the medium transmission rate, that is, percentage of unscattered light that reaches the camera from point $\mathbf{x}$; it is defined as $t(\mathbf{x}) = e^{-\beta \cdot d(\mathbf{x})}$, where β is the homogeneous scattering coefficient, and $d(\mathbf{x})$ is the distance of point $\mathbf{x}$ from the camera.

In the following sections, we describe the specific image formation models for underwater and how they can be used for image restoration, covering also the related data-driven approaches of *perceptual* image enhancement to facilitate improved visual perception by underwater robots.

9.2.1 Underwater Image Formation Models

In the actual underwater medium, forward scattering causes images to blur, while backscattering creates veiling light that leads to low contrast and a haze-like appearance. In particular, veiling light scattered into the line of sight is the main contributor to underwater image degradation, hence the forward scattering component is often ignored for simplicity [45]. Moreover, color components of different frequencies have different attenuation degrees when light passes through water medium. Taking these domain characteristics into account, the Jaffe–McGlamery imaging model [46, 47] is defined as:

$$\mathbf{U}_\lambda(\mathbf{x}) = \underbrace{\mathbf{I}_\lambda(\mathbf{x}) \cdot e^{-\beta_\lambda d(\mathbf{x})}}_{direct\ attenuation} + \underbrace{\mathbf{B}_\lambda \cdot (1 - e^{-\beta_\lambda d(\mathbf{x})})}_{backscattering}. \quad (9.2)$$

where, λ denotes the three different wavelength components of RGB channels in light. The transmission map of water medium $t_\lambda(\mathbf{x})$ is represented by depth map $d(\mathbf{x})$ and medium attenuation coefficient β_λ as: $t_\lambda(\mathbf{x}) = e^{-\beta_\lambda \cdot d(\mathbf{x})}$. It can be observed that the composition and structure of this equation is very similar to the atmo-

spheric scattering model (of Eq. 9.1). The essential difference is that the Jaffe–McGlamery model considers wavelength-dependency in the estimation of direct transmission and backscattering components.

Recently, a revised underwater image formation model proposed by Akkaynak *et al.* shows that, in the ocean, the effects of light absorption can be comparable to light scattering, or even dominate it depending on the type of water [1]. Moreover, the attenuation coefficient relies not only on the inherent water optical properties but also on the camera sensor, scene irradiance, and imaging range. Most importantly, the attenuation coefficients of backscattering and direct signal in Eq. 9.2 are different, which are given by: $\beta_\lambda^D = -\ln(\frac{\mathbf{U}_\lambda(\mathbf{z}) - \mathbf{B}_\lambda(\mathbf{z})}{\mathbf{I}_\lambda})/\mathbf{z}$, and $\beta_\lambda^B = -\ln(1 - \frac{\mathbf{B}_\lambda(\mathbf{z})}{\mathbf{B}_\lambda^\infty})/\mathbf{z}$; where $\beta_\lambda^{D(B)}$ is the attenuation coefficient of direct transmitted signal (backscattered signal) and $\mathbf{B}_\lambda^\infty$ is wideband veiling light; the backscattered light $\mathbf{B}_\lambda$ is a function of depth map $\mathbf{z}$ representing the distance of the target object from camera along line of sight. Based on these definitions, the imaging model, starting from Eq. 9.2, is revised as:

$$\mathbf{U}_\lambda(\mathbf{x}) = \mathbf{I}_\lambda(\mathbf{x}) \cdot e^{-\beta_\lambda^D(\mathbf{V}_D) \cdot \mathbf{z}} + \mathbf{B}_\lambda^\infty \cdot (1 - e^{-\beta_\lambda^B(\mathbf{V}_B) \cdot \mathbf{z}}). \quad (9.3)$$

where the vectors $\mathbf{V}_D$ and $\mathbf{V}_B$ denote the dependencies of the coefficients β_λ^D and β_λ^B over a number of factors including range, sensor response, and ambient light.

According to this revised model, degraded underwater images strongly depend on two parameters: backscattering and attenuation coefficient of direct signal, whose estimation method was mentioned by the follow-up work of Akkaynak and Treibitz [48]. The estimation process of backscatter is inspired by Dark Channel Prior (DCP) [44], which is modeled by:

$$\mathbf{B}_\lambda = \mathbf{B}_\lambda^\infty \cdot (1 - e^{-\beta_\lambda^B \cdot \mathbf{z}}) + \mathbf{I}_\lambda' \cdot e^{-\beta_\lambda^{D'} \cdot \mathbf{z}}, \quad (9.4)$$

where $\mathbf{B}_\lambda^\infty$, β_λ^B, $\mathbf{I}_\lambda'$, $\beta_\lambda^{D'}$ are estimated using nonlinear least squares fitting. The attenuation coefficients of direct signal is estimated using an optimization framework by illumination map, which is computed by the local color space averaging.

9.2.2 Physics-based Modeling: Color Correction and Image Recovery

When the atmospheric scattering model (Eq. 9.1) is applied for underwater image recovery, the direct transmission component $\mathbf{I}(\mathbf{x})\cdot t(\mathbf{x})$ is considered to be equivalent to *forward scattering*, while the air-light $\mathbf{B}\cdot(1 - t(\mathbf{x}))$ is assumed to have analogous effects as the *backscatter*. Both unknown variables $\mathbf{B}$ and $t(\cdot)$ are generally estimated by using the DCP concept [44]; DCP is based on the observation that most local image patches contain some dark areas, that is, at least one color channel (R/G/B) has some pixels with very low intensity. The transmission map can be obtained by applying this theory with *minimum* filtering; ambient light is then calculated by finding the value of the corresponding point with the highest brightness in hazy image. However, since red light suffers aggressive attenuation underwater, these assumptions do not hold for the red channel. To address this issue, UDCP (Underwater DCP) [49] ignores red channel and only applies DCP to the blue and green channels, while RDCP (Red DCP) [50] adjusts the coefficient of red channel so that its weights decrease rapidly with increasing depth. Moreover, Berman *et al.* [40] made the assumption that colors of a haze-free image are distributed globally within tight clusters in a RGB space. This assumption holds for a given water type and facilitates a fast linear estimation of the transmission map. However, relaxing the nonlinearity of underwater optical distortions contributes to poor enhancement results and limits the generalizability of this prior.

The Jaffe–McGlamery underwater imaging model (see Eq. 9.2) has been widely used to address these issues. Trucco *et al.* proposed a self-tuning image restoration filter [51], which ignores backscatter component and is ideally suitable for shallow-water, diffuse-light conditions with limited backscatter. However, underwater images are more affected by backscatter than forward scattering in most situations, hence such assumptions made by those early methods have limited applications. Therefore, researchers usually ignore forward scattering and only consider the influence of direct signal and backscatter. To restore clear latent images $\mathbf{I}_\lambda(\mathbf{x})$, they apply different efficient methods to estimate transmission map $t_\lambda(\mathbf{x})$ and global background light $\mathbf{B}_\lambda$. For instance, Chiang *et al.* estimated the transmission map according to the classical DCP algorithm to reduce haze from scattering and to compensate for each wavelength's attenuation discrepancy to restore color cast [36]. Galdran *et al.* estimated

the transmission map by RDCP method [50], which has fewer free parameters and can handle presence of artificial illumination in images. Hou *et al.* applied UDCP and quad-tree subdivision methods to estimate the transmission map and global background light, respectively; they also sped up efficiency of the model by alternating direction method of multipliers [52]. Yan *et al.* adopted a more detailed model with different attenuation coefficients [53] to split the image backscatter and the direct attenuation components to reconstruct color corrected images.

The revised imaging model (see Eq. 9.3) has been successfully used for underwater image restoration as well. In 2019, Akkaynak *et al.* formulated an underwater image restoration method named Sea-Thru [48]. As Figure 9.3 demonstrates, it can recover raw pixel colors by consistently removing water using RGB-D images. However, there are no readily available equivalent RGB-D sensors for underwater, given that infrared light strongly attenuates underwater; thus these methods' requirements of dense scene depth and water body properties as prior are not typically available in practice. Besides, they are not end-to-end deployable and tend to be computationally too demanding for real-time applications.

9.2.3 Data-driven Approaches: Image Enhancement and Restoration

A practical alternative is to approximate the underlying solution by learning-based methods, which demonstrated remarkable success in recent years. These approaches are mainly governed by large-scale supervised learning of deep Convolutional Neural Networks (CNNs) [54], Generative Adversarial Networks (GANs) [55], and Vision Transformers (ViTs) [56]. Their training objective is to learn sequences of nonlinear filters for approximating the transformation between the *distorted* and *enhanced* image domains [57–59]. Here, the assumption is that comprehensive databases contain enough image samples to learn the underlying distortion function represented by pixel-to-pixel mapping or visual style transfer.

1. **CNN-based Encoder-Decoder Models:** The deep CNN-based generative models significantly boost the SOTA performance in learning image-to-image translation (*e.g.*,, enhancement, style transfer) for both terrestrial [60, 61] and underwater imagery [62, 63]. These models typically have a U-shaped encoder–decoder structure [64]: the encoder extracts multiscale hierarchical features, which are then exploited by the decoder to generate the up-sampled and enhanced output. Domain knowledge about the optical distortions and imaging parameters are generally incorporated into the model design and/or loss function formulation. For instance, Water-Net model [57] uses a CNN-based encoder to generate confidence

FIGURE 9.3
The Sea-thru method (with courtesy from [48]) applies the revised imaging model (with courtesy from [1]) to accurately remove water from underwater images; an example of image color-corrected by running the publicly available implementation.

maps for white balance, histogram equalization, and gamma correction – which are then combined by a gated fusion network to decode and learn perceptual enhancement. Moreover, Jamadandi and Mudenagudi [65] used an encoder–decoder module with wavelet pooling and un-pooling; Liu *et al.* [62] employed a deep residual framework and incorporate sharpness loss; Hashisho *et al.* [66] formulated a U-shaped denoising auto-encoder; and Zhang *et al.* [67] incorporated a multiscale retinex component into their respective learning pipelines. Islam *et al.* [9] outlines several loss components for restoring color, contrast, and sharpness into a deep residual architecture for learning underwater image simultaneous enhancement and super-resolution (SESR) combined. Other models incorporate domain knowledge into the synthetic data generation alone. For instance, in UWCNN, Li *et al.* [68] used scene priors to combine the optical properties of light propagation with a physics-based underwater imaging model to generate paired samples, then apply a deep CNN model to learn perceptual enhancement.

2. **GAN-based Models:** GAN-based models attempt to improve the generalization performance of deep generative learning by employing a two-player min-max game [55], where an adversarial *discriminator* evaluates the *generator*-enhanced images compared to ground truth samples. This forces the generator to learn realistic enhancement while evolving with the discriminator toward equi-

librium. In a pioneering work, Li *et al.* [73] proposed the WaterGAN method, which takes RGB-D images as input and learns enhancement using synthetically generated underwater images from in-air images and depth pairings. In the UGAN model, Fabbri *et al.* [59] removed the dependency on depth pairings of in-air images and used only synthetic RGB image pairs for learning enhancement; they also demonstrated the effectiveness and learning behaviors of various GAN-based architectures (*e.g.,*, least-squared GAN [74], Wasserstein GAN [75], energy-based GAN [76]). Many other contemporary methods [70, 77, 78] use a similar principle, that is, CycleGAN [79]-based style transfer for synthetic paired sample generation, then adversarial training for learning perpetual image enhancement. Since paired training data is not always available, two-way GANs [79, 80] are often deployed for learning unpaired learning of perceptual image enhancement [81]. Furthermore, Islam *et al.* [2] and Ignatov *et al.* [82] showed that additional loss terms for preserving the high-level feature-based content improve the quality of image enhancement using GANs. Such approaches are computationally efficient and can be extended to filtering or video enhancement as well. As seen in Figure 9.4, several GAN-based underwater image enhancement models have reported impressive results from both paired [59, 73] and unpaired training [2].

3. **Transformer-based Models:** In recent years, researchers have explored the potential of

FIGURE 9.4
A few qualitative comparisons are shown for perceptual enhancement of underwater imagery by various SOTA models: Ancuti *et al.* (with courtesy from [69]), UGAN (with courtesy from [59]), UWGAN (with courtesy from [70]), Deep SESR (with courtesy from [9]), Water-Net (with courtesy from [57]), U-shaped Tx (with courtesy from [71]), and Auto-En Tx (with courtesy from [72].)

ViTs [56, 83] for underwater image enhancement with inspiring success [71, 84]. ViT abandons part of the convolution structure and models the global long-range relationship; its self-attention mechanism can capture the global context information of underwater images. In 2021, Peng *et al.* proposed a U-shaped Transformer [71], which integrates a channel-wise multiscale feature fusion transformer (CMSFFT) and a spatial-wise global feature modeling transformer (SGFMT) for learning underwater image enhancement. Guo *et al.* [84] implemented a harmonization Transformer to exploit long-range context to capture the global information toward image enhancement. Moreover, Tang *et al.* [72] employed a neural architecture search strategy to automatically search the optimal U-Net architecture, which can boost the learning capability of the deep transformer-based models for underwater image enhancement.

All the data-driven approaches discussed above (*i.e.*, based on CNN, GAN, and ViTs) have their strengths and weaknesses for training and inference on certain types of underwater scenes. A comprehensive analysis on these aspects with qualitative and quantitative evaluation is necessary and can greatly benefit the underwater imaging and robotics community. While some recent works offer small-scale benchmark datasets [57, 85], more comprehensive benchmark evaluation of perceptual enhancement algorithms and nonreference image quality assessment techniques have not been explored in depth in the literature.

9.2.4 Visual Filtering for Robot Perception

One major operational challenge for visually guided underwater robots is that despite using high-end cameras, visual sensing is often greatly affected by the abovementioned attenuation and scattering characteristics. These optical artifacts trigger nonlinear distortions in the captured images, which severely affect the performance of vision-based tasks such as detection and tracking, semantic segmentation, and visual servoing. Fast and accurate image enhancement techniques can alleviate these problems by restoring the perceptual and statistical qualities [3, 59, 67] of the distorted images in real time. Traditionally, simple physics-based filters and energy functions are applied for instantaneous color recovery [86]. The idea is to estimate the *transmission map* and *ambient* light in the scene [3, 87, 88] and then adjust the pixel intensities for denoising and color correction. Various *model-free* filters [67, 89] are also used to reduce noise and improve global contrast for improving underwater visual data.

Although the contemporary data-driven approaches offer SOTA image enhancement performance, their applicability as visual filters in robots' autonomy pipeline has been limited due to real-time operational constraints. Nevertheless, careful design choices and post-training model optimizations (*e.g.*,, pruning, quantization) [91] can facilitate their single-board implementations. For instance, Islam *et al.* [2] showed that inference-optimized GAN-based image enhancement models can provide over 48 FPS on NVIDIA Jetson AGX Xavier and over 25 FPS on Jetson TX2 devices. They also demonstrated that such image enhancement filters can significantly improve the visual perception performance of robots in standard tasks

FIGURE 9.5

A few instances of fast visual filtering by FUnIE-GAN (with courtesy from [2]) are shown on the left; improved performances of standard models for underwater object detection (with courtesy from [27]) and human pose estimation (with courtesy from [90]) are shown on the right.

such as detection, tracking, and human body pose estimation; see Figure 9.5. DeepSeaColor by Jamieson *et al.* [92] improves upon the Sea-Thru algorithm by eliminating several approximations and improving computational efficiency through the use of CNNs which are trained online. It uses one CNN for modeling the backscatter, and another one for attenuation, both of which are trained with self-supervision using the captured image and the range map. With efficient learning-based approximations, DeepSeaColor is able to achieve real-time inference speeds while offering comparable, and in some cases exceeding the performances of the Sea-Thru algorithm, enabling more robust real-time vision-based behaviors. Nevertheless, further research efforts are required for balancing the *performance-efficiency* trade-off in existing visual filters, especially when incorporating domain knowledge of the revised underwater imaging model [1] into lightweight learning pipelines.

9.3 Underwater Visual Perception

Object detection, classification, and semantic segmentation are well-studied problems in the domains of robot vision and deep learning [93, 94] for their usefulness in estimating scene geometry, inferring interactions and spatial relationships among objects, and salient object identification. They are particularly important for detailed scene understanding in autonomous driving by visually guided robots. Over the last decade, substantial contributions from both industrial and academic researchers have led to remarkable advancements of the SOTA methodologies [94, 95], which are largely propelled by various genres of deep learning-based models. For visually guided underwater robots, however, the existing solutions for object detection and semantic scene parsing are significantly less advanced. The practicalities and limitations are twofold. First, the visual content of underwater imagery is entirely different because of the domain-

specific object categories, background patterns, and optical distortion artifacts (see the discussions in Section 9.2); hence, the learning-based SOTA models trained on terrestrial data are not directly applicable. Secondly, there are very few underwater datasets to facilitate large-scale comprehensive training for general-purpose use. Despite the challenges, contemporary research works have made substantial inroads in applications such as fish detection or classification [7, 96], invasive fish tracking [6, 97], pipeline inspection [98, 99], coral-reef classification or coverage estimation [4, 100], visual saliency estimation [8, 101], and semantic scene segmentation [10, 102].

9.3.1 Object Detection, Classification, and Tracking

Visual detection and classification algorithms are traditionally used for target tracking in underwater exploration and navigation tasks [11]. Beyond software solutions (*i.e.*, object detector or classifier models), these applications generally require efficient *machine vision* systems integrated into an embedded device for real-time visual perception. For instance, Ortiz *et al.* [103] integrated a probabilistic line detector into an onboard Kalman filtering pipeline of AUVs for localizing submarine power cables. Xiang *et al.* [99] designed a custom vision system for coordinated formation control in pipeline inspection and following tasks. Traditional stereo vision systems are also used for tracking multiple fish by using histogram back-projection on low-contrast and low frame-rate stereo videos [104]. Furthermore, various model-free and model-based diver detection and tracking modules [27, 28, 105] are deployed in human–robot cooperative underwater missions. The model-free algorithms rely on spatial (*e.g.*,, color, texture) or frequency domain (*e.g.*,, swimming motion cues) filtering, while the model-based algorithms exploit appearance signatures for diver detection and tracking; further discussions are provided in Section 9.5.1. Various efficient systems have also been developed for autonomous fish tracking and recognition

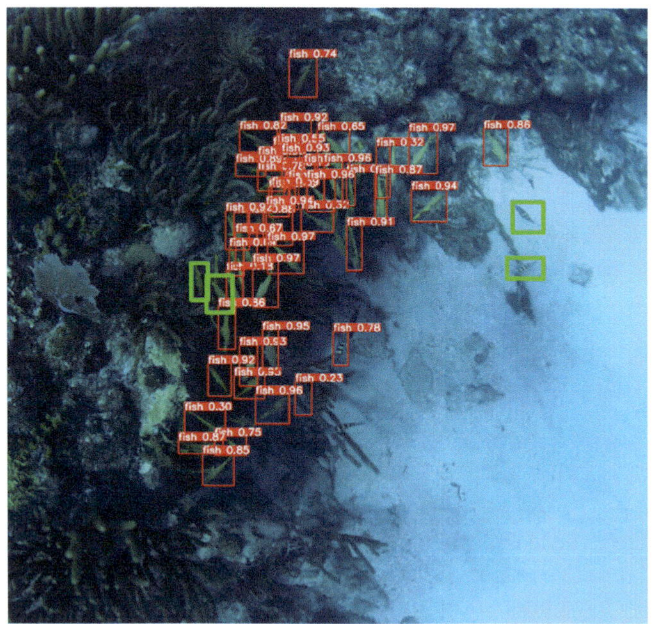
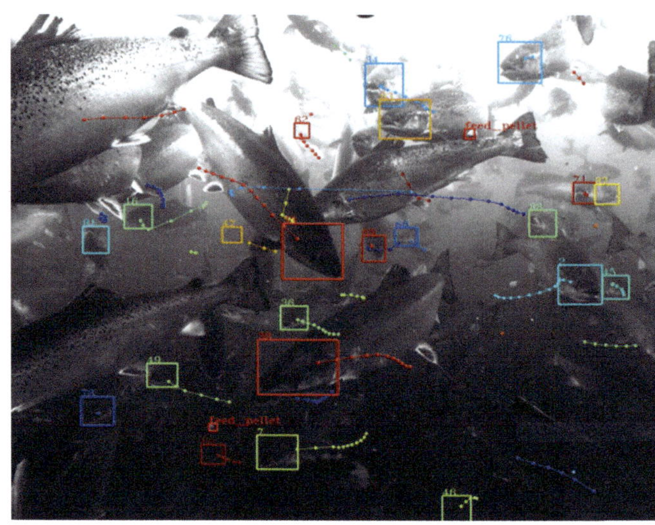

FIGURE 9.6
Real-time fish detection by MegaFishDetector (with courtesy from [106]) (left) and individual detection on tracking in video by *Tidal* (with courtesy from [107]) (right).

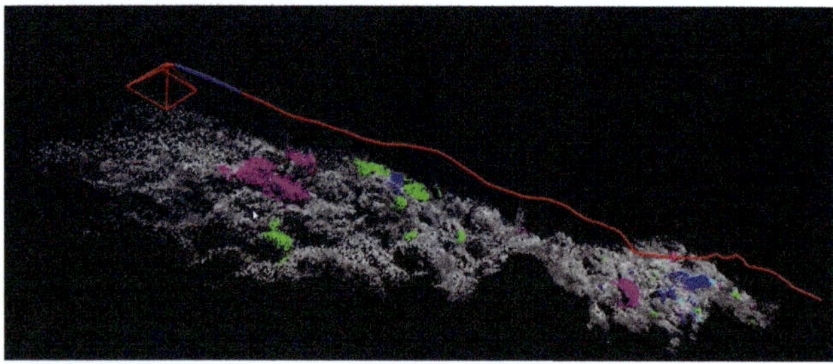
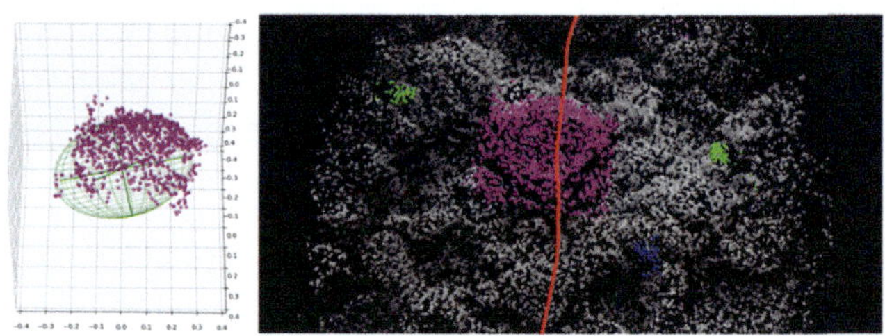

(a) Semantic map of coral reef

(b) Volume evaluation of a brain coral

FIGURE 9.7
The identification and tracking, semantic mapping, and volumetric modeling of coral reefs; the evaluation is performed on data collected in the Caribbean (Barbados) (a) Semantic map of coral reef (with courtesy from [5]), (b) Volume evaluation of a brain coral (with courtesy from [117].)

at a fast rate [107–109]. These systems typically fuse motion priors of fishes with their appearance-based features that are learned from image or video data; see Figure 9.6. Deep learning-based models are also utilized to detect other robots in order to assist with robot-to-robot relative localization [110, 111]. To this end, the region-based deep CNNs [108] and bounding-box regression models [97] are popular choices for fast inference on devices such as Jetson TX2/Nanos or Raspberry Pis. MegaFishDetector [106] currently provides the largest openly available fish detector, trained on multiple publicly available fish datasets.

Recognizing and counting corals is another area of interest. Over the years, various techniques for automating coral reef surveying have been developed for real-time en-vironmental assessments [112, 113]. Early work utilized point annotations [114, 115] for specific coral reef species on publicly available datasets [116]. More recent work combined a deep learning approach with VIO to identify and track the locations of different types of corals in the Caribbean producing a semantic map [5]; see Figure 9.7a. The features detected by this VIO system can be used for volumetric modeling of the detected corals [117], as shown in Figure 9.7b. Robotic operations produce visual data in a temporal sequence, and track detected objects across time enabling the generation of soft labels, which can be then utilized to improve detection rate [118]. Earlier work classified bleached (dead) and live corals [119].

While underwater object detection and tracking performances of existing systems are promising, several other visual perception capabilities remain unsolved; a few prominent open problems are (1) multimodal learning of fish shape descriptors with/without priors [7, 96], (2) fish counting and weight estimation [120, 121], (3) density tracking/estimation and health monitoring of invasive fish populations [6, 122], and (4) coral reef fish recognition and density estimation [123]. Along that line, vision-based coral reef classification [4, 124], coverage estimation [100], and health monitoring are also challenging open problems in marine sciences. The existing vision-based systems use data-driven models [125] to learn hierarchical image features, often with texture priors, for coral reef classification, mapping, and spectral analyses. Some of these systems have been deployed [126] in remote sensing platforms for mapping studies in the Great Barrier Reefs, Shark Bay, and the Florida Keys with reasonable success. However, there are significant knowledge gaps in the literature on bio-geo-physical understanding of coral reef habitats with high-resolution imagery coupled with benthic community data.

9.3.2 Visual Saliency Estimation and Attention Modeling

The most essential capability of visually guided underwater robots is to identify interesting and relevant image regions to make effective operational and navigational decisions on the go. As shown in Figure 9.8a, identifying *salient* pixels in image space facilitates visual attention modeling for general-purpose decision-making. The existing systems and solutions for visual saliency estimation can be categorically discussed from the perspectives of model adaptation [10, 127] and high-level robot tasks [128, 129]. The model-based techniques are particularly beneficial for fast visual search [127, 130], enhanced object detection [131, 132], and monitoring applications [15, 133]. For instance, Maldonado-Ramírez *et al.* [134] used ad hoc visual descriptors learned by a convolutional autoencoder to identify salient landmarks for fast place recognition. Koreitem *et al.* [127] used a bank of pre-specified image patches containing interesting objects or relevant scenes to learn a similarity operator that guides the robot's visual search in an unconstrained setting. Such similarity operators are essentially spatial saliency predictors which assign a degree of *relevance* to the visual scene based on the prior model-driven knowledge of what may constitute as salient, for example, coral reefs [4, 117], companion divers [27, 131], wrecks [10], and fish [96].

Model-free approaches are more feasible for autonomous exploratory applications compared to model-based ones [135, 136]. Classical approaches use binary morphology filters [137] or adopt feature contrast evaluation techniques to encode low-level image-based

(a) Salient object detection

(b) Semantic segmentation

FIGURE 9.8
A few underwater scenes and corresponding pixel annotations are shown for class-agnostic binary saliency estimation and semantic image segmentation; here, different color annotations represent various classes or object categories (a) Salient object detection (with courtesy from [8]), (b) Semantic segmentation (with courtesy from [10].)

features (*e.g.*, color, luminance, texture, object shapes) into super-pixel descriptors [138–140]. These low-dimensional representations are then exploited by heuristics or learning-based models to infer global saliency. For example, Girdhar *et al.* [128] formulated an online topic-modeling scheme that encodes visible features into a low-dimensional semantic descriptor, then adopt a probabilistic approach to compute a *surprise score* for the current observation based on the presence of high-level patterns in the scene. Kim *et al.* [141] introduced an online bag-of-words (BoW) scheme to measure intra- and inter-image saliency estimation for robust key-frame selection in SLAM-based navigation. Wang *et al.* [140] encoded multiscale image features into a topographical descriptor, then applied Bhattacharyya measure [142] to extract salient pixels by segmenting out the background.

Several research efforts [101, 143–145] have demonstrated inspiring results for object-level saliency estimation and visual attention modeling as well. In particular, heterogeneous deep residual learning-based attention models [8, 9] have reported over 45% faster processing in marine robotics use cases such as uninformed visual search and object localization. Nevertheless, no such studies exist for use cases such as visual question answering, video summarizing, and text-to-image generative modeling; similar to terrestrial applications, these research problems have the potential to revolutionize underwater robotics domain as well.

9.3.3 Semantic Segmentation and Scene Parsing

While a class-agnostic saliency estimation provides interesting foreground regions, pixel-wise scene parsing and the semantic segmentation of underwater imagery offer holistic information about the spatial distribution and interaction among the objects in the scene. Despite the advantages, these research problems have not been explored in depth for underwater robotic applications. Due to the difficulties in acquiring large-scale labeled data and the limited applicability of traditional transfer learning pipelines, the existing systems attempt to collect and annotate small-scale application-specific image data. Therefore, although the existing works [10, 102, 146] demonstrate promising results, they only consider a handful of object categories. For instance, Islam *et al.* [10] considered eight object categories that are important for human–robot cooperative underwater missions: robots, human divers, wrecks/ruins, aquatic plants, fish, reefs, and sea floor (see Figure 9.8b). Other datasets consider even fewer object categories such as marine debris (*i.e.*, various types of trash) [147] or ship hull defects (*i.e.*, corrosion, paint peel) [148]. With limited training samples per object category over only a few waterbody types, these datasets do not facilitate generalizable deep visual learning for underwater image segmentation tasks.

Researchers have tried to address these issues by using intelligent annotation tools, semi-supervised learning, and online learning schemes. For instance, Beijbom *et al.* [114] used automated point-annotation to expedite coverage estimation of the dominant benthic types in survey images of coral reefs. Zurowietz *et al.* [149] developed an online image annotation tool for environmental monitoring and exploration applications. Weakly supervised learning pipelines [150], often with synthetically generated images [151, 152], are also utilized for underwater image segmentation. Unsupervised methods such as fuzzy clustering and stochastic optimizations [153, 154] have shown promising results as well. Nevertheless, comprehensive applications of semantic scene parsing remain a challenge and call for significant advancements in near future.

ture, particularly for benthic habitat segmentation [155], photogrammetry [156], and 3D visualizations of bathymetric maps or coral reef surveys [124, 157].

9.4 Vision-based Underwater Robot Localization

Most of the underwater navigation algorithms are based on high-grade acoustic sensors such as Doppler Velocity Logs (DVL) and Ultra-Short Baseline (USBL) positioning systems, which are expensive and require dedicated surface vehicle support for paired communication [158]. For fully autonomous operation without a GPS or surface connectivity, underwater robots often solve the localization problem by coupling IMUs with a compass and pressure sensors. Such solutions, classified as *dead-reckoning*, suffer from unbounded drift and are not accurate enough for navigation in cluttered areas [159]. Vision-based SLAM (VSLAM) methods are thus preferred, which require exteroceptive sensors (*e.g.*,, lidar, sonar, camera) to measure the 3D structure of the environment. However, a lack of natural landmarks and noisy visual conditions in underwater scenes make it difficult to apply VSLAM and VIO methods [38, 160, 161]. The following discussions provide an overview of how these difficulties are addressed by SOTA vision-based state estimation and localization algorithms.

9.4.1 Vision-Based State Estimation

Real-time vision-based state estimation methods – which estimate state $\mathbf{x}_R$, including position and rotation, of an underwater robot R – fall into two main categories. First, there are *direct methods*, such as LSD-SLAM [162] and DSO [163, 164], which use pixel intensities to minimize the photometric error and they can produce dense maps. Despite their success in the terrestrial domain, benchmark work [38, 160] showed that direct methods can track only for a short time, because the *brightness constancy* assumption is violated due to the frequent lighting variation and appearance changes in underwater images. Second, there are *indirect methods* (*e.g.*,, ORB-SLAM3 [165]) – which detect features in each frame and track them over different frames, solving the structure from motion problem. Indirect methods are more popular for underwater SLAM being more robust in different conditions, especially when the used feature descriptors are (to some extent) illumination invariant.

For more robust and accurate state estimation, vision-based state estimation is integrated with other sensors such as inertial sensors. Sensor fusion has been performed with two types of approaches. One type is based

FIGURE 9.9
Custom made sensor suite mounted on a dual DPV. The mechanical scanning profiling sonar scans around the sensor while the cameras see in front. (With courtesy from [17].)

on *Kalman filter*, where inertial measurements are utilized in the propagation step, while visual features are considered in the update phase [166–168]. A second type formulates the problem as a nonlinear optimization [165, 169]. Kalman filter-based approaches are generally computationally efficient, thus suitable for embedded systems; instead the nonlinear least-square optimization-based methods are more computationally demanding as a sliding window of states is used. The latter type has been more prevalently adopted in the underwater domain as ORB-SLAM and subsequent versions showed overall good performance compared to other packages, although new advances in the Kalman filter-based methods [170] might lead to a change. As an example, we show here a recent optimization-based method tailored for underwater that included cameras, IMU, and a mechanical scanning profiling sonar [17]; see Figure 9.9. Adding the sonar helps in cases where the visibility is low as well as to recover accurate distance information. We highlight how different sensors can be combined in the optimization framework, focusing on the sonar error formulation, to provide a guidance on the design process to include a new sensor.

The formal definition of the optimization problem is as follows. The state vector contains the robot position $_W\mathbf{p}_I^T = [_Wp_x, _Wp_y, _Wp_z]^T$, the robot attitude expressed by the quaternion $\mathbf{q}_I^T$, and the linear velocity $_W\mathbf{v}_I^T$ (expressed in world coordinates); in addition, the state vector contains the gyroscopes bias $\mathbf{b}_g$ and the accelerometers bias $\mathbf{b}_a$. The error-state vector is defined in minimal coordinates, while the perturbation takes place in the tangent space. Thus, the state $\mathbf{x}_R$ and the error-state vector $\delta\boldsymbol{\chi}_R$ are represented as $\mathbf{x}_R = [_W\mathbf{p}_I^T, \mathbf{q}_I^T, _W\mathbf{v}_I^T, \mathbf{b}_g^T, \mathbf{b}_a^T]^T$ and $\delta\boldsymbol{\chi}_R = [\delta\mathbf{p}^T, \delta\mathbf{q}^T, \delta\mathbf{v}^T, \delta\mathbf{b}_g^T, \delta\mathbf{b}_a^T]^T$; the error for each component of the state vector is then expressed by a transformation between the tangent space and minimal

coordinates [171]. The joint nonlinear optimization cost function $J(\mathbf{x})$ for the reprojection error $\mathbf{e}_r$, the IMU error $\mathbf{e}_s$, and sonar error term $\mathbf{e}_t$ [17]:

$$J(\mathbf{x}) = \sum_{i=1}^{I=2} \sum_{k=1}^{K} \sum_{j \in \mathcal{J}(i,k)} \mathbf{e}_r^{i,j,k^T} \mathbf{P}_r^k \mathbf{e}_r^{i,j,k} + \sum_{k=1}^{K-1} \mathbf{e}_s^{k^T} \mathbf{P}_s^k \mathbf{e}_s^k$$
$$+ \sum_{k=1}^{K-1} \mathbf{e}_t^{k^T} \mathbf{P}_t^k \mathbf{e}_t^k \tag{9.5}$$

where i denotes the camera index, that is, left or right camera in a stereo camera system, with landmark index j observed in the kth camera frame. $\mathbf{P}_r^k$, $\mathbf{P}_s^k$, and $\mathbf{P}_t^k$ represent the information matrix of visual landmark, IMU, and sonar range measurement for the kth frame, respectively. The reprojection error function for the stereo camera system describes the difference between a keypoint measurement in camera coordinate frame and the corresponding landmark projection according to the stereo projection model. Each IMU error term combines all accelerometer and gyroscope measurements by the *IMU preintegration* between successive camera measurements and represents both the robot *pose*, *speed*, and *bias* errors between the predicted and the actual state.

The sonar measurements are used to correct the robot *pose* estimate as well as to optimize both visual and acoustic landmarks' position. Due to the low visibility of underwater environments, sonar provides features with more accurate scale when it is hard to find visual landmarks. Nevertheless, a particular challenge here is the temporal displacement between the two sensory measurements (*i.e.*, from camera and sonar). To address the above challenge, visual features detected in close proximity to the sonar return are generally grouped together and used to construct a *patch*. The distance between the sonar measurement and the visual patch is then used as an additional constraint. For computational efficiency, the sonar range correction only takes place when a new camera frame is added to the pose graph. Since sonar has a faster measurement rate than cameras, only the nearest *range* to the robot pose in terms of timestamp is used to calculate a small patch from *visual landmarks* around the *sonar landmarks*. Each sonar point detected, that is, $_W\mathbf{l}_S = [l_x, l_y, l_z]$ in world coordinates, is then compared to a patch of neighboring visual landmarks. More specifically, the sonar error term is expressed as [17]:

$$\hat{r} = \left\| _W\hat{\mathbf{p}}_I^k - \text{mean}(\mathcal{L}_S) \right\|, \tag{9.6}$$

where $\mathcal{L}_S$ is the subset of visual landmarks around the sonar landmark. Note that $\hat{r}$ is the difference between the sonar landmark and the centroid of the visual landmarks; the purpose of $\text{mean}(\mathcal{L}_S)$ is to filter out the noise in individual visual landmarks. Given the range measurement $\mathbf{z}_t^k$, the error term $\mathbf{e}_t^k(_W\mathbf{p}_I^k, \mathbf{z}_t^k)$ is used to correct the position $_W\mathbf{p}_I^k$. Rahman *et al.* [172] further introduced the loop closure (LC) by accurate trajectory estimation to facilitate robust mapping of underwater structures.

The key to include a new sensor is to define the error term and include it in the optimization framework. Next, we present an experimental analysis of a number of SLAM algorithms and highlight their operational considerations.

9.4.2 Comparison of Visual-Inertial SLAM Algorithms

In recent years, many Visual Odometry, VIO, and Visual SLAM systems have been developed [163, 173] using monocular, stereo, or multi-camera rigs mostly for indoor and outdoor environments that can be classified as indirect (feature-based), direct, or semi-direct methods, as mentioned in the previous subsection. Combining vision with an IMU solves the scale issue in monocular VSLAM, as it can be used to estimate the motion between camera frames. To name a few, currently, ORB-SLAM3 [165] is one of the most popular feature-based visual-inertial SLAM systems that use the BoW approach for LC and relocalization. VINS-Mono [174] also uses a similar principle but uses a *loosely* coupled sensor fusion method to align monocular vision with inertial measurement for estimator initialization. On the other hand, OKVIS [169] *tightly* integrates visual measurements with IMU readings for nonlinear optimization-based visual-inertial state estimation. SVIn2 [17], as described above, augments the OKVIS package with acoustic range data from a mechanical scanning sonar, which improved the trajectory estimation accuracy, especially when there is varying visibility underwater.

In Table 9.1, we present a performance comparison of recent SOTA visual-inertial SLAM systems: DM-VIO [175] (direct/optimization), Kimera-VIO [176] (indirect/optimization), OKVIS [169] (indirect/optimization), OpenVINS [170] (indirect/filtering), OpenVINS_LC [170], ORB-SLAM3 [165] (indirect/optimization), ORB-SLAM3_LC [165], ROVIO [167] (patch-based direct/filtering), SVIN_LC [17] (indirect/optimization), SVO [177] (semi-direct/optimization), SVO_LC [177], VINS-Fusion [178] (indirect/optimization), and VINS-Fusion_LC [178]. Three publicly available datasets proposed by Joshi *et al.* [38] are utilized; see Figure 9.2 for representative images. The packages presented here are newer versions from the ones used in the earlier comparison [38] and they exhibit better behavior than the aforementioned study. As seen in Table 9.1, *loop closure (LC)* provides a significant boost in accuracy. The *Sunken bus* dataset is a particularly challenging one where 7 out of 13 systems were able to track less than 30% of the total trajectory. The main reason relates to the sharp changes in brightness and contrast. While recent methods are more robust than earlier approaches, the performance of underwater state estimation can be further improved by including visual filtering (discussed in Section 9.2.4) in the pipeline. Efforts are still needed for a fully robust system that can recover and track in all different challenging underwater conditions. In addition, while some datasets have been provided [38, 179], they lack a ground truth comparable to benchmark datasets for out-of-the-water environments, making a complete quantitative evaluation not possible.

9.4.3 Beyond Traditional VIO

Recently, with the growing research on deep visual learning, both classical and data-driven methods have been tested and benchmarked by Billings *et al.* [180], showing that deep learning-based visual state estimation methods are capable of tracking even in textureless environments. However, these methods handled drift poorly, which the authors of that benchmark addressed by designing a visual-inertial fusion network that improves the

TABLE 9.1

Performance Comparison of Various Monocular Visual-Inertial SLAM Algorithms: (**I**) DM-VIO [175], (**II**) Kimera-VIO [176], (**III**) OKVIS [169], (**IV**) OpenVINS [170], (**V**) OpenVINS_LC [170], (**VI**) ORB-SLAM3 [165], (**VII**) ORB-SLAM3_LC [165], (**VIII**) ROVIO [167], (**IX**) SVIN_LC [17], (**X**) SVO [177], (**XI**) SVO_LC [177], (**XII**) VINS-Fusion [178], and (**XIII**) VINS-Fusion_LC [178]. For Each Dataset, the First Row Specifies the Translation RMSE After *sim3* Trajectory Alignment; the Second Row Is the Percentage of Time the Trajectory Was Tracked. In Blue the Lowest RMSE from a SLAM System in the Corresponding Dataset When the Tracking Percentage is Greater Than 90%.

	I	II	III	IV	V	VI	VII	VIII	IX	X	XI	XII	XIII
Cave dataset	0.51	2.34	1.08	1.21	0.36	0.42	**0.14**	0.41	0.30	0.31	0.25	0.35	0.17
(98.7m)	100%	49.5%	100%	99%	98.9%	99.9%	100%	100%	100%	96.8%	96.7%	99.9%	99.6%
Sunken bus	1.81	0.95	0.67	0.2	0.17	0.07	0.07	0.77	0.42	0.14	0.14	0.51	**0.31**
dataset (67.5m)	99.9%	33.0%	99.9%	21.0%	20.9%	20.4%	20.4%	99.9%	99.9%	20.2%	20.2%	99.7%	99.2%
Cemetery dataset	0.33	3.18	1.72	1.56	0.33	1.87	0.71	1.53	**0.31**	1.43	0.7	1.6	0.38
(95.7m)	9.6%	100.0%	99.9%	98.7%	97.6%	87.0%	83.8%	99.9%	97.5%	89.5%	81.0%	97.3%	96.8%

pose estimation accuracy. Another study from Burguera *et al.* [181] proposed an unsupervised deep neural network for detecting LCs and compared it in a real dataset with other classical methods. Their experiments reported better performance in re-identifying regions that were seen before.

In addition to the image enhancement techniques described earlier, some recent work has looked at explicitly classifying particulates suspended in the water captured by the image to reject corresponding keypoints [182]. Although a thorough quantitative evaluation is needed, qualitative results showed the ability of a SLAM system to track in scenes that otherwise would fail. Other works used application-specific knowledge about reliable features that can be used for tracking and reconstruction. For example, in cave-like environments, light that divers carry creates a boundary between illuminated and nonilluminated areas; features that fall on that boundary can be used for accurate dense 3D reconstruction [21, 22]. Moreover, specific setups often require SLAM systems to use multiple cameras. For example, Billings *et al.* [23] fused a forward-looking stereo system with a fisheye camera mounted on a robot manipulator. Another trend that can be observed is to fuse sensors that are specific to the underwater domain to improve the overall quality of the state estimation. For example, an inexpensive solution that included a monocular camera and a single-beam echosounder was proposed to resolve depth ambiguity [25]. Several other robust SLAM systems have been proposed by fusing acoustic odometry from a DVL, camera, IMU, and depth sensor data [20, 183] augmenting ORB-SLAM2 [184] and ORB-SLAM3 [165].

A different set of work has been focusing on imaging sensors, such as sidescan sonar or multibeam sonars, which return images used then for feature tracking [185–189]. Acoustic measurements however are sparser than camera data and noisy in certain conditions because of multi-path effects. To achieve a denser map, recent works have looked at *photometric stereo* approaches that use cameras in conjunction with lights to estimate the scene depth [190–192]. Photometric stereo is based on the insight that difference in light illumination can provide information about the *albedo* and *normal* of an object point [193, 194]. Such work assumed, however, that the robot is stationary. Recently, Roznere *et al.* [195] relaxed the stationarity assumption in the photometric stereo problem integrating the state estimate from a monocular SLAM system, achieving promising offline 3D reconstruction.

All methods discussed so far considered a single robot. When using multiple robots, because of the communication constraints in data rate underwater, state estimation techniques need to share data efficiently. Pfingsthorn *et al.* [196] described a strategy for identifying relevant camera images that need to be shared with other underwater robots and that can improve the quality of the map. Paull *et al.* [197] presented a cooperative SLAM framework, where the size of *packets* shared among AUVs is reduced by marginalization, which was tested on sidescan sonar data. Another recent work [198] devised a method for imaging sonar-based SLAM that shares only scene descriptors between robots, thus reducing the burden on the communication channel. Selecting an inexpensive sensor configuration and a robust SLAM system that can achieve real-time dense reconstruction, as well as techniques for multi-robot state estimation, is still an open question in the underwater domain.

9.5 Vision-based Underwater HRI

In underwater human–robot cooperative missions, a team of human divers and autonomous robots perform a set of common tasks [26, 199, 200] where a robot follows and interacts with its companion diver at certain stages. Such human-in-the-loop task execution is essential, because fully autonomous mission planning is challenging due to the lack of radio communication and global positioning information underwater. It reduces operational overhead by eliminating the necessity of teleoperation and enables underwater robots to dynamically configure their mission parameters by regularly interacting with their companion divers [32, 201]. Beyond teleoperation, various vision-based systems [30, 32, 202] are widely used for spontaneous diver-robot cooperation in underwater missions, which we discuss next.

9.5.1 Autonomous Diver Following

Autonomous diver-following by visually guided underwater robots involves different perception tasks, including detection, semantic segmentation, and body-pose estimation, as shown in Figure 9.10. Autonomous diver following is generally challenging due to several operational and sensing constraints. The major challenges lie in ensuring robust visual diver detection, fast onboard tracking, and the generation of smooth motion trajectories for autonomous following. In particular, divers' appearances and motion signatures to a follower robot vary greatly based on their swimming styles, choices of wearables, and relative orientations in the six-degrees-of-freedom (6-DOF) environment. Due to this immense variability, classical model-based detection and tracking algorithms fail to achieve good generalization performance [203, 204]. These techniques are prone to *target drift*, caused by the accumulation of detection errors in noisy visual conditions. The tracking instability can be

FIGURE 9.10
A few illustrations for deep learning-based visual diver detection: anatomical body-pose estimation (with courtesy from [90]), semantic object segmentation (With courtesy from [10]), and bounding box detection (With courtesy from [27].)

considerably reduced by incorporating a motion model based on human swimming cues in the frequency domain. Sattar *et al.* [31] showed that intensity variations in the spatio-temporal domain caused by a diver's swimming gait tend to generate high-energy responses in the 1–2 Hz frequency range. This inherent periodicity can be used as a cue for robust tracking in noisy visual conditions. This idea is further generalized in the MDPM tracker [205]; it fuses spatial-domain and frequency-domain features to track arbitrary swimming directions of a diver by using a hidden Markov model-based optimizer. While the MDPM tracker relaxes the directional constraint of the existing frequency-domain trackers, it does not take advantage of important features such as divers' appearance/wearables, and its performance is affected by their swimming trajectories, that is, straight-on or sideways trajectories.

The deep visual tracking-by-detection approaches can overcome these challenges by learning complex appearance-based models from large-scale and comprehensive data [27, 105, 206]. They are also considerably more robust to noise and image distortions compared to model-free algorithms, which are prone to target drift in noisy visual conditions. Various SOTA diver detection models are typically trained (off-line) on large underwater datasets and sometimes quantized and/or pruned in order to get faster inference by balancing the trade-offs between robustness and efficiency. A family of MobileNet [207] and YOLO [208] architectures are the most commonly used, while custom CNN-based models [27] are designed for diver tracking as well. The 2D bounding-box detectors [27] are generally paired with nonlinear box-trackers [105] in autonomous diver-following pipelines. Other forms of trackers based on diver body-pose estimation [111], pixel-level segmentation [209], and online appearance modeling [28] have been explored in the literature as well.

9.5.2 Diver-Robot Communication Languages

The ability to interact with companion divers and understand their instructions is essential for underwater robots operating in a cooperative mission. In traditional systems, a set of *fiducial markers* is used by human divers to communicate simple instructions to their companion robots. These systems typically use a predefined mapping between various markers and atomic instructions as *language rules*. The most commonly used fiducial markers are those with square, black-and-white patterns that are easily detectable as visual targets, such as AR-Tags [210] and April-Tags [211], among others. Circular markers with similar patterns such as the photomodeler coded targets and Fourier tags [212] have also been used in practice. Grammar-based diver-robot communication systems use more expressive language rules. For instance, RoboChat [32] allows programming both atomic action commands as well as procedural statements by using sequences of visual symbols encoded into AR-Tag markers. Each marker represents one symbolic `token`, and specific sequences of such `tokens` constitute various programs to modulate robot behavior. However, RoboChat language requires a separate marker for each token; hence, a large number of marker cards need to be securely carried during the mission, and specific cards are required to be searched and ordered to formulate a syntactically correct instruction; this whole process imposes a rather high cognitive load on divers. Moreover, the symbol-to-instruction mapping is not intuitive, which makes it inconvenient for rapidly programming a robot.

The first limitation is addressed by Xu *et al.* [202], in which a set of discrete motions of a pair of fiducial markers is interpreted as language rules. Different features such as shape, orientation, and size of these motions are extracted and mapped to various instructions.

FIGURE 9.11
Divers are instructing AUVs via AR-Tags using Robo-Chat (with courtesy from [32]) (left); via hand gestures using Robo-Chat-Gest (with courtesy from [26]) (right).

Since more information is embedded in each token, a large number of instructions can be supported by using only two fiducial markers. However, this method introduces additional computational overhead to track the markers' shape, orientation, and motion trajectory. These are extremely challenging in real-world scenarios as both the divers and robots are suspended in a 6-DOF environment. Besides, the symbol-to-instruction mapping remains counter-intuitive. RoboChatGest [29] addresses these limitations by designing a hand gesture-based diver-to-robot communication system that is syntactically simpler and more intuitive than existing grammar-based systems. Such visual languages typically have the following computational components [29, 30]: (1) a set of grammar: hand gesture to instruction mapping rules; (2) hand gesture recognition modules: deep visual learning-based light models; and (3) a fast instruction decoder: for ensuring robustness to noisy inputs. As shown in Figure 9.11, divers can reliably use visual languages for dynamic reconfiguration of motion parameters on-the-fly. This eliminates the need for a human operator on the surface for interpreting instructions to the robot, as done in many tethered operations [213].

Many approaches extend these ideas for multi-robot cooperation; a few most prominent systems are: inter-robot communication in the context of joint activities [214], robot communication via motion [215], and robot-to-robot relative pose estimation [111]. These systems generally use data-driven approaches to learn visual cues to detect/track relative pose and motion for effective inter-robot or robot-diver communication. Nevertheless, the practical utility of such systems for multi-human-robot cooperation has not yet been explored in-depth. As studied by Wu *et al.* [216], various environmental-related factors and operational constraints contribute to this, impacting the feasibility and usability of semi-autonomous robots alongside human companions in underwater missions. Some of the prominent factors are cognitive load, reaction time, adaptability, reaction time, implicit or explicit interaction modes, degree of autonomy, and other application-specific uncertainties.

9.6 High-Impact Applications and Future Research Directions

The computer vision and underwater robotics technologies described in the previous sections can foster numerous high-impact applications for our society but still require advancements to enable autonomous robots to achieve certain tasks more efficiently and effectively. Table 9.2 summarizes our earlier discussions and highlights a set of open research problems in the literature. In this section, we describe several applications where such technologies have been applied recently with inspiring success. Then, we also highlight the relevant open problems to fully enable such applications and future research opportunities.

9.6.1 Marine Life Monitoring and Conservation

Visually guided underwater robots and machine vision systems have a major role to play in marine ecosystem monitoring and ensuring the success of our conservation efforts. In particular, programs for coral reef restoration [220, 221], sea turtle conservation [222], and seagrass bed mapping and surveying [223, 224] would greatly benefit from improved sensing solutions and 3D visualization tools. Coral reefs significantly impact the marine ecosystem and are an important indicator of coastal oceanic health, climate change, and ocean acidification [225]. Besides, we need next-generation machine vision systems to learn how marine species such as sea turtles observe and get affected by artificial light pollution [222] so that better policies and conservation measures can be taken in coastal communities. Moreover, seagrass density mapping and *transect* surveys are still manually carried out by humans [224]; low-cost automated sensing platforms with dynamic visualization capabilities would potentially revolutionize our understanding of seagrass ecosystems such as its transitional loss characteristics, species distributions, density, and growth, as well as keystone species interactions.

However, the design and development of long-term visual monitoring systems for oceanic ecosystems are challenging due to the finite power budgets and expensive deployment/maintenance costs. Recent work by McCammon *et al.* [226] proposed to use ambient underwater soundscapes to adaptively *focus* visual sensing during

TABLE 9.2
We Identified Numerous Open Research Problems from Our Discussions in Sections 9.2–9.5; Here, We List a Few Prominent Ones for Each Domain: Underwater Imaging, Visual Perception, Robot Localization and Mapping, and Underwater HRI.

Research Domain	Open Research Problems
Underwater imaging	**1.** Find closed-form solutions to estimate transmission map and backscatter signals from underwater imagery; the SOTA methods use DCP [44] and other noisy approximates which provide sub-optimal solutions. **2.** Formulate real-time visual filtering algorithms based on the revised underwater image formation model [1]; the existing approaches are computationally too demanding for real-time robot vision [48]. **3.** Design unpaired learning pipelines [2] for accurate image enhancement and color restoration. **4.** Design loss functions or training objectives to preserve the geometric constraints of the scene (*e.g.*, relative camera motion, object distances) while learning underwater image enhancement or restoration.
Visual perception	**1.** Enable multimodal sensing and estimation for learning to count objects and recognize object shape, size, and weight distributions from videos [120, 123] or long-term monitoring cameras. **2.** Develop 3D object detection and density estimation capabilities for underwater robots; existing methods either estimate 2D bounding-box detection or rely on offline processing, and/or consider only a few object categories [115]. **3.** Design robust learning pipelines for visual question answering, video summarizing, and text-to-image generative modeling for underwater scenes; there are no such frameworks available in the literature. **4.** Develop robust weakly-supervised or self-supervised learning pipelines for image segmentation in benthic habitat classification [155], photogrammetry [156], robot navigation [10], and automatic coral reef surveying tasks [124]. The existing solutions are not generalizable beyond the training data for application-specific tasks.
Robot localization & mapping	**1.** Develop a comprehensive benchmark dataset for vision-based state estimation and VIO with accurate ground truth; such datasets (*e.g.*, KITTI [217]) will significantly flourish research in the underwater robotics domain. **2.** Design a VSLAM system with inexpensive sensors that can achieve real-time dense reconstruction of underwater scenes; while recent work made inspiring progress [17, 198], it still remains a challenging open problem. **3.** Solve the multi-view 2D feature correspondence problem for noisy underwater scenes. Optical image distortions severely affect traditional algorithms (*e.g.*,SIFT, ORB); addressing this issue with geometry-aware keypoint filtering will ensure the applicability of standard 3D reconstruction [22] or indirect VIO pipelines [160] to underwater data. **4.** Develop multi-robot state estimation algorithms for cooperative underwater exploration [218] and mapping [219].
Vision-based underwater HRI	**1.** Design efficient active learning pipelines for real-time diver detection and visual tracking [27]; existing methods either use static (pre-trained) models that fail to adapt and generalize to challenging unseen scenarios in real-time. **2.** Develop vision-based interaction frameworks for human–robot teams; that is, enable robot capabilities for following and interacting with a team of divers (and other robots) during cooperative underwater missions [201]. **3.** Design a unified human–robot interaction framework for both explicit and implicit interactions.

periods of high *bioactivity*. This technique uses an auto-periodic Gaussian process to model the acoustic bioactivity over different times of the day and then plans for the most information-rich sensing schedule. Moreover, the ability to follow marine animals is the key to understanding their natural behavior and ecosystem interactions. The Mesobot AUV [227] has been used to observe and track slow-moving animals in the mesopelagic zone, where Katija *et al.* [228] used a pre-trained animal-specific detector to enable robust tracking of siphonophores.

FIGURE 9.12
CUREE is designed to observe corals and marine animals using vision and passive acoustics.

Many marine species however are extremely rare, which makes it difficult to train supervised detection models for all species of interest. Cai *et al.* [229] have instead used a semi-supervised approach that learns a pose invariant appearance model of any arbitrary animal online and then uses the learned model for detection and tracking. Such semi-supervised approaches have been deployed in low-cost systems such as the CUREE robot [230] (see Figure 9.12), which is able to track a wide variety of animals such as barracuda (see Figure 9.13), jacks, stingrays, and jellyfish.

Fish and other marine species do not conform to their natural behavior in the presence of robots and other electromechanical devices [231, 232]. Therefore, close-proximity monitoring of their swimming, mating, and behavioral patterns becomes impossible with traditional robots and remote sensing platforms. The most alternative has been to use flippers or fins (*i.e.*, hydro-fins) instead of propulsion devices such as thrusters [231]. Robots such as CUREE [230] have explored the use of drifting to eliminate noise during measurements. While these systems can get rid of the propulsion noise, they cannot offer varying degrees of body undulation and/or fin oscillations for thrust generation at a power budget. Hence, their battery life and on-board processing powers are limited. More recent robotic systems deploy multiple control surfaces such as the caudal fins, pectoral fins, pelvic fins, dorsal fin, and cyclic hydraulic actuation with soft or elastic body materials to achieve fast and maneuverable propulsion [232, 233]. However, these robots are designed to be teleoperated by human divers for short-term data collection and inspection; they do not offer any autonomous features for vision-based navigation or long-term monitoring support. Therefore, more research efforts are needed for developing low-cost robotic solutions with biomimetic motion and agile swimming capabilities for applications such as long-term monitoring of high-risk

(a) Barracuda detection by a semi-supervised visual tracker

(b) The tracked trajectory

FIGURE 9.13
Visually guided underwater robots have been successfully used to track arbitrary animals without explicit prior training (with courtesy from [229].) Here, we illustrate an example of (a) the CUREE robot (with courtesy from [230]) successfully tracking a barracuda; and (b) its tracked trajectory, which shows that the animal is territorial and returning to its home base after a sortie to a nearby shoal.

ecosystems, invasive fish population health and behavioral aspects, artificial coral reefs and seagrass beds.

9.6.2 Fully Autonomous Structure Inspection and 3D Reconstruction

Subsea structures (*e.g.*, submarine cables, oil pipelines, pressure hulls) operate in physically demanding environments and are susceptible to various hazards with significant economic and environmental consequences. Several modes of failure need to be regularly monitored

FIGURE 9.14
BlueRoV2 observing an Aqua2 AUV mapping a shipwreck in the Caribbean (Barbados).

and inspected. With the recent revolution of deep visual learning and single-board supercomputers, an increasing number of vision-based monitoring/inspection systems are being deployed on AUVs, ROVs, and Autonomous Inspection Drones (AIDs), as well as in Underwater Wireless Sensor Networks (UWSNs) [234–236]. For autonomous operation, mobile robots need to robustly navigate around the structure [19, 141] and collect the necessary observations for detecting leaks/cracks, corrosion/erosion, or any physical displacements of the structure [237]. In addition to classical camera-based systems, sonar vision, optical, and radiographic technologies are deployed for long-term autonomous monitoring as well [99, 238]. Standard mapping and 3D reconstruction pipelines (*e.g.*, COLMAP [239] and OpenSfM [240]) are also used to map and visualize structural properties, particularly for marine archaeological studies. Besides, end-to-end learning-based planning algorithms have been proposed to navigate around shipwrecks [14] and coral reefs [15] while solely depending on the visual input. Contemporary methods also showed that deploying a team of robots at different distances from the structure enables detail mapping from the closest robots to high situational awareness for the robots further from the structure [241]; Figure 9.14 shows an illustration where an Aqua2 AUV is navigating over the Stavronikita shipwreck, Barbados, while a BlueRoV2 in autonomous mode observes it from a distance.

Traditionally, multi-beam echo sounders on AUVs/ROVs are used to create 3D bathymetric maps of areas of interests in the form of point cloud data [242]. In recent years, high-resolution cameras and hyperspectral imaging systems are used for bathymetry and orthomosaics as well [243]. These systems utilize Structure-from-Motion (SfM) photogrammetry to generate high-resolution and detailed 3D visualizations with millimeter-to-centimeter scale at a relatively low cost. However, image correction filters are needed to accurately adopt these libraries for underwater scenes [244, 245]. To this end, a prospective future research direction is to develop domain-specific end-to-end SfM pipelines to (learn to) generate accurate 3D reconstructions of underwater scenes.

9.6.3 Underwater Cave Mapping

Underwater caves are a crucial source of fresh water, provide important information for hydrogeology, and play a central role in advancing our knowledge of marine archaeology. Underwater cave mapping by humans is a labor-intensive, tedious, and above all, an extremely dangerous operation. For autonomous cave mapping by AUVs, one of the biggest challenges lies in vision-based estimation in the complete absence of ambient light, which results in constantly moving shadows due to the motion of the camera/light setup. In turn, the constantly moving shadows provide a denser set of features enabling semidense reconstructions [22]. Existing systems deploy stereo vision and utilize the *cone* of video-light intersection with the cave boundaries (*i.e.*, walls, floor, ceiling) to construct a *wire frame* outline of caves [246]. Recent results utilizing an inexpensive action camera such as GoPro 9 (see Figure 9.15a and b) and the visual-inertial component of SVIn [19, 219] demonstrated high accuracy compared to manual surveyed portions of an underwater cave; see Figure 9.15c.[1] The SOTA vision-based SLAM and motion estimation techniques demonstrate that imaging systems provide complementary sensing to sonar and inertial motion estimation pipelines and improve the mapping accuracy at a good update rate. Nevertheless, more research efforts are needed to achieve real-time low-power mapping capabilities, particularly for large subsea structures and unexplored caves with sparse feature landmarks, low illumination, and other motion constraints [223, 247]. Specifically, ensuring robust loop-closing and robot relocalization capabilities by fusing sonar, visual, inertial, and depth information is still a challenging open problem.

9.6.4 Aquaculture and Marine Ecology

Aquaculture and deep ocean agriculture (mariculture) farms [248] are considered the most promising solutions for meeting the ever-increasing food demands of the world population. Several infrastructures have been successful worldwide for harvesting fish (*e.g.*, carp, salmon, tuna) [249], crustaceans (*e.g.*, shrimp, crayfish), mollusks (*e.g.*, oysters [250], mussels), and aquatic plants (*e.g.*, seaweed, kelp). The inspiring success of underwater farming warrants scalable and smart technological solutions for rapid and innovative growth. Unfortunately, the existing

(a) Data collection while diving inside the Cueva del Agua, Spain

(b) A sample image from the dataset at the Cueva del Agua, Spain

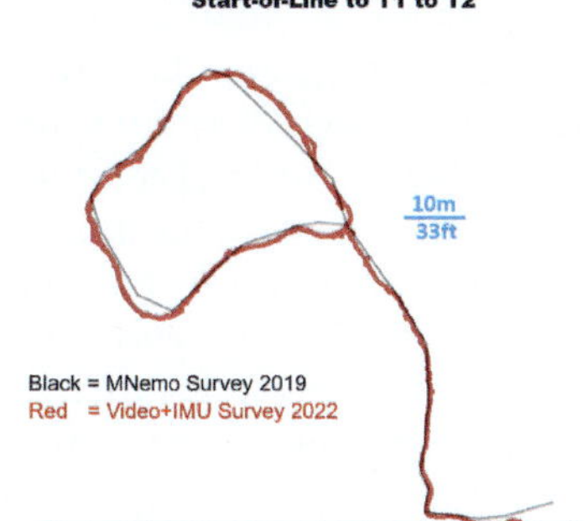

(c) Estimated trajectory together with manually measured ground truth from baseline

FIGURE 9.15
Underwater cave mapping: (a, b) a stereo GoPro setup with Kalden lights is utilized to collect visual-inertial data from an underwater cave in Spain; (c) the estimated trajectory matches the manual survey of the specific portion of the cave.

aquaculture and mariculture infrastructures are still rustic compared to traditional agricultural systems. A lack of intelligent system integration and data-driven feedback loops keeps the operating costs high, which in turn limits viable industrial expansions. Researchers in academia and industry have been trying to come up with better and more sustainable systems for autonomous monitoring and harvesting [251, 252]. The potential applications of robotics and computer vision are immense, both for automating the ecological sensing/monitoring processes in subsea vegetation pods [253] as well as offshore fish pens [254, 255]. In particular, novel AI and Internet of Things (AIoT)-based systems with integrated high-resolution cameras and optical sensors are being developed [254] for real-time monitoring of fish pens, from measuring oxygen levels to ensuring fish remain healthy by accurately maintaining feeding cycles. One of the major challenges is the low visibility together with the non-rigid structures of the aqua-pens [256]. Low-cost and scalable systems for real-time biomass estimation and other eco-parameter monitoring/control are also critically important for next-generation of underwater farming strategies.

Overall, the most desired intelligent features of futuristic industry-scale aquaculture farms are (1) interactive smart interfaces for dynamic eco-parameter visualization, (2) edge-cloud capabilities to dynamically control those eco-parameters by remote operators, and (3) automated pod/pen maintenance systems for feeding and harvesting cycles. These aspired capabilities call for scientific advancements in multimodal sensing and data-driven learning pipelines to understand various crop/fish-specific eco-parameters and other ecological characteristics. Other engineering challenges include efficient deployments of these AI features on edge devices with constraints on power, cost, and cloud connectivity.

9.7 Conclusion

We are in the midst of a technological revolution of computer vision, deep learning, and remote sensing platforms in underwater robotics applications. New hardware and software systems with improved solutions are emerging every day which is taking us closer to complete autonomy. In this chapter, we provided an overview of these SOTA systems and highlighted the emerging technologies. First, we presented the environmental and operational challenges of the existing solutions for prominent applications such as subsea inspection, monitoring, surveying, and bathymetric mapping. We highlighted their practical considerations for real-time deployments by AUVs/ROVs as well as in AIoT-based remote sensory devices. In particular, we focused on underwater imaging technologies and visual filtering pipelines that enable robust sensing and estimation in adverse underwater sensing conditions. We then discussed how the filtered perception pipelines facilitate autonomous identification, tracking, and semantic segmentation of important object categories such as coral reefs, various species of fish, robots or instruments, human divers, subsea structures. Subsequently, we discussed the latest underwater state estimation and localization systems of autonomous mobile robots for safe and efficient navigation. We presented an experimental review and highlighted the practical challenges of implementing visual-inertial SLAM for subsea structure inspection and cave mapping applications. We also mentioned how some of these challenges are addressed in human–robot cooperative missions by enabling guided autonomy through diver–robot interaction. Lastly, we highlighted several prospective research directions and high-impact applications of real-time robot vision for marine life monitoring and conservation, aquaculture, subsea mapping and inspection, and more. As several scientific and engineering challenges remain to be solved, we hope this overview will help researchers to identify critical problems and in developing next-generation technological solutions.

Acknowledgments

The findings presented in this chapter have been supported in part by the National Science Foundation under grants #1943205, #1919647, #2024741, #2024541 and #2024653, #2144624, #2330416; and the generous support of Halcyon Diving Systems and FLIR cameras. The authors would also like to acknowledge the help of the Woodville Karst Plain Project (WKPP), El Centro Investigador del Sistema Acuífero de Quintana Roo A.C. (CINDAQ), and Ricardo Constantino, Project Baseline in collecting data and providing access to challenging underwater caves.

Note

1. Manual survey by Ricardo Constantino, https://project baseline.org/

References

[1] D. Akkaynak and T. Treibitz, "A revised underwater image formation model," in *IEEE Conference on Computer Vision and Pattern Recognition (CVPR)*, 2018, pp. 6723–6732.

[2] M. J. Islam, Y. Xia, and J. Sattar, "Fast underwater image enhancement for improved visual perception," *IEEE Robotics and Automation Letters (RA-L)*, vol. 5, no. 2, pp. 3227–3234, 2020.

[3] M. Roznere and A. Quattrini Li, "Real-time model-based image color correction for underwater robots," in *2019 IEEE/RSJ International Conference on Intelligent Robots and Systems (IROS)*. IEEE, 2019, pp. 7191–7196.

[4] I. Alonso, M. Yuval, G. Eyal, T. Treibitz, and A. C. Murillo, "CoralSeg: Learning coral segmentation from sparse annotations," *Journal of Field Robotics (JFR)*, vol. 36, no. 8, pp. 1456–1477, 2019.

[5] M. Modasshir, S. Rahman, O. Youngquist, and I. Rekleitis, "Coral identification and counting with an autonomous underwater vehicle," in *IEEE International Conference on Robotics and Biomimetics (ROBIO)*, Kuala Lumpur, Malaysia, (Finalist of T. J. Tarn Best Paper in Robotics), Dec. 2018, pp. 524–529.

[6] S. B. Garner, A. M. Olsen, R. Caillouet, M. D. Campbell, and W. F. Patterson III, "Estimating reef fish size distributions with a mini remotely operated vehicle-integrated stereo camera system," *PLOS One*, vol. 16, no. 3, p. e0247985, 2021.

[7] M.-C. Chuang, J.-N. Hwang, K. Williams, and R. Towler, "Automatic fish segmentation via double local thresholding for Trawl-based underwater camera systems," in *IEEE International Conference on Image Processing*. IEEE, 2011, pp. 3145–3148.

[8] M. J. Islam, R. Wang, K. de Langis, and J. Sattar, "SVAM: Saliency-guided visual attention modeling by autonomous underwater robots," *In Review: IEEE Transactions on Pattern Analysis and Machine Intelligence (T-PAMI). ArXiv:2011.06252*, 2020.

[9] M. J. Islam, P. Luo, and J. Sattar, "Simultaneous enhancement and super-resolution of underwater imagery for improved visual perception," in *Robotics: Science and Systems (RSS)*, 2020.

[10] M. J. Islam, C. Edge, Y. Xiao, P. Luo, M. Mehtaz, C. Morse, S. S. Enan, and J. Sattar, "Semantic segmentation of underwater imagery: dataset and benchmark," in *IEEE/RSJ International Conference on Intelligent Robots and Systems (IROS)*. IEEE/RSJ, 2020.

[11] D. Lee, G. Kim, D. Kim, H. Myung, and H.-T. Choi, "Vision-based object detection and tracking for autonomous navigation of underwater robots," *Ocean Engineering*, vol. 48, pp. 59–68, 2012.

[12] M. Kalaitzakis, B. Cain, N. Vitzilaios, I. Rekleitis, and J. Moulton, "A marsupial robotic system for surveying and inspection of freshwater ecosystems," *Journal of Field Robotics*, vol. 38, no. 1, pp. 121–138, 2020.

[13] B. R. Kennedy, K. Cantwell, M. Malik, C. Kelley, J. Potter, K. Elliott, E. Lobecker, L. M. Gray, D. Sowers, M. P. White *et al.*, "The unknown and the unexplored: Insights into the pacific deep-sea following NOAA CAPSTONE expeditions," *Frontiers in Marine Science*, vol. 6, p. 480, 2019.

[14] N. Karapetyan, J. Johnson, and I. Rekleitis, "Human diver-inspired visual navigation: Towards coverage path planning of shipwrecks," *Marine Technology Society Journal, "Best of OCEANS 2020"*, vol. 55, no. 4, pp. 24–32, 2021.

[15] T. Manderson, J. C. G. Higuera, R. Cheng, and G. Dudek, "Vision-based autonomous underwater swimming in dense coral for combined collision avoidance and target selection," in *IEEE/RSJ International Conference on Intelligent Robots and Systems (IROS)*. IEEE, 2018, pp. 1885–1891.

[16] F. Shkurti, I. Rekleitis, and G. Dudek, "Feature tracking evaluation for pose estimation in underwater environments," in *2011 Canadian Conference on Computer and Robot Vision*. IEEE, 2011, pp. 160–167.

[17] S. Rahman, A. Quattrini Li, and I. Rekleitis, "Svin2: A multi-sensor fusion-based underwater slam system," *The International Journal of Robotics Research*, vol. 41, no. 11-12, pp. 1022–1042, 2022.

[18] A. Q. Li, A. Coskun, S. Doherty, S. Ghasemlou, A. Jagtap, M. Modasshir, S. Rahman, A. Singh, M. Xanthidis, J. O'Kane *et al.*, "Vision-based shipwreck mapping: On evaluating features quality and open source state estimation packages," in *OCEANS 2016 MTS/IEEE Monterey*. IEEE, 2016, pp. 1–10.

[19] S. Rahman, A. Quattrini Li, and I. Rekleitis, "Sonar visual inertial SLAM of underwater structures," in *IEEE International Conference on Robotics and Automation*, Brisbane, Australia, May 2018, pp. 5190–5196.

[20] S. Xu, T. Luczynski, J. S. Willners, Z. Hong, K. Zhang, Y. R. Petillot, and S. Wang, "Underwater visual acoustic slam with extrinsic calibration," in *2021 IEEE/RSJ International Conference on Intelligent Robots and Systems (IROS)*. IEEE, 2021, pp. 7647–7652.

[21] N. Weidner, S. Rahman, A. Quattrini Li, and I. Rekleitis, "Underwater cave mapping using stereo vision," in *IEEE International Conference on Robotics and Automation (ICRA)*, Singapore, May 2017, pp. 5709 – 5715.

[22] S. Rahman, A. Quattrini Li, and I. Rekleitis, "Contour based reconstruction of underwater structures using sonar, visual, inertial, and depth sensor," in *IEEE/RSJ International Conference on Intelligent Robots and Systems (IROS)*, Macau, Nov. 2019, pp. 8048–8053.

[23] G. Billings, R. Camilli, and M. Johnson-Roberson, "Hybrid visual slam for underwater vehicle manipulator systems," *IEEE Robotics and Automation Letters*, 2022.

[24] B. Yu, J. Wu, and M. J. Islam, "Udepth: Fast monocular depth estimation for visually-guided underwater robots," in *IEEE International Conference on Robotics and Automation (ICRA)*. IEEE, 2023.

[25] M. Roznere and A. Quattrini Li, "Underwater monocular image depth estimation using single-beam echosounder," in *2020 IEEE/RSJ International Conference on Intelligent Robots and Systems (IROS)*, 2020, pp. 1785–1790.

[26] M. J. Islam, M. Ho, and J. Sattar, "Understanding human motion and gestures for underwater human–robot collaboration," *Journal of Field Robotics (JFR)*, vol. 36, no. 5, pp. 851–873, 2019.

[27] M. J. Islam, M. Fulton, and J. Sattar, "Toward a generic diver-following algorithm: balancing robustness and efficiency in deep visual detection," *IEEE Robotics and Automation Letters (RA-L)*, vol. 4, no. 1, pp. 113–120, 2018.

[28] K. De Langis and J. Sattar, "Realtime multi-diver tracking and re-identification for underwater human-robot collaboration," in *2020 IEEE International Conference on Robotics and Automation (ICRA)*. IEEE, 2020, pp. 11 140–11 146.

[29] M. J. Islam, M. Ho, and J. Sattar, "Dynamic reconfiguration of mission parameters in underwater human-robot collaboration," in *IEEE International Conference on Robotics and Automation (ICRA)*, 2018, pp. 1–8.

[30] D. Chiarella, M. Bibuli, G. Bruzzone, M. Caccia, A. Ranieri, E. Zereik, L. Marconi, and P. Cutugno, "Gesture-based language for diver-robot underwater interaction," in *OCEANS 2015-Genova*. IEEE, 2015, pp. 1–9.

[31] J. Sattar and G. Dudek, "Underwater human-robot interaction via biological motion identification," in *Proceedings of the International Conference on Robotics: Science and Systems V, RSS*. Seattle, Washington, USA: MIT Press, June 2009, pp. 185–192.

[32] G. Dudek, J. Sattar, and A. Xu, "A visual language for robot control and programming: A human-interface study," in *IEEE International Conference on Robotics and Automation (ICRA)*. IEEE, 2007, pp. 2507–2513.

[33] M. Yang, J. Hu, C. Li, G. Rohde, Y. Du, and K. Hu, "An in-depth survey of underwater image enhancement and restoration," *IEEE Access*, vol. 7, pp. 123 638–123 657, 2019.

[34] R. Schettini and S. Corchs, "Underwater image processing: State of the art of restoration and image enhancement methods," *EURASIP journal on advances in signal processing*, vol. 2010, pp. 1–14, 2010.

[35] J. S. Jaffe, "Computer modeling and the design of optimal underwater imaging systems," *IEEE Journal of Oceanic Engineering*, vol. 15, no. 2, pp. 101–111, 1990.

[36] J. Y. Chiang and Y.-C. Chen, "Underwater image enhancement by wavelength compensation and dehazing," *IEEE Transactions on Image Processing*, vol. 21, no. 4, pp. 1756–1769, 2011.

[37] H. Feng, X. Yin, L. Xu, G. Lv, Q. Li, and L. Wang, "Underwater salient object detection jointly using improved spectral residual and fuzzy c-means," *Journal of Intelligent & Fuzzy Systems*, vol. 37, no. 1, pp. 329–339, 2019.

[38] B. Joshi, S. Rahman, M. Kalaitzakis, B. Cain, J. Johnson, M. Xanthidis, N. Karapetyan, A. Hernandez, A. Quattrini Li, N. Vitzilaios, and I. Rekleitis, "Experimental comparison of open source visual-inertial-based state estimation algorithms in the underwater domain," in *IEEE/RSJ International Conference on Intelligent Robots and Systems (IROS)*, Macau, Nov. 2019, pp. 7221–7227.

[39] S. Rahman, N. Karapetyan, A. Quattrini Li, and I. Rekleitis, "A modular sensor suite for underwater reconstruction," in *MTS/IEEE OCEANS - Charleston*.   IEEE, 2018, pp. 1–6.

[40] D. Berman, D. Levy, S. Avidan, and T. Treibitz, "Underwater single image color restoration using haze-lines and a new quantitative dataset," *arXiv preprint arXiv:1811.01343*, 2018.

[41] M. Bryson, M. Johnson-Roberson, O. Pizarro, and S. B. Williams, "True color correction of autonomous underwater vehicle imagery," *Journal of Field Robotics (JFR)*, vol. 33, no. 6, pp. 853–874, 2016.

[42] X. Ding, Y. Wang, J. Zhang, and X. Fu, "Underwater image dehaze using scene depth estimation with adaptive color correction," in *OCEANS 2017 - Aberdeen*, 2017, pp. 1–5.

[43] X. Xue, Z. Hao, L. Ma, Y. Wang, and R. Liu, "Joint luminance and chrominance learning for underwater image enhancement," *IEEE Signal Processing Letters*, vol. 28, pp. 818–822, 2021.

[44] K. He, J. Sun, and X. Tang, "Single image haze removal using dark channel prior," *IEEE Transactions on Pattern Analysis and Machine Intelligence*, vol. 33, no. 12, pp. 2341–2353, 2010.

[45] Y. Y. Schechner and N. Karpel, "Recovery of underwater visibility and structure by polarization analysis," *IEEE Journal of Oceanic engineering*, vol. 30, no. 3, pp. 570–587, 2005.

[46] X. Wu and H. Li, "A simple and comprehensive model for underwater image restoration," in *2013 IEEE International Conference on Information and Automation (ICIA)*, 2013, pp. 699–704.

[47] E. Trucco and A. Olmos-Antillon, "Self-tuning underwater image restoration," *IEEE Journal of Oceanic Engineering*, vol. 31, no. 2, pp. 511–519, 2006.

[48] D. Akkaynak and T. Treibitz, "Sea-Thru: A method for removing water from underwater images," in *IEEE Conference on Computer Vision and Pattern Recognition (CVPR)*, 2019, pp. 1682–1691.

[49] P. L. Drews, E. R. Nascimento, S. S. Botelho, and M. F. M. Campos, "Underwater depth estimation and image restoration based on single images," *IEEE Computer Graphics and Applications*, vol. 36, no. 2, pp. 24–35, 2016.

[50] A. Galdran, D. Pardo, A. Picón, and A. Alvarez-Gila, "Automatic red-channel underwater image restoration," *Journal of Visual Communication and Image Representation*, vol. 26, pp. 132–145, 2015.

[51] E. Trucco and A. T. Olmos-Antillon, "Self-tuning underwater image restoration," *IEEE Journal of Oceanic Engineering*, vol. 31, no. 2, pp. 511–519, 2006.

[52] G. Hou, Z. Pan, G. Wang, H. Yang, and J. Duan, "An efficient nonlocal variational method with application to underwater image restoration," *Neurocomputing*, vol. 369, pp. 106–121, 2019.

[53] S. Yan, X. Chen, Z. Wu, J. Wang, Y. Lu, M. Tan, and J. Yu, "Hybrur: A hybrid physical-neural solution for unsupervised underwater image restoration," *arXiv preprint arXiv:2107.02660*, 2021.

[54] P. Isola, J.-Y. Zhu, T. Zhou, and A. A. Efros, "Image-to-image translation with conditional adversarial networks," in *Proc. of the IEEE Conference on Computer Vision and Pattern Recognition (CVPR)*, 2017, pp. 1125–1134.

[55] I. Goodfellow, J. Pouget-Abadie, M. Mirza, B. Xu, D. Warde-Farley, S. Ozair, A. Courville, and Y. Bengio, "Generative adversarial nets," in *Advances in Neural Information Processing Systems (NIPS)*, 2014, pp. 2672–2680.

[56] A. Vaswani, N. Shazeer, N. Parmar, J. Uszkoreit, L. Jones, A. N. Gomez, Ł. Kaiser, and I. Polosukhin, "Attention is all you need," *Advances in Neural Information Processing Systems*, vol. 30, 2017.

[57] C. Li, C. Guo, W. Ren, R. Cong, J. Hou, S. Kwong, and D. Tao, "An underwater image enhancement benchmark dataset and beyond," in *IEEE Transactions on Image Processing (TIP)*. IEEE, 2019, pp. 4376–4389.

[58] X. Yu, Y. Qu, and M. Hong, "Underwater-GAN: Underwater image restoration via conditional generative adversarial network," in *International Conference on Pattern Recognition*. Springer, 2018, pp. 66–75.

[59] C. Fabbri, M. J. Islam, and J. Sattar, "Enhancing underwater imagery using generative adversarial networks," in *IEEE International Conference on Robotics and Automation (ICRA)*. IEEE, 2018, pp. 7159–7165.

[60] Z. Cheng, Q. Yang, and B. Sheng, "Deep colorization," in *Proceedings of the IEEE International Conference on Computer Vision (ICCV)*, 2015, pp. 415–423.

[61] B. Cai, X. Xu, K. Jia, C. Qing, and D. Tao, "DehazeNet: An end-to-end system for single image haze removal," *IEEE Transactions on Image Processing*, vol. 25, no. 11, pp. 5187–5198, 2016.

[62] P. Liu, G. Wang, H. Qi, C. Zhang, H. Zheng, and Z. Yu, "Underwater image enhancement with a deep residual framework," *IEEE Access*, vol. 7, pp. 94 614–94 629, 2019.

[63] W. Wang and J. Shen, "Deep visual attention prediction," *IEEE Transactions on Image Processing*, vol. 27, no. 5, pp. 2368–2378, 2017.

[64] O. Ronneberger, P. Fischer, and T. Brox, "U-Net: Convolutional networks for biomedical image segmentation," in *International Conference on Medical Image Computing and Computer-assisted Intervention*. Springer, 2015, pp. 234–241.

[65] A. Jamadandi and U. Mudenagudi, "Exemplar-based underwater image enhancement augmented by wavelet corrected transforms," in *IEEE Conference on Computer Vision and Pattern Recognition (CVPR) Workshops*, 2019, pp. 11–17.

[66] Y. Hashisho, M. Albadawi, T. Krause, and U. F. von Lukas, "Underwater color restoration using U-Net denoising autoencoder," in *2019 11th International Symposium on Image and Signal Processing and Analysis (ISPA)*. IEEE, 2019, pp. 117–122.

[67] S. Zhang, T. Wang, J. Dong, and H. Yu, "Underwater image enhancement via extended multi-scale retinex," *Neurocomputing*, vol. 245, pp. 1–9, 2017.

[68] C. Li, S. Anwar, and F. Porikli, "Underwater scene prior inspired deep underwater image and video enhancement," *Pattern Recognition*, vol. 98, p. 107038, 2020.

[69] C. O. Ancuti, C. Ancuti, C. De Vleeschouwer, and P. Bekaert, "Color balance and fusion for underwater image enhancement," *IEEE Transactions on image processing*, vol. 27, no. 1, pp. 379–393, 2017.

[70] Y. Guo, H. Li, and P. Zhuang, "Underwater image enhancement using a multiscale dense generative adversarial network," *IEEE Journal of Oceanic Engineering*, 2019.

[71] L. Peng, C. Zhu, and L. Bian, "U-shape transformer for underwater image enhancement," *arXiv preprint arXiv:2111.11843*, 2021.

[72] Y. Tang, T. Iwaguchi, H. Kawasaki, R. Sagawa, and R. Furukawa, "Autoenhancer: Transformer on U-Net architecture search for underwater image enhancement," in *Proceedings of the Asian Conference on Computer Vision*, 2022, pp. 1403–1420.

[73] J. Li, K. A. Skinner, R. M. Eustice, and M. Johnson-Roberson, "Watergan: Unsupervised generative network to enable real-time color correction of monocular underwater images," *IEEE Robotics and Automation letters*, vol. 3, no. 1, pp. 387–394, 2017.

[74] X. Mao, Q. Li, H. Xie, R. Y. Lau, Z. Wang, and S. Paul Smolley, "Least squares generative adversarial networks," in *Proceedings of the IEEE International Conference on Computer Vision (ICCV)*, 2017, pp. 2794–2802.

[75] M. Arjovsky, S. Chintala, and L. Bottou, "Wasserstein generative adversarial networks," in *International Conference on Machine Learning (ICML)*, 2017, pp. 214–223.

[76] J. Zhao, M. Mathieu, and Y. LeCun, "Energy-based generative adversarial network," *arXiv preprint arXiv:1609.03126*, 2017.

[77] C. Li, J. Guo, and C. Guo, "Emerging from water: Underwater image color correction based on weakly supervised color transfer," *IEEE Signal processing letters*, vol. 25, no. 3, pp. 323–327, 2018.

[78] N. Wang, Y. Zhou, F. Han, H. Zhu, and J. Yao, "Uwgan: Underwater gan for real-world underwater color restoration and dehazing," *arXiv preprint arXiv:1912.10269*, 2019.

[79] J.-Y. Zhu, T. Park, P. Isola, and A. A. Efros, "Unpaired image-to-image translation using cycle-consistent adversarial networks," in *Proceedings of the IEEE International Conference on Computer Vision (ICCV)*, 2017, pp. 2223–2232.

[80] Z. Yi, H. Zhang, P. Tan, and M. Gong, "DualGAN: Unsupervised dual learning for image-to-image translation," in *Proceedings of the IEEE International Conference on Computer Vision (ICCV)*, 2017, pp. 2849–2857.

[81] Y.-S. Chen, Y.-C. Wang, M.-H. Kao, and Y.-Y. Chuang, "Deep photo enhancer: Unpaired learning for image enhancement from photographs with GANs," in *Proceedings of the IEEE Conference on Computer Vision and Pattern Recognition (CVPR)*. IEEE, 2018, pp. 6306–6314.

[82] A. Ignatov, N. Kobyshev, R. Timofte, K. Vanhoey, and L. Van Gool, "DSLR-quality photos on mobile devices with deep convolutional networks," in *Proceedings of the IEEE International Conference on Computer Vision (ICCV)*, 2017, pp. 3277–3285.

[83] Z. Wang, X. Cun, J. Bao, W. Zhou, J. Liu, and H. Li, "Uformer: A general U-shaped transformer for image restoration," in *Proceedings of the IEEE/CVF Conference on Computer Vision and Pattern Recognition*, 2022, pp. 17 683–17 693.

[84] Z. Guo, D. Guo, Z. Gu, H. Zheng, B. Zheng, and G. Wang, "Unsupervised underwater image clearness via transformer," in *OCEANS 2022-Chennai*. IEEE, 2022, pp. 1–4.

[85] D. Berman, D. Levy, S. Avidan, and T. Treibitz, "Underwater single image color restoration using haze-lines and a new quantitative dataset," *IEEE Transactions on Pattern Analysis and Machine Intelligence*, 2020.

[86] L. A. Torres-Méndez and G. Dudek, "Color correction of underwater images for aquatic robot inspection," in *International Workshop on Energy Minimization Methods in Computer Vision and Pattern Recognition*. Springer, 2005, pp. 60–73.

[87] Y. Cho, J. Jeong, and A. Kim, "Model-assisted multiband fusion for single image enhancement and applications to robot vision," *IEEE Robotics and Automation Letters (RA-L)*, vol. 3, no. 4, pp. 2822–2829, 2018.

[88] Y. Cho and A. Kim, "Visibility enhancement for underwater visual slam based on underwater light scattering model," in *2017 IEEE International Conference on Robotics and Automation (ICRA)*. IEEE, 2017, pp. 710–717.

[89] H. Lu, Y. Li, and S. Serikawa, "Underwater image enhancement using guided trigonometric bilateral filter and fast automatic color correction," in *IEEE International Conference on Image Processing*. IEEE, 2013, pp. 3412–3416.

[90] Z. Cao, T. Simon, S.-E. Wei, and Y. Sheikh, "Realtime multi-person 2D pose estimation using part affinity fields," in *IEEE Conference on Computer Vision and Pattern Recognition (CVPR)*, 2017, pp. 7291–7299.

[91] P. Warden and D. Situnayake, *TinyML: Machine Learning with TensorFlow on Arduino, and Ultra-Low Power Micro-Controllers*. O'Reilly Media, 2019.

[92] S. Jamieson, J. P. How, and Y. Girdhar, "Deepseecolor: Realtime adaptive color correction for autonomous underwater vehicles via deep learning methods," arXiv:2303.04025, 2023.

[93] Z.-Q. Zhao, P. Zheng, S.-t. Xu, and X. Wu, "Object detection with deep learning: A review," *IEEE Transactions on Neural Networks and Learning Systems*, vol. 30, no. 11, pp. 3212–3232, 2019.

[94] A. Garcia-Garcia, S. Orts-Escolano, S. Oprea, V. Villena-Martinez, and J. Garcia-Rodriguez, "A review on deep learning techniques applied to semantic segmentation," *arXiv preprint arXiv:1704.06857*, 2017.

[95] F. Perazzi, J. Pont-Tuset, B. McWilliams, L. Van Gool, M. Gross, and A. Sorkine-Hornung, "A benchmark dataset and evaluation methodology for video object segmentation," in *IEEE*

Conference on Computer Vision and Pattern Recognition (CVPR), 2016, pp. 724–732.

[96] M. Ravanbakhsh, M. R. Shortis, F. Shafait, A. Mian, E. S. Harvey, and J. W. Seager, "Automated fish detection in underwater images using shape-based level sets," *Photogrammetric Record*, vol. 30, no. 149, pp. 46–62, 2015.

[97] M. Sung, S.-C. Yu, and Y. Girdhar, "Vision based real-time fish detection using convolutional neural network," in *OCEANS 2017-Aberdeen*. IEEE, 2017, pp. 1–6.

[98] G. L. Foresti and S. Gentili, "A vision based system for object detection in underwater images," *International Journal of Pattern Recognition and Artificial Intelligence*, vol. 14, no. 02, pp. 167–188, 2000.

[99] X. Xiang, B. Jouvencel, and O. Parodi, "Coordinated formation control of multiple autonomous underwater vehicles for pipeline inspection," *International Journal of Advanced Robotic Systems*, vol. 7, no. 1, p. 3, 2010.

[100] O. Beijbom, P. J. Edmunds, D. I. Kline, B. G. Mitchell, and D. Kriegman, "Automated annotation of coral reef survey images," in *IEEE Conference on Computer Vision and Pattern Recognition (CVPR)*. IEEE, 2012, pp. 1170–1177.

[101] Z. Chen, Y. Sun, Y. Gu, H. Wang, H. Qian, and H. Zheng, "Underwater object segmentation integrating transmission and saliency features," *IEEE Access*, vol. 7, pp. 72 420–72 430, 2019.

[102] Y. Zhou, J. Wang, B. Li, Q. Meng, E. Rocco, and A. Saiani, "Underwater scene segmentation by deep neural network," *UK Robotics and Autonomous Systems*, 2019.

[103] A. Ortiz, M. Simó, and G. Oliver, "A vision system for an underwater cable tracker," *Machine Vision and Applications*, vol. 13, no. 3, pp. 129–140, 2002.

[104] M.-C. Chuang, J.-N. Hwang, K. Williams, and R. Towler, "Tracking live fish from low-contrast and low-frame-rate stereo videos," *IEEE Transactions on Circuits and Systems for Video Technology*, vol. 25, no. 1, pp. 167–179, 2014.

[105] F. Shkurti, W.-D. Chang, P. Henderson, M. J. Islam, J. C. G. Higuera, J. Li, T. Manderson, A. Xu, G. Dudek, and J. Sattar, "Underwater multi-robot convoying using visual tracking by detection," in *IEEE/RSJ International Conference on Intelligent Robots and Systems (IROS)*. IEEE, 2017.

[106] D. Yang, L. Cai, S. Jamieson, and Y. Girdhar, "Biological hotspot mapping in coral reefs with robotic visual surveys," *arXiv*, 2023.

[107] X.Company, "The tidal project," https://x.company/projects/tidal/, 2020, accessed: 20-12-2022.

[108] A. Salman, S. A. Siddiqui, F. Shafait, A. Mian, M. R. Shortis, K. Khurshid, A. Ulges, and U. Schwanecke, "Automatic fish detection in underwater videos by a deep neural network-based hybrid motion learning system," *ICES Journal of Marine Science*, vol. 77, no. 4, pp. 1295–1307, 2020.

[109] H. Qin, X. Li, J. Liang, Y. Peng, and C. Zhang, "Deepfish: Accurate underwater live fish recognition with a deep architecture," *Neurocomputing*, vol. 187, pp. 49–58, 2016.

[110] B. Joshi, M. Modasshir, T. Manderson, H. Damron, M. Xanthidis, A. Quattrini Li, I. Rekleitis, and G. Dudek, "Deepurl: Deep pose estimation framework for underwater relative localization," in *IEEE/RSJ International Conference on Intelligent Robots and Systems (IROS)*, Las Vegas, NV, 2020, pp. 1777–1784.

[111] M. J. Islam, J. Mo, and J. Sattar, "Robot-to-robot relative pose estimation using humans as markers," *Autonomous Robots*, vol. 45, no. 4, pp. 579–593, 2021.

[112] H. Singh, R. Armstrong, F. Gilbes, R. Eustice, C. Roman, O. Pizarro, and J. Torres, "Imaging coral I: imaging coral habitats with the seabed AUV," *Subsurface Sensing Technologies and Applications*, vol. 5, no. 1, pp. 25–42, 2004.

[113] M. Grasmueck, G. P. Eberli, D. A. Viggiano, T. Correa, G. Rathwell, and J. Luo, "Autonomous underwater vehicle (AUV) mapping reveals coral mound distribution, morphology, and oceanography in deep water of the straits of Florida," *Geophysical Research Letters*, vol. 33, no. 23, 2006.

[114] O. Beijbom *et al.*, "Towards automated annotation of benthic survey images: Variability of human experts and operational modes of automation," *PLoS ONE*, vol. 10, no. 7, p. e0130312, 2015.

[115] M. Modasshir, A. Quattrini Li, and I. Rekleitis, "Mdnet: Multi-patch dense network for coral classification," in *MTS/IEEE OCEANS - Charleston*. IEEE, 2018, pp. 1–6.

[116] O. Beijbom, P. Edmunds, C. Roelfsema, J. Smith, D. Kline, B. Neal, M. Dunlap, V. Moriarty, T. Fan,

C. Tan, S. Chan, T. Treibitz, A. Gamst, B. Mitchell, and D. Kriegman, "Data from: Towards automated annotation of benthic survey images: variability of human experts and operational modes of automation," PloS ONE 2015.

[117] M. Modasshir, S. Rahman, and I. Rekleitis, "Autonomous 3D semantic mapping of coral reefs," in *12th Conference on Field and Service Robotics (FSR)*, Tokyo, Japan, Aug. 2019, pp. 365–379.

[118] M. Modasshir and I. Rekleitis, "Augmenting coral reef monitoring with an enhanced detection system," in *IEEE International Conference on Robotics and Automation*, Paris, France, 2020, pp. 1874–1880.

[119] T. Manderson, J. Li, N. Dudek, D. Meger, and G. Dudek, "Robotic coral reef health assessment using automated image analysis," *Journal of Field Robotics*, vol. 34, no. 1, pp. 170–187, 2017.

[120] D. A. Konovalov, A. Saleh, D. B. Efremova, J. A. Domingos, and D. R. Jerry, "Automatic weight estimation of harvested fish from images," in *2019 Digital Image Computing: Techniques and Applications (DICTA)*. IEEE, 2019, pp. 1–7.

[121] J. Martinez-de Dios, C. Serna, and A. Ollero, "Computer vision and robotics techniques in fish farms," *Robotica*, vol. 21, no. 3, pp. 233–243, 2003.

[122] H. E. Harris, A. Q. Fogg, M. S. Allen, R. N. Ahrens, and W. F. Patterson, "Precipitous declines in Northern Gulf of Mexico invasive lionfish populations following the emergence of an ulcerative skin disease," *Scientific Reports*, vol. 10, no. 1, pp. 1–17, 2020.

[123] S. Villon, D. Mouillot, M. Chaumont, E. S. Darling, G. Subsol, T. Claverie, and S. Villéger, "A deep learning method for accurate and fast identification of coral reef fishes in underwater images," *Ecological Informatics*, vol. 48, pp. 238–244, 2018.

[124] B. M. Hopkinson, A. C. King, D. P. Owen, M. Johnson-Roberson, M. H. Long, and S. M. Bhandarkar, "Automated classification of three-dimensional reconstructions of coral reefs using convolutional neural networks," *LoS ONE*, vol. 15, no. 3, p. e0230671, 2020.

[125] A. Mahmood, M. Bennamoun, S. An, F. Sohel, F. Boussaid, R. Hovey, G. Kendrick, and R. B. Fisher, "Deep learning for coral classification," in *Handbook of Neural Computation*. Elsevier, 2017, pp. 383–401.

[126] C. Roelfsema, S. Phinn, S. Jupiter, J. Comley, and S. Albert, "Mapping coral reefs at reef to reef-system scales, 10s–1000skm2, using object-based image analysis," *International Journal of Remote Sensing*, vol. 34, no. 18, pp. 6367–6388, 2013.

[127] K. Koreitem, F. Shkurti, T. Manderson, W.-D. Chang, J. C. G. Higuera, and G. Dudek, "One-shot informed robotic visual search in the wild," in *2020 IEEE/RSJ International Conference on Intelligent Robots and Systems (IROS)*. IEEE, 2020, pp. 5800–5807.

[128] Y. Girdhar, P. Giguere, and G. Dudek, "Autonomous adaptive exploration using realtime online spatiotemporal topic modeling," *International Journal of Robotics Research (IJRR)*, vol. 33, no. 4, pp. 645–657, 2014.

[129] Z. U. Rehman, "Salient Object Detection from Underwater Image," Ph.D. dissertation, Capital University, 2019.

[130] M. Johnson-Roberson, O. Pizarro, and S. Williams, "Saliency ranking for benthic survey using underwater images," in *International Conference on Control Automation Robotics & Vision*. IEEE, 2010, pp. 459–466.

[131] J. Zhu, S. Yu, L. Gao, Z. Han, and Y. Tang, "Saliency-based diver target detection and localization method," *Mathematical Problems in Engineering*, vol. 2020, 2020.

[132] D. L. Rizzini, F. Kallasi, F. Oleari, and S. Caselli, "Investigation of vision-based underwater object detection with multiple datasets," *International Journal of Advanced Robotic Systems*, vol. 12, no. 6, p. 77, 2015.

[133] M. Modasshir and I. Rekleitis, "Enhancing coral reef monitoring utilizing a deep semi-supervised learning approach," in *IEEE International Conference on Robotics and Automation (ICRA)*. IEEE, 2020, pp. 1874–1880.

[134] A. Maldonado-Ramírez and L. A. Torres-Mendez, "Learning ad-hoc compact representations from salient landmarks for visual place recognition in underwater environments," in *International Conference on Robotics and Automation (ICRA)*. IEEE, 2019, pp. 5739–5745.

[135] Y. Girdhar and G. Dudek, "Modeling curiosity in a mobile robot for long-term autonomous exploration and monitoring," *Autonomous Robots*, vol. 40, no. 7, pp. 1267–1278, 2016.

[136] I. Rekleitis, G. Dudek, and E. Milios, "Multi-robot collaboration for robust exploration," *Annals of Mathematics and Artificial Intelligence*, vol. 31, no. 1-4, pp. 7–40, 2001.

[137] D. R. Edgington, K. A. Salamy, M. Risi, R. Sherlock, D. Walther, and C. Koch, "Automated event detection in underwater video," in *Oceans*, vol. 5. IEEE, 2003, pp. 2749–2753.

[138] A. Maldonado-Ramírez and L. A. Torres-Méndez, "Robotic visual tracking of relevant cues in underwater environments with poor visibility conditions," *Journal of Sensors*, vol. 2016, 2016.

[139] N. Kumar, H. K. Sardana, and S. Shome, "Saliency-based shape extraction of objects in unconstrained underwater environment," *Multimedia Tools and Applications*, vol. 78, no. 11, pp. 15 121–15 139, 2019.

[140] H. B. Wang, X. Dong, J. Shen, X. W. Wu, and Z. Chen, "Saliency-based adaptive object extraction for color underwater images," in *Applied Mechanics and Materials*, vol. 347. Trans Tech Publ, 2013, pp. 3964–3970.

[141] A. Kim and R. M. Eustice, "Real-time Visual SLAM for Autonomous underwater hull inspection using visual saliency," *IEEE Transactions on Robotics (TRO)*, vol. 29, no. 3, pp. 719–733, 2013.

[142] A. Bhattcharyya, "On a measure of divergence between two statistical populations defined by their probability distributions," *Bulletin Calcutta Math Society*, vol. 1, no. 35, pp. 99–110, 1943.

[143] M. Jian, Q. Qi, J. Dong, Y. Yin, and K.-M. Lam, "Integrating QDWD with pattern distinctness and local contrast for underwater saliency detection," *Journal of Visual Communication and Image Representation*, vol. 53, pp. 31–41, 2018.

[144] Y. Zhu, B. Hao, B. Jiang, R. Nian, B. He, X. Ren, and A. Lendasse, "Underwater image segmentation with co-saliency detection and local statistical active contour model," in *OCEANS*. IEEE, 2017, pp. 1–5.

[145] X. Li, J. Hao, M. Shang, and Z. Yang, "Saliency segmentation and foreground extraction of underwater image based on localization," in *OCEANS*. IEEE, 2016, pp. 1–4.

[146] F. Liu and M. Fang, "Semantic segmentation of underwater images based on improved deeplab," *Journal of Marine Science and Engineering*, vol. 8, no. 3, p. 188, 2020.

[147] J. Hong, M. Fulton, and J. Sattar, "Trashcan: A semantically-segmented dataset towards visual detection of marine debris," *arXiv preprint arXiv:2007.08097*, 2020.

[148] M. Waszak, A. Cardaillac, B. Elvesæter, F. Rødølen, and M. Ludvigsen, "Semantic segmentation in underwater ship inspections: Benchmark and data set," *IEEE Journal of Oceanic Engineering*, 2022.

[149] M. Zurowietz, D. Langenkämper, B. Hosking, H. A. Ruhl, and T. W. Nattkemper, "Maia—a machine learning assisted image annotation method for environmental monitoring and exploration," *PLoS ONE*, vol. 13, no. 11, p. e0207498, 2018.

[150] A. B. Labao and P. C. Naval, "Weakly-labelled semantic segmentation of fish objects in underwater videos using a deep residual network," in *Asian Conference on Intelligent Information and Database Systems*. Springer, 2017, pp. 255–265.

[151] A. Smith, J. Coffelt, and K. Lingemann, "A deep learning framework for semantic segmentation of underwater environments," in *OCEANS 2022, Hampton Roads*. IEEE, 2022, pp. 1–7.

[152] M. O'Byrne, V. Pakrashi, F. Schoefs, and B. Ghosh, "Semantic segmentation of underwater imagery using deep networks trained on synthetic imagery," *Journal of Marine Science and Engineering*, vol. 6, no. 3, p. 93, 2018.

[153] X. Li, J. Song, F. Zhang, X. Ouyang, and S. U. Khan, "MapReduce-based fast fuzzy C-means algorithm for large-scale underwater image segmentation," *Future Generation Computer Systems*, vol. 65, pp. 90–101, 2016.

[154] B. Arain, C. McCool, P. Rigby, D. Cagara, and M. Dunbabin, "Improving underwater obstacle detection using semantic image segmentation," in *International Conference on Robotics and Automation (ICRA)*. IEEE, 2019, pp. 9271–9277.

[155] H. Mohamed, K. Nadaoka, and T. Nakamura, "Automatic semantic segmentation of benthic habitats using images from towed underwater camera in a complex shallow water environment," *Remote Sensing*, vol. 14, no. 8, p. 1818, 2022.

[156] H. Zhang, A. Gruen, and M. Li, "Deep learning for semantic segmentation of coral images in underwater photogrammetry," *ISPRS Annals of the Photogrammetry, Remote Sensing and Spatial Information Sciences*, vol. 2, pp. 343–350, 2022.

[157] A. King, S. M. Bhandarkar, and B. M. Hopkinson, "A comparison of deep learning methods for semantic segmentation of coral reef survey images," in *Proceedings of the IEEE Conference on Computer Vision and Pattern Recognition Workshops*, 2018, pp. 1394–1402.

[158] Y. R. Petillot, G. Antonelli, G. Casalino, and F. Ferreira, "Underwater robots: From remotely operated vehicles to intervention-autonomous underwater vehicles," *IEEE Robotics & Automation Magazine*, vol. 26, no. 2, pp. 94–101, 2019.

[159] L. Paull, S. Saeedi, M. Seto, and H. Li, "Auv navigation and localization: A review," *IEEE Journal of Oceanic Engineering*, vol. 39, no. 1, pp. 131–149, 2013.

[160] A. Quattrini Li, A. Coskun, S. M. Doherty, S. Ghasemlou, A. S. Jagtap, M. Modasshir, S. Rahman, A. Singh, M. Xanthidis, J. M. O'Kane, and I. Rekleitis, "Experimental comparison of open source vision based state estimation algorithms," in *International Symposium of Experimental Robotics (ISER)*, Tokyo, Japan, Mar. 2016.

[161] K. Köser and U. Frese, "Challenges in underwater visual navigation and slam," in *AI Technology for Underwater Robots*.   Springer, 2020, pp. 125–135.

[162] J. Engel, T. Schöps, and D. Cremers, "LSD-SLAM: Large-scale direct monocular SLAM," in *European Conference on Computer Vision (ECCV)*, 2014, pp. 834–849.

[163] J. Engel, V. Koltun, and D. Cremers, "Direct sparse odometry," *IEEE Transactions on Pattern Analysis and Machine Intelligence*, vol. 40, no. 3, pp. 611–625, 2017.

[164] J. Mo, M. J. Islam, and J. Sattar, "Fast direct stereo visual slam," *IEEE Robotics and Automation Letters*, vol. 7, no. 2, pp. 778–785, 2021.

[165] C. Campos, R. Elvira, J. J. Gómez, J. M. M. Montiel, and J. D. Tardós, "ORB-SLAM3: An accurate open-source library for visual, visual-inertial and multi-map SLAM," *IEEE Transactions on Robotics*, vol. 37, no. 6, pp. 1874–1890, 2021.

[166] A. I. Mourikis and S. I. Roumeliotis, "A multistate constraint Kalman filter for vision-aided inertial navigation," in *Proceedings of the IEEE International Conference on Robotics and Automation (ICRA)*, 2007, pp. 3565–3572.

[167] M. Bloesch, S. Omari, M. Hutter, and R. Siegwart, "Robust visual inertial odometry using a direct EKF-based approach," in *2015 IEEE/RSJ International Conference on Intelligent Robots and Systems (IROS)*, 2015.

[168] F. Shkurti, I. Rekleitis, M. Scaccia, and G. Dudek, "State estimation of an underwater robot using visual and inertial information," in *IEEE/RSJ International Conference on Intelligent Robots and Systems (IROS)*, San Francisco, CA, US, Sep. 2011, pp. 5054–5060.

[169] S. Leutenegger, S. Lynen, M. Bosse, R. Siegwart, and P. Furgale, "Keyframe-based visual–inertial odometry using nonlinear optimization," *The International Journal of Robotics Research*, vol. 34, no. 3, pp. 314–334, 2015.

[170] P. Geneva, K. Eckenhoff, W. Lee, Y. Yang, and G. Huang, "OpenVINS: A research platform for visual-inertial estimation," in *Proceedings of the IEEE International Conference on Robotics and Automation*, Paris, France, 2020.

[171] C. Forster, L. Carlone, F. Dellaert, and D. Scaramuzza, "On-manifold preintegration for real-time visual–inertial odometry," *IEEE Transactions on Robotics*, vol. 33, no. 1, pp. 1–21, 2016.

[172] S. Rahman, A. Quattrini Li, and I. Rekleitis, "An underwater SLAM system using sonar, visual, inertial, and depth sensor," in *IEEE/RSJ International Conference on Intelligent Robots and Systems (IROS)*, Macau, (IROS ICROS Best Application Paper Award. Finalist), Nov. 2019, pp. 1861–1868.

[173] C. Cadena, L. Carlone, H. Carrillo, Y. Latif, D. Scaramuzza, J. Neira, I. Reid, and J. J. Leonard, "Past, present, and future of simultaneous localization and mapping: Toward the robust-perception age," *IEEE Transactions on Robotics*, vol. 32, no. 6, pp. 1309–1332, 2016.

[174] T. Qin, P. Li, and S. Shen, "VINS-Mono: A robust and versatile monocular visual-inertial state estimator," *IEEE Transactions on Robotics*, vol. 34, no. 4, pp. 1004–1020, 2018.

[175] L. von Stumberg and D. Cremers, "DM-VIO: Delayed marginalization visual inertial odometry," *IEEE Robotics and Automation Letters*, vol. 7, no. 2, pp. 1408–1415, 2022.

[176] A. Rosinol, M. Abate, Y. Chang, and L. Carlone, "Kimera: An open-source library for real-time metric-semantic localization and mapping," in *Proceedings of the IEEE International Conference*

on Robotics and Automation (ICRA), 2020. [Online]. Available: https://github.com/MIT-SPARK/Kimera

[177] C. Forster, Z. Zhang, M. Gassner, M. Werlberger, and D. Scaramuzza, "Svo: Semidirect visual odometry for monocular and multicamera systems," *IEEE Transaction Robot.*, vol. 33, no. 2, pp. 249–265, 2017.

[178] T. Qin, S. Cao, J. Pan, and S. Shen, "A general optimization-based framework for global pose estimation with multiple sensors," arXiv:1901.03642, 2019.

[179] M. Ferrera, V. Creuze, J. Moras, and P. Trouvé-Peloux, "AQUALOC: An underwater dataset for visual–inertial–pressure localization," *The International Journal of Robotics Research*, vol. 38, no. 14, pp. 1549–1559, 2019.

[180] B. Teixeira, H. Silva, A. Matos, and E. Silva, "Deep learning for underwater visual odometry estimation," *IEEE Access*, vol. 8, pp. 44 687–44 701, 2020.

[181] A. Burguera and F. Bonin-Font, "An unsupervised neural network for loop detection in underwater visual slam," *Journal of Intelligent & Robotic Systems*, vol. 100, no. 3, pp. 1157–1177, 2020.

[182] L. M. Hodne, E. Leikvoll, M. Yip, A. L. Teigen, A. Stahl, and R. Mester, "Detecting and suppressing marine snow for underwater visual slam," in *Proceedings of the IEEE/CVF Conference on Computer Vision and Pattern Recognition*, 2022, pp. 5101–5109.

[183] E. Vargas, R. Scona, J. S. Willners, T. Luczynski, Y. Cao, S. Wang, and Y. R. Petillot, "Robust underwater visual slam fusing acoustic sensing," in *IEEE International Conference on Robotics and Automation (ICRA)*. IEEE, 2021, pp. 2140–2146.

[184] R. Mur-Artal and J. D. Tardós, "Orb-slam2: An open-source slam system for monocular, stereo, and RGB-D cameras," *IEEE Transactions on Robotics*, vol. 33, no. 5, pp. 1255–1262, 2017.

[185] J. Folkesson, J. Leonard, J. Leederkerken, and R. Williams, "Feature tracking for underwater navigation using sonar," in *IEEE/RSJ International Conference on Intelligent Robots and Systems (IROS)*, 2007.

[186] D. Ribas, P. Ridao, and J. Neira, *Underwater SLAM for Structured Environments Using an Imaging Sonar*. Springer, 2010, vol. 65.

[187] A. Mallios, P. Ridao, D. Ribas, M. Carreras, and R. Camilli, "Toward autonomous exploration in confined underwater environments," *Journal of Field Robotics*, vol. 33, no. 7, pp. 994–1012, 2016.

[188] P. Ozog, G. Troni, M. Kaess, R. M. Eustice, and M. Johnson-Roberson, "Building 3D mosaics from an autonomous underwater vehicle, doppler velocity log, and 2D imaging sonar," in *IEEE International Conference on Robotics and Automation (ICRA)*, 2015.

[189] E. Westman, A. Hinduja, and M. Kaess, "Feature-based SLAM for imaging sonar with under-constrained landmarks," in *ICRA*, 2018.

[190] C. Tsiotsios, M. E. Angelopoulou, T.-K. Kim, and A. J. Davison, "Backscatter compensated photometric stereo with 3 sources," in *CVPR*, 2014.

[191] C. Tsiotsios, T. Kim, A. Davison, and S. Narasimhan, "Model effectiveness prediction and system adaptation for photometric stereo in murky water," *Computer Vision and Image Understanding*, vol. 150, pp. 126–138, 2016.

[192] Z. Murez, T. Treibitz, R. Ramamoorthi, and D. Kriegman, "Photometric stereo in a scattering medium," in *Proceedings of the IEEE International Conference on Computer Vision (ICCV)*, 2015.

[193] R. J. Woodham, "Photometric method for determining surface orientation from multiple images," *Optical Engineering*, vol. 19, no. 1, pp. 139 – 144, 1980.

[194] O. Drbohlav and M. Chaniler, "Can two specular pixels calibrate photometric stereo?" in *ICCV*, vol. 2, 2005, pp. 1850–1857.

[195] M. Roznere, P. Mordohai, and A. Quattrini Li, "Towards photometric stereo for non-stationary underwater robots: High-resolution model accuracy vs. real-time application," in *ICRA2022 Robotic Perception and Mapping: Emerging Techniques Workshop*, 2022.

[196] M. Pfingsthorn, A. Birk, and H. Bülow, "An efficient strategy for data exchange in multi-robot mapping under underwater communication constraints," in *2010 IEEE/RSJ International Conference on Intelligent Robots and Systems*. IEEE, 2010, pp. 4886–4893.

[197] L. Paull, G. Huang, M. Seto, and J. J. Leonard, "Communication-constrained multi-AUV cooperative slam," in *2015 IEEE International Conference on Robotics and Automation (ICRA)*. IEEE, 2015, pp. 509–516.

[198] J. McConnell, Y. Huang, P. Szenher, I. Collado-Gonzalez, and B. Englot, "Draco-slam: Distributed robust acoustic communication-efficient slam for imaging sonar equipped underwater robot teams," in *2022 IEEE/RSJ International Conference on Intelligent Robots and Systems (IROS)*. IEEE, 2022, pp. 8457–8464.

[199] N. Weidner, S. Rahman, A. Q. Li, and I. Rekleitis, "Underwater cave mapping using stereo vision," in *2017 IEEE International Conference on Robotics and Automation (ICRA)*. IEEE, 2017, pp. 5709–5715.

[200] J. Sattar, G. Dudek, O. Chiu, I. Rekleitis, P. Giguere, A. Mills, N. Plamondon, C. Prahacs, Y. Girdhar, M. Nahon *et al.*, "Enabling autonomous capabilities in underwater robotics," in *IEEE/RSJ International Conference on Intelligent Robots and Systems (IROS)*. IEEE, 2008, pp. 3628–3634.

[201] M. J. Islam, J. Hong, and J. Sattar, "Person-following by autonomous robots: A categorical overview," *The International Journal of Robotics Research (IJRR)*, vol. 38, no. 14, pp. 1581–1618, 2019.

[202] A. Xu, G. Dudek, and J. Sattar, "A natural gesture interface for operating robotic systems," in *IEEE International Conference on Robotics and Automation (ICRA)*. IEEE, 2008, pp. 3557–3563.

[203] J. Sattar and G. Dudek, "On the performance of color tracking algorithms for underwater robots under varying lighting and visibility," in *IEEE International Conference on Robotics and Automation (ICRA)*. IEEE, 2006, pp. 3550–3555.

[204] J. Sattar and G. Dudek, "Robust Servo-control for underwater robots using banks of visual filters," in *IEEE International Conference on Robotics and Automation (ICRA)*. IEEE, 2009, pp. 3583–3588.

[205] M. J. Islam and J. Sattar, "Mixed-domain biological motion tracking for underwater human-robot interaction," in *IEEE International Conference on Robotics and Automation (ICRA)*. IEEE, 2017, pp. 4457–4464.

[206] K. Koreitem, J. Li, I. Karp, T. Manderson, F. Shkurti, and G. Dudek, "Synthetically trained 3D visual tracker of underwater vehicles," in *OCEANS 2018 MTS/IEEE Charleston*. IEEE, 2018, pp. 1–7.

[207] A. G. Howard, M. Zhu, B. Chen, D. Kalenichenko, W. Wang, T. Weyand, M. Andreetto, and H. Adam, "MobileNets: Efficient convolutional neural networks for mobile vision applications," *arXiv preprint arXiv:1704.04861*, 2017.

[208] J. Redmon and A. Farhadi, "YOLO9000: Better, faster, stronger," *arXiv preprint arXiv:1612.08242*, 2016.

[209] T. Agarwal, M. Fulton, and J. Sattar, "Predicting the future motion of divers for enhanced underwater human-robot collaboration," in *2021 IEEE/RSJ International Conference on Intelligent Robots and Systems (IROS)*. IEEE, 2021, pp. 5379–5386.

[210] M. Fiala, "ARTag, a fiducial marker system using digital techniques," in *2005 IEEE Computer Society Conference on Computer Vision and Pattern Recognition (CVPR)*, vol. 2. IEEE, 2005, pp. 590–596.

[211] E. Olson, "AprilTag: A robust and flexible visual fiducial system," in *Robotics and Automation (ICRA), 2011 IEEE International Conference on*. IEEE, 2011, pp. 3400–3407.

[212] J. Sattar, E. Bourque, P. Giguere, and G. Dudek, "Fourier Tags: Smoothly degradable fiducial markers for use in human-robot interaction," in *Canadian Conference on Computer and Robot Vision (CRV)*. IEEE, 2007, pp. 165–174.

[213] G. Dudek, M. Jenkin, C. Prahacs, A. Hogue, J. Sattar, P. Giguere, A. German, H. Liu, S. Saunderson, A. Ripsman *et al.*, "A visually guided swimming robot," in *IEEE/RSJ International Conference on Intelligent Robots and Systems (IROS)*. IEEE, 2005, pp. 3604–3609.

[214] K. Koreitem, J. Li, I. Karp, T. Manderson, and G. Dudek, "Underwater communication using full-body gestures and optimal variable-length prefix codes," in *2019 International Conference on Robotics and Automation (ICRA)*. IEEE, 2019, pp. 8018–8025.

[215] M. Fulton, C. Edge, and J. Sattar, "Robot communication via motion: Closing the underwater human-robot interaction loop," in *2019 International Conference on Robotics and Automation (ICRA)*. IEEE, 2019, pp. 4660–4666.

[216] X. Wu, R. E. Stuck, I. Rekleitis, and J. M. Beer, "Towards a framework for human factors in underwater robotics," in *Human Factors and Ergonomics Society International Annual Meeting*, vol. 59, Los Angeles, CA, USA, Sep. 2015, pp. 1115–1119.

[217] A. Geiger, P. Lenz, C. Stiller, and R. Urtasun, "Vision meets robotics: The kitti dataset," *The International Journal of Robotics Research*, vol. 32, no. 11, pp. 1231–1237, 2013.

[218] M. Xanthidis, B. Joshi, J. M. O'Kane, and I. Rekleitis, "Multi-robot exploration of underwater structures," in *IFAC Conference on Control Applications in Marine Systems, Robotics and Vehicles*, 2022, accepted.

[219] B. Joshi, M. Xanthidis, S. Rahman, and I. Rekleitis, "High definition, inexpensive, underwater mapping," in *IEEE International Conference on Robotics and Automation (ICRA)*, Philadelphia, PA, USA, 2022, pp. 1113–1121.

[220] NOAA, "The National Coral Reef Monitoring Program (NCRMP)," https://www.coris.noaa.gov/monitoring/, 200, accessed: 2-2-2021.

[221] O. Hoegh-Guldberg, P. J. Mumby, A. J. Hooten, R. S. Steneck, P. Greenfield, E. Gomez *et al.*, "Coral reefs under rapid climate change and ocean acidification," *Science*, vol. 318, no. 5857, pp. 1737–1742, 2007.

[222] A. E. Windle, D. S. Hooley, and D. W. Johnston, "Robotic vehicles enable high-resolution light pollution sampling of sea turtle nesting beaches," *Frontiers in Marine Science*, vol. 5, p. 493, 2018.

[223] P. L. Negre, F. Bonin-Font, and G. Oliver, "Cluster-based loop closing detection for underwater slam in feature-poor regions," in *2016 IEEE International Conference on Robotics and Automation (ICRA)*. IEEE, 2016, pp. 2589–2595.

[224] A. Vasilijevic, N. Miskovic, Z. Vukic, and F. Mandic, "Monitoring of seagrass by lightweight AUV: A posidonia oceanica case study surrounding murter island of croatia," in *22nd Mediterranean Conference on Control and Automation*. IEEE, 2014, pp. 758–763.

[225] D. O. Obura, G. Aeby, N. Amornthammarong, W. Appeltans, N. Bax, J. Bishop, R. E. Brainard, S. Chan, P. Fletcher, T. A. Gordon *et al.*, "Coral reef monitoring, reef assessment technologies, and ecosystem-based management," *Frontiers in Marine Science*, vol. 6, p. 580, 2019.

[226] S. Mccammon, T. A. Mooney, and Y. Girdhar, "Adaptive online sampling of periodic processes with application to coral reef acoustic abundance monitoring," in *Proceedings of the IEEE/RSJ International Conference on Intelligent Robots and Systems (IROS)*, 2022, p. 8.

[227] D. R. Yoerger, A. F. Govindarajan, J. C. Howland, J. K. Llopiz, P. H. Wiebe, M. Curran, J. Fujii, D. Gomez-Ibanez, K. Katija, B. H. Robison, B. W. Hobson, M. Risi, and S. M. Rock, "A hybrid underwater robot for multidisciplinary investigation of the ocean twilight zone," *Science Robotics*, vol. 6, 6 2021.

[228] K. Katija, P. L. D. Roberts, J. Daniels, A. Lapides, K. Barnard, M. Risi, B. Y. Ranaan, B. G. Woodward, and J. Takahashi, "Visual tracking of deepwater animals using machine learning-controlled robotic underwater vehicles," 2021.

[229] L. Cai, N. McGuire, R. Hanlon, T. A. Mooney, and Y. Girdhar, "Semi-supervised visual tracking of marine animals using autonomous underwater vehicles," *International Journal of Computer Vision (IJCV)*, 2023.

[230] Y. Girdhar, N. McGuire, L. Cai, S. Jamieson, S. McCammon, B. Claus, J. E. S. Soucie, J. E. Todd, and T. A. Mooney, "CUREE: A curious underwater robot for ecosystem exploration," in *Proceedings of the IEEE International Conference on Robotics and Automation*, 2023.

[231] M. Ay, D. Korkmaz, G. Ozmen Koca, C. Bal, Z. H. Akpolat, and M. C. Bingol, "Mechatronic design and manufacturing of the intelligent robotic fish for bio-inspired swimming modes," *Electronics*, vol. 7, no. 7, p. 118, 2018.

[232] R. K. Katzschmann, A. De Maille, D. L. Dorhout, and D. Rus, "Cyclic hydraulic actuation for soft robotic devices," in *2016 IEEE/RSJ International Conference on Intelligent Robots and Systems (IROS)*. IEEE, 2016, pp. 3048–3055.

[233] R. Williamson, "Mit Sofi: A Study in Fabrication, Target Tracking, and Control of Soft Robotic Fish," Ph.D. dissertation, Massachusetts Institute of Technology, 2022.

[234] K. Johansen, "Taking subsea operations to the next level by applying machine vision to perform autonomous inspections," in *Offshore Technology Conference*. OnePetro, 2022.

[235] M. Jacobi and D. Karimanzira, "Underwater pipeline and cable inspection using autonomous underwater vehicles," in *2013 MTS/IEEE OCEANS-Bergen*. IEEE, 2013, pp. 1–6.

[236] K. M. Awan, P. A. Shah, K. Iqbal, S. Gillani, W. Ahmad, and Y. Nam, "Underwater wireless sensor networks: A review of recent issues and

challenges," *Wireless Communications and Mobile Computing*, vol. 2019, 2019.

[237] H. Zhang, S. Zhang, Y. Wang, Y. Liu, Y. Yang, T. Zhou, and H. Bian, "Subsea pipeline leak inspection by autonomous underwater vehicle," *Applied Ocean Research*, vol. 107, p. 102321, 2021.

[238] Z. Huang, L. Wan, M. Sheng, J. Zou, and J. Song, "An underwater image enhancement method for simultaneous localization and mapping of autonomous underwater vehicle," in *2019 3rd International Conference on Robotics and Automation Sciences (ICRAS)*, 2019, pp. 137–142.

[239] J. L. Schönberger and J.-M. Frahm, "Structure-from-motion revisited," in *Conference on Computer Vision and Pattern Recognition (CVPR)*, 2016.

[240] Mapillary, "Opensfm: Open-source structure from motion library," https://github.com/mapillary/OpenSfM, 2020, accessed: 10-10-2022.

[241] M. Xanthidis, B. Joshi, M. Roznere, W. Wang, N. Burgdorfer, A. Quattrini Li, P. Mordohai, S. Nelakuditi, and I. Rekleitis, "Mapping of underwater structures by a team of autonomous underwater vehicles," in *International Symposium of Robotics Research*, 2022, accepted.

[242] M. Reggiannini and O. Salvetti, "Seafloor analysis and understanding for underwater archeology," *Journal of Cultural Heritage*, vol. 24, pp. 147–156, 2017.

[243] G. A. Hatcher, J. A. Warrick, A. C. Ritchie, E. T. Dailey, D. G. Zawada, C. Kranenburg, and K. K. Yates, "Accurate bathymetric maps from underwater digital imagery without ground control," *Frontiers in Marine Science*, vol. 7, p. 525, 2020.

[244] R. Chen, Z. Cai, and W. Cao, "Mffn: An underwater sensing scene image enhancement method based on multiscale feature fusion network," *IEEE Transactions on Geoscience and Remote Sensing*, vol. 60, no. 1, pp. 1–12, 2022.

[245] A. Jordt, "Underwater 3D reconstruction based on physical models for refraction and underwater light propagation," Ph.D. dissertation, 2014.

[246] N. Weidner, "Underwater Cave Mapping and Reconstruction Using Stereo Vision," Master's thesis, Computer Science and Engineering Department, University of South Carolina, Columbia, SC, 2017.

[247] S. Rahman, "A Multi-Sensor Fusin Based Underwater SLAM System," Ph.D. dissertation, Computer Science and Engineering Department, University of South Carolina, Columbia, SC, 2020.

[248] A. G. Tacon, "Trends in global aquaculture and aquafeed production: 2000–2017," *Reviews in Fisheries Science & Aquaculture*, vol. 28, no. 1, pp. 43–56, 2020.

[249] E. Kelasidi, B. Su, W. Caharija, M. Føre, M. Pedersen, and K. Frank, "Autonomous monitoring and inspection operations with UUVS in fish farms," *IFAC-PapersOnLine*, vol. 55, no. 31, pp. 401–408, 2022.

[250] X. Lin, N. Jha, M. Joshi, N. Karapetyan, Y. Aloimonos, and M. Yu, "Oystersim: Underwater simulation for enhancing oyster reef monitoring," in *OCEANS 2022, Hampton Roads*. IEEE, 2022, pp. 1–6.

[251] M. Sun, X. Yang, and Y. Xie, "Deep learning in aquaculture: A review," *Journal of Computers*, vol. 31, no. 1, pp. 294–319, 2020.

[252] D. Li, J. Bao *et al.*, "Research progress on key technologies of underwater operation robot for aquaculture." *Transactions of the Chinese Society of Agricultural Engineering*, vol. 34, no. 16, pp. 1–9, 2018.

[253] G. Dini, E. Princi, S. Gamberini, and L. Gamberini, "Nemo's garden: Growing plants underwater," in *OCEANS 2016 MTS/IEEE Monterey*. IEEE, 2016, pp. 1–6.

[254] I. S. Inc., "Aquaculture intelligence: Advanced aquaculture monitoring tools for real-time visibility in fish farms," https://www.innovasea.com/aquaculture-intelligence/, 2020, accessed: 12-12-2022.

[255] Y. Girdhar, "Aquaculture monitoring system and method," Patent, 2022.

[256] C. Schellewald, A. Stahl, and E. Kelasidi, "Vision-based pose estimation for autonomous operations in aquacultural fish farms," *IFAC-PapersOnLine*, vol. 54, no. 16, pp. 438–443, 2021.

10

Applications of Computer Vision in Entertainment and Media Industry

Mahmudul Hasan, Kishan Shamsundar Athrey, Arfeen Khalid, Danfeng Xie, Ehsan Younessian, and Tony Braskich

10.1 Introduction

Media and[1] entertainment have a profound impact on human lives. From providing news and information to entertaining and engaging audiences, the industry has played a significant role in shaping culture and society. The entertainment industry has come a long way since the first commercial movie was screened for an audience at a theater in Paris in 1895 [1] and the first television show was broadcasted by British Broadcasting Corporation (BBC) in 1926 [2]. Today, the entertainment industry is not limited to only film and television. Its many other segments come in different forms and mediums, such as news and print media, streaming platforms, sports broadcasting, gaming, social media, theme parks, etc. [3] as graphically depicted in Figure 10.1.

As the lives of modern people get more manageable with the advancements in technology, they have more free time to seek entertainment in their daily lives. People in the USA spend on average eight hours daily on entertainment and media content [4]. Due to technological advances, creating and distributing media content has become easy. People are spending time and money on entertainment more than ever. With the support of high-speed network availability and easy-to-access media platforms, users are accessing unlimited entertainment on the go. This has led to the rapid growth of the industry worldwide. According to market research reports, global revenues of the media and entertainment industry were $2 trillion (USA) in 2020 and are expected to reach nearly $2.6 trillion by 2024. This rapid growth is primarily because people switch from traditional media channels like cable and radio to digital or mobile streaming applications [5].

The competition is high across media and entertainment companies due to rapid growth and demand for content. Content providers and distribution companies are facing increasing pressure to deliver high-quality media content and entertainment services; otherwise, they cannot attract and retain customers to drive business value. These changing landscapes are leading companies to resort to artificial intelligence (AI) and computer vision (CV) technology-based solutions. AI/CV in the media and entertainment industry enables companies to build digital solutions that help service providers deliver quality services and enhance their customer experiences. Many professionals in the entertainment industry are adopting AI and unlocking new digital approaches to create and deliver high-quality content to customers.

AI comprises many exciting and emerging fields, among which CV is the most challenging and pioneering. CV deals with computational methods to help computers understand and interpret the content of digital images and videos [6]. It enables computer systems to automatically see, identify, and understand the visual world, simulating human vision using computational methods. Artificial vision remains one of the most challenging fields, especially due to the enormous complexity of the high-dimensional physical world. Human vision is based on a lifetime of learning with context to identify specific scenes, objects, faces, and beyond. Similarly, modern artificial vision technology uses machine learning and deep learning methods to train computers on how to recognize specific patterns such as objects, faces, or people in visual scenes. CV tasks include methods for acquiring, processing, analyzing, and understanding digital images, and extraction of high-dimensional data from the real world to produce numerical or symbolic information so that downstream applications easily use them for decision-making.

Content creators and publishers have started to take advantage of recent advancements in CV across many sectors [7]. Newly invented highly accurate CV and machine learning models created enormous opportunities for creating high-quality content and rapid distribution. In this chapter, we aim to provide a comprehensive overview of different CV-based applications in the entertainment industry with respect to recent advances. In Figure 10.2,

DOI: 10.1201/9781003328957-10

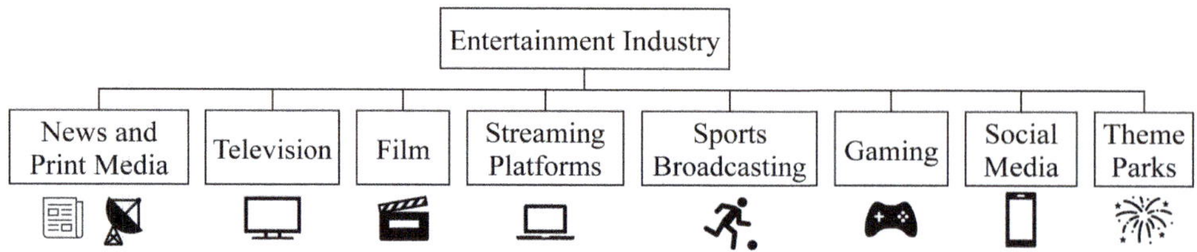

FIGURE 10.1
A comprehensive breakdown of the entertainment industry into different segments.

we provide a rough enumeration of CV methods that are relevant to the entertainment industry. For the sake of later discussion, we will briefly discuss the following subfields of CV – recognition, motion analysis, scene understanding, generative methods, and image restoration with example applications from the entertainment industry.

The CV problems belonging to the recognition category identify whether visual data contains some specific objects, patterns, features, or activities. The best algorithm for such tasks is based on either convolutional neural networks (CNNs) [8] or image transformers [9]. Illustrations of their capabilities can be seen in some CV challenges such as ImageNet Large Scale Visual Recognition [10], Common Objects in Context (COCO) [11], etc. These are examples of benchmark datasets for image classification and object detection tasks with millions of images. We briefly discuss some recognition tasks below.

- **Image classification:** A model sees an image or short video as the input and can predict whether this input belongs to certain classes with some level of confidence [12]. For example, a social media company may want to categorize an image as safe or unsafe for the public.

- **Object detection:** This task identifies and localizes objects belonging to some classes in an image or video [13]. For example, a media company may want to detect and blur sensitive content such as blood and gory scenes.

- **Instance classification:** An individual instance of an object is recognized in some medium. Examples include identification of a specific person's face or fingerprint, identification of a specific vehicle, or identification of a specific product [14].

- **Content retrieval:** Retrieving all images or videos in a larger database that have certain characteristics [15]. The content can be specified in different ways, for example, in terms of similarity relative to a target image by utilizing reverse image search techniques, or in terms of high-level search criteria given as text input. For example, a news broadcasting company may want to find all similar past instances of a particular news story.

- **Pose estimation:** Estimating the position or orientation of a specific object relative to the camera [16]. In sports analysis videos, coaches may want to

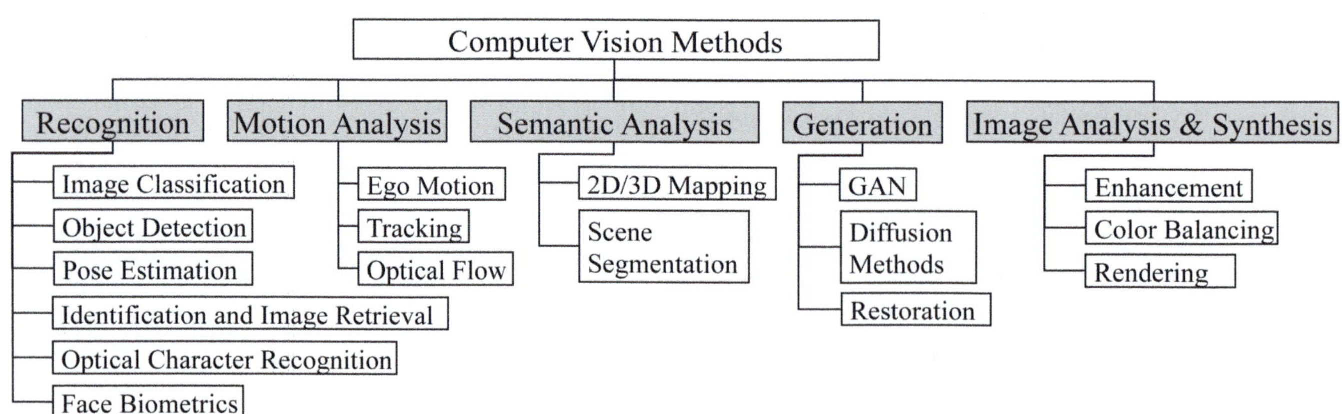

FIGURE 10.2
A basic list of subfields in computer vision that are applicable to the entertainment and media sectors. This list is not comprehensive as our emphasis is specifically on the entertainment industry.

analyze the pose of specific players during games for future improvement plans.

- **Optical character recognition (OCR):** Identifying either handwritten or printed characters and graphics in images [17, 18]. Text extracted this way from the images and video in the wild are used as important metadata, which further fuels other applications such as sports score tracking.

- **Facial biometrics:** Identifying faces, recognizing celebrities, recognizing emotion, estimating age, recognizing gender, and similar tasks based on facial features [19].

Motion estimation deals with a set of problems where the relative position, path, or velocity of a set of points or objects are determined from a sequence of images in either 2D or 3D scenes. Some motion estimation tasks are discussed as follows.

- **Egomotion:** Determining the 2D or 3D rigid motion such as rotation and translation of a system with respect to an environment [20]. A sequence of images produced by the system's own camera is used to determine the motion. It has uses in virtual reality (VR) to understand the position in an environment.

- **Tracking:** Following the movements of a smaller set of interest points or objects, for example, vehicles, objects, humans, or other organisms in an image sequence. For example, in sports video analysis, we may want to track the position of the ball and players in the field [21].

- **Optical flow:** Determining, for each point in the image, how that point is moving relative to the image plane. This motion is a result both of how the corresponding point is moving in the scene and how the camera is moving relative to the scene [22].

Scene understanding attempts to analyze objects in context with respect to the 3D structure of the scene, its layout, and the spatial, functional, and semantic relationships between objects.

- **Scene reconstruction:** Given one or more images of a scene, or a video, scene reconstruction aims at computing a 3D model of the scene. In the simplest case, the model can be a set of 3D points [23]. It has applications in game development.

- **Scene segmentation:** This task labels each pixel or contour with specific scene categories [24]. For example, it can be used to extract important metadata for media content understanding and retrieval.

Generative methods are the most advanced set of algorithms that generate new data or instances with or without any prompt. Vision-based generative models take an image, sketch, random number, or text as the input and generate high-quality pictures and short videos.

- **Generative adversarial networks (GANs)** [25, 26] train a generative model by framing the problem as a supervised learning problem with two subtasks – the generator that generates new images, and the discriminator that tries to classify images as either real or fake. The two models are trained together in an adversarial manner, until the discriminator model is fooled about half the time, meaning the generator model is generating plausible examples.

- **Stable diffusion** [27] is a state-of-the-art text-to-image generation algorithm that uses a process called diffusion to generate images. Diffusion works by training an artificial neural network to reverse a process of adding random noise to an image. Once trained, the neural network can take an image made up of random pixels and turn it into an image that matches the text prompt.

Image restoration and enhancement [28–30] aims to remove noise such as sensor noise or motion blur from images. The simplest possible approach for noise removal is various types of filters such as low-pass filters or median filters. More sophisticated methods assume a model of how the local image structures look to distinguish them from noise. Nowadays, various deep learning-based approaches are being used for image restoration and enhancement. The film industry has greatly benefited from such image analysis to produce high-quality movies.

Rendering is a significant CV application that greatly influences the movie and video gaming sectors. It involves the generation of a final image or a series of images from a three-dimensional (3D) scene. This conversion process transforms a virtual scene, comprising models, textures, lighting, and camera angles, into two-dimensional (2D) images suitable for display on a screen or for use in a video. In the movie industry, rendering is used extensively in computer-generated imagery (CGI) to create realistic or fantastical visual effects, animated characters, and virtual environments. It allows filmmakers to bring their imaginative visions to life and create immersive worlds that enhance storytelling.

Multimedia content providers in the entertainment industry have been adopting CV in many different applications. We arranged these applications into multiple segments based on their similarities as shown in Figure 10.3. However, there might be overlaps between these applications in terms of problem definition and the nature of the solution. Some of these segments fall into a broader

Applications of Computer Vision in the Entertainment Industry

<table>
<tr><td>Content Generation</td><td>Content Moderation</td><td>Sports</td><td>Gaming and AR/VR/MR</td></tr>
<tr><td>
o Image/art generation
o Video generation
o Preview/thumbnail creation
o Rotoscoping and Rotomation
o Computer generated imagery
o Film editing
</td><td>
o Sensitive content detection
o Adult content detection
o Deepfake detection
</td><td>
o Performance analysis
o Pose analysis
o Risk Assessment
o Graphical enhancement
o Event detection
</td><td>
o Environment design
o Character design
o Motion and tracking
o Lighting and shading
o AR/VR/MR
o Rendering
</td></tr>
</table>

<table>
<tr><td>Experience Enhancement</td><td>Streaming</td><td>Information Retrieval</td><td>Advertisement</td></tr>
<tr><td>
o Intro/recap detection
o Commercial detection
o Duplicate content detection
o Audience analysis
o Automatic sign language
o Descriptive audio
</td><td>
o Quality management

Recommendation
o Visual recommendation
o Improved personalization
</td><td>
o On-screen information
o Content recognition
o Product recognition
o Metadata curation and indexing
</td><td>
o Contextual advertisement
o Brand alignment and safety
o Dynamic product placement
</td></tr>
</table>

FIGURE 10.3

A list of applications of CV in entertainment and media. The top row corresponds to applications related to content generation and the bottom is for the experience.

category of content creation such as content generation, content moderation, sports, gaming, and AR/VR. The rest of the segments fall into the experience category such as information retrieval, advertisement, experience enhancement, streaming, and recommendation. We will briefly introduce these segments here and leave the detail in their respective dedicated sections.

Content creation is the foundation of the entertainment industry and is being greatly benefited by CV these days [31]. Multimedia companies are using advanced CV tools to enhance and produce visual content automatically. Recently developed generative deep learning models have the ability to generate high-quality images and art, as well as short videos. Various media companies and streaming platforms are also utilizing CV to automatically create personalized previews and thumbnails that can attract and retain the interest of their customer. During the post-production process of film editing, CV has been particularly beneficial in rotoscoping, rotomation, and model-based animation, as opposed to CGI. Additionally, CV techniques are being used for media and news summarization and the creation of short-form content. In Section 10.2, we will delve into content-generation applications in more detail.

In the current era of streaming and social media, where individuals can create and distribute content for public consumption, content moderation and compliance are widely debated issues. This rapid growth of content has introduced serious challenges such as fake and tempered content to social networks and media companies. To address this issue and ensure that content is safe for a wide range of audiences, CV-based methods are being utilized to rapidly and cost-effectively detect sensitive, adult, and deepfake content. This automated process is highly efficient and saves a significant amount of manual effort. For further details on content moderation, please refer to Section 10.3.

The large amount of available content has made it challenging for consumers to find their desired content. To address this issue, companies are utilizing various types of recommendation engines to present relevant content to users. By providing personalized content, companies not only attract new users but also retain existing ones. In the past, recommendation engines relied on textual tags, genres, and user data for recommendations. However, CV-based recommendation systems use more detailed visual metadata, such as actions, places, and objects, which greatly enhances personalization and leads to greater customer satisfaction. We will discuss more on CV-based recommendation in Section 10.4.

The entertainment industry is generating an unprecedented volume of content every day such as news, films, books, music and songs, advertisements, social media posts, reviews, and user-generated videos. For example, in YouTube alone, stats [32] show that one hour of video is uploaded to YouTube every second. This creates considerable challenges when it comes to efficient content labeling, search, and retrieval processes, especially in the case of video and audio. CV is used for generating useful metadata for content indexing, understanding, and improved visual search. It also helps to provide on-screen information, content recognition, and product recognition. In Section 10.5, we will take a closer look at various CV-based information retrieval applications.

Advertisement placement research has gained significant attention in recent years due to the importance of content monetization. Traditional advertisement placement lacks context and, as a result, has limited reach. In contrast, contextual advertisement enables ad placement based on the semantics of content and environment, which allows for deeper consumer engagement and increased revenue. CV tools and techniques have proven to be valuable resources for extracting important context information that can be utilized for advertisement placement. Companies are using CV for brand alignment to deliver brand promises and avoid disappointing customers. Additionally, CV is being used for dynamic product placement, where brands or ads are placed dynamically inside TV shows and movies based on the context of the scene. We will talk more about these topics in Section 10.6.

Entertainment companies are facing stiff competition and are therefore investing in research to improve the content consumption experience for both new and existing customers. To achieve this, they are utilizing AI and CV tools to enhance features like Intro/Recap detection, Credit detection for binge-watching, and commercial advertisement detection for digital video recorder (DVR) systems. Additionally, streaming platforms are employing duplicate content detection techniques to keep their databases noise-free. Social media companies are using augmented reality (AR) and CV to enhance content and make photos and videos more engaging. Accessibility tools such as audience analysis, automatic sign language, and descriptive audio generation are being developed using CV. Section 10.7 will provide more information on various CV-based experience enhancement applications.

Streaming companies are increasingly concerned with how their media content is viewed and perceived by their audiences, but manual surveillance of this content is not feasible due to its sheer volume. To address this issue, streaming companies are utilizing CV-based techniques for streaming quality management. We will provide a detailed analysis of streaming quality in Section 10.8.

CV plays a significant role in game development, where it is utilized to automate several time-consuming tasks, including 3D mapping, character design, and environment design. The latest trend in this field is augmented and virtual reality (VR), and CV is also employed in developing AR/VR/MR solutions such as hand/body tracking, motion capture, 3D mapping, and emotion recognition. These applications and how CV is revolutionizing this sector will be explored in depth in Section 10.9.

The sports broadcasting industry is huge, with billions of viewers tuning in to watch their favorite teams play. Automation is being used to enhance the viewing experience and create sports content more efficiently. AI and CV tools are being utilized for tasks such as event detection, automatic highlight generation, ball tracking, goal-line technology, and scoring. These applications not only improve the viewing experience but also promote fairness in sports. In Section 10.10, we will provide a more detailed analysis of CV applications in the sports industry.

10.2 Content Generation

10.2.1 Image Generation

Current state-of-the-art content generation methods utilize deep generative models (DGM) for high-quality images. It had broad implications in the media industry as it can generate valuable image or video art efficiently while reducing costs. Applications such as StyleGAN [33], VQGAN[34], DALLE-2 [35], Imagen[36], and Stable Diffusion [37] have drawn great attention from the public. Current widely used image generation models can be categorized into two categories: GAN [25, 26] and diffusion models [38, 39]. GANs have more mature technology and have been widely adopted in the media industry, while diffusion models have gained popularity in very recent years but with very fast development and promising results. In this section, we provide an overview of GAN-based and diffusion-based image generation methods.

1. **Generative adversarial network:** GAN is a framework that uses deep learning to estimate generative models via an adversarial process, as shown in Figure 10.4. In this framework, two models are trained simultaneously – a generative model G that models the data distribution and a discriminative model D that estimates the probability of whether a sample came from the generator G or the training data. The training process for G is to maximize the probability to fool discriminator D. This is a zero-sum game and can be formulated as a two-player mini-max problem:

$$\min_{G} \max_{D} V(D, G) = E_{x \sim p_{data}(x)}[log D(x)]$$
$$+ E_{x \sim p_z(z)}[log(1 - D(G(z)))]$$
$$(10.1)$$

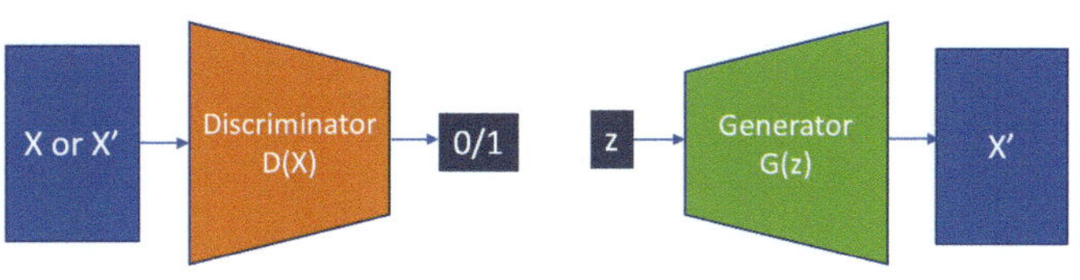

FIGURE 10.4
A vanilla architecture of a generative adversarial network.

where $p_{data}(x)$ is the data distribution, $p_z(z)$ is the distribution of a noise variable. By optimizing this formula, GAN is able to achieve a unique solution, with G successful model the training distribution, and D cannot distinguish whether the data is from G or training data. Detailed description and analysis of GANs can be found in [26, 40, 41].

GANs achieved great performance on many image generation tasks [33, 34, 42, 43] in terms of FID [44], Inception Score [45] and Precision [46] metrics. They have been widely used in the marketing and media industry. Rosebud AI[2] has used GAN to generate custom images of fashion models who do not exist. Generated Media Inc.[3] uses StyleGAN [33] to generate synthetic human faces of different ages, gender, and ethnicity. Reallusion Inc.[4] uses GAN to generate faces for 3D graphic models for video games or animation. Mobile applications such as FaceApp[5] and ZAO[6] use GAN to allow users to edit a person's facial appearance or even swap faces. Furthermore, start-ups such as DataGen use GAN to generate synthetic datasets for training.

However, GAN suffers from some drawbacks. One noticeable drawback is that GANs may generate less diverse images [47, 48] and may suffer from mode collapse compared to other image generation methods [33, 49, 50]. The other drawback is that GANs are often hard to train, which makes it difficult to scale up and transfer to new domains [48, 51].

2. **Diffusion models:** Diffusion models [35–39] have recently drawn great interest as they achieve state-of-the-art performance in many applications, including text-to-image generation, text-to-video generation, super-resolution, and image synthesis. They have a relatively simple training scheme so that the training of very large models at scale becomes possible. Applications such as DALL-E 2 [35], Stable Diffusion [37], and Imagen [36] have drawn broad interest from the public. Companies such as Google, Meta, OpenAI, and Stability AI [35–37, 52] are using diffusion models for their content-creating applications.

Diffusion models come from the idea of the diffusion process as shown in Figure 10.5 - by adding random noises to an image in many steps, the image gradually becomes pure noise. By training a diffusion model to reverse the diffusion process, that is, removing noises gradually in many steps, we are able to generate a clean image from pure noise.

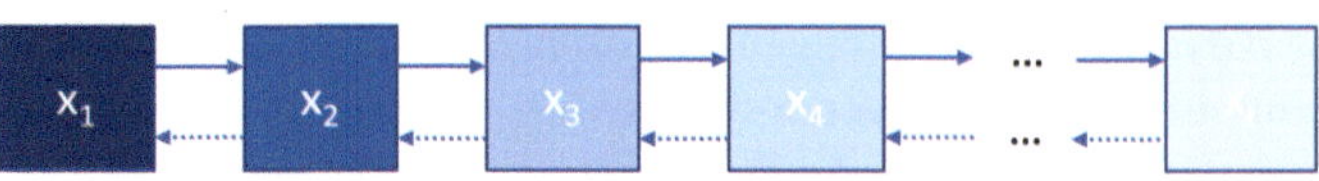

FIGURE 10.5
Simplified architecture of a diffusion model.

The forward diffusion process can be defined as

$$q(x_t|x_{t-1}) = \mathcal{N}(x_t|\sqrt{1 - \beta_t}x_{t-1}, \beta_t I) \qquad (10.2)$$

where x_t and x_{t-1} are images at steps t and $t - 1$, respectively. $\beta_t \in (0, 1)$ is the strength of noises in step t and $\mathcal{N}(\cdot)$ is a Gaussian distribution. We can regard the forward diffusion process as the encoding process.

The diffusion model is trained to learn the reversal of the forward diffusion process, which can be called the decoding process. Usually, a neural network is used to approximate the decoding process, which can be defined as:

$$p(x_{t-1}|x_t) = \mathcal{N}(x_{t-1}|\mu_\theta(x_t, t), \Sigma_\theta(x_t, t)) \qquad (10.3)$$

By learning the reversal of the diffusion process, diffusion models are able to generate diverse and high-quality images given pure noise, which is called unconditional image generation. By pairing class c with input x, the diffusion model can be trained using the pair (x, c) for conditional image generation. Conditional image generation has drawn more interest from the public as it has powerful functionalities such as text-to-image generation, image inpainting, super-resolution, etc. Conditional image generation applications such as DALL-E 2, Stable Diffusion, and Imagen have gained great popularity and impact on the media industry.

DALL-E 2 [35] is developed by OpenAI using CLIP [53] as a prior embedding encoder. CLIP has shown its ability to learn robust image representations containing semantics and style. DALL-E 2 leveraged these representations for text-to-image generation by using two-stage diffusion models. First, given a text prompt y, the CLIP model is able to encode y into image embedding z_i using a diffusion model or an auto-regression (AR) model. Then, another diffusion model is used as a decoder $p(x|z_i)$ to produce images x conditioned on input embedding z_i.

Imagen [36] is a text-to-image generation model developed by Google. Instead of using the CLIP model that is trained on image-text pair, Imagen uses a pre-trained transformer-based large language model as a text encoder [54], which is only trained on text data. It also proposed dynamic thresholding and efficient U-Net for generating more photorealistic and detailed images. Imagen achieves state-of-the-art in text-to-image tasks in terms of COCO FID [44] and DrawBench [36] metrics.

Stable diffusion [37] using latent diffusion model to conduct image generation. Images are first encoded into latent space using pre-trained autoencoders [34]. Then, the diffusion process and the training happen in the latent space. It also applied a cross-attention mechanism into the U-Net [55] backbone model for more flexible conditional image generation. By using latent diffusion and cross-attention mechanism, stable diffusion achieves highly competitive performance on various tasks including text-to-image generation, unconditional image synthesis, and super-resolution, while using significantly fewer computational resources compared to pixel-based diffusion models. Stable diffusion is widely used as it is relatively lightweight and is open-sourced.

In recent years, the use of text-to-image technologies in the media business has shown tremendous promise. These AI-powered platforms have the potential to drastically change how media material is created and consumed, opening up an abundance of new opportunities for content production, automation, and personalization [56]. One important application of text-to-image technology in the media sector is to create storyboards for film and television, which significantly reduces the need for manual artwork as well as speeds up the pre-production process. Furthermore, based on specific textual prompts, these technologies can enable the generation of personalized visual content for digital marketing campaigns, providing a high degree of personalization and automation in content creation [57].

Text-to-image technologies have also found compelling use cases within the field of news and journalism. These AI systems can be harnessed to transform news articles into visualizations and infographics, a feature that could potentially boost reader involvement and understanding. Additionally, in instances where there is a lack of suitable photographic content or the available imagery is inadequate, these technologies can step in to generate images, thus widening the scope for visual narratives in journalism [58].

Furthermore, text-to-image technologies are also used in AI profile picture generators[59] to produce realistic and unique profile pictures. These technologies employ GANs or diffusion models trained on large datasets of human faces to generate original, life-like images. In particular, these models understand the nuanced details that contribute to human facial features, including facial structure, skin texture, hair types, and even lighting conditions. As a result, the generated profile pictures are often indistinguishable from real human photographs. These AI-generated images can be useful in a variety of contexts, including anonymous user profiles, avatars for gaming or VR, or placeholders in web and interface design.

Another noteworthy application of AI in the field of visual design is AI movie poster generators. These technologies are capable of generating unique and eye-catching movie posters based on a variety of inputs, such as movie titles, genres, plot summaries, or even specific themes or moods. Trained on extensive databases of existing movie posters, these models understand the visual elements and design principles that characterize effective and compelling movie posters. Consequently, they can produce custom movie posters that not only capture the essence of a film but also adhere to industry standards and trends. The potential applications of this technology extend beyond just movie posters, with possibilities in promotional materials, book covers, album art, and more [60].

10.2.2 Video Generation

Early works on video generation [61–64] are majorly concentrated on generating simple videos such as human actions and moving digits. Sync-DRAW [61] first proposed text-to-video generation using a variational autoencoder with recurrent attention. Pan et al. [62] extended GAN-based image generation to text-to-video generation.

Currently, text-to-video generation models [65, 66] achieved great improvement due to the advancement of large pre-trained language models [53, 54] and diffusion models [36]. Autoregressive transformers [67–70] have shown their enhancement in video generation. GODIVA [69] and NÜWA [70] are able to generate more diverse videos given text prompts. However, their methods may suffer from inconsistency between frames as their video frames are simply generated in a simple chronological order without strong temporal alignment.

Diffusion-based text-to-video generation model capable of generating high-quality and diverse video with a high degree of control. Diffusion-based text-to-video generation model includes Meta's Make-A-Video [65] and Google's Imagen Video [35, 66]. Ho et al. [71] first proposed a video diffusion model that uses a 3D U-Net as the backbone of the diffusion model and was trained directly on text-video pairs for video generation. Meta's Make-A-Video proposed an efficient way for video generation – instead of training on text-video pairs, Meta's model proposed a spatial-temporal pipeline to generate videos with high resolution and high frame rate. This pipeline first generates keyframes using the diffusion model given text prompts. Then, multiple spatial-temporal super-resolution models cascaded together to interpolate frames between keyframes as well as improve spatial resolution. Google's Imagen Video model [66] designed a pipeline that cascades seven video diffusion models. The pipeline contains one base video diffusion model, three spatial super-resolution diffusion models, and three temporal super-resolution diffusion models. These models can be trained in parallel thus reducing training time.

Text-to-video technology's growing capabilities are enabling an upsurge of creative applications in the media industry. Music video creation is one notable application. These technologies can be used to generate one-of-a-kind visual tales that correspond to song lyrics, paving the way for more personalized and innovative music video material [72].

Text-to-video technologies are proven to be useful instruments for video editing in post-production. Editors can supply written descriptions of the necessary alterations, and the technology can embed these changes into the video content, speeding the editing process and perhaps lowering the time and effort required [73, 74].

Furthermore, text-to-video technologies have the potential to be very useful in education and corporate training. They can be used to create interesting educational movies or customized training materials from textual descriptions or scripts. This technology has the potential to improve student learning experiences and provide individualized training for employees in a variety of industries [75].

Finally, in the field of interactive storytelling, these technologies are enabling viewers to participate in narratives in novel ways. By providing textual inputs, viewers can adjust the path of a tale, and the technology can generate accompanying video content in real time. This interactive element increases viewer engagement and provides a one-of-a-kind, personalized narrative experience[76].

10.2.3   Computer-Generated Imagery

CGI has transformed the media industry, notably film and television production, by enabling the creation of realistic and fanciful sequences that would otherwise be impossible or prohibitively expensive to capture in real life [77, 78]. With the incorporation of AI, the use of CGI has gotten more advanced, enabling the generation and manipulation of digital information.

The application of AI in CGI has resulted in numerous improvements in a variety of sectors. In the entertainment industry, AI has been used to age performers up or down, allowing films to be completed even after an actor's death. This is becoming increasingly prevalent as the industry uses AI developments to improve the realism of CGI on human faces. The industry has also witnessed the development of "deep fakes" of performers who aren't even physically in a film, as well as the usage of AI for object detection and CV in motion capture. This has taken film and video game production to the next level, allowing the development of AR and VR games and movies with in-game material and NPCs that behave more humanly and responsively than traditional coding [79].

The incorporation of AI into software products such as Adobe Photoshop and their generative AI model Fire-

fly has also transformed the way designers operate [80]. Creators can use generative AI to add, erase, or replace portions of an image and extend the edges of an image at the speed of their imagination. The Eliminate Tool, a Photoshop AI-powered function, can eliminate undesired elements from photographs while keeping the integrity of neighboring objects and creating seamless transitions over complex and varied backdrops.

AI generative technologies in the domain of animation have enabled independent creatives to make high-quality cartoons in less time and at a lesser cost than traditional animation studios. The YouTube show The Corridor Crew[81], for example, used AI approaches to turn live action into animation, resulting in an automated animation process that eliminated the problem of visual noise. As a consequence, a polished animated short film was created for a fraction of the cost and time required for traditional animation[82].

10.2.4   Automated Preview and Thumbnail Generation

Automated preview and thumbnail creation is the process of generating short and meaningful video previews and thumbnail images using CV and machine learning techniques. These previews and thumbnails can provide users with a quick and personalized representation of the content in a video, which can help increase engagement and drive more views. Automated preview and thumbnail creation can be particularly useful for video sharing and streaming platforms such as YouTube, Vimeo, Netflix [83], Prime Video, and others. It makes the content more interesting and encourages them to click through to view the full content. It also helps improve the browsing experience for users by providing personalized insights into the content. It saves time and resources for content creators, as they can automate the process of generating previews and thumbnails rather than creating them manually.

Automated preview and thumbnail creation requires analyzing the video content in order to identify key moments and scenes, rank these scenes according to some measures, and then selecting and presenting the most relevant parts as previews or thumbnails [84]. CV-based techniques such as object detection, scene recognition, action classification, motion analysis, and image processing are used to extract useful information from the video and generate visually appealing and informative previews and thumbnails. User feedback is another important technique used in automated preview and thumbnail generation. Content creators can collect data on which previews and thumbnails are most engaging by analyzing metrics such as click-through rates, watch durations, and user interactions. This data can then be used to optimize the selection and presentation of video content, ensuring that users are presented with previews and thumbnails that are most likely to capture their attention and encourage them to click through to view the full content. End-to-end deep learning based frameworks [85–87] are also being used to generate previews and thumbnails, allowing for faster and more accurate previews and thumbnails.

10.2.5 Rotoscoping and Rotomation

Rotoscoping and rotomation are widely used in the film industries to create realistic and appealing animations and special effects. Rotoscoping is the process of tracing live-action footage frame-by-frame to create an animation sequence. This is used to create special effects, such as the realistic movement of animated characters or objects to live-action footage. On the other hand, rotomation involves filming actors in front of a green screen and then replacing the background with CGI. This technique is often used to create complex visual effects and realistic action sequences.

Both of these techniques require a combination of artistic skill and technical expertise. They are both time-consuming and labor-intensive. In recent years, advances in CV and machine learning [88–90] have made it possible to automate some aspects of rotoscoping and rotomation, such as object tracking and image segmentation. This can help speed up the process and reduce the workload for artists and animators.

10.2.6 Film editing and Post-producing

The use of CV-based techniques is increasing in film editing and post-production. Film editing is a labor-intensive task that requires both artistic touch and efficiency. However, certain aspects of film editing and post-production can be automated using CV and machine learning.

1. **Automated shot selection:** CV can save time and resources for filmmakers and editors by quickly identifying the best takes from a film shoot based on preset criteria. For example, the CV algorithm could be programmed to select shots with the best framing, focus, lighting, and performance by the actors. This can help speed up the editing process and ensure that the final product is of the highest quality. Research work in [91] proposed a video shot selection framework based on face pose.

2. **Scene detection:** CV is being used to automatically detect scene changes and segment footage into individual scenes, which makes it easier for editors to organize and work with large amounts of footage [92].

3. **Color grading:** Movies are shoot with different cameras in diverse environments and lighting conditions. It makes color grading an important post-production step [93]. CV is being used to automatically identify color balance issues in the footage and adjust the color and tone.

4. **Object removal:** CV can be used to automatically remove unwanted objects from the footage, which can save time and resources for filmmakers. Research work in [94] proposed a deep GAN for artifact removal.

10.3 Content Moderation

10.3.1 Sensitive Content Detection

As media becomes increasingly prevalent on the internet, users are exposed to a wide range of content that ranges from professional to user-generated material across multiple platforms. This content can include violent, sexually explicit, racist, hate speech-promoting, terrorist propaganda, child exploitation, and other types of inappropriate content. To ensure a safe and welcoming environment for users, media platforms employ both human moderators and automated systems to detect and eliminate sensitive content.

In the automated system, visually sensitive content is detected using a variety of CV techniques. For instance, Meta, formerly known as Facebook, uses a variety of techniques to detect sensitive content on their platforms -

1. Machine learning classifiers are trained to detect various visually sensitive content.

2. A large user-reported repository of visual content is matched using TMK [95] CV technique.

3. Whole post integrity classifier takes a multimodal approach to moderate the content present in it.

CV techniques have been used for many years to flag visually sensitive content. The type of technique depends on the complexity, media type, and scale of the visual content being monitored. The techniques have ranged from simple feature-based algorithms to complex deep-learning algorithms. State-of-the-art research as described below explores different spatial and temporal architectures in deep learning as detecting sensitive content in images and videos requires different spatial and temporal understanding. For instance, convolutional models and image transformer models are generally spatial models and are suitable for images. However, videos contain a temporal aspect to them, the models should accommodate the temporal data. In such cases, the authors use LSTM, GRU, and other sequential models [96] along with a convolutional network [8] or a modified version of transformers for the videos such as Swin [97], and video transformers [9]. These techniques are very useful and are generally the backbone of many industry-grade content moderation tools. However, these models are as good as the data they are trained on and hence can't generalize well to all possible scenarios of occurrences of sensitive content. Hence, they must rely on user-reported content.

User-reported content makes up a huge volume of cases that get flagged as sensitive content and must be removed from circulation on their platforms. This poses a different set of challenges as the scale and diversity of the flagged content increases; it becomes harder to update and maintain machine learning models to reliably detect these in a short period of time. This is where CV-based matching techniques play an important role. Techniques such as TMK [95] are perceptual hashing algorithms to create the fingerprints of the images and videos reported by the users and they are scaled to search their entire platform. Every image or video on their platform gets fingerprinted and gets matched against the flagged content. The TMK algorithm is robust to different image and video transformations such as translation, rotation, and time shifts. This provides a very reliable and scalable way to make sure the platform is safe and inclusive for the users.

10.3.2 Adult Content Detection

Adult content detection primarily involves detecting violent images, videos, nudity, and sexually explicit content. These can appear on social media platforms, and professionally generated content such as movies, TV shows, etc. The presence of such content in most cases may not be desirable especially on social media or streaming platforms to make it safer for users of all ages. On the other hand, such content may appear as a part of the storyline in professionally generated content. Depending on the distribution destination and target audience, such scenes need to be moderated automatically, given the scale of the content.

Several companies offer adult content detection solutions that use CV techniques, such as Microsoft's Azure Content Moderator,[7] Google's Vision API,[8] Amazon's Rekognition,[9] etc. These services use a combination of machine learning and CV techniques to detect and filter out adult content in real time and can be customized to meet the specific needs of different industries and platforms. The CV techniques used to detect these are very similar to those used in sensitive content detection, where we use a combination of traditional machine learning and advanced deep learning models to detect them in videos and images.

10.3.3 Deepfake Detection

Deepfakes refer to fake multimedia content, which has been digitally altered or synthetically created using deep neural networks. The threat of deepfakes lies in their potential to spread misinformation and undermine trust in media and institutions. With the ability to create realistic-looking videos of people saying or doing things they never actually did, deepfakes can be used to manipulate public opinion, interfere with political elections, or cause reputational harm to individuals.

The most common techniques to create deepfake manipulations are identity swap, face reenactment, attribute manipulation, and entire face synthesis. Typically, the deepfake generation and detection are performed using deep learning architectures such as CNNs, GANs, and other generative neural networks. We can reverse engineer the techniques used to generate them and develop detection methods. This AI arms race between the attackers/offenders/adversaries (i.e., deepfake generation methods) and the defenders (i.e., deepfake detection methods) is what makes the domain challenging as the generation techniques keep evolving and improving.

Identity swap or face swap generation includes replacing a person's face with another person's face. Face reenactment involves changing the facial expression of the individual. Attribute manipulation or face retouching is basically altering certain face attributes such as skin tone, age, and gender and entire face synthesis consists of generating entirely nonexistent face images.

A variety of techniques have been proposed to detect these in the literature. For instance, Koopman et al. [98] analyzed photo response non-uniformity (PRNU) for detection. Warping artifacts, eye blinking, optical flow, heart rate, image quality, local image textures, long short-term memory (LSTM), compression artifacts, metric learning, etc. are also used. Recently, Liu et al. [99] proposed a block shuffling learning method to detect deepfakes, where the image is divided into blocks, and using random shuffling intra-block and inter-block-based features are extracted. Matern et al. [100] used visual features with logistic regression and MLP, Rossler et al. [101] used mesoscopic, steganalysis, and CNN features; Sabir et al. [102] used RNN; In contrast, Zhao et al. [103] designed a spatiotemporal network, which can utilize complementary global and local information. In particular, this framework uses a spatial module for the global information, and the local information module extracts features from the patches selected by attention layers. Bharati et al. [104] proposed a Deep Boltzmann machine-based framework to detect face retouches. Mc-Closkey et al. [105] presented a color cues-based system to detect GAN-generated images. Deepfake detection is being increasingly benefited by state-of-the-art CV models like the EfficientNet and vision transformers, which have shown improved performance across various benchmarks [106].

10.4 Recommendation

The growth of streaming video services offering thousands of movies and TV series has led to the need for recommendation systems (RSs) that suggests relevant videos to targeted users. Their primary goal is to reduce the negative impacts of the over-choice imposed on users, which can make finding a video of interest challenging, time-consuming, and exhausting. Analyzing large video catalogs, RS aims to present users with a personalized subset of videos, usually organized in different menus with particular themes. Algorithms used for RS can be broadly categorized into collaborative filtering (CF), content-based filtering (CBF), and hybrid [107].

CF solutions rely on usage data of all users, either watch history or explicit feedback such as rating scores, to identify preference patterns and predict the impression of unseen video on a user [108]. CBF solutions leverage the content information, typically represented as embedding vectors [109]. For instance, in the entertainment domain, video properties can be keywords or concepts in subtitles or on-screen texts, colors or visual concepts from keyframes, motion and activities detected in a video segment, or audio features and concepts from a music piece. Hybrid recommender systems combine two previous approaches, using a fusion approach [110]. In this chapter, the main focus is on CBF solutions using the audio-visual characteristics of the video content and how they are leveraged in hybrid solutions.

10.4.1 Content-based Filtering RS Solutions

For content analysis, most CBF studies use short forms of media derived from movies, such as trailers, movie clips, posters, or a few select frames to reduce computational expense or since the full movies are not freely available. To enhance movie recommendation quality, Chen et al. [111] used visual clues extracted from movie posters and selected movie frames. They merged the two tasks of visual feature extraction using a pre-trained CNN and movie recommendation into a unified optimization process called unified visual contents matrix factorization (UVMF). Deldjoo et al. [112] proposed a CBF movie recommendation system that leverages a set of low-level

visual features defined as shot length, color variation, lighting key, and motion vectors to describe movies. For instance, comedy movies usually include bright colors, while horror movies are made of dark ones. Several improvements have been made to the proposed movie RS in [113–115], including experiments with MPEG-7, pre-trained CNNs, and esthetic features. A comprehensive set of content features (audio, visual, and textual) was investigated by Du et al. [116] to develop RS for movies that effectively handle missing multimodal features. They also introduced a priority-aware late fusion method to boost the quality of movie recommendations. Almeida et al. [117] investigated various deep learning features that represent the visual appearance, audio, and motion information from video content to tackle the cold-start recommendation problem. In addition, different fusion methods are explored to evaluate how well these multimodal features can be fused to fully exploit their complementary information. They found that recommendations generated with deep learning audio features and action-centric deep learning features outperform MFCC and state-of-the-art iDT features.

Exploring the emotional aspect of the movies, Canini et al. [118] proposed an affective framework to extract audio-visual features such as dominant color, color energy, lighting key, sound energy, low-energy ratio, MFCCs, and movie grammar descriptors such as illuminated color, shot length, and shot type transition rate allowing movie scenes to be compared based on their emotive differences. Roy and Guntuku [119] introduced an emotion-based visual movie recommender system named visual-CLiMF (Collaborative Less-is-More Filtering). Visual-CLiMF is built upon the original CLiMF approach [120] and further combines visual clues with user watch history to learn a latent representation of emotional factors.

10.4.2 Hybrid RS Solutions

Research work in [121] proposed a hybrid movie recommendation framework called VideoTopic that utilizes both visual and textual features of movies and uses LDA to represent movie contents as well as user interests based on topics extracted from user watch history. For the textual features, they found actors and directors more effective compared to other movie metadata such as summary and writer. For the visual information, SIFT visual features are extracted from the keyframes of trailers. The second stage involves finding movies with minimal topic distribution differences based on user watch histories. The VideoTopic outperforms other methods using unimodal textual or visual features with respect to accuracy metrics reported on MovieLens 1M test dataset. Deldjoo et al. [122] proposed a hybrid RS called collaborative-filtering enriched CBF (CFeCBF) by leveraging canonical correlation analysis to combine audio and visual features together with movie metadata such as genre and cast into a unique representation referred to as the Movie Genome. In order to recommend new movies for which no usage data is available, they trained a CF model using existing movies and applied it to the movie genome. Evaluation results indicate that multimodal recommendations generated by the proposed CFeCBF model outperform plain CBF baseline using only genre or cast to describe a movie. Using multimodal content relevance and user feedback, Mei et al. [123] proposed VideoReach, a hybrid RS for online videos that do not require a sufficient collection of user profiles. From online video content and context, different multimodal clues are extracted, including visual and audio tracks as well as associated texts such as queries, keywords, and surrounding text. Also, they consider that different parts of videos have varying degrees of interest for users, as well as different features and modalities contribute to the overall relevance of the video. With relevance feedback and attention fusion, they designed a unified framework for VideoReach that incorporates both multimodal relevance and user feedback seamlessly.

Going beyond content analysis and user–content interactions, scholars have investigated a range of knowledge bases, such as semantic knowledge, taxonomic knowledge, social knowledge, and domain-specific knowledge, to improve recommendation performance. Song et al. [124] explore RS using knowledge graphs to address data sparsity and cold start issues. While existing methods focus on learning representations for matching users and items, this study proposes generating recommendations through meaningful paths in the knowledge graph. By formulating the problem as a sequential decision process and training a policy function with policy gradient methods, the proposed approach not only provides effective recommendations but also offers valuable explanations, as demonstrated in experiments with real-world datasets. Wang et al. [125] explore the use of knowledge graphs (KG) to enhance recommendation systems by considering high-order relations between items through attributes. The proposed method, knowledge graph attention network (KGAT), effectively models these relations by recursively propagating embeddings and employing an attention mechanism. Empirical results demonstrate that KGAT outperforms existing KG-based recommendation methods in terms of accuracy and interpretability. Wang et al. [126] discuss the development of Graph Learning based Recommender Systems (GLRS), which utilize graph learning approaches to model users' preferences and item characteristics for recommendations. GLRS differs from other recommendation approaches by leveraging explicit or implicit connections in graphs. This chapter provides a systematic review of GLRS, including their characterization, challenges, progress, and new research directions, highlighting how they extract knowledge from graph-based representations to improve recommendation accuracy, reliability, and explainability.

10.5 Metadata Curation and Information Retrieval

CV plays an important role in curating metadata and retrieving valuable information from media content. Broadcasting and streaming media come with three different modalities – audio, visual, and text. Among them, visual modality contains the densest information. Efficient curation of this meaningful information will tremendously help improve customer experience and develop innovative features that, in turn, help improve customer churn rate. In this section, we will discuss some innovative customer-focusing features that were made possible by state-of-the-art CV.

10.5.1 On-screen Information Retrieval

TV media contains so much meaningful information. We can automatically extract this information only by employing CV based techniques. There are recently developed frameworks available that are being used to recognize celebrities, identify cast, recognize on-screen text, etc. One such application is a feature called X-Ray [127] in Amazon Prime Videos. It enables curious audiences to know more about the on-screen casts along with other information and interact with them. This type of application is the outcome of several decades of face recognition research [128, 129]. Among them, the most influential work is FaceNet [130]; it directly learns a mapping from face images to a compact Euclidean space where distances directly correspond to a measure of face similarity. This method is the illumination and poses invariant that makes it possible to efficiently apply in large-scale face verification and recognition system.

In sports broadcasting, the application of CV gets more interesting. OCR is being used to identify current scores in the on-screen scoreboard in sports like Soccer, Cricket, F1, Golf, and NFL. This automatic identification of scores made it possible to generate automated highlights and important clips and moments. "Key Plays" [131] feature in the streaming app Peacock is one such great CV application. Apart from understanding the scoreboard, more explicit information such as news ticker, credit scroll, etc. are present on-screen that can be extracted using OCR. Recently developed large CNN-based machine learning models [132, 133] made it possible to accurately recognize on-screen characters. These end-to-end trainable models are fast enough to apply them in real-time and also clever enough to recognize characters in the presence of diverse fonts and sizes.

10.5.2 Automatic Content Recognition

Automatic content recognition (ACR) is a technology built into smart TVs, streaming devices, or set-top boxes to identify what is currently playing on the screen. ACR captures audio or visual signals from the screen and then identifies the content such as the name of the show, movie, commercial, etc. using sophisticated content-matching techniques. ACR is capable of identifying content in almost any form of media such as streaming shows, linear TV, live sports, on-demand videos, movies played from DVD, etc. ACR is being used to measure content viewership, ad performance, targeted ads, and personalized content recommendations. Using ACR media companies not only improves the customer experience but also generates a revenue stream. Nowadays, most smart TVs are equipped with ACR capability, for example, Vizio [134] a smart TV maker is profiting from selling such TVs.

CV is an essential part of ACR as visual signal provides the most significant and distinguishable clues. The most commonly used method for ACR is visual fingerprinting. Some parts of the video or the frames are represented in a compact form that contains unique characteristics. These representations or fingerprints are matched against the reference fingerprints in a database. Upon matching, associated meta information such as the name of the show, program id, etc. is retrieved from the database, which is then used in downstream applications. Some fingerprinting techniques are color-based [135], random transform [136], deep neural network-based [137], etc.

10.5.3 Product Recognition

Automated product recognition on TV screens is indeed a crucial step in bringing T-commerce into our daily lives. With the ability to recognize products featured in television shows, commercials, or other programming, viewers can easily make purchases using their remote controls while sitting on their couches. Vision-based product recognition involves using CV algorithms to identify consumer goods in videos. This technology has come a long way in recent years thanks to advancements in deep learning and neural networks. CNNs, for example, have proven to be particularly effective at identifying objects and features in images and videos, making them well-suited for product recognition tasks.

Google Lens [138] is a great example of how this technology can be applied in real-world scenarios. By using CV algorithms, Google Lens can identify products in images and provide users with information about those products, such as where to buy them or how much they cost. This technology has the potential to revolutionize the way we shop, making it easier and more convenient than ever before. However, there are still some challenges that need to be addressed before T-commerce can become a reality. One of the biggest challenges is developing algorithms that can accurately recognize products in a wide range of settings, including different lighting conditions, camera angles, and product orientations. Another challenge is developing a reliable and secure payment system that viewers can use to make purchases from their TV screens. Research works in [139–141] have addressed this challenging problem with significant accuracy. These research works, along with many others, demonstrate the potential of CV and deep learning in automated product recognition for T-commerce. As technology continues to advance, we can expect to see even more innovative approaches to this challenging problem.

10.5.4 Content Indexing and Metadata Curation

Media content indexing and metadata curation are essential processes for managing and organizing large-volume multimedia content. The metadata not only includes information such as title, description, keywords, date, and other relevant details but also incorporates important visual clues such as objects, actions, concepts, etc. that help identify and categorize the media content. By creating accurate and comprehensive visual metadata for multimedia content, it becomes easier to search, retrieve, and manage the content. Media content indexing and metadata curation also enable efficient content discovery and distribution, which is particularly important for digital media platforms such as online video or audio streaming services. In addition to facilitating content management, media content indexing and metadata curation can also help improve user experience by providing relevant recommendations, personalized content, and better search results. These processes involve advanced technologies such as CV and machine learning to analyze and extract meaningful information from multimedia content.

10.6 Advertisement

10.6.1 Contextual Advertisement

Both video streaming platforms and linear TV broadcasters generate revenue by selling advertising spots to advertisement partners. Contextual advertising is a new phenomenon in this sphere, which is a type of targeted advertising that is based on the context of the content being viewed by a user [142, 143]. In contextual advertising, CV-based algorithms are used to analyze video content, and then display ads that are the most relevant to the topic of the content. For example, if a user is watching a car racing movie, contextual advertising may display ads for cars. On the other hand, displaying a car ad after a car crash is not always ideal. In this way, contextual advertising provides more relevant and personalized ads to users, while also providing a higher rate of return for advertisers, since they are more likely to reach an audience that is already interested in the product or service being advertised.

CV is being used to make contextual advertising more relevant and promising.

1. **Image analysis:** CV algorithms such as image classification [12], object detection [13], text recognition [133], etc. can be used to extract relevant metadata to be used as important clues for contextual advertisement.

2. **Video analysis:** CV-based techniques can be used to identify the optimal placement of ads in long-form video content such as TV shows and movies. Ads can be placed in pre-roll, mid-roll, and post-roll positions, and the choice of placement can have a significant impact on how the ad and the corresponding brand are perceived by viewers [144]. CV techniques can be used to analyze the video content and identify natural breaking points, such as scene changes or other transitions, where ads can be inserted without disrupting the flow of the content. This can help to ensure that ads are placed in optimal locations that are less likely to be perceived as intrusive or disruptive by viewers [145].

10.6.2 Brand Alignment

Brand alignment is a crucial aspect of advertising and marketing in the media industry. Brand alignment is the process of ensuring that all aspects of a company's brand, including its messaging, visual identity, overall reputation, and branding elements, such as its logo, colors, and typography are consistent and cohesive across all marketing and communication channels. Manually ensuring brand alignment is a nearly impossible task due to the vast horizon of reach these days. However, CV can assist here by automatically identifying the colors and font types used in a company's branding materials and analyzing whether they are consistent across different platforms.

10.6.3 Brand Safety

Brand safety refers to the measures that companies take to ensure that their ads are displayed and they have a presence in safe and appropriate environments. This includes monitoring and controlling the context in which their ads are displayed in order to avoid associating their brand with inappropriate or offensive content. CV can help by analyzing images and videos on web pages and social media platforms to identify potentially unsafe or inappropriate content and prevent ads from being displayed in those contexts. CV algorithms can also recognize logos associated with a company's brand, which can be used to monitor and protect against trademark infringement or unauthorized use of the company's branding materials.

10.6.4 Dynamic Product Placement

Dynamic product placement is a new phenomenon in advertising and marketing where digital products are placed into existing video content [146]. The technology uses CV algorithms to identify appropriate locations within the video content, and then dynamically inserts the product into the video during playback. This technique enables advertisers to tailor their product placements to the content being viewed, providing more relevant and effective advertising to viewers. Dynamic product placement can also be used in live broadcasts, where product placements can be updated in real-time based on the content of the broadcast. By integrating product placement seamlessly into the video content, advertisers can create a more immersive and engaging advertising experience for viewers.

10.7 Experience Enhancement

Research studies show that the majority of customers are generally willing to pay more when it comes to getting a better experience [147]. Therefore, a tremendous amount of value can be added to the overall revenue generation of media companies by offering rich and engaging content experiences to their users. These days, media companies are putting in a lot of effort to increase user engagement. With the recent popularity of streaming services, people are constantly looking for better content experiences. This situation puts substantial pressure on content providers to continue enhancing their content consumption experience strategies in order to retain current customers as well as attract new customers. In the following subsections, we discuss some of the most popular CV-based experience enhancement applications.

10.7.1 Intro/Outro/Recap Detection

The viewership of streaming media has increased tremendously in the past decade. With the growing number of streaming services, there has been a serious decline in the number of users watching broadcast television. The younger generation prefers digital content that gives them a more personalized experience as well as the flexibility to stream media whenever they like and onto the device of their choice.

One of the many media consumption enhancement experiences is the capability to binge-watch an episodic TV show or movie series. This allows users to skip the redundant segments (intro sequence and end credits sequence) of their favorite TV shows when they want to continuously watch a series of episodes in one sitting or want to jump to any post-credits scenes or the next movie. Recent studies [148, 149] show that the majority of viewers in the USA prefer binge-watching as their regular manner of consuming entertainment content. Having the capability to skip these segments, users can not only save time but also enjoy the uninterrupted natural continuation of the story. In addition to intro and outro segments, some users might also prefer to skip the recap of the previous episode when binge-watching.

In order to engage more customers, it becomes essential to automatically detect time markers for the intro, outro, and recap detection in episodic TV shows as well as movie series. This can automate the process of providing skip functionality to a huge collection of videos owned by streaming platforms. It can also drastically reduce the time and cost of hiring human annotators to manually go through each video and identify binge markers.

Active research has been going on in this domain and different CV techniques have been explored to auto-detect binge markers. Nematollahi et al. [150] use blank screen transitions and histograms of shot boundaries to develop a framework that identifies intro and outro segments of a show. Other approaches [151, 152] investigate the use of visual fingerprints and histograms of similar frames across multiple video assets to mark the beginning and end of intro sequences. Recent research [153] proposes an end-to-end deep learning framework to detect intro and recap segments in TV shows. They employ a multimodal approach by extracting visual and audio features from videos and using a Bi-LSTM to capture the long and short-term dependencies among extracted features over time.

There are also a number of commercial solutions [154–156] available that leverage visual fingerprinting and probabilistic AI in order to provide safe content skipping to facilitate binge-watching for its customers.

10.7.2　Commercial Detection

Another major application of media enhancement experience is the automatic detection of commercials/ads in broadcast TV as well as streaming videos. It is a well-known fact that advertising is a multibillion dollar industry that provides a big opportunity to traditional broadcasters as well as OTT companies to increase their revenues by taking advantage of the latest technological developments in AI and particularly CV.

CV-powered automated ad detection generates start and end time markers of where a commercial appears within the content. These ad markers are then used to replace existing ads with targeted and personalized ads, thus enhancing the overall viewer experience [157]. Real-time automated ad detection in live streams also effectively eliminates dependence on broadcasters to supply ad markers. DVR systems use ad markers to offer ad skip functionality to their customers for an uninterrupted and smooth watch experience [158]. For content that is stored in databases, cutting out commercials can reduce storage requirements.

For advertisers, commercial detection is important as it helps them verify the fulfillment of contracts, that is, helps them to keep track of exactly when and how many times their ads are being broadcast.

The CV research community has been actively looking into the domain of commercial detection. Li et al. [159] propose an SVM-based classifier that is trained on shot features extracted by using a deep CNN to detect commercials in TV broadcasting. Another recent research [160] aims to detect ads in video broadcasts by employing a two-stream audio-visual CNN combining both audio and visual features. Other researchers [161, 162] explore a novel approach to detect commercial blocks by distinguishing the presence of TV channel logos from other types of logos appearing during commercials using visual features such as edge detection, color analysis, and HOG features. Khan et al. [163] employ color histograms and SURF feature descriptors to detect ads in an unsupervised manner.

10.7.3　Duplicate Content Detection

The exponential growth in the production of multimedia digital content poses a huge challenge to online video hosting and sharing services which are responsible for storing, indexing, and analyzing this massive amount of data. If we consider YouTube alone, stats [32] show that one hour of video is uploaded to YouTube every second. In this scenario, it is highly likely that there are multiple copies of a single video being uploaded by more than one user resulting in the presence of duplicate content. To address this issue, there is a need for a system that can automatically detect duplicate content to help optimize storage requirements. Duplicate content detection also helps in video search by showing more diverse results instead of multiple copies of the same video. It can also help provide copyright protection to content creators by detecting and flagging copied content.

A major obstacle encountered when developing a solution for duplicate content detection is the scalability of the proposed solution. Given the vast amount of multimedia data including countless hours of videos, the solution to search for duplicate content needs to be highly efficient. Fortunately, technological advancements in the field of AI/CV have provided multiple solutions to tackle this problem. Recent research in [164] aims to leverage intermediate convolution layer features of a deep CNN to generate compact video representations. A second deep neural network is then trained to compute the similarity between two candidate videos using their respective compact representations. Chen et al. [165] used simple visual features but parallelize their detection architecture to achieve fast and accurate results. A more recent attempt [166] learns both intra- and inter-frame relations when computing frame-to-frame similarity matrices between two videos. Han et al. [167] take a step further and propose an approach to jointly learn spatial similarity, temporal similarity, and partial alignment of video pairs as a multitask learning problem. Another recent research [168] eliminates the need for annotated data by using contrastive learning to learn video representations from a large set of unlabeled videos. They also employ a transformer-based architecture to aggregate frame-level representation features into clip-level which are then used to compute similarity scores between videos.

10.7.4 Audio Description

Audio description (AD) can be defined as narrations of different visual aspects portrayed in a video such as a movie or a TV show [169, 170]. In other words, these narrations explain what is being shown on screen including names of characters, their physical actions, facial expressions, costumes, and props as well as scene background, scene changes, and any text that is displayed on the screen. The AD helps the blind and visually impaired community to better understand the content of a video and enables them to enjoy multimedia content like sighted people. ADs are becoming increasingly popular in streaming platforms, which now provide an option to enable it for many of its movies and TV shows making it accessible to a wider audience.

Manual generation of ADs is an expensive and time-consuming task. Advances in CV and Machine Learning have made it possible to automate or semiautomate the process of generating ADs for multimedia videos. Recent research in [171] proposes a single framework to automate the generation of ADs for videos by performing three tasks – predicting AD insertion time, AD generation, and AD optimization. Their approach works by first dividing a given video into clips (that do not have voiceovers) and then extracting keyframes from each clip. The clips are then supplied to an attention-based model that consists of a sentence localizer and a caption generator. The generated ADs are further refined for the best experience. Another recent research [172] combines different CV-based deep learning techniques such as face detection, facial expression detection, and audio synthesizers to analyze a given video and generate ADs.

Researchers have also started exploring the possibility of generating ADs for video games [173] to make them more accessible to the blind and visually impaired community. Additionally, state-of-the-art research in the area of using transformers for dense video captioning [174–177] is also very much related to and valuable for the AD generation task.

10.7.5 Automatic Sign Language

Sign language is used by the deaf–mute community all around the world as their primary means of communication. It is a visually rich language that involves the movement of hands, arms, and other body parts along with facial expressions to express a deaf–mute person's thoughts. Just like other spoken natural languages, sign languages have their own grammatical structures. With the advancement in AI, particularly CV, it is now possible to automatically recognize, translate, and produce sign language so that the deaf–mute community can fluidly communicate with the rest of the world.

1. **Sign language recognition (SLR) and translation (SLT):** It is a major challenge for people with hearing or speaking disabilities to communicate with normal people as normal people are generally not familiar with sign languages. Hence, there is a need for robust methods that can recognize sign language gestures and translate them to convey the message to normal people. Various CV techniques have been employed to bridge this gap by automatically recognizing the gestures in sign language videos and essentially interpreting what a hearing or speaking-impaired person is trying to communicate.

   The goal of SLR is to interpret the gestures in sign videos and generate a mid-level representation for each gesture (known as gloss). SLT then aims at translating glosses into textual or spoken words. Recent research [178] shows that transformers-based architectures can be used to jointly learn sign language recognition and translation as a single task. This removes the dependency on generating mid-level gloss representations and their proposed model is trainable in an end-to-end manner. More recent research in this domain has started to explore multilingual sign language translation (MSLT) task [179] which seeks to develop a single model that can translate between multiple signs and spoken languages.

An interesting application in the sign language domain is the retrieval of sign language videos with free-form textual queries [180], that is, search for a particular video from a given large database of sign language videos using a written piece of free-form text. Instead of doing traditional keyword matching, the authors achieve this task by learning a joint embedding space between text and videos.

2. **Sign language production (SLP):** SLP refers to the automatic generation of sign language gestures (3D sign pose sequences in the form of images or videos) from written text or spoken voice at a level comparable to a human translator.

   Recent work by Saunders et al. [181, 182] uses Progressive Transformers, a combination of Symbolic Encoders and Progressive Decoders, to learn the mapping between spoken language and skeleton pose sequences. Another recent work [183] proposes a sign language production method that directly converts spoken language sequences to photo-realistic continuous sign gesture videos. Such systems play a huge role in bridging the communication gap between hearing-impaired and hearing communities.

   Conversational User Interfaces (CUIs; such as Alexa that generally take spoken language as input) are becoming a very useful household item these days. Some CUIs also offer text-based input but there is still a need for these devices to support sign language input and output. Researchers have started exploring this domain [184] to help the deaf–mute community interact with CUIs using sign language.

10.8   Streaming

10.8.1   Streaming Quality Analysis

Video streaming has become a crucial element of contemporary entertainment due to the widespread use of on-demand services, whether it be professionally produced content or content created by users. Many people globally enjoy watching their preferred sports, movies, shows, and gaming videos on different streaming platforms. Nonetheless, the streaming experience's quality can differ significantly based on multiple factors, such as internet speed, device compatibility, and the streaming service's video compression technology. Hence, both content providers and viewers prioritize the need to guarantee a high-quality streaming experience regardless of the type of content. Streaming video can suffer from defects introduced during recording, encoding, decoding, and/or transmission.

At the encoding stage, streaming video requires compression using standards, such as H.264/AVC, HEVC, and VP9, to stream at acceptable bitrates. However, overcompression of videos can lead to different artifacts that reduce the quality of the video perceived by the user. These artifacts also known as perceptual artifacts can be classified into spatial and temporal defects. Spatial artifacts are created by block-based video coding schemes due to block partitioning and quantization. These artifacts include blurring, blocking, ringing, basis pattern effect, and color bleeding. Temporal artifacts refer to those distortion effects that are not observed when the video is paused but during video playback. These artifacts include Flickering, mosquito noise, Jerkiness, texture floating, etc. Temporal artifacts are more challenging to quantify as they keep evolving with video and coding techniques.

Visual quality is estimated typically in two different ways – full reference-based (FR) quality assessment and No-reference (NR)-based quality assessment. As the name suggests, FR-based requires a source video as a reference to compare the quality of the video in question. NR assessment does not require the source video as a reference to compare.

Researchers have come up with various quality metrics to quantify the degradation in the quality of the video because of the perceptual artifacts. For instance, historically, for full reference VQA, metrics such as peak signal-to-noise ratio (PSNR), structural similarity index (SSIM), multi-scale structure similarity index (MS-SSIM), SSIMplus, etc. were proposed. Very recently, Netflix proposed Video Multimethod Assessment Fusion (VMAF) [185] as an alternative to measuring subjective quality by combining multiple elementary quality metrics. The basic rationale is that each elementary metric may have its own strengths and weaknesses with respect to the source content characteristics, type of artifacts, and degree of distortion. By "fusing" elementary metrics into a final metric using a machine-learning algorithm – a Support Vector Machine (SVM) regressor – which assigns weights to each elementary metric, the final metric could preserve all the strengths of the individual metrics and deliver a more accurate final score. Recently, Zhang et al. [186] have proposed deep learning-based architectures to measure the quality of the video frames, which they refer to as perceptual metrics. They extract features from different deep learning models at different depths and use them to compare against the reference video features.

They conclude that perceptual similarity is an emergent property shared across deep visual representations and can be an effective perceptual metric.

For NR-VQA, deep learning-based techniques have recently been utilized. One of the first methods utilizing deep learning was SACONVA [187], which extracted feature vectors from video data via a 3D shearlet transform. These features were mapped onto quality scores using logistic regression and a CNN. Wang et al. [188] combined deep spatial and temporal features for perceptual quality prediction. Specifically, spatial features were obtained through the pooling of a CNN's activation functions. Further, the standard deviations of motion vectors were considered temporal features. Next, two predictions were obtained from these two sets of features, and they were combined by using a Bayes classifier for video quality prediction. Agarla et al. [189] proposed an approach in which the image quality attributes, that is, sharpness, graininess, lightness, and color saturation, of video frames, were estimated first by using the deep features of a CNN. Based on these attributes, frame-level quality scores were estimated. Finally, a recurrent neural network was trained for video quality estimation by using the previously predicted frame-level scores as training data. The two-level video quality model (TVLQM) proposed by Korhonen et al. [190] first computed low-complexity features from the entire video sequence before the extraction of high-complexity features. Guan et al. [191] implemented a visual attention module that obtained frame-level perceptual quality scores.

10.9   Gaming and Augmented Intelligence (AR, VR, and MR)

10.9.1   Game Development

CV is being used increasingly in game development as it enables developers to create more immersive and interactive gaming experiences. Game development is a complex labor-intensive process, and some of it can be easily automated using CV. We listed some uses of CV in the gaming industry as follows.

1. **Environment design and 3D mapping:** CV can be used to scan real-world environments and convert them into 3D models that can be used in game development [192–194]. It can also be used to create procedural generation algorithms that can automatically generate unique and realistic environments for the game. In addition, CV can help to create dynamic environments that respond to the player's movements and actions, providing

a more interactive and engaging gameplay experience.

2. **Character design:** CV-based algorithms are used to scan and create realistic 3D models of characters [195, 196]. It also helps to recognize and track facial expressions and apply them to in-game characters for realistic animation and interactions. GANs can also be used to generate new faces for game characters [197].

3. **Object detection and tracking:** CV can be used to recognize and track objects in the game environment [198], which can help create more realistic interactions between the player and the environment. For example, a CV system can track the player's movements and adjust the environment to match their position.

4. **Lighting and shading:** CV can be used to simulate realistic lighting and shading effects in the game environment [199, 200]. By analyzing real-world lighting conditions and creating algorithms based on this data, CV can create more immersive and visually stunning game environments.

5. **Rendering:** This is essential in the video game industry, enabling the creation of visually appealing and interactive game environments. Real-time rendering techniques are employed to generate images on the fly multiple times per second as players explore the virtual world for smooth and responsive gameplay. This dynamic process involves complex CV algorithms and optimizations to balance visual quality and computational efficiency.

The significance of cutting-edge image-based rendering is increasingly recognized, with a rise in popularity. This approach utilizes differentiable rendering techniques [201], which enables the development of a unified CV model that can generate 2D images from 3D scene descriptions.

10.9.2   Augmented Intelligence – AR/VR/MR

In AR, CV is used to recognize and track objects in the real world, such as markers or images, and overlay virtual content on top of them [202]. CV can also help to align and adjust virtual objects in real-time based on the real-world environment, providing a seamless AR experience.

In VR, CV can be used to track the user's head and hand movements, allowing them to interact with virtual objects in a natural way [203]. It can also be used to create realistic simulations of real-world objects and environments, enhancing the immersive experience.

In Mixed Reality (MR), CV is used to merge the real and virtual worlds in a way that the virtual objects interact with real-world objects in a realistic manner [204]. This requires sophisticated CV algorithms that can recognize and track the 3D structure of the real-world environment and map virtual objects onto it.

10.10 Sports

Sports have the ability to connect people and communities. It is the common language of people from all sectors irrespective of their age, gender, and race. Due to technological advances and the rise of mass media, lots of resources are being invested in sports to arrange global tournaments, broadcasting, performance analysis, etc. AI and CV have recently started to play important roles in sports. For example, sports performance assessment is a big thing these days, which was previously mainly of interest to coaches and sports scientists. But now it has applications in broadcasting media, which make interesting performance statistics available to viewers. CV-guided graphics and animations are making sports more enjoyable to the audience. Many advanced CV-based technologies are being used in important decision-making during gameplay. Nevertheless, CV methods have a huge potential in many aspects of sports ranging from the automatic annotation of broadcasting footage, better understanding of sports injuries, enhanced viewing experience, sports event detection, automatic highlight generation, audience analysis, enhanced graphical effects, etc. The media industry is catching up with the recent advances in CV and delivering sports highlights during live broadcasts. Next, we will review some state-of-the-art research work focusing on CV in sports.

10.10.1 Sports Performance Assessment

Performance assessments in sports are being performed through automatic player identification and tracking, pose estimation, training assistance, etc. Player identification is the first essential step to several downstream tasks. Some recent research works [205–211] proposed different useful techniques to identify players in sports videos. Research work in [205] introduces a transformer network for recognizing players through their jersey numbers in broadcast National Hockey League (NHL) videos. A semi-supervised approach is proposed in [206] that requires few game-specific annotations to effectively identify and track players in the video. Work in [207] proposed a teacher student–like framework that takes the advantage of massively available unlabeled sports videos to improve player and ball tracking. An unsupervised approach to classify players in hockey according to their team affiliation is proposed in [209]. Research work in [210] shows a single fish-eye camera in enough for real-time detection of

players on a football field. A multi-tasking network is proposed in [211] that delivers jersey number recognition using a pose-guided network. A pose-based automatic performance analysis framework is proposed in [212] for diving, where is it very difficult for the judges to evaluate performance.

10.10.2 Pose Analysis

Pose analysis is the key component for individual player performance analysis and coaching assistance. Despite the progress in this field, it is still a challenging problem in sports. Several research works [212–216] addressed this problem in sports videos. A multi-person 3D pose estimation and tracking method is proposed in [213] based on 2D poses from multiple cameras. Human pose analysis in an aquatic training scenario is proposed in [214]. A transformer-based pose estimation method is proposed in [216] that is capable of detecting arbitrary key points in limbs.

10.10.3 Sports Risk Assessment and Injury Prediction

Sports risk assessment and injury prediction are relatively new fields that have been made possible by recent advancements in CV and deep learning. One important area of research in this field is head injury assessment in sports such as rugby, where it can be difficult for officials to accurately assess the severity of a head injury on the field. Research works such as [217] and [218] have addressed this issue using video analytics and deep learning techniques. These methods involve analyzing video footage of the game to identify potential head injuries and assess their severity. By automating this process, officials can make more accurate and informed decisions about player safety and potentially reduce the risk of serious head injuries. Similar work has been done in other sports as well, such as MLB pitchers, where early detection of injuries is important for ensuring player safety and preventing long-term damage. In [219], deep learning techniques were used to analyze video footage of MLB pitchers and identify early signs of injury, such as changes in pitching mechanics or arm motion. By detecting these changes early, coaches and medical staff can take preventative measures to reduce the risk of injury and ensure the long-term health of the players.

10.10.4 Event Detection

Event detection is an important prior step for automatic sports highlight generation, video indexing, etc. Many event detection frameworks has been proposed across many sports such as soccer [220–225], hockey [215, 225–227], aquatic sports [228, 229], athletics [230], etc. A temporally aware feature poling method is proposed in

[220] for action spotting in soccer videos. Research work in [221] proposed an experimental study that combines multimodal audio and video streams for action spotting in soccer videos. Instead of going through trout of traditional atomic action detection, research work in [222] performs group activity detection and with-ball interaction in soccer video. A CV-based pass count possession statistics generation method is proposed in [223], which learns the dual task of pass localization and team identification using a reinforcement learning agent. CV has been widely used in hockey video understanding in challenging tasks such as puck localization and multi-task event recognition [226], temporal hockey action recognition [215], automatic play segmentation [227], interesting event detection [225], interaction classification [208], etc. CV has been used in other sports such as for motion event detection [230] in athletics, event classification in diving, [228], and waterline detection in canoe sprint video analysis [229], real-time temporal and spatial video analysis in table tennis [231], actionness and game states recognition in basketball [232], etc.

10.10.5   Graphical Enhancement

Graphical enhancement plays a significant role in improving the viewing experience of sports events, and CV techniques are increasingly being used to develop innovative and engaging sports graphics. One area of research in this field is sports field registration, which involves accurately aligning the graphics overlays with the physical sports field. The research work mentioned in [233] proposes a solution for sports field registration, which could have a wide range of applications in sports broadcasting. Accurate field registration can facilitate augmented sports tactics analysis, where graphics can be overlaid onto the field to highlight player movements and team strategies. It can also enable virtual advertisement insertion, where ads can be seamlessly integrated into the broadcast feed. Additionally, accurate field registration can facilitate true-view replay, where viewers can watch a 3D reconstruction of the game from any angle. Overall, CV techniques have great potential for enhancing the sports viewing experience, and ongoing research in this field is likely to lead to even more innovative and engaging sports graphics in the future.

10.11   Ethical Implications of AI/CV in the Entertainment Industry

Broadly speaking, the field of AI ethics is evolving along with the technological advances of AI. Its purpose is to avoid harm to users or society overall that might come from misuse of, poor or misguided implementation of, or an unintended consequence of an AI system [234]. The set of issues covered by AI ethics discussions is as broad as the varieties of AI in use today, but in this section, we focus on ethical issues specific to CV in the entertainment industry.

CV applications, in general, can lead to several risks, including privacy violations, security breaches, spoofing or adversarial input, and the like [235]. While many of these may not directly apply to entertainment media applications, where CV is typically employed on industry-produced content, as opposed to, say, security cameras, there are several important issues to consider. As discussed earlier, CV is used in entertainment media to detect and identify people, objects, products, and more, in support of several applications. These detections, if flawed, can lead to serious user experience issues.

As an extreme example, in 2015, Google Photos rolled out an image labeling feature, giving user-uploaded photos an automatic label. But, disturbingly, this feature labeled several photos of Black persons as "gorillas" [236]. The culprit is not machine learning or AI algorithms themselves, but "the sets of images on which these systems are trained" [236]. Studies over the last several years have highlighted these shortcomings. Buolamwini and Gebru [237] studied a face database released by the US Government specifically designed to be geographically diverse, but found it to be overwhelmingly comprised of lighter-skinned subjects and males and additionally found that commercial gender classifiers performed quite poorly on darker-skinned female subjects. Han and Jain [238] similarly reviewed the Labeled Faces in the Wild dataset, which they noted had a "significantly biased race distribution."

Further, while imbalances in the face and person image datasets may have the most obvious consequences, well-known "comprehensive" image datasets, such as ImageNet, are themselves constrained, "ultimately closed systems, limited in their coverage of non-Western contexts and temporally bounded" [239]. These imbalances can impact entertainment media applications in more pervasive and subtle ways. For example, systems that can more effectively recognize objects and concepts in Western contexts will fail in contextual advertising and recommendation applications employed on content from other regions.

In addition to applications utilizing CV for recognition and identification, content generation applications are also subject to the negative effects of bias. Super-resolution applications may, in one example, unintentionally "whiten" a well-recognized and highly photographed Black man [240]. Generative AI in media merits special attention, even when bias in data sets is controlled. A 2021 documentary about the television star Anthony Bourdain, who died in 2018, crafted three lines of dialogue using generative AI to mimic Bourdain's voice but did not disclose this to its audience. The backlash was fierce [241], illustrating the sensitivity of the public to AI even when carefully controlled.

Given these potential pitfalls, a framework must be established to guide the development and use of AI in entertainment media applications. Companies in the industry may establish AI ethics guidelines for their employees, while an increasing set of laws also govern what can be done [234]. Indeed, as Bergquist argues in [242], ethics must be incorporated with AI development because:

- AI is still undergoing rapid development, particularly in the media industry.

- Laws such as the EU General Data Protection Regulation (GDPR) impose restrictions on its use.

- Failing to combat issues, such as bias, can be quite costly to a company.

- Media applications bring unique challenges, such as copyright considerations.

- Practices demanded by ethical standards enhance performance (i.e., a biased CV algorithm is a poorly performing one).

Preferably, an organization or company building an ethical framework for the use of AI in its products will have support from its leadership. Specifically, as Ammanath [243] illustrates, responsibly employing AI requires enterprises to ensure executive leaders prioritize ethics in their AI strategies, establish an oversight board for AI applications, build a diverse workforce to ensure a variety of perspectives, and establish processes for evaluating AI in the development cycle.

Similarly, Bergquist's proposed AI ethics pipeline [242] begins with organizational leadership and continues with incorporating ethics awareness and decisions into product design. Following these, the decisions then belong to the researchers and developers, who must:

- Understand the training data to actively combat bias,

- Understand the problem to be solved, avoid assumptions that could lead to biased behavior, and

- Communicate to stakeholders about biases and tradeoffs that may be present in their system [242].

Given the rapid advancement of Generative AI technology, its use in media likely merits its own guidelines. In [244], the authors argue that deepfakes and Generative AI, like other technology, may be abused through unethical use. Therefore, they propose a three-part framework for assessing Generative AI use:

- **Consent:** has explicit consent been obtained from all those depicted in the original media, and from those who created it?

- **Respect:** does the generated content fail to respect those depicted?

- **Deception:** if the generated media may be perceived as original (not generated), what is the potential harm to the viewer, or to society?

Both ethical and legal considerations (e.g., Europe's GDPR) increasingly require explainability in AI, but this can also provide benefits to the organization, rather than being a burdensome addition. Indeed, "fostering trust in AI means in part demystifying how it works, and the variety of stakeholders surrounding AI deployment have different needs for explainability" [245]. Explainability is itself an area of significant research and development. Two widely cited methods for explainability are:

- Local Interpretable Model-Agnostic Explanations (LIME), which addresses explainability through a local approximation of the model for a given prediction, which then identifies the features relevant to the prediction. [246]

- Shapely Additive Explanations (SHAP), which uses a game-theoretic approach to measure marginal contributions of features. [247]

Model and component reuse has been a key tenet of our ability to scale to several applications. When developers prepare components for internal or external distribution, they typically include at least some documentation detailing how the component is to be used. In our experience, though, this documentation focuses more on the practical aspects of running the software and integrating it into a larger system. However, Mitchell et al. introduce and advocate the use of Model Cards, a documentation framework to accompany ML models with information that can combat the misuse that can lead to biased or unintended results [248]. Model Cards contain contextual information, describing the intended use and factors (such as groups of evaluation data), which could affect performance. Additionally, there are performance benchmarks, including both a description of the metrics used, as well as a quantitative analysis that is disaggregated across factors to demonstrate the factors' effect on performance [248].

Despite our efforts to establish guardrails for ML models, models may still perform in unexpected ways. A well-known article [249] demonstrated that early deep neural networks could be fooled with unrecognizable images and label them with very high certainty. A common problem is that many ML models are "trained in a world in which everything... belongs to one of a few categories," [250] but the domain of possible inputs to the system is vast compared to the domain of the training set. These models are, in effect, forced to make a choice, even when uncertain.

Traditionally, ML models used in practice have not had a true measure of uncertainty. Instead, "the probability vector obtained at the end of the pipeline (the softmax output) is often erroneously interpreted as model confidence. A model can be uncertain in its predictions even with a high softmax output" [251]. Recent efforts have focused on measuring and reporting uncertainty in deep neural networks. Bayesian Neural Networks, which are probabilistic, are computationally intractable as a solution for measuring uncertainty but have led to practical solutions for measuring uncertainty [250]. Practitioners employing ML in entertainment applications may find that uncertainty estimates are among the best tools for eliminating bias and other unintended behaviors.

10.12   Conclusion

In this chapter, we provided a comprehensive overview of applications of CV in the media industry. This was a challenging task as most media companies do not publish their work in peer-reviewed journals and conferences. We aimed to achieve this by reviewing state-of-the-art research works, blogs, and news sources that have both present and future potential applications. Some media companies publish their work via technical reports in their respective web blogs. We summarized these works in this chapter and cited them when appropriate.

We covered most of the aspects of the media industry with regard to AI and CV. These aspects included but were not limited to both the media creation and experience side of the industry. We discussed many recent AI/CV-based approaches for media production in the domain of content generation, moderation, information retrieval, and sports. On the other hand, on the experience side, we discussed applications in the domain of advertisement, experience, streaming, and recommendation. We hope this comprehensive chapter will provide research direction to people who want to conduct groundbreaking research in the media domain as well as to people in the media industry who want to provide the best media consumption experience to their customers.

Notes

1. Authors are the members of Media Analytics research team at Comcast AI Technologies, Philadelphia, PA, USA.

2. https://www.pixelvibe.com/

3. https://generated.photos/

4. https://www.reallusion.com/

5. https://www.faceapp.com/

6. https://zaodownload.com/

7. https://learn.microsoft.com/en-us/azure/cognitive-services/computer-vision/concept-detecting-adult-content

8. https://cloud.google.com/vision/docs/detecting-safe-search

9. https://docs.aws.amazon.com/rekognition/latest/dg/moderation.html

References

[1] History.com. (2022, October) First commercial movie screened. [Online]. Available: https://www.history.com/this-day-in-history/first-commercial-movie-screened/

[2] J. L. Baird. (1933) Television in 1932 (BBC annual report 1933). [Online]. Available: https://www.bairdtelevision.com/television-in-1932-bbc-annual-report-1933.html

[3] A. Stein and B. B. Evans, *An introduction to the entertainment industry*. Peter Lang, 2009.

[4] statista.com. (2023, January) Media use in the U.S. - statistics & facts. [Online]. Available: https://www.statista.com/topics/1536/media-use/#topicOverview

[5] Zippia. (2022, May) 20 trending U.S. media and entertainment industry statistics [2022]. [Online]. Available: https://www.zippia.com/advice/media-and-entertainment-industry-statistics/

[6] D. A. Forsyth and J. Ponce, *Computer vision: A modern approach*. Prentice hall professional technical reference, 2002.

[7] F. Tsalakanidou, S. Papadopoulos, V. Mezaris, I. Kompatsiaris, B. Gray, D. Tsabouraki, M. Kalogerini, F. Negro, M. Montagnuolo, J. d. Vos *et al.*, "The AI4media project: Use of next-generation artificial intelligence technologies for media sector applications," in *IFIP International Conference on Artificial Intelligence Applications and Innovations*. Springer, 2021, pp. 81–93.

[8] K. He, X. Zhang, S. Ren, and J. Sun, "Deep residual learning for image recognition," in *Proceedings*

of the IEEE Conference on Computer Vision and Pattern Recognition, 2016, pp. 770–778.

[9] N. Parmar, A. Vaswani, J. Uszkoreit, L. Kaiser, N. Shazeer, A. Ku, and D. Tran, "Image transformer," in *International Conference on Machine Learning*. PMLR, 2018, pp. 4055–4064.

[10] O. Russakovsky, J. Deng, H. Su, J. Krause, S. Satheesh, S. Ma, Z. Huang, A. Karpathy, A. Khosla, M. Bernstein *et al.*, "Imagenet large scale visual recognition challenge," *International Journal of Computer Vision*, vol. 115, no. 3, pp. 211–252, 2015.

[11] T.-Y. Lin, M. Maire, S. Belongie, J. Hays, P. Perona, D. Ramanan, P. Dollár, and C. L. Zitnick, "Microsoft coco: Common objects in context," in *Computer Vision–ECCV 2014: 13th European Conference, Zurich, Switzerland, September 6-12, 2014, Proceedings, Part V 13*. Springer, 2014, pp. 740–755.

[12] W. Wang, Y. Yang, X. Wang, W. Wang, and J. Li, "Development of convolutional neural network and its application in image classification: A survey," *Optical Engineering*, vol. 58, no. 4, p. 040901, 2019.

[13] L. Liu, W. Ouyang, X. Wang, P. Fieguth, J. Chen, X. Liu, and M. Pietikäinen, "Deep learning for generic object detection: A survey," *International Journal of Computer Vision*, vol. 128, no. 2, pp. 261–318, 2020.

[14] M. Ye, J. Shen, G. Lin, T. Xiang, L. Shao, and S. C. Hoi, "Deep learning for person re-identification: A survey and outlook," *IEEE Transactions on Pattern Analysis and Machine Intelligence*, vol. 44, no. 6, pp. 2872–2893, 2021.

[15] A. Latif, A. Rasheed, U. Sajid, J. Ahmed, N. Ali, N. I. Ratyal, B. Zafar, S. H. Dar, M. Sajid, and T. Khalil, "Content-based image retrieval and feature extraction: A comprehensive review," *Mathematical Problems in Engineering*, vol. 2019, 2019.

[16] X. Zhu and D. Ramanan, "Face detection, pose estimation, and landmark localization in the wild," in *2012 IEEE Conference on Computer Vision and Pattern Recognition*. IEEE, 2012, pp. 2879–2886.

[17] N. Islam, Z. Islam, and N. Noor, "A survey on optical character recognition system," *arXiv preprint arXiv:1710.05703*, 2017.

[18] S. Long, X. He, and C. Yao, "Scene text detection and recognition: The deep learning era," *International Journal of Computer Vision*, vol. 129, no. 1, pp. 161–184, 2021.

[19] K. Sundararajan and D. L. Woodard, "Deep learning for biometrics: A survey," *ACM Computing Surveys (CSUR)*, vol. 51, no. 3, pp. 1–34, 2018.

[20] V. Casser, S. Pirk, R. Mahjourian, and A. Angelova, "Unsupervised monocular depth and egomotion learning with structure and semantics," in *Proceedings of the IEEE/CVF Conference on Computer Vision and Pattern Recognition Workshops*, 2019, pp. 0–0.

[21] E. Ristani and C. Tomasi, "Features for multitarget multi-camera tracking and re-identification," in *Proceedings of the IEEE Conference on Computer Vision and Pattern Recognition*, 2018, pp. 6036–6046.

[22] S. Baker, D. Scharstein, J. Lewis, S. Roth, M. J. Black, and R. Szeliski, "A database and evaluation methodology for optical flow," *International Journal of Computer Vision*, vol. 92, no. 1, pp. 1–31, 2011.

[23] C. Hane, C. Zach, A. Cohen, R. Angst, and M. Pollefeys, "Joint 3D scene reconstruction and class segmentation," in *Proceedings of the IEEE Conference on Computer Vision and Pattern Recognition*, 2013, pp. 97–104.

[24] J. Fu, J. Liu, H. Tian, Y. Li, Y. Bao, Z. Fang, and H. Lu, "Dual attention network for scene segmentation," in *Proceedings of the IEEE/CVF Conference on Computer Vision and Pattern Recognition*, 2019, pp. 3146–3154.

[25] I. Goodfellow, J. Pouget-Abadie, M. Mirza, B. Xu, D. Warde-Farley, S. Ozair, A. Courville, and Y. Bengio, "Generative adversarial networks," *Communications of the ACM*, vol. 63, no. 11, pp. 139–144, 2020.

[26] A. Creswell, T. White, V. Dumoulin, K. Arulkumaran, B. Sengupta, and A. A. Bharath, "Generative adversarial networks: An overview," *IEEE Signal Processing Magazine*, vol. 35, no. 1, pp. 53–65, 2018.

[27] R. Rombach, A. Blattmann, D. Lorenz, P. Esser, and B. Ommer, "High-resolution image synthesis with latent diffusion models," 2021.

[28] J. Liang, J. Cao, G. Sun, K. Zhang, L. Van Gool, and R. Timofte, "Swinir: Image restoration using swin transformer," in *Proceedings of the IEEE/CVF International Conference on Computer Vision*, 2021, pp. 1833–1844.

[29] H. Chen, X. He, L. Qing, Y. Wu, C. Ren, R. E. Sheriff, and C. Zhu, "Real-world single image super-resolution: A brief review," *Information Fusion*, vol. 79, pp. 124–145, 2022.

[30] G. Li, Y. Yang, X. Qu, D. Cao, and K. Li, "A deep learning based image enhancement approach for autonomous driving at night," *Knowledge-Based Systems*, vol. 213, p. 106617, 2021.

[31] Netflix Technology Blog. (2022, September) New series: Creating media with machine learning. [Online]. Available: https://netflixtechblog.com/new-series-creating-media-with-machine-learning-5067ac110bcd

[32] jeffbullas.com. (1017, October) 35 mind numbing Youtube facts, figures and statistics – infographic. [Online]. Available: https://www.jeffbullas.com/35-mind-numbing-youtube-facts-figures-and-statistics-infographic/

[33] T. Karras, S. Laine, and T. Aila, "A style-based generator architecture for generative adversarial networks," in *Proceedings of the IEEE/CVF Conference on Computer Vision and Pattern Recognition*, 2019, pp. 4401–4410.

[34] P. Esser, R. Rombach, and B. Ommer, "Taming transformers for high-resolution image synthesis," in *Proceedings of the IEEE/CVF Conference on Computer Vision and Pattern Recognition*, 2021, pp. 12 873–12 883.

[35] A. Ramesh, P. Dhariwal, A. Nichol, C. Chu, and M. Chen, "Hierarchical text-conditional image generation with clip latents," *arXiv preprint arXiv:2204.06125*, 2022.

[36] C. Saharia, W. Chan, S. Saxena, L. Li, J. Whang, E. Denton, S. K. S. Ghasemipour, B. K. Ayan, S. S. Mahdavi, R. G. Lopes *et al.*, "Photorealistic text-to-image diffusion models with deep language understanding," *arXiv preprint arXiv:2205.11487*, 2022.

[37] R. Rombach, A. Blattmann, D. Lorenz, P. Esser, and B. Ommer, "High-resolution image synthesis with latent diffusion models," in *Proceedings of the IEEE/CVF Conference on Computer Vision and Pattern Recognition*, 2022, pp. 10 684–10 695.

[38] J. Ho, A. Jain, and P. Abbeel, "Denoising diffusion probabilistic models," *Advances in Neural Information Processing Systems*, vol. 33, pp. 6840–6851, 2020.

[39] A. Nichol, P. Dhariwal, A. Ramesh, P. Shyam, P. Mishkin, B. McGrew, I. Sutskever, and M. Chen, "Glide: Towards photorealistic image generation and editing with text-guided diffusion models," *arXiv preprint arXiv:2112.10741*, 2021.

[40] H. Huang, P. S. Yu, and C. Wang, "An introduction to image synthesis with generative adversarial nets," *arXiv preprint arXiv:1803.04469*, 2018.

[41] I. Goodfellow, "Nips 2016 tutorial: Generative adversarial networks," *arXiv preprint arXiv:1701.00160*, 2016.

[42] T. Karras, S. Laine, M. Aittala, J. Hellsten, J. Lehtinen, and T. Aila, "Analyzing and improving the image quality of stylegan," in *Proceedings of the IEEE/CVF Conference on Computer Vision and Pattern Recognition*, 2020, pp. 8110–8119.

[43] A. Brock, J. Donahue, and K. Simonyan, "Large scale gan training for high fidelity natural image synthesis," *arXiv preprint arXiv:1809.11096*, 2018.

[44] M. Heusel, H. Ramsauer, T. Unterthiner, B. Nessler, and S. Hochreiter, "Gans trained by a two time-scale update rule converge to a local nash equilibrium," *Advances in Neural Information Processing Systems*, vol. 30, 2017.

[45] T. Salimans, I. Goodfellow, W. Zaremba, V. Cheung, A. Radford, and X. Chen, "Improved techniques for training gans," *Advances in Neural Information Processing Systems*, vol. 29, 2016.

[46] T. Kynkäänniemi, T. Karras, S. Laine, J. Lehtinen, and T. Aila, "Improved precision and recall metric for assessing generative models," *Advances in Neural Information Processing Systems*, vol. 32, 2019.

[47] C. Nash, J. Menick, S. Dieleman, and P. W. Battaglia, "Generating images with sparse representations," *arXiv preprint arXiv:2103.03841*, 2021.

[48] A. Q. Nichol and P. Dhariwal, "Improved denoising diffusion probabilistic models," in *International Conference on Machine Learning*. PMLR, 2021, pp. 8162–8171.

[49] T. Miyato, T. Kataoka, M. Koyama, and Y. Yoshida, "Spectral normalization for generative adversarial networks," *arXiv preprint arXiv:1802.05957*, 2018.

[50] A. Brock, T. Lim, J. M. Ritchie, and N. Weston, "Neural photo editing with introspective

adversarial networks," *arXiv preprint arXiv:1609. 07093*, 2016.

[51] P. Dhariwal and A. Nichol, "Diffusion models beat gans on image synthesis," *Advances in Neural Information Processing Systems*, vol. 34, pp. 8780–8794, 2021.

[52] O. Gafni, A. Polyak, O. Ashual, S. Sheynin, D. Parikh, and Y. Taigman, "Make-a-scene: Scene-based text-to-image generation with human priors," *arXiv preprint arXiv:2203.13131*, 2022.

[53] A. Radford, J. W. Kim, C. Hallacy, A. Ramesh, G. Goh, S. Agarwal, G. Sastry, A. Askell, P. Mishkin, J. Clark *et al.*, "Learning transferable visual models from natural language supervision," in *International Conference on Machine Learning.* PMLR, 2021, pp. 8748–8763.

[54] C. Raffel, N. Shazeer, A. Roberts, K. Lee, S. Narang, M. Matena, Y. Zhou, W. Li, P. J. Liu *et al.*, "Exploring the limits of transfer learning with a unified text-to-text transformer." *Journal of Machine Learning Research*, vol. 21, no. 140, pp. 1–67, 2020.

[55] O. Ronneberger, P. Fischer, and T. Brox, "U-Net: Convolutional networks for biomedical image segmentation," in *International Conference on Medical Image Computing and Computer-assisted Intervention.* Springer, 2015, pp. 234–241.

[56] FXHome. (2023, February) AI tools for filmmakers. [Online]. Available: https://www.desktop-documentaries.com/ai-tools-for-filmmakers.html

[57] Desktop-documentaries.com. (2023, May) Utilizing AI for documentary production. [Online]. Available: https://www.desktop-documentaries.com/ai-for-documentary-filmmakers.html

[58] N. Routley. (2023, February) Infographic: Generative ai explained by ai. [Online]. Available: https://www.visualcapitalist.com/generative-ai-explained-by-ai/

[59] N. Ruiz, Y. Li, V. Jampani, Y. Pritch, M. Rubinstein, and K. Aberman, "Dreambooth: Fine tuning text-to-image diffusion models for subject-driven generation," *arXiv preprint arxiv:2208.12242*, 2022.

[60] M. T. Miller. (2023, May) AI-generated movie posters are truly wild and disconcerting. [Online]. Available: https://nerdist.com/article/ai-generated-movie-posters-artificial-intelligence-robomojo/

[61] G. Mittal, T. Marwah, and V. N. Balasubramanian, "Sync-draw: Automatic video generation using deep recurrent attentive architectures," in *Proceedings of the 25th ACM International Conference on Multimedia*, 2017, pp. 1096–1104.

[62] Y. Pan, Z. Qiu, T. Yao, H. Li, and T. Mei, "To create what you tell: Generating videos from captions," in *Proceedings of the 25th ACM International Conference on Multimedia*, 2017, pp. 1789–1798.

[63] Y. Liu, X. Wang, Y. Yuan, and W. Zhu, "Cross-modal dual learning for sentence-to-video generation," in *Proceedings of the 27th ACM International Conference on Multimedia*, 2019, pp. 1239–1247.

[64] T. Gupta, D. Schwenk, A. Farhadi, D. Hoiem, and A. Kembhavi, "Imagine this! scripts to compositions to videos," in *Proceedings of the European Conference on Computer Vision (ECCV)*, 2018, pp. 598–613.

[65] U. Singer, A. Polyak, T. Hayes, X. Yin, J. An, S. Zhang, Q. Hu, H. Yang, O. Ashual, O. Gafni *et al.*, "Make-a-video: Text-to-video generation without text-video data," *arXiv preprint arXiv:2209.14792*, 2022.

[66] J. Ho, W. Chan, C. Saharia, J. Whang, R. Gao, A. Gritsenko, D. P. Kingma, B. Poole, M. Norouzi, D. J. Fleet *et al.*, "Imagen video: High definition video generation with diffusion models," *arXiv preprint arXiv:2210.02303*, 2022.

[67] R. Rakhimov, D. Volkhonskiy, A. Artemov, D. Zorin, and E. Burnaev, "Latent video transformer," *arXiv preprint arXiv:2006.10704*, 2020.

[68] W. Yan, Y. Zhang, P. Abbeel, and A. Srinivas, "VideoGPT: Video generation using VQ-VAE and transformers," *arXiv preprint arXiv:2104.10157*, 2021.

[69] C. Wu, L. Huang, Q. Zhang, B. Li, L. Ji, F. Yang, G. Sapiro, and N. Duan, "GODIVA: Generating open-domain videos from natural descriptions," *arXiv preprint arXiv:2104.14806*, 2021.

[70] C. Wu, J. Liang, L. Ji, F. Yang, Y. Fang, D. Jiang, and N. Duan, "NUWA: Visual synthesis pre-training for neural visual world creation," in *European Conference on Computer Vision.* Springer, 2022, pp. 720–736.

[71] J. Ho, T. Salimans, A. Gritsenko, W. Chan, M. Norouzi, and D. J. Fleet, "Video diffusion models," *arXiv preprint arXiv:2204.03458*, 2022.

[72] L. Loic. (2022, July) The 9 best AI video generators. [Online]. Available: https://www.makeuseof.com/best-ai-video-generators-text-to-video/

[73] P. Esser, J. Chiu, P. Atighehchian, J. Granskog, and A. Germanidis, "Structure and content-guided video synthesis with diffusion models," arXiv:2302.03011, 2023.

[74] RunwayResearch. (2023, May) Reimagining creativity with artificial intelligence. [Online]. Available: https://research.runwayml.com

[75] M. Kharbach. (2023, May) 6 of the best ai tools to make videos from text (text to video). [Online]. Available: https://www.educatorstechnology.com/2023/02/6-of-best-ai-tools-to-make-videos-from-2.html

[76] Charisma.ai. (2023, May) Build conversational characters powered by AI. [Online]. Available: https://charisma.ai

[77] A. Conner-Simons. (2018, September) An AI for CGI. [Online]. Available: https://www.csail.mit.edu/news/ai-cgi

[78] CGIKite. (2022, September) AI for CGI: Advancements and opportunities. [Online]. Available: https://cgikite.com/fr/tpost/l5nb7koti1-ai-for-cgi-advancements-and-opportunitie

[79] SabrePC. (2023, February) 5 recent amazing AI advancements. [Online]. Available: https://www.sabrepc.com/blog/Deep-Learning-and-AI/5-recent-amazing-ai-advancements

[80] P. Clark. (2023, May) Adobe supercharges photoshop with firefly generative AI. [Online]. Available: https://blog.adobe.com/en/publish/2023/05/23/photoshop-new-features-ai-contextual-presets

[81] C. Crew. (2023, February) Did we just change animation forever? [Online]. Available: https://www.youtube.com/watch?v=_9LX9HSQkWo

[82] A. Schwanke. (2023, April) Move over, disney. AI is giving youtubers the tools to make your next animation obsession. [Online]. Available: https://www.techradar.com/features/move-over-disney-ai-is-giving-youtubers-the-tools-to-make-your-next-animation-obsession

[83] Netflix Technology Blog. (2018, November) Delivering meaning with previews on web. [Online]. Available: https://netflixtechblog.com/delivering-meaning-with-previews-on-web-3cedc0341b9e

[84] Netflix Technology Blog. (2017, December) Artwork personalization at Netflix. [Online]. Available: https://netflixtechblog.com/artwork-personalization-c589f074ad76

[85] M. Gygli, Y. Song, and L. Cao, "Video2GIF: Automatic generation of animated GIFs from video," in *Proceedings of the IEEE Conference on Computer Vision and Pattern Recognition (CVPR)*, June 2016.

[86] K. Zhang, W.-L. Chao, F. Sha, and K. Grauman, "Video summarization with long short-term memory," in *Computer Vision–ECCV 2016: 14th European Conference, Amsterdam, The Netherlands, October 11–14, 2016, Proceedings, Part VII 14*. Springer, 2016, pp. 766–782.

[87] E. Apostolidis, E. Adamantidou, A. I. Metsai, V. Mezaris, and I. Patras, "Video summarization using deep neural networks: A survey," *Proceedings of the IEEE*, vol. 109, no. 11, pp. 1838–1863, 2021.

[88] S. Saboo, F. Lefebvre, and V. Demoulin, "Deep learning and interactivity for video rotoscoping," in *2020 IEEE International Conference on Image Processing (ICIP)*. IEEE, 2020, pp. 643–647.

[89] W. Li, F. Viola, J. Starck, G. J. Brostow, and N. D. Campbell, "Roto++ accelerating professional rotoscoping using shape manifolds," *ACM Transactions on Graphics (TOG)*, vol. 35, no. 4, pp. 1–15, 2016.

[90] J. Stringer, N. Sundararajan, J. Pina, S. Kar, N. L. Dabby, A. Lin, L. Bermudez, and S. Hilmarsdottir, "Rotomation: AI powered rotoscoping at LAIKA," in *Special Interest Group on Computer Graphics and Interactive Techniques Conference Talks*, 2020, pp. 1–2.

[91] Z. Yang, H. Ai, B. Wu, S. Lao, and L. Cai, "Face pose estimation and its application in video shot selection," in *Proceedings of the 17th International Conference on Pattern Recognition, 2004. ICPR 2004.*, vol. 1. IEEE, 2004, pp. 322–325.

[92] A. Rao, L. Xu, Y. Xiong, G. Xu, Q. Huang, B. Zhou, and D. Lin, "A local-to-global approach to multi-modal movie scene segmentation," in *Proceedings of the IEEE/CVF Conference on Computer Vision and Pattern Recognition*, 2020, pp. 10 146–10 155.

[93] P. Postma and B. Chorley, "Color grading with color management," *SMPTE Motion Imaging Journal*, vol. 125, no. 8, pp. 69–74, 2016.

[94] L. Galteri, L. Seidenari, M. Bertini, and A. Del Bimbo, "Deep generative adversarial compression artifact removal," in *Proceedings of the IEEE International Conference on Computer Vision*, 2017, pp. 4826–4835.

[95] J. Dalins, C. Wilson, and D. Boudry, "PDQ & TMK + PDQF – a test drive of Facebook's perceptual hashing algorithms," arXiv:1912.07745, 2019.

[96] G. Varol, I. Laptev, and C. Schmid, "Long-term temporal convolutions for action recognition," *IEEE Transactions on Pattern Analysis and Machine Intelligence*, vol. 40, no. 6, pp. 1510–1517, 2017.

[97] Z. Liu, Y. Lin, Y. Cao, H. Hu, Y. Wei, Z. Zhang, S. Lin, and B. Guo, "Swin transformer: Hierarchical vision transformer using shifted windows," in *Proceedings of the IEEE/CVF International Conference on Computer Vision*, 2021, pp. 10 012–10 022.

[98] M. Koopman, A. M. Rodriguez, and Z. Geradts, "Detection of deepfake video manipulation," in *The 20th Irish Machine Vision and Image Processing Conference (IMVIP)*, 2018, pp. 133–136.

[99] S. Liu, Z. Lian, S. Gu, and L. Xiao, "Block shuffling learning for deepfake detection," *arXiv preprint arXiv:2202.02819*, 2022.

[100] F. Matern, C. Riess, and M. Stamminger, "Exploiting visual artifacts to expose deepfakes and face manipulations," in *2019 IEEE Winter Applications of Computer Vision Workshops (WACVW)*. IEEE, 2019, pp. 83–92.

[101] A. Rossler, D. Cozzolino, L. Verdoliva, C. Riess, J. Thies, and M. Nießner, "Faceforensics++: Learning to detect manipulated facial images," in *Proceedings of the IEEE/CVF International Conference on Computer Vision*, 2019, pp. 1–11.

[102] E. Sabir, J. Cheng, A. Jaiswal, W. AbdAlmageed, I. Masi, and P. Natarajan, "Recurrent convolutional strategies for face manipulation detection in videos," *Interfaces (GUI)*, vol. 3, no. 1, pp. 80–87, 2019.

[103] T. Zhao, X. Xu, M. Xu, H. Ding, Y. Xiong, and W. Xia, "Learning self-consistency for deepfake detection," in *Proceedings of the IEEE/CVF International Conference on Computer Vision*, 2021, pp. 15 023–15 033.

[104] A. Bharati, R. Singh, M. Vatsa, and K. W. Bowyer, "Detecting facial retouching using supervised deep learning," *IEEE Transactions on Information Forensics and Security*, vol. 11, no. 9, pp. 1903–1913, 2016.

[105] S. McCloskey and M. Albright, "Detecting GAN-generated imagery using color cues," *arXiv preprint arXiv:1812.08247*, 2018.

[106] D. A. Coccomini, N. Messina, C. Gennaro, and F. Falchi, "Combining efficientnet and vision transformers for video deepfake detection," in *Image Analysis and Processing–ICIAP 2022: 21st International Conference, Lecce, Italy, May 23–27, 2022, Proceedings, Part III*. Springer, 2022, pp. 219–229.

[107] Y. Deldjoo, M. Schedl, P. Cremonesi, and G. Pasi, "Recommender systems leveraging multimedia content," *ACM Computing Surveys*, vol. 53, no. 5, sep 2020. [Online]. Available: https://doi.org/10.1145/3407190

[108] Y. Koren and R. Bell, *Advances in Collaborative Filtering*. Boston, MA: Springer US, 2011, pp. 145–186. [Online]. Available: https://doi.org/10.1007/978-0-387-85820-3_5

[109] Y. Hu, Y. Koren, and C. Volinsky, "Collaborative filtering for implicit feedback datasets," in *2008 Eighth IEEE International Conference on Data Mining*, 2008, pp. 263–272.

[110] E. Çano, "Hybrid recommender systems: A systematic literature review," *Intelligent Data Analysis*, vol. 21, pp. 1487–1524, 11 2017.

[111] X. Chen, P. Zhao, J. Xu, Z. Li, L. Zhao, Y. Liu, V. S. Sheng, and Z. Cui, "Exploiting visual contents in posters and still frames for movie recommendation," *IEEE Access*, vol. 6, pp. 68 874–68 881, 2018.

[112] Y. Deldjoo, M. Elahi, P. Cremonesi, F. Garzotto, P. Piazzolla, and M. Quadrana, "Content-based video recommendation system based on stylistic visual features," *Journal on Data Semantics*, vol. 5, pp. 1–15, 06 2016.

[113] Y. Deldjoo, M. G. Constantin, H. Eghbal-Zadeh, B. Ionescu, M. Schedl, and P. Cremonesi, "Audio-visual encoding of multimedia content for enhancing movie recommendations," in *Proceedings of the 12th ACM Conference on Recommender Systems*, ser. RecSys '18. New York, NY, USA: Association for Computing Machinery, 2018, p. 455–459. [Online]. Available: https://doi.org/10.1145/3240323.3240407

[114] Y. Deldjoo, M. Elahi, and P. Cremonesi, "Using visual features and latent factors for movie recommendation," in *CBRecSys@RecSys*, 2016.

[115] Y. Deldjoo, M. Elahi, M. Quadrana, and P. Cremonesi, "Using visual features based on MPEG-7 and deep learning for movie recommendation," *International Journal of Multimedia Information Retrieval*, vol. 7, pp. 1–13, 11 2018.

[116] X. Du, H. Yin, L. Chen, Y. Wang, Y. Yang, and X. Zhou, "Personalized video recommendation using rich contents from videos," *IEEE Transactions on Knowledge and Data Engineering*, vol. 32, no. 3, p. 492–505, mar 2020. [Online]. Available: https://doi.org/10.1109/TKDE.2018.2885520

[117] A. Almeida, J. P. de Villiers, A. De Freitas, and M. Velayudan, "The complementarity of a diverse range of deep learning features extracted from video content for video recommendation," *Expert Systems with Applications*, vol. 192, p. 116335, 2022. [Online]. Available: https://www.sciencedirect.com/science/article/pii/S095741742101633X

[118] L. Canini, S. Benini, and R. Leonardi, "Affective recommendation of movies based on selected connotative features," *IEEE Transactions on Circuits and Systems for Video Technology*, vol. 23, no. 4, pp. 636–647, 2013.

[119] S. Roy and S. C. Guntuku, "Latent factor representations for cold-start video recommendation," in *Proceedings of the 10th ACM Conference on Recommender Systems*, 2016, pp. 99–106.

[120] Y. Shi, A. Karatzoglou, L. Baltrunas, M. Larson, N. Oliver, and A. Hanjalic, "CLiMF: Learning to maximize reciprocal rank with collaborative less-is-more filtering," in *Proceedings of the Sixth ACM Conference on Recommender Systems*, ser. RecSys '12. New York, NY, USA: Association for Computing Machinery, 2012, p. 139–146. [Online]. Available: https://doi.org/10.1145/2365952.2365981

[121] Q. Zhu, M.-L. Shyu, and H. Wang, "Videotopic: Content-based video recommendation using a topic model," in *2013 IEEE International Symposium on Multimedia*, 2013, pp. 219–222.

[122] Y. Deldjoo, M. F. Dacrema, M. G. Constantin, H. Eghbal-zadeh, S. Cereda, M. Schedl, B. Ionescu, and P. Cremonesi, "Movie genome: Alleviating new item cold start in movie recommendation," *User Modeling and User-adapted Interaction*, vol. 29, no. 2, pp. 291–343, 2019.

[123] T. Mei, B. Yang, X.-S. Hua, and S. Li, "Contextual video recommendation by multimodal relevance and user feedback," *ACM Transactions on Information Systems*, vol. 29, no. 2, apr 2011. [Online]. Available: https://doi.org/10.1145/1961209.1961213

[124] W. Song, Z. Duan, Z. Yang, H. Zhu, M. Zhang, and J. Tang, "Ekar: An explainable method for knowledge aware recommendation," *arXiv e-prints*, p. arXiv:1906.09506, Jun. 2019.

[125] X. Wang, X. He, Y. Cao, M. Liu, and T.-S. Chua, "KGAT: Knowledge graph attention network for recommendation," in *Proceedings of the 25th ACM SIGKDD International Conference on Knowledge Discovery & Data Mining*, ser. KDD '19. New York, NY, USA: Association for Computing Machinery, 2019, p. 950–958. [Online]. Available: https://doi.org/10.1145/3292500.3330989

[126] S. Wang, L. Hu, Y. Wang, X. He, Q. Sheng, M. Orgun, L. Cao, F. Ricci, and P. Yu, "Graph learning based recommender systems: A review," in *Proceedings of the Thirtieth International Joint Conference on Artificial Intelligence, IJCAI 2021*, ser. IJCAI International Joint Conference on Artificial Intelligence, Z.-H. Zhou, Ed. International Joint Conferences on Artificial Intelligence, 2021, pp. 4644–4652, 30th International Joint Conference on Artificial Intelligence, IJCAI 2021 ; Conference date: 19-08-2021 Through 27-08-2021.

[127] amazon.com. (2023, February) Prime video x-ray. [Online]. Available: https://www.amazon.com/adlp/xray/

[128] W. Zhao, R. Chellappa, P. J. Phillips, and A. Rosenfeld, "Face recognition: A literature survey," *ACM Computing Surveys (CSUR)*, vol. 35, no. 4, pp. 399–458, 2003.

[129] M. Wang and W. Deng, "Deep face recognition: A survey," *Neurocomputing*, vol. 429, pp. 215–244, 2021.

[130] F. Schroff, D. Kalenichenko, and J. Philbin, "FaceNet: A unified embedding for face recognition and clustering," in *Proceedings of the IEEE Conference on Computer Vision and Pattern Recognition*, 2015, pp. 815–823.

[131] engadget.com. (2023, February) Peacock's latest update includes a 'key plays' feature for premier league games. [Online]. Available: https://www.engadget.com/peacock-ui-update-183247828.html

[132] I. Krylov, S. Nosov, and V. Sovrasov, "Open images v5 text annotation and yet another mask text spotter," in *Asian Conference on Machine Learning*. PMLR, 2021, pp. 379–389.

[133] B. Shi, X. Bai, and C. Yao, "An end-to-end trainable neural network for image-based sequence recognition and its application to scene text recognition," *IEEE Transactions on Pattern Analysis and Machine Intelligence*, vol. 39, no. 11, pp. 2298–2304, 2016.

[134] theverge.com. (2023, February) Vizio's profit on ads, subscriptions, and data is double the money it makes selling TVs. [Online]. Available: https://www.theverge.com/2021/11/10/22773073/vizio-acr-advertising-inscape-data-privacy-q3-2021

[135] M. A. Gavrielides, E. Sikudova, and I. Pitas, "Color-based descriptors for image fingerprinting," *IEEE Transactions on Multimedia*, vol. 8, no. 4, pp. 740–748, 2006.

[136] J. S. Seo, J. Haitsma, T. Kalker, and C. D. Yoo, "A robust image fingerprinting system using the radon transform," *Signal Processing: Image Communication*, vol. 19, no. 4, pp. 325–339, 2004.

[137] Y. Li, D. Wang, and L. Tang, "Robust and secure image fingerprinting learned by neural network," *IEEE Transactions on Circuits and Systems for Video Technology*, vol. 30, no. 2, pp. 362–375, 2019.

[138] Google. (2023, February) Google lens. [Online]. Available: https://lens.google/#shopping

[139] L. Karlinsky, J. Shtok, Y. Tzur, and A. Tzadok, "Fine-grained recognition of thousands of object categories with single-example training," in *Proceedings of the IEEE Conference on Computer Vision and Pattern Recognition*, 2017, pp. 4113–4122.

[140] Y. Wei, S. Tran, S. Xu, B. Kang, M. Springer *et al.*, "Deep learning for retail product recognition: Challenges and techniques," *Computational Intelligence and Neuroscience*, vol. 2020, 2020.

[141] N. Zhang, T. Mei, X.-S. Hua, L. Guan, and S. Li, "Tap-to-search: Interactive and contextual visual search on mobile devices," in *2011 IEEE 13th International Workshop on Multimedia Signal Processing*, 2011, pp. 1–5.

[142] A. Broder, M. Fontoura, V. Josifovski, and L. Riedel, "A semantic approach to contextual advertising," in *Proceedings of the 30th Annual International ACM SIGIR Conference on Research and Development in Information Retrieval*, 2007, pp. 559–566.

[143] K. Zhang and Z. Katona, "Contextual advertising," *Marketing Science*, vol. 31, no. 6, pp. 980–994, 2012.

[144] A. Carlitz, *The Impact of Video Advertisement Placement and Video Advertisement Transparency on Video Advertisement Avoidance: Pre-Rolls, Mid-Rolls, and Psychological Reactance*. Ohio University, 2020.

[145] M. M. Islam, M. Hasan, K. S. Athrey, T. Braskich, and G. Bertasius, "Efficient movie scene detection using state-space transformers," *arXiv preprint arXiv:2212.14427*, 2022.

[146] techcrunch.com. (2022, May) Virtual product placement ads are coming to Amazon Prime video and Peacock. [Online]. Available: https://techcrunch.com/2022/05/17/virtual-product-placement-ads-are-coming-to-amazon-prime-video-and-peacock/

[147] SuperOffice. (2023, February) How a customer experience strategy helps scale revenue growth. [Online]. Available: https://www.superoffice.com/blog/customer-experience-strategy/

[148] J. A. Starosta and B. Izydorczyk, "Understanding the phenomenon of binge-watching—A systematic review," *International Journal of Environmental Research and Public Health*, vol. 17, no. 12, p. 4469, 2020.

[149] YouGovAmerica. (2023, February) Most Americans binge watch TV shows. [Online]. Available: https://today.yougov.com/topics/entertainment/articles-reports/2017/09/13/58-americans-binge-watch-tv-shows

[150] M. Nematollahi and X.-P. Zhang, "Automatic video intro and outro detection on internet television," in *Proceedings of the First International Workshop on Internet-Scale Multimedia Management*, 2014, pp. 43–46.

[151] Netflix Technology Blog. (2015, April) Extracting contextual information from video assets. [Online]. Available: https://netflixtechblog.com/extracting-contextual-information-from-video-assets-ee9da25b6008

[152] T. Redaelli and J. Ekedahl, "Automated intro detection for tv series (dissertation)," 2020. [Online]. Available: http://urn.kb.se/resolve?urn=urn:nbn:se:kth:diva-269419

[153] X. Hao, K. Chettiar, B. Cheung, V. Germano, and R. Hamid, "Intro and recap detection for movies and TV series," in *Proceedings of the IEEE/CVF Winter Conference on Applications of Computer Vision*, 2021, pp. 167–176.

[154] vionlabs.com. (2023, February) Vionlabs press release 2021 Feb 2. [Online]. Available: https://www.vionlabs.com/post/vionlabs-press-release-2021-feb-2

[155] Amazon Rekognition. (2023, February) Amazon Rekognition video for media analysis. [Online]. Available: https://aws.amazon.com/rekognition/media-analysis/

[156] Cognitive Mill. (2023, February) End credits detection. [Online]. Available: https://cognitivemill.com/solutions/end-credits-detection/

[157] The Bridge. (2017, August) Automated ad detection is a game-changer for ott content aggregators. [Online]. Available: https://www.thebroadcastbridge.com/content/entry/9293/automated-ad-detection-is-a-game-changer-for-ott-content-aggregators

[158] B. Satterwhite and O. Marques, "Automatic detection of TV commercials," *IEEE Potentials*, vol. 23, no. 2, pp. 9–12, 2004.

[159] M. Li, Y. Guo, and Y. Chen, "CNN-based commercial detection in TV broadcasting," in *Proceedings of the 2017 VI International Conference on Network, Communication and Computing*, 2017, pp. 48–53.

[160] S. Minaee, I. Bouazizi, P. Kolan, and H. Najafzadeh, "Ad-Net: Audio-visual convolutional neural network for advertisement detection in videos," *arXiv preprint arXiv:1806.08612*, 2018.

[161] A. Gomes, M. P. Queluz, and F. Pereira, "Automatic detection of TV commercial blocks: A new approach based on digital on-screen graphics classification," in *2017 11th International Conference on Signal Processing and Communication Systems (ICSPCS)*. IEEE, 2017, pp. 1–6.

[162] P. Carvalho, A. Pereira, and P. Viana, "Automatic TV logo identification for advertisement detection without prior data," *Applied Sciences*, vol. 11, no. 16, p. 7494, 2021.

[163] N. A. Khan, U. Amin, M. Umer *et al.*, "Unsupervised commercials identification in videos," *International Journal of Advanced Computer Science and Applications*, vol. 8, no. 2, 2017.

[164] G. Kordopatis-Zilos, S. Papadopoulos, I. Patras, and Y. Kompatsiaris, "Near-duplicate video retrieval with deep metric learning," in *Proceedings of the IEEE International Conference on Computer Vision Workshops*, 2017, pp. 347–356.

[165] Y. Chen, W. He, Y. Hua, and W. Wang, "CompoundEyes: Near-duplicate detection in large scale online video systems in the cloud," in *IEEE INFOCOM 2016-The 35th Annual IEEE International Conference on Computer Communications*. IEEE, 2016, pp. 1–9.

[166] G. Kordopatis-Zilos, S. Papadopoulos, I. Patras, and I. Kompatsiaris, "ViSiL: Fine-grained spatiotemporal video similarity learning," in *Proceedings of the IEEE/CVF International Conference on Computer Vision*, 2019, pp. 6351–6360.

[167] Z. Han, X. He, M. Tang, and Y. Lv, "Video similarity and alignment learning on partial video copy detection," in *Proceedings of the 29th ACM International Conference on Multimedia*, 2021, pp. 4165–4173.

[168] X. He, Y. Pan, M. Tang, Y. Lv, and Y. Peng, "Learn from unlabeled videos for near-duplicate video retrieval," in *Proceedings of the 45th International ACM SIGIR Conference on Research and Development in Information Retrieval*, 2022, pp. 1002–1011.

[169] Disabilities, Opportunities, Internetworking, and Technology. (2021, April) What is audio description? [Online]. Available: https://www.washington.edu/doit/what-audio-description

[170] nomensa. (2023, February) What is audio description? [Online]. Available: https://www.nomensa.com/blog/what-audio-description

[171] Y. Wang, W. Liang, H. Huang, Y. Zhang, D. Li, and L.-F. Yu, "Toward automatic audio description generation for accessible videos," in *Proceedings of the 2021 CHI Conference on Human Factors in Computing Systems*, 2021, pp. 1–12.

[172] I. R. Filho, F. Honorato, J. W. Lucena, J. P. Teixeira, and T. Maritan, "An approach for automatic description of characters for blind people," in *Proceedings of the Brazilian Symposium on Multimedia and the Web*, 2021, pp. 53–56.

[173] S. C. Sansalone, C. M. Culver, and M. A. Sukhai, "Audio description in video games research in progress," in *Proceedings of the 2022 CONF-IRM*, 2022.

[174] Q. Zhang, Y. Song, and Q. Jin, "Unifying event detection and captioning as sequence generation via pre-training," in *European Conference on Computer Vision*. Springer, 2022, pp. 363–379.

[175] T. Wang, R. Zhang, Z. Lu, F. Zheng, R. Cheng, and P. Luo, "End-to-end dense video captioning with parallel decoding," in *Proceedings of the IEEE/CVF International Conference on Computer Vision*, 2021, pp. 6847–6857.

[176] Z. Liu, X. Wu, and Y. Yu, "Multi-task video captioning with a stepwise multimodal encoder," *Electronics*, vol. 11, no. 17, p. 2639, 2022.

[177] A. Rohrbach, M. Rohrbach, S. Tang, S. Joon Oh, and B. Schiele, "Generating descriptions with grounded and co-referenced people," in *Proceedings of the IEEE Conference on Computer Vision and Pattern Recognition*, 2017, pp. 4979–4989.

[178] N. C. Camgoz, O. Koller, S. Hadfield, and R. Bowden, "Sign language transformers: Joint end-to-end sign language recognition and translation," in *Proceedings of the IEEE/CVF Conference on Computer Vision and Pattern Recognition*, 2020, pp. 10 023–10 033.

[179] A. Yin, Z. Zhao, W. Jin, M. Zhang, X. Zeng, and X. He, "MLSLT: Towards multilingual sign language translation," in *Proceedings of the IEEE/CVF Conference on Computer Vision and Pattern Recognition*, 2022, pp. 5109–5119.

[180] A. Duarte, S. Albanie, X. Giró-i Nieto, and G. Varol, "Sign language video retrieval with free-form textual queries," in *Proceedings of the IEEE/CVF Conference on Computer Vision and Pattern Recognition*, 2022, pp. 14 094–14 104.

[181] B. Saunders, N. C. Camgoz, and R. Bowden, "Progressive transformers for end-to-end sign language production," in *European Conference on Computer Vision*. Springer, 2020, pp. 687–705.

[182] B. Saunders, N. C. Camgoz, and R. Bowden, "Mixed signals: Sign language production via a mixture of motion primitives," in *Proceedings of the IEEE/CVF International Conference on Computer Vision*, 2021, pp. 1919–1929.

[183] B. Saunders, N. C. Camgoz, and R. Bowden, "Signing at scale: Learning to co-articulate signs for large-scale photo-realistic sign language production," in *Proceedings of the IEEE/CVF Conference on Computer Vision and Pattern Recognition*, 2022, pp. 5141–5151.

[184] A. Glasser, V. Mande, and M. Huenerfauth, "Accessibility for deaf and hard of hearing users: Sign language conversational user interfaces," in *Proceedings of the 2nd Conference on Conversational User Interfaces*, 2020, pp. 1–3.

[185] Netflix Technology Blog. (2016, Jun) Toward a practical perceptual video quality metric. [Online]. Available: https://netflixtechblog.com/toward-a-practical-perceptual-video-quality-metric-653f208b9652

[186] R. Zhang, P. Isola, A. A. Efros, E. Shechtman, and O. Wang, "The unreasonable effectiveness of deep features as a perceptual metric," in *Proceedings of the IEEE Conference on Computer Vision and Pattern Recognition*, 2018, pp. 586–595.

[187] Y. Li, L.-M. Po, C.-H. Cheung, X. Xu, L. Feng, F. Yuan, and K.-W. Cheung, "No-reference video quality assessment with 3D shearlet transform and convolutional neural networks," *IEEE Transactions on Circuits and Systems for Video Technology*, vol. 26, no. 6, pp. 1044–1057, 2015.

[188] Z. Wang, H. R. Sheikh, and A. C. Bovik, "No-reference perceptual quality assessment of JPEG compressed images," in *Proceedings. International Conference on Image Processing*, vol. 1. IEEE, 2002, pp. I–I.

[189] M. Agarla, L. Celona, and R. Schettini, "An efficient method for no-reference video quality assessment," *Journal of Imaging*, vol. 7, no. 3, p. 55, 2021.

[190] J. Korhonen, "Two-level approach for no-reference consumer video quality assessment," *IEEE Transactions on Image Processing*, vol. 28, no. 12, pp. 5923–5938, 2019.

[191] X. Guan, F. Li, Y. Zhang, and P. C. Cosman, "End-to-end blind video quality assessment based on visual and memory attention modeling," *IEEE Transactions on Multimedia*, 2022.

[192] B. Schwarz, "Mapping the world in 3D," *Nature Photonics*, vol. 4, no. 7, pp. 429–430, 2010.

[193] A. Hornung, K. M. Wurm, M. Bennewitz, C. Stachniss, and W. Burgard, "OctoMap: An efficient probabilistic 3D mapping framework based on octrees," *Autonomous Robots*, vol. 34, pp. 189–206, 2013.

[194] J. McCormac, A. Handa, A. Davison, and S. Leutenegger, "SemanticFusion: Dense 3D semantic mapping with convolutional neural networks," in *2017 IEEE International Conference on Robotics and automation (ICRA)*. IEEE, 2017, pp. 4628–4635.

[195] B. C. Ko, "A brief review of facial emotion recognition based on visual information," *Sensors*, vol. 18, no. 2, p. 401, 2018.

[196] Y. Feng, F. Wu, X. Shao, Y. Wang, and X. Zhou, "Joint 3D face reconstruction and dense alignment with position map regression network," in *Proceedings of the European Conference on Computer Vision (ECCV)*, 2018, pp. 534–551.

[197] A. Kammoun, R. Slama, H. Tabia, T. Ouni, and M. Abid, "Generative adversarial networks for face generation: A survey," *ACM Computing Surveys*, vol. 55, no. 5, pp. 1–37, 2022.

[198] M. Jung, H. Yang, and K. Min, "Improving deep object detection algorithms for game scenes," *Electronics*, vol. 10, no. 20, 2021. [Online]. Available: https://www.mdpi.com/2079-9292/10/20/2527

[199] S. McAuley, S. Hill, N. Hoffman, Y. Gotanda, B. Smits, B. Burley, and A. Martinez, "Practical physically-based shading in film and game production," in *ACM SIGGRAPH 2012 Courses*, 2012, pp. 1–7.

[200] O. Nalbach, E. Arabadzhiyska, D. Mehta, H.-P. Seidel, and T. Ritschel, "Deep shading: Convolutional neural networks for screen space shading," in *Computer Graphics Forum*, vol. 36, no. 4. Wiley Online Library, 2017, pp. 65–78.

[201] H. Kato, D. Beker, M. Morariu, T. Ando, T. Matsuoka, W. Kehl, and A. Gaidon, "Differentiable rendering: A survey," *arXiv preprint arXiv:2006.12057*, 2020.

[202] I. Rabbi and S. Ullah, "A survey on augmented reality challenges and tracking," *Acta graphica: znanstveni časopis za tiskarstvo i grafičke komunikacije*, vol. 24, no. 1-2, pp. 29–46, 2013.

[203] F. Mueller, D. Mehta, O. Sotnychenko, S. Sridhar, D. Casas, and C. Theobalt, "Real-time hand tracking under occlusion from an egocentric RGB-D sensor," in *Proceedings of the IEEE International Conference on Computer Vision*, 2017, pp. 1154–1163.

[204] M. Speicher, B. D. Hall, and M. Nebeling, "What is mixed reality?" in *Proceedings of the 2019 CHI Conference on Human Factors in Computing Systems*, 2019, pp. 1–15.

[205] K. Vats, W. McNally, P. Walters, D. A. Clausi, and J. S. Zelek, "Ice hockey player identification via transformers and weakly supervised learning," in *Proceedings of the IEEE/CVF Conference on Computer Vision and Pattern Recognition*, 2022, pp. 3451–3460.

[206] A. Maglo, A. Orcesi, and Q.-C. Pham, "Efficient tracking of team sport players with few game-specific annotations," in *Proceedings of the IEEE/CVF Conference on Computer Vision and Pattern Recognition*, 2022, pp. 3461–3471.

[207] R. Vandeghen, A. Cioppa, and M. Van Droogenbroeck, "Semi-supervised training to improve player and ball detection in soccer," in *Proceedings of the IEEE/CVF Conference on Computer Vision and Pattern Recognition*, 2022, pp. 3481–3490.

[208] F. Askari, R. Ramaprasad, J. J. Clark, and M. D. Levine, "Interaction classification with key actor detection in multi-person sports videos," in *Proceedings of the IEEE/CVF Conference on Computer Vision and Pattern Recognition*, 2022, pp. 3580–3588.

[209] M. Koshkina, H. Pidaparthy, and J. H. Elder, "Contrastive learning for sports video: Unsupervised player classification," in *Proceedings of the IEEE/CVF Conference on Computer Vision and Pattern Recognition*, 2021, pp. 4528–4536.

[210] A. Cioppa, A. Deliege, N. U. Huda, R. Gade, M. Van Droogenbroeck, and T. B. Moeslund, "Multimodal and multiview distillation for real-time player detection on a football field," in *Proceedings of the IEEE/CVF Conference on Computer Vision and Pattern Recognition Workshops*, 2020, pp. 880–881.

[211] H. Liu and B. Bhanu, "Pose-guided R-CNN for jersey number recognition in sports," in *Proceedings of the IEEE/CVF Conference on Computer Vision and Pattern Recognition Workshops*, 2019, pp. 0–0.

[212] M. Nekoui, F. O. T. Cruz, and L. Cheng, "Falcons: Fast learner-grader for contorted poses in sports," in *Proceedings of the IEEE/CVF Conference on Computer Vision and Pattern Recognition Workshops*, 2020, pp. 900–901.

[213] L. Bridgeman, M. Volino, J.-Y. Guillemaut, and A. Hilton, "Multi-person 3D pose estimation and tracking in sports," in *Proceedings of the IEEE/CVF Conference on Computer Vision and Pattern Recognition Workshops*, 2019, pp. 0–0.

[214] D. Zecha, M. Einfalt, and R. Lienhart, "Refining joint locations for human pose tracking in sports videos," in *Proceedings of the IEEE/CVF Conference on Computer Vision and Pattern Recognition Workshops*, 2019, pp. 0–0.

[215] Z. Cai, H. Neher, K. Vats, D. A. Clausi, and J. Zelek, "Temporal hockey action recognition via pose and optical flows," in *Proceedings of the IEEE/CVF Conference on Computer Vision and Pattern Recognition Workshops*, 2019, pp. 0–0.

[216] K. Ludwig, D. Kienzle, and R. Lienhart, "Recognition of freely selected keypoints on human limbs," in *Proceedings of the IEEE/CVF Conference on Computer Vision and Pattern Recognition*, 2022, pp. 3531–3539.

[217] N. Nonaka, R. Fujihira, M. Nishio, H. Murakami, T. Tajima, M. Yamada, A. Maeda, and J. Seita, "End-to-end high-risk tackle detection system for rugby," in *Proceedings of the IEEE/CVF Conference on Computer Vision and Pattern Recognition*, 2022, pp. 3550–3559.

[218] Z. Martin, S. Hendricks, and A. Patel, "Automated tackle injury risk assessment in contact-based sports-a rugby union example," in *Proceedings of the IEEE/CVF Conference on Computer Vision and Pattern Recognition*, 2021, pp. 4594–4603.

[219] A. Piergiovanni and M. S. Ryoo, "Early detection of injuries in MLB pitchers from video," in *Proceedings of the IEEE/CVF Conference on Computer Vision and Pattern Recognition Workshops*, 2019, pp. 0–0.

[220] S. Giancola and B. Ghanem, "Temporally-aware feature pooling for action spotting in soccer broadcasts," in *Proceedings of the IEEE/CVF Conference on Computer Vision and Pattern Recognition*, 2021, pp. 4490–4499.

[221] B. Vanderplaetse and S. Dupont, "Improved soccer action spotting using both audio and video streams," in *Proceedings of the IEEE/CVF Conference on Computer Vision and Pattern Recognition Workshops*, 2020, pp. 896–897.

[222] R. Sanford, S. Gorji, L. G. Hafemann, B. Pourbabaee, and M. Javan, "Group activity detection from trajectory and video data in soccer," in *Proceedings of the IEEE/CVF Conference on Computer Vision and Pattern Recognition Workshops*, 2020, pp. 898–899.

[223] S. Sarkar, D. P. Mukherjee, and A. Chakrabarti, "Watch and act: Dual interacting agents for automatic generation of possession statistics in soccer," in *Proceedings of the IEEE/CVF Conference on Computer Vision and Pattern Recognition*, 2022, pp. 3560–3568.

[224] R. Li and B. Bhanu, "Fine-grained visual dribbling style analysis for soccer videos with augmented dribble energy image," in *Proceedings of the IEEE/CVF Conference on Computer Vision and Pattern Recognition Workshops*, 2019, pp. 0–0.

[225] K. Vats, M. Fani, P. Walters, D. A. Clausi, and J. Zelek, "Event detection in coarsely annotated sports videos via parallel multi-receptive field 1D convolutions," in *Proceedings of the IEEE/CVF Conference on Computer Vision and Pattern Recognition Workshops*, 2020, pp. 882–883.

[226] K. Vats, M. Fani, D. A. Clausi, and J. Zelek, "Puck localization and multi-task event recognition in broadcast hockey videos," in *Proceedings of the IEEE/CVF Conference on Computer Vision and Pattern Recognition*, 2021, pp. 4567–4575.

[227] H. Pidaparthy, M. H. Dowling, and J. H. Elder, "Automatic play segmentation of hockey videos," in *Proceedings of the IEEE/CVF Conference on Computer Vision and Pattern Recognition*, 2021, pp. 4585–4593.

[228] G. Kanojia, S. Kumawat, and S. Raman, "Attentive spatio-temporal representation learning for diving classification," in *Proceedings of the IEEE/CVF Conference on Computer Vision and Pattern Recognition Workshops*, 2019, pp. 0–0.

[229] M.-S. von Braun, P. Frenzel, C. Kading, and M. Fuchs, "Utilizing mask R-CNN for waterline detection in canoe sprint video analysis," in *Proceedings of the IEEE/CVF Conference on Computer Vision and Pattern Recognition Workshops*, 2020, pp. 876–877.

[230] M. Einfalt and R. Lienhart, "Decoupling video and human motion: Towards practical event detection in athlete recordings," in *Proceedings of the IEEE/CVF Conference on Computer Vision and Pattern Recognition Workshops*, 2020, pp. 892–893.

[231] R. Voeikov, N. Falaleev, and R. Baikulov, "TTNet: Real-time temporal and spatial video analysis of table tennis," in *Proceedings of the IEEE/CVF Conference on Computer Vision and Pattern Recognition Workshops*, 2020, pp. 884–885.

[232] J. Quiroga, H. Carrillo, E. Maldonado, J. Ruiz, and L. M. Zapata, "As seen on TV: Automatic basketball video production using gaussian-based actionness and game states recognition," in *Proceedings of the IEEE/CVF Conference on Computer Vision and Pattern Recognition Workshops*, 2020, pp. 894–895.

[233] Y.-J. Chu, J.-W. Su, K.-W. Hsiao, C.-Y. Lien, S.-H. Fan, M.-C. Hu, R.-R. Lee, C.-Y. Yao, and H.-K. Chu, "Sports field registration via keypoints-aware label condition," in *Proceedings of the IEEE/CVF Conference on Computer Vision and Pattern Recognition*, 2022, pp. 3523–3530.

[234] D. Mehan, *Artificial Intelligence: Ethical, Social and Security Impacts for the Present and the Future.* Ely, Cambridgeshire, United Kingdom: IT Governance Ltd, 2022, ch. 4.

[235] M. Skirpan and T. Yeh, "Designing a moral compass for the future of computer vision using speculative analysis," in *2017 IEEE Conference on Computer Vision and Pattern Recognition Workshops (CVPRW)*, 2017, pp. 1368–1377.

[236] B. Christian, *The Alignment Problem: Machine Learning and Human Values.* New York, NY: Norton and Company, 2020, ch. 1, pp. 29–30.

[237] J. Buolamwini and T. Gebru, "Gender shades: Intersectional accuracy disparities in commercial gender classification," in *Proceedings of the 1st Conference on Fairness, Accountability and Transparency*, ser. Proceedings of Machine Learning Research, S. A. Friedler and C. Wilson, Eds., vol. 81. PMLR, 23–24 Feb 2018, pp. 77–91. [Online]. Available: https://proceedings.mlr.press/v81/buolamwini18a.html

[238] H. Han and A. K. Jain, "Age, gender and race estimation from unconstrained face images," in *MSU Technical Report MSU-CSE-14-5*, 2014.

[239] D. Raji, E. Denton, E. M. Bender, A. Hanna, and A. Paullada, "AI and the everything in the whole wide world benchmark," in *Proceedings of the Neural Information Processing Systems Track on Datasets and Benchmarks*, J. Vanschoren and S. Yeung, Eds., vol. 1, 2021. [Online]. Available: https://datasets-benchmarks-proceedings.neurips.cc/paper/2021/file/084b6fbb10729ed4da8c3d3f5a3ae7c9-Paper-round2.pdf

[240] J. Vincent. What a machine learning tool that turns Obama white can (and can't) tell us about AI bias. [Online]. Available: https://www.theverge.com/21298762/face-depixelizer-ai-machine-learning-tool-pulse-stylegan-obama-bias

[241] H. Rosner, "The ethics of a deepfake anthony bourdain voice," *The New Yorker*, July 2021. [Online]. Available: https://www.newyorker.com/culture/annals-of-gastronomy/the-ethics-of-a-deepfake-anthony-bourdain-voice

[242] Y. Bergquist, "AI ethics: A framework for the media industry," *Medium*, May 2022. [Online]. Available: https://medium.com/@ybergquist/ai-ethics-a-framework-for-the-media-industry-2a5cf2719e13

[243] B. Ammanath, *Trustworthy AI.* Hoboken, NJ: John Wiley & Sons, Inc., 2022, ch. 10.

[244] B. Lyon and M. Tora, *Exploring Deepfakes.* Packt Publishing Ltd., Livery Place, 35 Livery Street, Birmingham, B3 2PB, UK.: Packt Publishing, 2023, ch. 2.

[245] B. Ammanath, *Trustworthy AI.* Hoboken, NJ: John Wiley & Sons, Inc., 2022, ch. 5.

[246] M. T. Ribeiro, S. Singh, and C. Guestrin, "Why should I trust you?": Explaining the predictions of any classifier," in *Proceedings of the 22nd ACM SIGKDD International Conference on Knowledge Discovery and Data Mining*, ser. KDD '16. New York, NY, USA: Association for Computing Machinery, 2016, p. 1135–1144. [Online]. Available: https://doi.org/10.1145/2939672.2939778

[247] S. M. Lundberg and S.-I. Lee, "A unified approach to interpreting model predictions," in *Proceedings of the 31st International Conference on Neural Information Processing Systems*, ser. NIPS'17. Red Hook, NY, USA: Curran Associates Inc., 2017, p. 4768–4777.

[248] M. Mitchell, S. Wu, A. Zaldivar, P. Barnes, L. Vasserman, B. Hutchinson, E. Spitzer, I. D. Raji, and T. Gebru, "Model cards for model reporting," in *Proceedings of the Conference on Fairness, Accountability, and Transparency.* ACM, jan 2019. [Online]. Available: https://doi.org/10.1145%2F3287560.3287596

[249] A. Nguyen, J. Yosinski, and J. Clune, "Deep neural networks are easily fooled: High confidence predictions for unrecognizable images," in *2015 IEEE Conference on Computer Vision and Pattern Recognition (CVPR)*, 2015, pp. 427–436.

[250] B. Christian, *The Alignment Problem: Machine Learning and Human Values.* New York, NY: Norton and Company, 2020, ch. 9, pp. 281–286.

[251] Y. Gal, "Uncertainty in deep learning," Ph.D. dissertation, University of Cambridge, 2016.

11

Quality Assessment in Media and Entertainment: Challenges and Trends

Abhinau K. Venkataramanan, Zaixi Shang, Joshua P. Ebenezer, Meixu Chen, Zhengzhong Tu, and Alan C. Bovik

Assessing and optimizing the perceptual quality of media plays an important role in enhancing the subjective experiences of hundreds of millions of television, streaming, and social media viewers. Quality assessment methods may be broadly categorized into "subjective" and "objective" methods. Subjective quality assessment consists of conducting studies where participants manually rate the quality of media. On the other hand, objective quality models consist of algorithms designed to accurately predict subjective quality. In this chapter, we will discuss the challenges in assessing the quality of popular visual media, including streaming video, virtual reality, high-dynamic-range, high-framerate, and user-generated video. In particular, we will discuss subjective databases and state-of-the-art quality models that have been developed to address the problem of quality assessment. Finally, we will discuss future trends and open problems that are interesting to tackle.

11.1 Standard Streaming Videos

Video-On-Demand (VOD) and streaming videos comprise the most ubiquitous video modality consumed by users on the Internet. VOD includes a broad range of platforms such as Netflix, Amazon Prime Video, Hulu, Disney+, HBO Max, etc. In this chapter, we will use VOD to refer to professionally generated content, which is typically produced with high-quality equipment that deliver pristine source contents. User-generated content will be discussed separately in Section 11.2 and is considered a separate video modality in this chapter, even when it is "streamed" on the Internet.

In order to stream videos under the bandwidth constraints of the Internet, lossy compression is used to significantly reduce file sizes. However, since compression is lossy, compressed videos are "distorted" versions of their respective source contents. Common distortions associated with video compression are blocking artifacts, temporal incoherency, and color bleeding.

In this section, we will first review video compression and popular SOTA codecs. Then, we will discuss rate-distortion optimization (RDO), which refers to the common paradigm used to "optimally" encode videos. Then, we will provide an overview of commonly used subjective quality databases that study the impact of compression on subjective quality. Finally, we will discuss state-of-the-art (SOTA) quality assessment algorithms for compressed videos and outline interesting future work.

11.1.1 Overview of Video Compression

The widespread use of streaming and social media platforms has led to an explosion in the amount of video content that is being generated and consumed on the Internet. In order to reliably store and deliver such large volumes of bandwidth-hungry video content, lossy compression is a necessity. In general, the term "lossy compression" refers to "encoding" videos in order to reduce their file sizes and/or bandwidth requirements, without guaranteeing perfect reconstruction at the decoder.

Modern video compression algorithms may be broadly summarized as comprising the following sequence of operations, whose order may vary.

1. Converting pixel values to an entropy-reducing or sparsifying transform domain, such as the Discrete Cosine Transform [1] or the Integer Transform [2].

2. Coding "intra" frames by computing residuals between neighboring blocks from the same frame.

3. Coding "inter" frames using block motion estimation and compensation to compute motion-compensated residuals.

DOI: 10.1201/9781003328957-11

4. Quantization of transformed residuals, which is the source of "loss" that may be tuned based on user preference.

5. Entropy coding of transformed residuals, motion vectors, and other data using algorithms such as Context Adaptive Variable Length Coding (CAVLC) [3] and Context Adaptive Binary Arithmetic Coding (CABAC) [4] to yield the compressed bitstream.

As described above, the amount of quantization determines the amount of "loss" in the bitstream, thereby affecting the file size/compression ratio/bitrate. The amount of compression (and so, quantization) is typically tuned by selecting values of the "quantization parameter" (QP), "constant rate factor" (CRF), or the "bitrate."

Popular video compression algorithms include H.264/Advanced Video Coding (AVC) [3], H.265/High Efficiency Video Coding (HEVC) [5], and the recent H.266/Versatile Video Coding (VVC) [6], which have all been developed by the Moving Picture Experts Group (MPEG). In addition to such proprietary algorithms, the popularity of open-source compression algorithms has grown in recent years with the development of algorithms such as VP9 [7] by Google and AOMedia Video 1 (AV1) [8] by the Alliance for Open Media (AOM).

11.1.2 Scaling and Rate-distortion Optimization

As mentioned in Section 11.1.1, compression algorithms are equipped with a user-selected parameter (such as CRF, QP, or directly setting the bitrate) to control the trade-off between the bitrate of the encoded video and the amount of "loss," which can be characterized by distortion functions. Traditionally, metrics such as Mean Square Error (MSE) or peak signal-to-noise ratio (PNSR) have been used to measure compression-induced distortion. However, with the development of perceptual video quality algorithms, models such as Structural Similarity (SSIM) [9, 10] and Visual Multimethod Assessment Fusion (VMAF) [11] have become widely adopted in the video industry.

In practice, videos are typically encoded at a lower encoding resolution compared to their native resolution in order to exploit spatial redundancy and to achieve higher compression ratios. This is particularly useful for source videos that do not contain a great amount of detail. For example, a 1080p source content may be scaled down to 480p prior to encoding by downscaling and compression. At the user's end, the encoded bitstream is decoded and upscaled to 1080p for display. This process is illustrated in Figure 11.1.

While the relationship between bitrate and distortion is generally monotonic for a given content, the perceptual

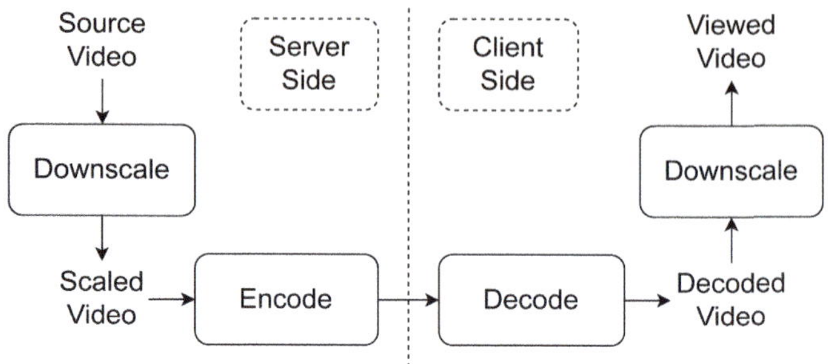

FIGURE 11.1

Combining scaling and encoding for optimal compression.

effects of compression significantly vary with and depend on content, and the combined effects of scaling and further complicates analysis and modeling of distortion. RDO algorithms such as the Dynamic Optimizer [12] tackle this problem by performing a convex-hull analysis to identify a set of "Pareto-optimal" encoding "recipes" (such as a combination of encoding scale and bitrate) from the given set. The Pareto-optimal set essentially consists of a set of encoding recipes that may be ordered in terms of their rates and distortions, thereby restoring monotonicity. During deployment, the best encoding recipe is chosen based on a rate/distortion constraint.

While the traditional approach to scaling involves the use of classical interpolation algorithms such as Bicubic/Lanczos interpolation, modern approaches use Deep Neural Networks (DNNs) to learn scaling algorithms from data by optimizing the rate-distortion tradeoff. Deep Video Precoding [13] and Deep Perceptual Optimization (DPO) [14] are two recent advancements in this nascent area of research. In addition to the aforementioned algorithms, "progressive" neural network architectures have been used to support fractional scale factors [15].

11.1.3 Subjective Quality Databases

The effect of video compression on perceptual quality has been an area of extensive research over the last decade. As a result, several subjective databases of compressed HD videos exist in the literature. In this section, we will describe some exemplary subjective databases that have aided the development of powerful quality assessment algorithms. A summary of the databases described below is provided in Table 11.1.

1. **Bristol Vision Institute Codec Comparison Database (BVI-CC):** The BVI-CC [16] database compares the performance of three popular video codecs, namely HEVC, VVC, and AV1, when compressing High-Definition (HD) and Ultra-High-Definition (UHD) content. The database consists of nine source

TABLE 11.1
Summary of Subjective Video Databases

Database	Source Videos	Test Videos
BVI-CC [16]	9	198
LIVE-SJTU A/VQA [17]	14	336
VQEG-HDTV [18]	45	828
LIVE Streaming [19]	45	245

sequences subjected to both scaling and compression distortions, leading to a total of 198 distorted test videos.

The study found that according to human viewers, AV1 and HEVC performed similarly on average, while VVC generally outperformed both AV1 and HEVC. In addition, the study investigated the effect of the Dynamic Optimizer, which was found to improve the performance of HEVC over AV1.

2. **LIVE-SJTU Audio and Video Quality Assessment Database (LIVE-SJTU A/VQA)**: The LIVE-SJTU A/VQA [17] database studies the joint impact of audio and video compression on subjective quality, which is representative of streaming use-cases. The database consists of 14 reference and 336 distorted sequences, generated by applying both video compression (with and without scaling) using the HEVC codec, and audio compression using the Advanced Audio Coding (AAC) codec.

The study found that subjects reported higher difficulty in assessing audio quality compared to video quality, with joint audio–video quality being reported as the most difficult. Furthermore, the authors observed that subjective scores varied non-monotonically with combined audio and video bitrates, thereby illustrating the complexity of the multimodal quality assessment task.

3. **Video Quality Experts Group - High Definition Television Database (VQEG-HDTV)**: VQEG-HDTV [18] is a collection of five publicly available databases created by VQEG to validate objective quality models that may be used to evaluate the perceptual quality of HD videos. In total, the set of databases consists of 45 diverse source contents compressed using the MPEG-2 [20] and AVC codecs at various bitrates to yield 828 test contents. In addition to compression, the database also includes simulated IP network errors.

4. **LIVE Livestreaming Database**: The LIVE Livestreaming database [19] is a large subjective database that studies the impact of several streaming-related distortions such as compression, aliasing, judder, and frame drops. The database consists of 45 HD and UHD source contents, distorted to produce 245 test contents. The compression artifacts were produced using the AVC codec applied at four quality levels.

The study observed that UHD videos were typically rated higher than HD videos. Furthermore, aliasing due to scaling affected the quality of HD videos to most, while judder had the greatest impact on the subjective quality of UHD videos.

11.1.4 Objective Quality Assessment

In this section, we review three classes of objective quality models, namely, full-reference (FR), no-reference (NR), and bitstream-based quality models.

11.1.4.1 Full-Reference Quality Assessment

The journey of modern image and video quality assessment (I/VQA) begins with the development of the SSIM [9] index. Prior to SSIM, the standard measure of "distortion" was the PSNR, which depends logarithmically on the mean squared error (MSE). However, PSNR was found to correlate poorly with subjective judgments, which is the primary goal of objective quality models.

SSIM introduced a much-needed perceptual element to the measurements of distortions by using similarity metrics that embody Weber masking of both luminance and contrast. Weber masking refers to the phenomenon that human perception is sensitive to relative changes, rather than absolute changes. As a result, in the presence of a large "background value," such as a high luminance or contrast, small changes in the value are not perceptible, that is, are "masked." SSIM and multi-scale SSIM (MS-SSIM) have successfully been used in the quality control of most cable, satellite, and fiber optic Internet streaming television providers and by most social media outlets. Recently, a variant of SSIM called Enhanced SSIM [10] has been shown to achieve even higher accuracy at lower computational cost.

Another breakthrough in quality assessment occurred with the development of the Visual Information Fidelity (VIF) quality model. VIF was the first algorithm to use powerful statistical models of natural scenes, modified to model distortions, to conduct FR quality assessment. VIF uses a Gaussian Scale Mixture (GSM) to model the statistics of bandpass wavelet coefficients, and a local Gaussian channel to model distortions. Versions of VIF have been deployed to conduct the quality control of videos on the Internet, most notably by contributing most of the features in the SOTA VMAF quality model. Furthermore, the philosophy behind VIF has been used to develop reduced-reference quality models such as ST-RRED [21] and SpEED-QA [22].

The Detail Loss Metric (DLM) [23] is another quality model that has emerged as a powerful tool to measure the quality of compressed videos. This is particularly because the primary spatial artifact introduced during compression is blocking, which leads to loss in detail that DLM is designed to capture. DLM uses a dyadic wavelet decomposition of input frames and a model of the contrast sensitivity function (CSF) of the Human Visual System (HVS) to decompose distortions into detail losses and additive impairments. Due to its high performance, DLM contributes one part of the SOTA VMAF quality model.

As alluded to above, the SOTA video quality models used by the streaming industry are MS-SSIM and VMAF, which was jointly developed by Netflix and academic collaborators. The core principle of VMAF is to fuse smaller "atom" quality models to obtain a better-performing quality model. VMAF can be viewed as an ensemble model consisting of "weaker" quality models. In particular, VMAF uses DLM, VIF computed at four scales on a Gaussian pyramid, and a temporal difference feature to capture Motion, as its atom features. VMAF and its anti-hacking variant [24] have been widely adopted in the industry for quality assessment, RDO, and codec comparison.

11.1.4.2 *No-Reference Quality Models*

All the aforementioned algorithms are said to be "full-reference" since they assume the availability of a "reference video" for each test video. The reference video is a "pristine" version of the test video that is used as a gold standard against which the test video is evaluated. On the server-side, this assumption generally holds since the pristine source video is used to generate the encoded videos. As a result, the majority of quality algorithms used in the streaming video industry are FR.

Nevertheless, "no-reference" quality models have been an active area of research, finding applications in moderating the quality of source contents prior to encoding, and for any client-side quality assessment. As the name suggests, "no-reference" models operate on test videos directly without using a paired reference video. The classical approach to NR quality assessment has been using natural scene statistics (NSS).

"Natural" images have been observed to exhibit regular statistical properties when subjected to band-pass decompositions, and distorted videos exhibit deviations from ideal behavior. Modeling the statistics of band-pass coefficients yields quality-aware features that are used to predict the subjective quality of videos. This approach has formed the basis of popular NR image quality models such as DIIVINE [25], BRISQUE [26], and NIQE [27]. When assessing the quality of videos, these models are applied frame-wise and the quality scores are aggregated across frames.

Extending NSS-based quality models to video typically involves introducing temporal information in the form of spatio-temporal band-pass decompositions [28], or using optical flow/motion vectors. Quality models that utilize motion information have studied the use of optical flow statistics [29, 30], motion-coherency [31], and localized space-time slices called "chips" [32].

11.1.4.3 *Bitstream-based Quality Models*

A recent novel approach to assessing the quality of compressed videos is "bitstream-based" quality assessment. While traditional FR and NR models operate on decoded frames, that is, pixel values, bitstream-based models operate directly on information extracted from the encoded bitstream. This alleviates the additional computational burden of decoding videos prior to quality assessment. Common "features" obtained from the bitstream include values of frame resolution, frame rate, frame type (I, B, or P-frame), encoded frame size, bitrate, and QP.

The SOTA for bitstream-based quality assessment is the ITU P.1204.3 [33] standard, which uses a combination of a parametric model and a machine learning model to predict quality scores.

11.1.5 Challenges and Trends

Despite the success of FR models such as VMAF, there remains a variety of challenges to be tackled. Firstly, the SOTA FR model VMAF demands a high computational load due to the calculation of a heterogeneous set of atom features. Given the development of hardware encoders for popular codecs, quality assessment (using VMAF) has emerged as the bottleneck operation in RDO. In order to address these concerns and to advance the SOTA, the FUNQUE [34] model has been proposed. FUNQUE "unifies" the atom quality features by computing them from a common HVS-aware decomposition. The HVS-aware decomposition provides for an 10% improvement in correlation against subjective scores compared to VMAF, while the computation sharing reduces the computational load by a factor of 8!

Despite the advances made by FUNQUE, both VMAF and FUNQUE operate only on the luma (grayscale) channel and are almost entirely applied framewise. The only temporal information that has been included in the model is the temporal difference between successive reference frames. A common solution is to introduce optical-flow-based quality models and apply quality models on various color channels [35, 36]. However, such methods often drastically increase the computational burden of quality assessment. So, incorporating color and temporal quality information into SOTA FR models in an efficient manner has proven to be an open practical challenge.

Furthermore, the deep-learning revolution has arrived in the quality assessment space, with the development of models such as LPIPS [37], NIMA [38], Patch-VQ [39], and CONTRIQUE [40]. The current SOTA deep architecture for vision tasks is the Transformer [41], which "tokenizes" images into patches and applies attention mechanisms to identify salient features. The success of Transformers in computer vision suggests that their application to quality assessment could break new ground. A common challenge that hinders the widespread adoption of deep models is the computational burden of applying them at scale. However, the development of hardware-accelerated deep learning platforms and large server farms with dedicated tools such as Graphical Processing Units (GPUs) and Tensor Processing Units (TPUs) aims to bridge the gap between research and the industry.

Finally, in practice, watching most videos involves listening to the audio, which encounters many of the same compression/transmission challenges that video faces. As a result, end users often simultaneously encounter distorted audio and video. However, the effect of audio on subjective quality is less understood when compared to videos. In fact, human subjects report greater difficulty in rating audio and audio-visual quality when compared to visual quality. Due to these factors, despite recent advances [17], multimodal audio-visual quality assessment remains a challenging problem.

11.2 User-Generated Content (UGC) Videos

In January 1996, Bill Gates stated in a published article that "content is king" [42]. The television industry during the past century has arguably validated the value of his words, but it has become even more true in the 2020s as we enter a new phase of Internet where overwhelming user creation weighs over professional artifacts made by traditional media. The rise of social media platforms such as Facebook, YouTube, TikTok, coupled with the increased availability of affordable mobile capturing devices, has led to UGC being created, uploaded, and shared almost instantly and ubiquitously across the globe everyday. Among the vast amount of shared video contents,

UGC videos undoubtedly account for the largest portion of Internet streams. It has been reported that YouTube viewers watch over 1 billion hours of video every day and generate billions of views [43]. Thus, it is of great importance for streaming companies to understand and analyze this new type of content, and accordingly, design optimized UGC processing pipelines to improve the encoding and streaming efficiency of UGC videos.

11.2.1 Introduction to UGC Video Quality

VQA is one of the core research topics in video compression and streaming pipelines that are designed to monitor, control, and calibrate video processing algorithms. Most VQA models may be classified into two categories: full-reference models (FR-VQA) and no-reference or blind (NR-VQA or BVQA) metrics. FR-VQA models compare the perceptual difference against pristine videos, while BVQA models directly predict the quality without references. For UGC videos, the "pristine" video is never available, and hence, only NR-VQA models can be applied, which makes the UGC-VQA problem [44] very challenging. Moreover, unlike relatively simple synthesized distortions simulated in traditional VQA databases, UGC video distortions arise from realistic user captures. Thus, UGC videos suffer from unknown, highly diverse, and commingled impairments. These authentic and intermixed visual defects include, but are not limited to, bad lighting conditions, capturing by shaky hands, imperfect shooting devices, processing and editing effects, compression and streaming artifacts, etc. Figure 11.2 shows some examples of UGC videos having various degrees of degradation. Indeed, apart from the technical distortions such as noise, blur, ringing, and banding that have been studied for decades, there exist even more complicated and subjective factors that affect how viewers perceive regarding aesthetics, interestingness, and content preference. Finally, many current video databases dedicated for UGC quality prediction are far smaller in size compared to mainstream video databases [45]. Moreover, none of them are capable of modeling the full complexity and diversity of real-world UGC distortions. Due to these challenging aspects, predicting the perceptual quality for UGC videos has remained largely unsolved for several years. So, it is of great interest to both industry and academia to investigate this most general and impactful problem in video quality research.

11.2.2 Subjective UGC Video Quality Databases

Subjective video quality research usually requires extensive resources invested in time- and label-consuming human studies on a selected video corpus to obtain reliable subjective data. These human-labeled videos are then

(a)

(b)

(c)

(d)

FIGURE 11.2
Exemplar frames of UGC video contents sampled from the LIVE-VQC database (with courtesy from [46].) These natural videos captured from mobile devices are subject to (a-d) unknown, highly diverse video distortions, and making it harder for objective video quality models to accurately predict their perceived qualities.

used to build a video quality database for the training, benchmarking, and calibration of video quality prediction (VQA) models. Traditional subjective video quality databases are typically constructed by first selecting a small set of high-quality sources, dubbed "pristine" or "reference," then synthetically distorting them using simulated approaches. Some examples of databases constructed in this manner are LIVE-VQA [47], EPFL-PoliMI [48], CSIQ [49], and MCL [50]. These databases are limited in their diversity of contents and distortions, making them incapable of modeling more complicated, real-world UGC impairments.

Recently, researchers have started to build "in-the-wild" databases that include authentically distorted contents captured under realistic settings. We list and briefly summarize these efforts as follows.

1. **CVD2014 [51].** The Camera Video Database (CVD2014) is perhaps the first authentically distorted video quality database in the literature. It consists of 234 480p and 720p videos recorded using 78 different cameras that are prone to distortions induced by video acquisition. However, the database only covers five unique scenes.

2. **LIVE-Qualcomm Mobile In-Capture Database [52].** The LIVE-Qualcomm database proposed in 2017 is another camera-captured database containing 208 realistic consumer contents with smartphones in full HD (1080p) resolution. Compared to CVD2014, this dataset contains more than 50 different scenes, thus having better diversity in contents.

3. **KoNViD-1k [53].** The KoNViD-1k dataset is a subjective UGC video quality database consisting of 1,200 public-domain video sequences. The 1,200 sequences were sampled from a large public video dataset named YFCC100m [54], based on a selected feature space of blur, contrast, colorfulness, Spatial

Information (SI), Temporal Information (TI), and NIQE scores [27]. All the videos were trimmed and resized to 540p for eight seconds.

4. **LIVE-VQC [46].** The LIVE-VQC database is another large-scale UGC database containing videos that were captured using a diverse set of mobile devices. It consists of 585 contents that span various resolutions, layouts, and framerates. Subjective scores were collected from 4,776 unique participants via online crowdsourcing on Amazon Mechanical Turk.

5. **YouTube UGC [55].** In 2020, Google released a large-scale UGC dataset called YouTube UGC, whose videos were sampled from uncurated YouTube uploads. It contains a total of 1,380 unique 20-second-long video clips, having resolutions ranging from 360p to 4k and framerates ranging from 15 to 60. Over 8,000 subjects were hired to rate these videos in a crowdsourcing environment.

6. **KonVid-150k [56].** The KoNViD-150k dataset follows similar design principles as KonViD-1k, but was orders-of-magnitude larger. It consists of two subsets, a coarsely annotated set of 153,841 videos having five quality ratings each, and a densely annotated set of 1,596 videos with over 89 ratings each. Notably, the first subset of KonVid-150k can be used as weakly labeled pre-training set for various downstream VQA tasks.

7. **LIVE-FB LSVQ [39].** LSVQ is yet another large-scale "in-the-wild" UGC database containing 39,000 real-world videos and 117k space-time localized video patches, labeled by 5.5 million human perceptual quality annotations. It is constructed from web-scale crawled video sources including Internet Archive [57] and YFCC-100M [54] to make the dataset more representative of real-world content.

8. **LIVE-YT Gaming [58].** LIVE-YouTube Gaming dataset is the first UGC video dataset dedicated for a specific type of UGC – gaming contents. It consists of 600 real-streaming UGC gaming videos spanning 59 various types of games.

11.2.3 Objective UGC Video Quality Assessment

The goal of subjective video quality studies or databases is to help develop automatic objective VQA metrics. As described before, UGC video quality problems can only be solved by NR-VQA or BVQA algorithms since no "pristine" video is available for these UGC uploads. Thus, here we briefly review the evolution of BVQA models, from traditional feature-based approaches to recent deep learning-based models.

Traditional feature-based BVQA models. Almost all the earliest blind quality prediction models were developed for a single or at most two types of distortions. Some examples are models for analyzing noise [59], blockiness [60], blur [61], ringing [62], banding artifacts [63] in streamed, distorted videos, or to assess multiply concurrent compression and transmission defects [64, 65], to name a few. More recently, general-purpose BVQA models that are capable of capturing versatile distortions have been developed based on NSS [66] and its deduced simple but highly effective low-level features [26, 27] equipped with machine learning techniques. Among these NSS-based models, BRISQUE [26] and NIQE [27] operate in the divisively normalized image domain, BLIINDS [67] examines statistics in the DCT domain, and Video-BLIINDS [31] extended BLIINDS to analyze DCT transforms of frame differences. TLVQM [29] leverages a bag of well-defined quality-related features extracted from multiple visual spaces and then trains a shallow support vector regressor. VIDEVAL [44] is the most recent SOTA BVQA model, and it uses an ensemble of features based on NSS inspired by multiple prior BVQA models. VIDEVAL has been shown to achieve strong performance on UGC databases.

Deep learning-based BVQA models. Despite performing well on recent UGC-VQA databases, these "handcrafted" feature-based models have been observed to underperform on recently released larger-scale datasets such as LSVQ. On the other hand, deep convolutional neural networks (ConvNet) that are purely data-driven have gained strength rapidly. VSFA [68] is a strong CNN-based model that applies a pre-trained ResNet backbone as a feature extractor and integrates the deep spatial features using recurrent layers and temporal pooling. PatchVQ [39] models the relationship between local and global perceptual quality from space, time, and space-time video patches. Finally, PatchVQ spatiotemporally pools quality scores to make the final predictions. Li et al. [69] leveraged transfer learning for the BVQA task by mixed dataset pre-training on both authentic IQA databases and large-scale action recognition sets. Some researchers have also proposed taking advantage of both shallow (statistical) features and deep-learning features to make the best prediction, such as CNN-TLVQM [70], RAPIQUE [71], and GAME-VQP [72].

11.2.4 Challenges and Future Trends

Although promising progress has been made on the UGC-VQA problem, several challenges remain. Firstly, compared to mainstream vision tasks such as image/video classification [73, 74], existing video databases dedicated to video quality are quite limited in scale, thus unable to fully exploit the power of deep learning approaches. Building massive-scale, diverse, and unbiased video datasets for video quality and compression research continues to be a critical research topic. Secondly, current deep learning-based models, despite showing standout performance, require pretty heavy floating-point operations. Thus, in their current form, they are neither efficient nor practical to be widely deployed. Exploring accurate and also efficient VQA models that can make predictions almost real-time for streaming services is still very challenging and largely unsolved. Finally, thanks to the evolution of smartphones and portable cameras, users can capture, curate, and share advanced media like high dynamic range (HDR), high frame rate (HFR), and even virtual reality (VR) sources. Most current VQA research involving these types of media modalities mainly focuses on small-scale compression video studies, whereas future trends in UGC video quality research must extensively embrace such "advanced-for-now" but "normal-to-be" multimedia sources.

11.3 High Dynamic Range Video

The human vision system (HVS) can perceive luminance levels between 10^{-6} cd/m^2 and 10^8 cd/m^2 using various mechanical, photochemical, and neuronal adaptive processes [75]. However, traditional imaging and display systems produce content having much narrower ranges of luminance values than the vision system is able to perceive. These older content formats are commonly referred to as Standard Dynamic Range (SDR). HDR is a set of techniques that extend the ranges of luminances and colors that can be represented and displayed. HDR has seen increasing adoption over the past few years, particularly in the area of streaming and video hosting, where services such as Amazon Prime, Netflix, and YouTube now offer content in HDR. HDR is also used as the default standard for UHD Blu-Rays. Major TV manufacturers such as LG, Samsung, and Panasonic support HDR content, and manufacturers such as Lenovo and Apple have also recently released laptops that can display HDR content. HDR is now part of live broadcast and film production workflows and is progressing rapidly into an industry standard.

11.3.1 Challenges Facing the Adoption of HDR

The challenges of HDR VQA arise from the distinct features of HDR videos. HDR videos contain much wider

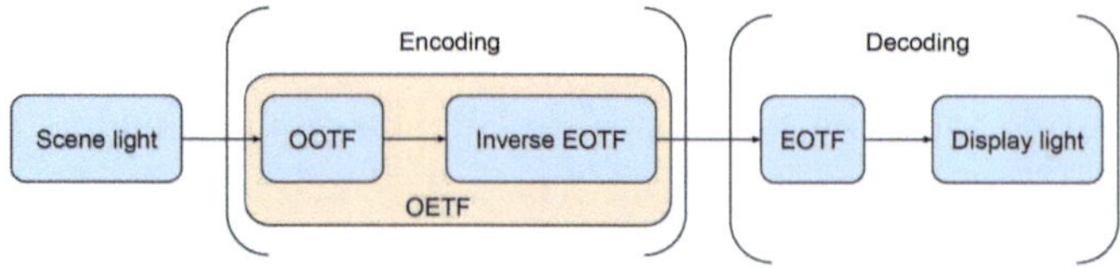

FIGURE 11.3
A representation of the PQ system. The initial step involves the transformation of the scene light into an electrical signal through the application of the OETF, followed by the conversion of the electrical signal into display light utilizing the EOTF.

luminance ranges than SDR. HDR video content represents peak brightness up to 10,000 nits (1 nit = 1 candela/meter^2), enabling the display of brighter highlights. Its lower limits are also below those of SDR content, enabling the display of darker black and deeper shadows. In order to represent such a wide range of luminance with limited data volume, while introducing minimal perceivable distortion due to quantization, the HLG and PQ electro-optical transfer functions (EOTFs) are designed to replace the legacy gamma EOTF defined in BT.1886 [76], which was originally designed to be compatible with old cathode ray tube (CRT) displays, as shown in Figure 11.3. Furthermore, the ranges of colors that can be represented by HDR video are also broader and richer than those captured by SDR. Modern HDR videos use the Rec. 2020 [77] color primaries, which cover 75.8% of the CIE 1931 color space, much larger than the Rec. 709/sRGB color gamut [78], which covers 35.6% of the CIE 1931 color space.

The use of different nonlinear transfer functions in HDR can greatly change the visibility and severity of compression distortions. The relationship between the intensity of the light as the physical magnitude measured on the display and how bright it appears to a human observer has long been known to be not linear [79]. The perceived brightness heavily depends on both the image stimulus and the viewing environment, including the image background, the peak luminance and the dynamic range of the display [80, 81]. As a result, the same magnitude of contrast may appear very different under different circumstances, making it difficult to predict the perception of distortions.

The increases in bit depth and the use of nonlinear transfer functions in HDR provide different challenges for the task of HDR VQA. Most existing VQA models can only operate on 8-bit luminance and color data, let alone account for HDR transfer functions and expanded color gamuts. For example, one of the most successful VQA models, the Video Multimethod Assessment Fusion (VMAF) algorithm [11] can be applied to 10-bit data, but it does not take into account the extended luminance range or transfer function of HDR10. However, there are a few VQA models that address the compression of HDR videos, even working with 10-bit data. These models will be discussed later in the section.

Finally, one more challenge is the scarcity of high-quality subjective VQA databases. As the HDR standards HDR10, HDR10+, and Dolby Vision are still fairly recent, publicly available HDR content is difficult to obtain, and there are few subjective VQA databases dedicated to HDR. Several existing databases have limited usefulness because they have been rendered obsolete by the latest HDR standards. Several other databases do comply with new HDR standards but are not publicly available due to copyright issues.

11.3.2 Publicly Available Subjective Quality Databases and Studies

1. **DML-HDR Video Database**: The DML-HDR video database [82] is designed to evaluate various HDR image and video quality metrics. The database consists of five videos captured by a RED SCARLET-X camera. The videos are approximately ten seconds long with a framerate of 30 frames per second (fps) and resolution of 2048×1080. The videos are stored in ".hdr" and 12 bit ".yuv" raw files using the Rec. BT.709 color primaries. The videos were then compressed using H.264/AVC and HEVC, displayed on a custom-built HDR display system, and viewed by 17 subjects.

2. **HDR-VQM Video Database**: The HDR-VQM Video Dataset [83] is a subjective database that contains ten source HDR video sequences with a resolution of 1920×1080 and the frame rate are 25 fps. These videos were compressed at eight different bitrates using a backward-compatible compression method. The videos were tone mapped to an 8-bit SDR version and compressed. When the videos were decompressed, they were inverse tone mapped to HDR. The videos were displayed on an HDR display with a peak luminance of 4,000 nits in a relatively bright environment ($200 \ cd/m^2$).

3. **Laboratory for Image and Video Engineering (LIVE) HDR Database)**: The LIVE HDR Database [84] is a subjective database that targets HDR10 videos. It contains 310 videos from 31 HDR10 source contents. All the source videos have a resolution of 3840×2160 and frame rates of 50 or 60 fps. The average duration of videos in the database is approximately eight seconds. The source sequences were processed by ten different compression and resolution combinations using the HEVC codec, simulating standard industrial practices. A subjective quality study was conducted using the database with 66 subjects. The subjective study was conducted under two different illumination conditions to test the effects of ambient illumination on the perceived quality of HDR content. It is worth noting that the LIVE HDR Database is the only publicly available HDR VQA database that complies with contemporary HDR standards.

Other than these three databases, there are a few other HDR subjective quality studies that have not made their data publicly available. Pan *et al.* [85] created a VQA database using 12 pristine HDR videos compressed with 12 quantization parameters (QP) levels. The videos are generated by both PQ and HLG and the BT.2020 color space, compressed using the AVS2 codec, developed by the Audio and Video Coding Standard (AVS) Working Group of China. A subjective quality study was conducted with 22 subjects using this dataset. Baroncini *et al.* [86] conducted a study of 12 compressed HDR videos evaluated by 40 human subjects. Rerabek *et al.* [87] conducted a study of 20 HDR videos with the resolution of 944×1080, with four distortion levels. Athar *et al.* [88] conducted a subjective study of 14 HDR10 content compressed using H.264 and HEVC to generate 140 distorted videos viewed and rated by 51 subjects.

11.3.3 Video Quality Algorithms

Existing VQA algorithms mostly consider only 8-bit data and without much consideration regarding the nonlinear transform. Some VQA algorithms that have color components use color space-conversion methods that are specific to Rec. BT.709 color spaces. This makes most of them inapplicable to HDR VQA scenarios. In this section, a few VQA algorithms dedicated to the HDR VQA problem are reviewed.

11.3.3.1 HDR-VDP

HDR-VDP is a modification proposed to extend the Visual Difference Predictor (VDP) to improve the prediction of perceivable differences in HDR applications [89]. HDR-VDP operates on linear luminance, and it relies on three stages to model the behavior of the optics and retina. HDR-VDP models the light scattering in the human eye using the Optical Transfer Function (OTF), scales the amplitude of the signal in units of Just Noticeable Differences (JND) to account for the nonlinear response of the HVS, and finally filters the data using a model of the CSF. Subsequently, improved versions of HDR-VDP have been proposed by including models of several luminance conditions, improved pooling methods, and better calibrations [90, 91].

11.3.3.2 *PU-SSIM and PU-PSNR*

Perceptually Uniform (PU)-SSIM and PU-PSNR have been proposed as an extension to the popular quality metrics [92]. A nonlinear transform has been proposed, called the perceptually uniform (PU) encoding, to ensure that the distortion visibility is approximately uniform across all encoded values. By using PU encoding to transform luminance values into perceptually uniform pixel values, SDR VQA algorithms can be directly applied to HDR images and videos.

11.3.3.3 *HDR-VQM*

HDR-VQM is an objective HDR video quality measure based on signal pre-processing, transformation, and subsequent frequency-based decomposition [83]. It uses the PU transform from PU-SSIM and PU-PSNR as a pre-processing step, then extracts the error signal in each subband in log-Gabor filters, and then the final result is obtained via a spatial-temporal pooling step.

11.3.3.4 *HDR-VMAF and HDR-ChipQA*

HDR-VMAF [84] and HDR-ChipQA [93] are two VQA methods based on a similar idea as PU-SSIM and PU-PSNR. They rely on nonlinear transforms to extend the base SDR algorithm to HDR-enhanced regions, that is, the bright and dark areas in the frames. HDR-VMAF is a FR method, while HDR-ChipQA is a NR method that uses statistical quality-aware features that are extracted in parallel from the original frames and the transformed frames.

11.4 High Frame Rate Video

Considerable advancement has been made in recent years in video technologies that improve the quality of experience (QoE) for end users. These improved capture, storage, and display technologies have facilitated the adoption of high resolution 4K videos, HDR videos, and HFR videos. The standard frame rate for movies is 24 fps and progressive television formats are typically 30 fps, while HFR videos are displayed at 50 fps or more. The increased frame rate of HFR videos alleviates temporal distortions such as motion blur and stutter. Powerful GPUs have made it possible to view high-resolution, HFR videos on consumer devices, and expansions in network capabili-

ties now allow customers to stream live events in a HFR format. Quality assessment of HFR videos is an important and open problem in order to stream and store these videos efficiently and improve QoE. In this section, we will discuss subjective quality databases and objective quality algorithms created for HFR.

11.4.1 Challenges Facing Adoption of HFR

A number of studies [94–96] have shown that for some contents quality saturates as frame rate is increased, but this is highly content dependent and HFR indeed makes a positive and significant difference in quality for high motion content. The perceptual benefits of HFR content are still relatively unknown due to a dearth of HFR content, and consumers are so accustomed to the aesthetics of SFR content that HFR content may be negatively received as being "too smooth" or having a "soap opera" effect, as happened in the case of the movie release of the "The Hobbit." However, this is a question of esthetics, and since display and capture technologies continue to develop the desire to view content that is as close to reality as possible will almost certainly lead to faster adoption of HFR. In fact, some studies [97, 98] have found that frame rates of above 900 fps may be necessary for certain types of contents to imitate optical reality. The prohibitive cost of producing and distributing HFR content and esthetic concerns remains a significant challenge to its adoption in the present.

Increasing the frame rate of videos forces encoders to compress a greater number of bits, and hence the trade-off between bitrate budgets and quality is an important one to consider. Very few NR or FR VQA models exist that are specifically designed to predict the quality of videos with different frame rates, but such models are necessary for automated, content-dependent, and data-driven decisions on the best combinations of resolution, frame rate, and bit depth for a fixed bitrate budget. The lack of these models poses a significant challenge to incorporating HFR into production workflows in an efficient and cost-friendly way that still delivers the best QoE.

11.4.2 Subjective Quality Databases

In this subsection, recent publicly available subjective quality databases that study the effect of frame rate on video quality are reviewed.

11.4.2.1 *LIVE YT HFR*

The LIVE YT HFR [99] database consists of 480 videos having six different frame rates. The temporally down-sampled videos were obtained from the source videos by frame dropping to 24 fps, 30 fps, 60 fps, 82 fps, and 98 fps. All of the videos were by varying the CRF in the

FIGURE 11.4
Screenshots of contents from the LIVE YTHFR database: (a) Runner, (b) 3 Runners, (c) Flips, (d) Hurdles, (e) Longjump, (f) bobblehead, (g) books, (h) bouncyball. (Figure is used with courtesy from [99].)

VP9 encoder. Each video was viewed by a minimum of 42 subjects. The videos in this database were displayed at their native frame rate by adjusting the refresh rate of the display to match the frame rate for each video. Screenshots of some contents from this database are shown in Figure 11.4.

The analysis of the subjective scores yielded the conclusion that the frame rate has a considerable impact on human subjective judgments of video quality but that the extent of the preference towards higher frame rates was content-dependent. Higher frame rates were strongly preferred for videos with significant camera motion. Another important conclusion drawn from this study was that the improvement in subjective quality diminishes above 60 fps.

11.4.2.2   *ETRI LIVE STSVQ*

The ETRI LIVE STSVQ [100] database contains 437 videos generated by spatial downsampling, temporal downsampling, and compression of 15 4K 10 bit source sequences. The resolutions of the downsampled videos are 1080p, 720p, and 540p and the frame rates represented in the database are 30 fps, 60 fps, and 120 fps, respectively. The videos were compressed using the x265 encoder at 5 bitrates. The temporally downsampled videos were created by frame dropping and were temporally interpolated linearly to the frame rate of their source videos before they were displayed, in contrast to LIVE YTHFR where the videos were displayed at their native frame rates. Each video was rated by 34 subjects.

The study concluded that as the bitrate budget was reduced, spatial and temporal downsampling led to higher quality scores as compression artifacts are minimized

when there is less video data to compress. However, the trade-off between spatial and temporal downsampling is highly content-dependent. For contents with rapid and large motions, the temporal downsampling can cause drops in quality even at lower bitrates and performing spatial downsampling alone may produce better quality judgments at limited bitrate budgets. For contents with little or no motion but a large amount of spatial detail, the rate-distortion curve would favor temporal downsampling over spatial downsampling at low bitrates.

11.4.2.3   *AVT-VQDB-UHD-1*

AVT-VQDB-UHD-1 [101] Test 4 consists of 120 videos with resolutions ranging from 360p to 2160p and frame rates ranging from 15 fps to 60 fps. The videos were compressed using HEVC. Each video was watched by at least 24 participants. The study showed that leading FR VQA models such as VMAF performed poorly on videos with very low framerates compared to the source video and that low frame rates had a strong negative impact on quality scores, but quality saturated as framerate was increased.

11.4.2.4   *BVI HFR*

BVI HFR [102] contains 22 source sequences and their temporally downsampled versions. The frame rates represented in the database are 15 fps, 30 fps, 60 fps, and 120 fps. The temporally downsampled videos were generated by frame-averaging. Frame dropping may result in strobing artefacts, while frame-averaging can cause motion blur. Frame-averaged videos require fewer bits than frame-dropped videos because edges lose their sharpness. The authors found in a mini-study that frame-averaged

TABLE 11.2
Performance Comparison of STGREED Against Different FR Algorithms on the LIVE-YT-HFR Database

	SROCC ↑	PLCC ↑	RMSE ↓
PSNR	0.7802	0.7481	7.75
SSIM [9]	0.5566	0.5418	9.99
MS-SSIM [104]	0.4135	0.5512	10.01
FSIM [105]	0.6528	0.6332	9.34
ST-RRED [106]	0.4516	0.6073	9.58
SpEED [107]	0.6051	0.5206	10.28
FRQM [108]	0.5133	0.5017	10.38
VMAF [11]	0.7782	0.7419	8.10
deepVQA [109]	0.4331	0.3996	10.87
STGREED	**0.8822**	**0.8869**	**5.48**

videos were preferred over frame-dropped videos for very low frame rates. Fifty-one subjects participated in this study. The study showed that increasing frame rates had a greater impact on visual quality when camera motion was present.

11.4.3   Video Quality Algorithms

Existing FR VQA algorithms such as VMAF and SSIM can only compare videos that have the same number of frames and can hence only be used to evaluate video quality across frame rates after the SFR video is temporally upsampled to the same frame rate as the HFR source video. Such models have been observed to perform poorly against human subjective opinions. We briefly review algorithms that have specifically designed for for assessing quality across frame rates.

11.4.3.1   *STGREED*

STGREED [103] is a FR VQA algorithm that is based on principles of natural video statistics. The statistics of temporally band-passed video coefficients are modelled as generalized Gaussian distributions. The entropies of these coefficients are computed, and it was observed that the entropies were relatively constant for the same frame rate for different bitrates but varied across frame rates. An entropic difference term was designed that would remove the frame rate bias while accounting for quality variations if the distorted video had the same frame rate as the source video. STGREED's performance on LIVE YTHFR is shown in Table 11.2 against that of other algorithms, using Spearman's Rank Ordered Correlation Coefficient (SROCC), Pearson's Linear Correlation Coefficient (PLCC), and the Root Mean Square Error (RMSE) between the predictions and the subjective scores as the metrics.

11.4.3.2   *VSTR*

The Video Space-Time Regularity (VSTR) model [110] computes displaced frame differences of bandpassed coefficients and chooses the displacement that minimizes the KL distance between a $N(0,1)$ distribution and the distribution of the displaced frame difference. The statistics of the most-Gaussian frame difference are modelled and entropic differences are computed between the reference and distorted videos to quantify quality. VSTR established that the statistics of pristine videos are regular and reliably Gaussian along directions of motion and less predictable along other directions and for distorted videos.

11.4.3.3   *FAVER*

Framerate-Aware Video Evaluator w/o Reference (FAVER) [111] is an NR VQA model designed for variable frame rate videos. FAVER models the statistics of temporally and spatially band-passed videos as GGDs. The temporal filtering is similar to that used in ST-GREED. Statistical features are extracted from luma and chromatic difference channels for QA. FAVER achieves the highest performance on the YTHFR database, with a correlation of approximately 0.6 against human quality scores. The relatively low value of the SOTA shows the difficulty and challenging nature of NR VQA for videos with different frame rates.

11.5   Virtual Reality

VR and its applications have attracted significant and increasing attention. However, the requirements of much larger file sizes, different storage formats, and immersive viewing conditions pose significant challenges to the goals of acquiring, transmitting, compressing and displaying high quality VR content. Towards meeting these challenges, it is important to understand the distortions that arise when delivering VR content and the effect they have on the perceived quality of displayed VR content.

11.5.1   Challenges Facing Adoption of VR

VR images are usually captured using a 360 camera equipped with multiple lenses that capture the entire 360 degrees of a scene. The recent Insta360 Titan is a professional 360 camera with eight 200° fisheye lenses that can capture both 2D and 3D images of resolution up to 11K. After the images are captured simultaneously by separate lenses, they are stitched together to generate a spherical image.

Unlike traditional viewing conditions where people watch images and videos on flat-panel computer and mobile displays, VR offers a more immersive viewing environment. Since the image can cover the entire viewing space, users are free to view the image in every direction. Usually, only a small portion of the image is displayed as they gaze in any given direction, so the content that a user sees is highly dependent on the spatial distribution of image content, the object being fixated on, and the spatial distribution of visual attention. The free-viewing of high resolution, immersive VR implies significant data volume, which leads to challenges when storing, transmitting and rendering the images that can affect the viewing quality. Therefore, it is important to be able to analyze and predict the perceptual quality of immersive VR image.

11.5.2 Subjective Quality Databases

In this subsection, recent publicly available subjective quality databases that study VR images and videos are reviewed.

11.5.2.1 *LIVE 3D VR IQA Database*

The LIVE 3D VR IQA Database [112] contains 450 distorted 3D VR images obtained from 15 pristine images modified by six types of distortions. The images were evaluated by 42 subjects using an HTC Vive headset with Tobii eye-tracking. This dataset includes traditional distortions like Gaussian noise, Gaussian blur, downsampling, as well as VR-specific stitching distortions. The dataset also includes recent compression distortions, including VP9 and H.265, to study and model the way they compress and perceptually distort VR images.

The study concluded that applying a reprojection weight to modify traditional IQA methods can help their performance on VR images. The study also showed that training on the subject data of these 3D VR images was an important step of the NR models to capture the unique perceptual peculiarities of the distorted VR image viewing experience. From the gaze maps of the study, it was shown that there exists an equator bias when viewing VR images. Subjects were more likely to view the center of the image (center bias), but this also depended on the content and whether there were objects of interest near the center. Sample images from the dataset are shown in Figure 11.5.

11.5.2.2 *IVQAD 2017 Database*

The IVQAD 2017 database [113] contains 10 raw videos and 150 distorted videos degraded by three quality factors: bit rate, frame rate, and resolution. To simulate quality degradation, three resolutions are picked as 4096 × 2048, 2048 × 1024, and 1024 × 512, and under each resolution, different bit rates and frame rates are selected to simulate different bandwidth. All the videos are encoded with MPEG-4 and 13 subjects were involved in this study. The study concluded that resolution and frame rate have almost the same degree of influence on quality assessment, whereas bit rate has the least influence.

11.5.2.3 *VRQ-TJU Database*

The VRQ-TJU database [114] contains 13 reference 3D VR videos subjected to two distortions: H.264 compression and JPEG 2000 compression, both symmetrically and asymmetrically on the original videos. In total, 104 symmetrically distorted videos and 260 asymmetrically distorted videos were obtained and evaluated by 30 subjects.

11.5.2.4 *LIVE-FBT-FCVR Database*

The LIVE-FBT-FCVR database [115] consists of two separate databases, one for 2D foveated compressed videos and the other one for 3D foveated compressed videos. Each database includes 10 reference videos and 180 foveated videos, which were processed by three levels of foveation on the pristine videos. In the 2D study, each video was of resolution 7680 × 3840 and was viewed by 36 subjects, while in the 3D study, each video was of resolution 5376 × 5376 and rated by 34 subjects. Both studies were conducted on top of a foveated video player having a motion-to-photon latency of around 50ms.

The study showed that in the 2D study, subjective quality was more affected by peripheral quality, while in the 3D study, subjective quality was largely affected by foveal quality.

11.5.3 VR Quality Algorithms

It is important to develop algorithms to assess the quality of VR content based on its characteristics. In this section, we briefly introduce different types of VR quality assessment algorithms, namely quality algorithms for VR non-foveated content and VR foveated QA.

VR Quality Algorithms. Several VR-specific QA models have been proposed over the years, with the increasing interest in VR. The very first attempts towards developing VR-specific algorithms were based on existing algorithms for flat panel displays such as SSIM and PSNR. Yu *et al.* [116] proposed a spherical PSNR model called S-PSNR that averages quality over all viewing directions. In [117], a craster parabolic projection-based PSNR (CPP-PSNR) VR-IQA model was introduced. Xu *et al.* [118] proposed two kinds of perceptual VQA (P-VQA) methods: a non content–based PSNR (NCP-PSNR) algorithm and a content-based

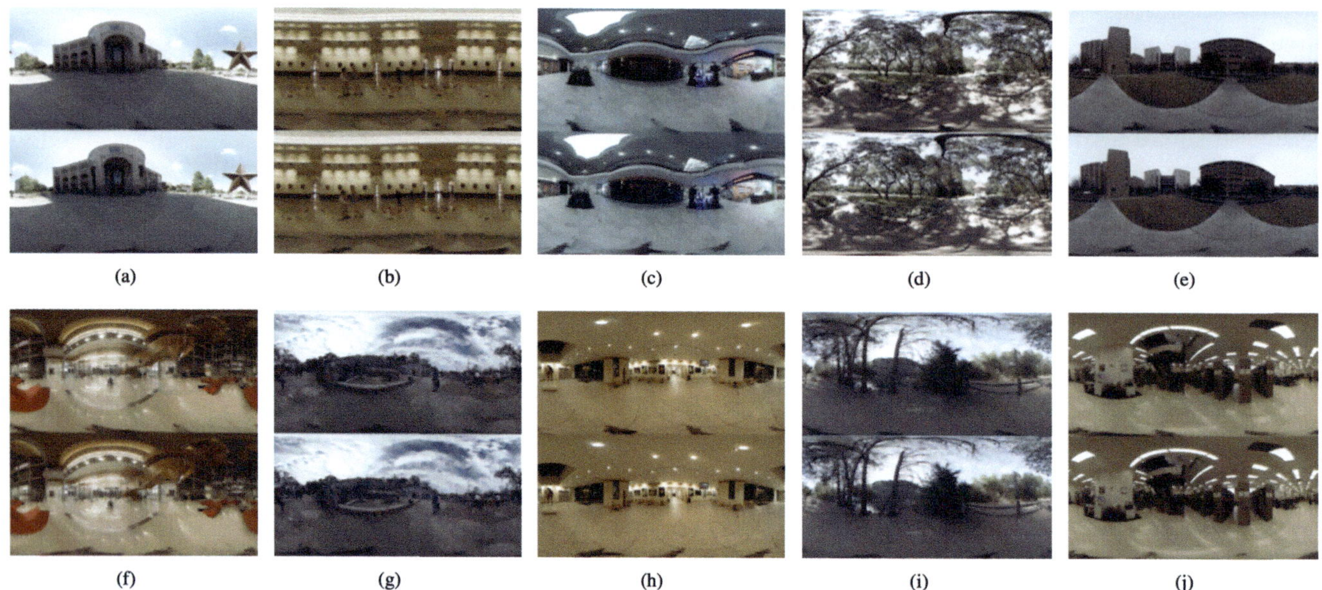

FIGURE 11.5
Screenshots of contents (a-j) from the LIVE 3D VR IQA database. (Figure used with courtesy from [112].)

PSNR (CP-PSNR) method. WS-PSNR[119] is yet another PSNR based VR-IQA method, which reweights pixels according to their location in space. SSIM has also been extended in a similar manner, as exemplified by S-SSIM [120].

Yang *et al.* [121] proposed a content-aware algorithm designed specifically to assess stitched VR images, by combining a geometric error metric with a locally constructed guided IQA method. An NR-IQA method designed to assess stitched panoramic images using convolutional sparse coding and compound feature selection was proposed in [122]. Given the explosive popularity of deep learning, many more recent VR-IQA methods have been learned to analyze immersive images and videos, often achieving impressive results. For example, in [114], the authors deployed an end-to-end 3D convolutional neural network to predict the quality of VR videos without reference. In [123] and [124], the power of adversarial learning was utilized to successfully predict the quality of images.

VR Foveated Quality Algorithms. The Foveated Wavelet Image Quality Index (FWQI) utilizes wavelets to extract position-dependent spatial quality information [125]. Several factors are taken into consideration, including the spatial CSF, which is used to determine local visual cutoff frequencies, which guides modeling of human visual sensitivity across the available wavelet subbands, when combined with assumptions on viewing distance and the display resolution. Lee *et al.* [126] proposed a foveal signal-to-noise ratio (FSNR) to evaluate the quality of picture or video streams. In this method,

a foveated image is obtained by a foveated coordinate transformation on the original image(s) whose quality is to be assessed. Riomac-Drlje *et al.* [127] proposed a foveated mean squared error (FMSE) that models the effects of spatial acuity reduction due to motion. Another model called the foveation-based content Adaptive Structural SIMilarity index (FA-SSIM), which is based on the popular IQA model SSIM [128] combines SSIM with a foveation-based sensitivity function. You *et al.* [129] proposed a full reference attention-driven foveated video quality metric (AFViQ) that accounts for the localization of fixations in images and videos. A recently developed NR FVQA model called Space-Variant BRISQUE (SVBRISQUE) achieves SOTA performance using NSS features and a neural noise model to predict the quality of immersive videos [115].

11.6 Conclusion

In this chapter, we have discussed various video modalities in media and entertainment, including popular modes such as video streaming and social media, and emerging modalities such as HDR and HFR video, and 3D VR. While all visual media rely on the same core perceptual principles, understanding domain-specific artifacts and distortions requires the development of dedicated subjective databases, which are in turn used to develop objective quality models. These models are then used to validate and enhance the quality of videos that are delivered to the end user.

Moreover, we have identified key challenges and opportunities for future research in these various domains. Some challenges are domain-specific, such as the lack of color-aware or framerate-aware quality models for streaming and HFR videos. On the other hand, we have also identified broader challenges that affect the field of VQA as a whole. One such challenge is the relatively small size of subjective databases, which is a consequence of the difficulty in obtaining subjective ratings. Recently, large-scale databases such as KonVid-150k and LIVE-FB LSVQ have been developed for user-generated social media content (UGC), but still more progress is required to usher in the deep learning era for VQA. At the same time, the high data requirements of emerging video modalities such as HDR, HFR, and VR video demand the development of efficient encoding and quality assessment algorithms. This presents an interesting opportunity to turn a challenge in one domain (high data volume) into a solution in another domain (higher data requirements for quality assessment).

In summary, the development of perceptual quality models has fueled the explosion of video streaming and consumption over the last decade, and the future holds exciting possibilities and interesting challenges for emerging visual media.

References

[1] N. Ahmed, T. Natarajan, and K.R. Rao. Discrete cosine transform. *IEEE Transactions on Computers*, C-23(1):90–93, 1974.

[2] H.S. Malvar, A. Hallapuro, M. Karczewicz, and L. Kerofsky. Low-complexity transform and quantization in h.264/avc. *IEEE Transactions on Circuits and Systems for Video Technology*, 13(7):598–603, 2003.

[3] Thomas Wiegand. Draft ITU-T recommendation and final draft international standard of joint video specification (ITU-T Rec. H. 264— ISO/IEC 14496-10 AVC). *JVT-G050*, 2003.

[4] D. Marpe, H. Schwarz, and T. Wiegand. Context-based adaptive binary arithmetic coding in the H.264/AVC video compression standard. *IEEE Transactions on Circuits and Systems for Video Technology*, 13(7):620–636, 2003.

[5] G. J. Sullivan, J. Ohm, W. Han, and T. Wiegand. Overview of the high efficiency video coding (HEVC) standard. *IEEE Transactions on Circuits and Systems for Video Technology*, 22(12):1649–1668, 2012.

[6] B. Bross, Y. Wang, Y. Ye, S. Liu, J. Chen, G. J. Sullivan, and J. R. Ohm. Overview of the versatile video coding (VVC) standard and its applications. *IEEE Transactions on Circuits and Systems for Video Technology*, 31(10):3736–3764, 2021.

[7] D. Mukherjee, J. Bankoski, A. Grange, J. Han, J. Koleszar, P. Wilkins, Y. Xu, and R. Bultje. The latest open-source video codec VP9 - An overview and preliminary results. In *Picture Coding Symposium (PCS)*, pages 390–393, 2013.

[8] Y. Chen, D. Murherjee, J. Han, A. Grange, Y. Xu, Z. Liu, S. Parker, C. Chen, H. Su, U. Joshi, C. Chiang, Y. Wang, P. Wilkins, J. Bankoski, L. Trudeau, N. Egge, J. Valin, T. Davies, S. Midtskogen, A. Norkin, and P. de Rivaz. An overview of core coding tools in the AV1 Video Codec. In *Picture Coding Symposium (PCS)*, pages 41–45, 2018.

[9] Z. Wang, A. C. Bovik, H. R. Sheikh, and E. P. Simoncelli. Image quality assessment: From error visibility to structural similarity. *IEEE Transactions on Image Processing*, 13(4):600–612, 2004.

[10] A. K. Venkataramanan, C. Wu, A. C. Bovik, I. Katsavounidis, and Z. Shahid. A Hitchhiker's guide to structural similarity. *IEEE Access*, 9:28872–28896, 2021.

[11] Z. Li, A. Aaron, I. Katsavounidis, A. Moorthy, and M. Manohara. Toward a practical perceptual video quality metric, *The Netflix Tech Blog*, Apr 2017.

[12] I. Katsavounidis. Dynamic optimizer - A perceptual video encoding optimization framework. *The Netflix Tech Blog*, 2018.

[13] E. Bourtsoulatze, A. Chadha, I. Fadeev, V. Giotsas, and Y. Andreopoulos. Deep video precoding, arXiv:1908.00812, 2019.

[14] A. Chadha, R. Anam, I. Fadeev, V. Giotsas, and Y. Andreopoulos. Escaping the complexity-bitrate-quality barriers of video encoders via deep perceptual optimization. In *Applications of Digital Image Processing XLIII*, volume 11510, pages 38 – 52. International Society for Optics and Photonics, SPIE, 2020.

[15] L. Chen, C. G. Bampis, Z. Li, J. Sole, and A. C. Bovik. A progressive architecture for learned fractional downsampling. In *2021 Picture Coding Symposium (PCS)*, pages 1–5, 2021.

[16] A. Katsenou, F. Zhang, M. Afonso, G. Dimitrov, and D. R. Bull. BVI-CC: A dataset for research on video compression and quality assessment. *Frontiers in Signal Processing*, 2, 2022.

[17] X. Min, G. Zhai, J. Zhou, M. C. Q. Farias, and A. C. Bovik. Study of subjective and objective quality assessment of audio-visual signals. *IEEE Transactions on Image Processing*, 29:6054–6068, 2020.

[18] Report on the validation of video quality models for high definition video content, 2010.

[19] Z. Shang, J. P. Ebenezer, Y. Wu, H. Wei, S. Sethuraman, and A. C. Bovik. Study of the subjective and objective quality of high motion live streaming videos. *IEEE Transactions on Image Processing*, 31:1027–1041, 2022.

[20] A. Puri. Video coding using the MPEG-2 compression standard. In Barry G. Haskell and Hsueh-Ming Hang, editors, *Visual Communications and Image Processing '93*, volume 2094, pages 1701 – 1713. International Society for Optics and Photonics, SPIE, 1993.

[21] R. Soundararajan and A. C. Bovik. Video quality assessment by reduced reference spatio-temporal entropic differencing. *IEEE Transactions on Circuits and Systems for Video Technology*, 23(4):684–694, 2013.

[22] C. G. Bampis, P. Gupta, R. Soundararajan, and A. C. Bovik. SpEED-QA: Spatial efficient entropic differencing for image and video quality. *IEEE Signal Processing Letters*, 24(9):1333–1337, 2017.

[23] S. Li, F. Zhang, L. Ma, and K. N. Ngan. Image quality assessment by separately evaluating detail losses and additive impairments. *IEEE Transactions on Multimedia*, 13(5):935–949, 2011.

[24] Z. Li, K. Swanson, C. Bampis, L. Krasula, and A. Aaron. Toward a better quality metric for the video community. *The Netflix Tech Blog*, page 2, 2020.

[25] A. K. Moorthy and A. C. Bovik. Blind image quality assessment: From natural scene statistics to perceptual quality. *IEEE Transactions on Image Processing*, 20(12):3350–3364, 2011.

[26] A. Mittal, A. K. Moorthy, and A. C. Bovik. No-reference image quality assessment in the spatial domain. *IEEE Transactions on Image Processing*, 21(12):4695–4708, 2012.

[27] A. Mittal, R. Soundararajan, and A. C. Bovik. Making a "completely blind" image quality analyzer. *IEEE Signal Processing Letters*, 20(3):209–212, 2013.

[28] S. V. Reddy Dendi and S. S. Channappayya. No-reference video quality assessment using natural spatiotemporal scene statistics. *IEEE Transactions on Image Processing*, 29:5612–5624, 2020.

[29] J. Korhonen. Two-level approach for no-reference consumer video quality assessment. *IEEE Transactions on Image Processing*, 28(12):5923–5938, 2019.

[30] K. Manasa and S. S. Channappayya. An optical flow-based no-reference video quality assessment algorithm. In *2016 IEEE International Conference on Image Processing (ICIP)*, pages 2400–2404, 2016.

[31] M. A. Saad, A. C. Bovik, and C. Charrier. Blind prediction of natural video quality. *IEEE Transactions on Image Processing*, 23(3):1352–1365, 2014.

[32] J. P. Ebenezer, Z. Shang, Y. Wu, H. Wei, S. Sethuraman, and A. C. Bovik. ChipQA: No-reference video quality prediction via space-time chips. *IEEE Transactions on Image Processing*, 30:8059–8074, 2021.

[33] R. R. R. Rao, S. Göring, P. List, W. Robitza, B. Feiten, U. Wüstenhagen, and A. Raake. Bitstream-based model standard for 4K/UHD: ITU-T P.1204.3 — Model details, evaluation, analysis and open source implementation. In *2020 Twelfth International Conference on Quality of Multimedia Experience (QoMEX)*, pages 1–6, 2020.

[34] A. K. Venkataramanan, C. Stejerean, and A. C. Bovik. Funque: Fusion of unified quality evaluators, arXiv:2202.11241, 2022.

[35] F. Zhang, A. Katsenou, C. Bampis, L. Krasula, Z. Li, and D. Bull. Enhancing VMAF through new feature integration and model combination. In *2021 Picture Coding Symposium (PCS)*, pages 1–5, 2021.

[36] L. Chen, C. G. Bampis, Z. Li, J. Sole, and A. C. Bovik. Perceptual video quality prediction emphasizing chroma distortions. *IEEE Transactions on Image Processing*, 30:1408–1422, 2021.

[37] R. Zhang, P. Isola, A. A. Efros, E. Shechtman, and O. Wang. The unreasonable effectiveness of deep features as a perceptual metric, arXiv:1801.03924, 2018.

[38] H. Talebi and P. Milanfar. NIMA: Neural image assessment. *IEEE Transactions on Image Processing*, 27(8):3998–4011, 2018.

[39] Z. Ying, M. Mandal, D. Ghadiyaram, and A. Bovik. Patch-VQ: 'patching up' the video quality problem. In *2021 IEEE/CVF Conference on Computer Vision and Pattern Recognition (CVPR)*. IEEE, jun 2021.

[40] P. C. Madhusudana, N. Birkbeck, Y. Wang, B. Adsumilli, and A. C. Bovik. Image quality assessment using contrastive learning, In Proceedings of the IEEE Winter Applications and Computer Vision Workshops (WACVW), 2021.

[41] A. Dosovitskiy, L. Beyer, A. Kolesnikov, D. Weissenborn, X. Zhai, T. Unterthiner, M. Dehghani, M. Minderer, G. Heigold, S. Gelly, J. Uszkoreit, and N. Houlsby. An image is worth 16x16 words: Transformers for image recognition at scale, arXiv:2010.11929, 2020.

[42] B. Gates. Content is king. *Retrieved October*, 29:2017, 1996.

[43] M. Mohsin. 10 youtube statistics that you need to know in 2021, May 2022.

[44] Z. Tu, Y. Wang, N. Birkbeck, B. Adsumilli, and A. C. Bovik. Ugc-vqa: Benchmarking blind video quality assessment for user generated content. *IEEE Transactions on Image Processing*, 30:4449–4464, 2021.

[45] S. Abu-El-Haija, N. Kothari, J. Lee, P. Natsev, G. Toderici, B. Varadarajan, and S. Vijayanarasimhan. Youtube-8m: A large-scale video classification benchmark. *arXiv preprint arXiv:1609.08675*, 2016.

[46] Z. Sinno and A. C. Bovik. Large-scale study of perceptual video quality. *IEEE Transactions on Image Processing*, 28(2):612–627, 2019.

[47] K. Seshadrinathan, R. Soundararajan, A. C. Bovik, and L. K Cormack. Study of subjective and objective quality assessment of video. *IEEE transactions on Image Processing*, 19(6):1427–1441, 2010.

[48] F. De Simone, M. Tagliasacchi, M. Naccari, S. Tubaro, and T. Ebrahimi. A H.264/AVC video database for the evaluation of quality metrics. In *2010 IEEE International Conference on Acoustics, Speech and Signal Processing*, pages 2430–2433, 2010.

[49] P. V. Vu and D. M. Chandler. ViS3: An algorithm for video quality assessment via analysis of spatial and spatiotemporal slices. *Journal of Electronic Imaging*, 23(1):013016, 2014.

[50] J. Y. Lin, R. Song, C. Wu, T. Liu, H. Wang, and C-C J. Kuo. MCL-V: A streaming video quality assessment database. *Journal of Visual Communication and Image Representation*, 30:1–9, 2015.

[51] M. Nuutinen, T. Virtanen, M. Vaahteranoksa, T. Vuori, P. Oittinen, and J. Häkkinen. CVD2014—A database for evaluating no-reference video quality assessment algorithms. *IEEE Transactions on Image Processing*, 25(7):3073–3086, 2016.

[52] D. Ghadiyaram, J. Pan, A. C. Bovik, A. K. Moorthy, P. Panda, and K. Yang. In-capture mobile video distortions: A study of subjective behavior and objective algorithms. *IEEE Transactions on Circuits and Systems for Video Technology*, 28(9):2061–2077, 2018.

[53] V. Hosu, F. Hahn, M. Jenadeleh, H. Lin, H. Men, T. Szirányi, S. Li, and D. Saupe. The Konstanz natural video database (KoNViD-1k). In *2017 Ninth International Conference on Quality of Multimedia Experience (QoMEX)*, pages 1–6, 2017.

[54] B. Thomee, D. A. Shamma, G. Friedland, B. Elizalde, K. Ni, D. Poland, D. Borth, and L. Li. YFCC100M: The new data in multimedia research. *Communications of the ACM*, 59(2):64–73, 2016.

[55] Y. Wang, S. Inguva, and B. Adsumilli. Youtube UGC dataset for video compression research. In *2019 IEEE 21st International Workshop on Multimedia Signal Processing (MMSP)*, pages 1–5. IEEE, 2019.

[56] F. Götz-Hahn, V. Hosu, H. Lin, and D. Saupe. KonVid-150k: A dataset for no-reference video quality assessment of videos in-the-wild. *IEEE Access*, 9:72139–72160, 2021.

[57] Internet Archive: Digital Library of Free & Borrowable Books, Movies, Music & Wayback Machine.

[58] X. Yu, Z. Ying, N. Birkbeck, Y. Wang, B. Adsumilli, and A. C Bovik. Subjective and objective analysis of streamed gaming videos. *arXiv preprint arXiv:2203.12824*, 2022.

[59] C. Chen, M. Izadi, and A. Kokaram. A perceptual quality metric for videos distorted by spatially correlated noise. In *Proceedings of the 24th ACM International Conference on Multimedia*, pages 1277–1285, 2016.

[60] Z. Wang, A.C. Bovik, and B.L. Evan. Blind measurement of blocking artifacts in images. In *Proceedings 2000 International Conference on Image Processing (Cat. No.00CH37101)*, volume 3, pages 981–984, 2000.

[61] P. Marziliano, F. Dufaux, S. Winkler, and T. Ebrahimi. A no-reference perceptual blur metric. In *Proceedings. International Conference on Image Processing*, volume 3, pages III–III, 2002.

[62] X. Feng and J. P. Allebach. Measurement of ringing artifacts in JPEG images. In Jan P. Allebach and Hui Chao, editors, *Digital Publishing*, volume 6076, pages 74 – 83. International Society for Optics and Photonics, SPIE, 2006.

[63] Z. Tu, J. Lin, Y. Wang, B. Adsumilli, and A. C. Bovik. Bband index: A no-reference banding artifact predictor. In *ICASSP 2020-2020 IEEE International Conference on Acoustics, Speech and Signal Processing (ICASSP)*, pages 2712–2716. IEEE, 2020.

[64] J. E. Caviedes and F. Oberti. No-reference quality metric for degraded and enhanced video. In *Visual Communications and Image Processing 2003*, volume 5150, pages 621–632. SPIE, 2003.

[65] C. Keimel, T. Oelbaum, and K. Diepold. No-reference video quality evaluation for high-definition video. In *2009 IEEE International Conference on Acoustics, Speech and Signal Processing*, pages 1145–1148, 2009.

[66] D. L. Ruderman. The statistics of natural images. *Network: Computation in Neural Systems*, 5(4):517, 1994.

[67] M. A. Saad, A. C. Bovik, and C. Charrier. Blind image quality assessment: A natural scene statistics approach in the DCT domain. *IEEE Transactions on Image Processing*, 21(8):3339–3352, 2012.

[68] D. Li, T. Jiang, and M. Jiang. Quality assessment of in-the-wild videos. In *Proceedings of the 27th ACM International Conference on Multimedia*, pages 2351–2359, 2019.

[69] B. Li, W. Zhang, M. Tian, G. Zhai, and X. Wang. Blindly assess quality of in-the-wild videos via quality-aware pre-training and motion perception. *IEEE Transactions on Circuits and Systems for Video Technology*, 2022.

[70] J. Korhonen, Y. Su, and J. You. Blind natural video quality prediction via statistical temporal features and deep spatial features. In *Proceedings of the 28th ACM International Conference on Multimedia*, pages 3311–3319, 2020.

[71] Z. Tu, X. Yu, Y. Wang, N. Birkbeck, B. Adsumilli, and A. C Bovik. RAPIQUE: Rapid and accurate video quality prediction of user generated content. *IEEE Open Journal of Signal Processing*, 2:425–440, 2021.

[72] X. Yu, Z. Tu, N. Birkbeck, Y. Wang, B. Adsumilli, and A. C Bovik. Perceptual quality assessment of UGC gaming videos. *arXiv preprint arXiv:2204.00128*, 2022.

[73] O. Russakovsky, J. Deng, H. Su, J. Krause, S. Satheesh, S. Ma, Z. Huang, A. Karpathy, A. Khosla, M. Bernstein, et al. Imagenet large scale visual recognition challenge. *International Journal of Computer Vision*, 115(3):211–252, 2015.

[74] A. Karpathy, G. Toderici, S. Shetty, T. Leung, R. Sukthankar, and L. Fei-Fei. Large-scale video classification with convolutional neural networks. In *Proceedings of the IEEE Conference on Computer Vision and Pattern Recognition (CVPR)*, June 2014.

[75] T. Kunkel, S. Daly, S. Miller, and J. Froehlich. Perceptual design for high dynamic range systems. In *High Dynamic Range Video*, pages 391–430. Elsevier, 2016.

[76] ITU. BT.1886 : Reference electro-optical transfer function for flat panel displays used in HDTV studio production. Technical report, Intl. Telecomm. Union, 2011.

[77] BT.2020 : Parameter values for ultra-high definition television systems for production and international programme exchange, 2015.

[78] ITU. BT.709 : Parameter values for the HDTV standards for production and international programme exchange. Technical report, Intl. Telecomm. Union, 2011.

[79] T. N. Cornsweet and H. M. Pinsker. Luminance discrimination of brief flashes under various conditions of adaptation. *The Journal of Physiology*, 176(2):294, 1965.

[80] M. Bertalmío. Chapter 5 - Brightness perception and encoding curves. In Marcelo Bertalmío, editor, *Vision Models for High Dynamic Range and Wide Colour Gamut Imaging*, Computer Vision and Pattern Recognition, pages 95–129. Academic Press, 2020.

[81] V. A. Billock and B. H. Tsou. To honor fechner and obey stevens: Relationships between psychophysical and neural nonlinearities. *Psychological Bulletin*, 137(1):1, 2011.

[82] Ma. Azimi, A. Banitalebi-Dehkordi, Y. Dong, M .T. Pourazad, and P. Nasiopoulos. Evaluating the performance of existing full-reference quality metrics on high dynamic range (HDR) video content. *TBD*, 2022.

[83] Manish Narwaria, Matthieu Perreira Da Silva, and Patrick Le Callet. HDR-VQM: An objective quality measure for high dynamic range video. *Signal Processing: Image Communication*, 35:46–60, 2015.

[84] Zaixi Shang, Joshua Peter Ebenezer, Yongjun Wu, Hai Wei, Sriram Sethuraman, and Alan C. Bovik. HDR-VMAF: Study of subjective and objective quality assessment of hdr videos. *submitted to IEEE Transactions on Image Processing*, 31, 2022.

[85] X. Pan, J. Zhang, S. Wang, S. Wang, Y. Zhou, W. Ding, and Y. Yang. HDR video quality assessment: Perceptual evaluation of compressed HDR video. *Journal of Visual Communication and Image Representation*, 57:76–83, 2018.

[86] V. Baroncini, K. Andersson, A.K. Ramasubramonian, and G. Sullivan. Verification test report for HDR/WCG video coding using HEVC main 10 profile. In *Proceedings of the JCTVC-X1018 24th JCT-VC Meeting*, 2016.

[87] M. Rerabek, P. Hanhart, P. Korshunov, and T. Ebrahimi. Subjective and objective evaluation of HDR video compression. In *9th International Workshop on Video Processing and Quality Metrics for Consumer Electronics (VPQM)*, 2015.

[88] S. Athar, T. Costa, K. Zeng, and Z. Wang. Perceptual quality assessment of UHD-HDR-WCG videos. In *IEEE International Conference on Image Processing*, pages 1740–1744, 2019.

[89] Rafal Mantiuk, Scott J. Daly, Karol Myszkowski, and Hans-Peter Seidel. Predicting visible differences in high dynamic range images: Model and its calibration. In Bernice E. Rogowitz, Thrasyvoulos N. Pappas, and Scott J. Daly, editors, *Human Vision and Electronic Imaging X*, volume 5666, pages 204 – 214. International Society for Optics and Photonics, SPIE, 2005.

[90] Rafal Mantiuk, Kil Joong Kim, Allan G. Rempel, and Wolfgang Heidrich. HDR-VDP-2: A calibrated visual metric for visibility and quality predictions in all luminance conditions. *ACM Transactions on Graphics*, 30(4), Jul 2011.

[91] Manish Narwaria, Matthieu Perreira Da Silva, Patrick Le Callet, and Romuald Pepion. On improving the pooling in HDR-VDP-2 towards better HDR perceptual quality assessment. In Bernice E. Rogowitz, Thrasyvoulos N. Pappas, and Huib de Ridder, editors, *Human Vision and Electronic Imaging XIX*, volume 9014, pages 143 – 151. International Society for Optics and Photonics, SPIE, 2014.

[92] Tunç O. Aydın, Rafal Mantiuk, and Hans-Peter Seidel. Extending quality metrics to full luminance range images. In Bernice E. Rogowitz and Thrasyvoulos N. Pappas, editors, *Human Vision and Electronic Imaging XIII*, volume 6806, pages 109 – 118. International Society for Optics and Photonics, SPIE, 2008.

[93] Joshua Peter Ebenezer, Zaixi Shang, Yongjun Wu, Hai Wei, Sriram Sethuraman, and Alan C. Bovik. HDR-chipQA: No-reference quality assessment for high dynamic range videos. *submitted to IEEE Transactions on Image Processing*, 31, 2022.

[94] R. Selfridge, K. C. Noland, and M. Hansard. Visibility of motion blur and strobing artefacts in video at 100 frames per second. In *Proceedings of the 13th European Conference on Visual Media Production (CVMP 2016)*, pages 1–10, 2016.

[95] P. J. Bex, G. K. Edgar, and A. T. Smith. Multiple images appear when motion energy detection fails. *Journal of Experimental Psychology: Human Perception and Performance*, 21(2):231, 1995.

[96] R. M. Nasiri, J. Wang, A. Rehman, S. Wang, and Z. Wang. Perceptual quality assessment of high frame rate video. In *2015 IEEE 17th International Workshop on Multimedia Signal Processing (MMSP)*, pages 1–6. IEEE, 2015.

[97] A. Mackin, K. C. Noland, and D. R. Bull. High frame rates and the visibility of motion artifacts. *SMPTE Motion Imaging Journal*, 126(5):41–51, 2017.

[98] A. B. Watson. High frame rates and human vision: A view through the window of visibility. *SMPTE Motion Imaging Journal*, 122(2):18–32, 2013.

[99] P. C. Madhusudana, X. Yu, N. Birkbeck, Y. Wang, B. Adsumilli, and A. C. Bovik. Subjective and objective quality assessment of high frame rate videos. *IEEE Access*, 9:108069–108082, 2021.

[100] D. Y. Lee, S. Paul, C. G. Bampis, H. Ko, J. Kim, S. Y. Jeong, B. Homan, and A. C. Bovik. A subjective and objective study of space-time subsampled video quality. *IEEE Transactions on Image Processing*, 2021.

[101] R. R. R. Rao, S. Göring, W. Robitza, B. Feiten, and A. Raake. AVT-VQDH-UHD-1: A large scale video quality database for UHD-1. In *2019 IEEE International Symposium on Multimedia (ISM)*, pages 17–177. IEEE, 2019.

[102] A. Mackin, F. Zhang, and D. R. Bull. A study of high frame rate video formats. *IEEE Transactions on Multimedia*, 21(6):1499–1512, 2018.

[103] P. C Madhusudana, N. Birkbeck, Y. Wang, B. Adsumilli, and A. C. Bovik. ST-greed: Space-time generalized entropic differences for frame rate dependent video quality prediction. *IEEE Transactions on Image Processing*, 30:7446–7457, 2021.

[104] Zhou Wang, Eero P Simoncelli, and Alan C Bovik. Multiscale structural similarity for image quality assessment. In *Asilomar Conference on Signals, Systems & Computers*, volume 2, pages 1398–1402 Vol.2, Nov 2003.

[105] Lin Zhang, Lei Zhang, Xuanqin Mou, and David Zhang. FSIM: A feature similarity index for image quality assessment. *IEEE Transactions on Image Processing*, 20(8):2378–2386, 2011.

[106] R. Soundararajan and A. C. Bovik. Video quality assessment by reduced reference spatio-temporal entropic differencing. *IEEE Transactions on Circuits and Systems for Video Technology*, 23(4):684–694, April 2013.

[107] C. G. Bampis, P. Gupta, R. Soundararajan, and A. C. Bovik. SpEED-QA: Spatial efficient entropic differencing for image and video quality. *IEEE Signal Processing Letters*, 24(9):1333–1337, Sep. 2017.

[108] F. Zhang, A. Mackin, and D. R. Bull. A frame rate dependent video quality metric based on temporal wavelet decomposition and spatiotemporal pooling. In *IEEE International Conference on Image Processing*, pages 300–304, Sep. 2017.

[109] Woojae Kim, Jongyoo Kim, Sewoong Ahn, Jinwoo Kim, and Sanghoon Lee. Deep video quality assessor: From spatio-temporal visual sensitivity to a convolutional neural aggregation network. In *European Conference on Computer Vision*, pages 219–234, September 2018.

[110] D. Y. Lee, H. Ko, J. Kim, and A. C. Bovik. Video quality model for space-time resolution adaptation. In *2020 IEEE 4th International Conference on Image Processing, Applications and Systems (IPAS)*, pages 34–39, 2020.

[111] Q. Zheng, Z. Tu, Y. Fan, X. Zeng, and A. C. Bovik. No-reference quality assessment of variable frame-rate videos using temporal bandpass statistics. In *ICASSP 2022 - 2022 IEEE International Conference on Acoustics, Speech and Signal Processing (ICASSP)*, pages 1795–1799, 2022.

[112] M. Chen, Y. Jin, T. Goodall, X. Yu, and A. C. Bovik. Study of 3D virtual reality picture quality. *IEEE Journal of Selected Topics in Signal Processing*, 14(1):89–102, 2019.

[113] H. Duan, G. Zhai, X. Yang, D. Li, and W. Zhu. IVQAD 2017: An immersive video quality assessment database. In *2017 International Conference on Systems, Signals and Image Processing (IWSSIP)*, pages 1–5. IEEE, 2017.

[114] J. Yang, T. Liu, B. Jiang, H. Song, and W. Lu. 3D panoramic virtual reality video quality assessment based on 3D convolutional neural networks. *IEEE Access*, 6:38669–38682, 2018.

[115] Y. Jin, M. Chen, T. Goodall, A. Patney, and A. C Bovik. Subjective and objective quality assessment of 2D and 3D foveated video compression in virtual reality. *IEEE Transactions on Image Processing*, 30:5905–5919, 2021.

[116] M. Yu, H. Lakshman, and B. Girod. A framework to evaluate omnidirectional video coding schemes. In *2015 IEEE International Symposium on Mixed and Augmented Reality*, pages 31–36. IEEE, 2015.

[117] V. Zakharchenko, K. P. Choi, and J. H. Park. Quality metric for spherical panoramic video. In *Optics and Photonics for Information Processing X*, volume 9970, page 99700C. International Society for Optics and Photonics, 2016.

[118] M. Xu, C. Li, Z. Chen, Z. Wang, and Z. Guan. Assessing visual quality of omnidirectional videos. *IEEE Transactions on Circuits and Systems for Video Technology*, 2018.

[119] Y. Sun, A. Lu, and L. Yu. Weighted-to-spherically-uniform quality evaluation for omnidirectional video. *IEEE Signal Processing Letters*, 24(9):1408–1412, 2017.

[120] S. Chen, Y. Zhang, Y. Li, Z. Chen, and Z. Wang. Spherical structural similarity index for objective omnidirectional video quality assessment. In *2018 IEEE International Conference on Multimedia and Expo (ICME)*, pages 1–6. IEEE, 2018.

[121] L. Yang, Z. Tan, Z. Huang, and G. Cheung. A content-aware metric for stitched panoramic image quality assessment. In *Proceedings of the IEEE International Conference on Computer Vision*, pages 2487–2494, 2017.

[122] S. Ling, G. Cheung, and P. Le Callet. No-reference quality assessment for stitched panoramic images using convolutional sparse coding and compound feature selection. In *2018 IEEE International Conference on Multimedia and Expo (ICME)*, pages 1–6. IEEE, 2018.

[123] H. T. Lim, H. G. Kim, and Y. M. Ra. VR IQA net: Deep virtual reality image quality assessment using adversarial learning. In *2018 IEEE International Conference on Acoustics, Speech and Signal Processing (ICASSP)*, pages 6737–6741. IEEE, 2018.

[124] H. G. Kim, H. T. Lim, and Y. M. Ro. Deep virtual reality image quality assessment with human perception guider for omnidirectional image. *IEEE Transactions on Circuits and Systems for Video Technology*, 2019.

[125] Z. Wang, A. C. Bovik, L. Lu, and J. L. Kouloheris. Foveated wavelet image quality index. 4472:42–53, 2001.

[126] S. Lee, M. S Pattichis, and A. C. Bovik. Foveated video quality assessment. *IEEE Transactions on Multimedia*, 4(1):129–132, 2002.

[127] S. Rimac-Drlje, M. Vranješ, and D. Žagar. Foveated mean squared error - A novel video quality metric. *Multimedia Tools and Applications*, 49(3):425–445, 2010.

[128] S. Rimac-Drlje, G. Martinović, and B. Zovko-Cihlar. Foveation-based content adaptive structural similarity index. In *2011 18th International Conference on Systems, Signals and Image Processing (IWSSIP)* pages 1–4, 2011.

[129] J. You, T. Ebrahimi, and A. Perkis. Attention driven foveated video quality assessment. *IEEE Transactions on Image Processing*, 23(1):200–213, 2014.

12

Immersive User Experiences: Trends and Challenges of Using XR Technologies

Vasudev Bhaskaran and Upal Mahbub

12.1 Introduction

Immersive experiences [52] and extended reality (XR) are two terms that are often used interchangeably to describe technologies and techniques that simulate real or imagined environments enabling users to interact with them in a more natural and intuitive way. Immersive experiences refer to any framework that creates a sense of presence and immersion in a simulated environment. This can include

1. **Assisted reality:** View a virtual display within the immediate field of view (FOV) to give access to relevant information and be able to be hands-free while viewing,

2. **Augmented reality (AR):** Overlay virtual objects in a real-world environment and interact with the virtual objects,

3. **Augmented virtuality:** Real objects projected and controlled in a virtual world as seen by the viewer, and

4. **Virtual reality (VR):** Where the user is fully immersed in the virtual world.

Note that Mixed Reality (MR) encompasses AR and Augmented Virtuality. Other immersive techniques include 360-degree video, haptic feedback, and spatial audio. Extended reality, or XR, is a term that encompasses all of these immersive technologies, as well as any other technology that extends our reality in some way. This can include technologies like telepresence[35], which enable people to interact with each other remotely as if they were in the same room, or even technologies that allow us to control machines and robots using our own movements and gestures.

The goal of immersive experiences and XR is to create more engaging, intuitive, and interactive experiences for users, whether that's in entertainment[1], gaming[21], education[43], training[48], or other applications. By simulating environments and allowing users to interact with them in a more natural way, these technologies can create a captivating and memorable experience that can help to drive engagement, retention, and even learning outcomes. As these technologies continue to evolve and become more widely adopted, we can expect to see a wide range of new applications and use cases emerge in the future.

XR has a long history covering nearly two centuries [12, 45, 54] as illustrated in Figure 12.1. The depiction shows that in recent years there has been an acceleration in the development of XR technologies embodied by the many HMDs available to the consumer. This acceleration in the realization of the HMDs is primarily due to the advancements in low-power computing and the development of dedicated display systems for HMDs that support high-resolution sensors for accurately imaging the real world at high refresh rates. Additionally, the rapid progress in communications infrastructure such as 5G contributed to the effort by enabling very low latency communications at high bandwidth which made it possible to offload computing from the head-mounted devices (HMD) to the edge device or the cloud for a better power distribution without affecting the immersive use experiences adversely [2]. Moreover, the advancement in the capabilities of the silicon in the HMD for an acceleration of the "perception" workloads in XR at very low power enabled running multiple XR perception tasks concurrently. Perception is the human process of understanding the surroundings through physical sensation based on sensory inputs from the ears, nose, skin, and tongue. In XR, inputs for visual perception usually come from an array of cameras and depth sensors, whereas other sensors such as IMUs, pressure sensors, and microphones provide input for other perception tasks. In the rest of this chapter, we will focus on the perception workloads in XR as to what they are, how are they used, and what are the challenges in the development of these workloads as we drive to lower power and have more capabilities in the XR system.

DOI: 10.1201/9781003328957-12

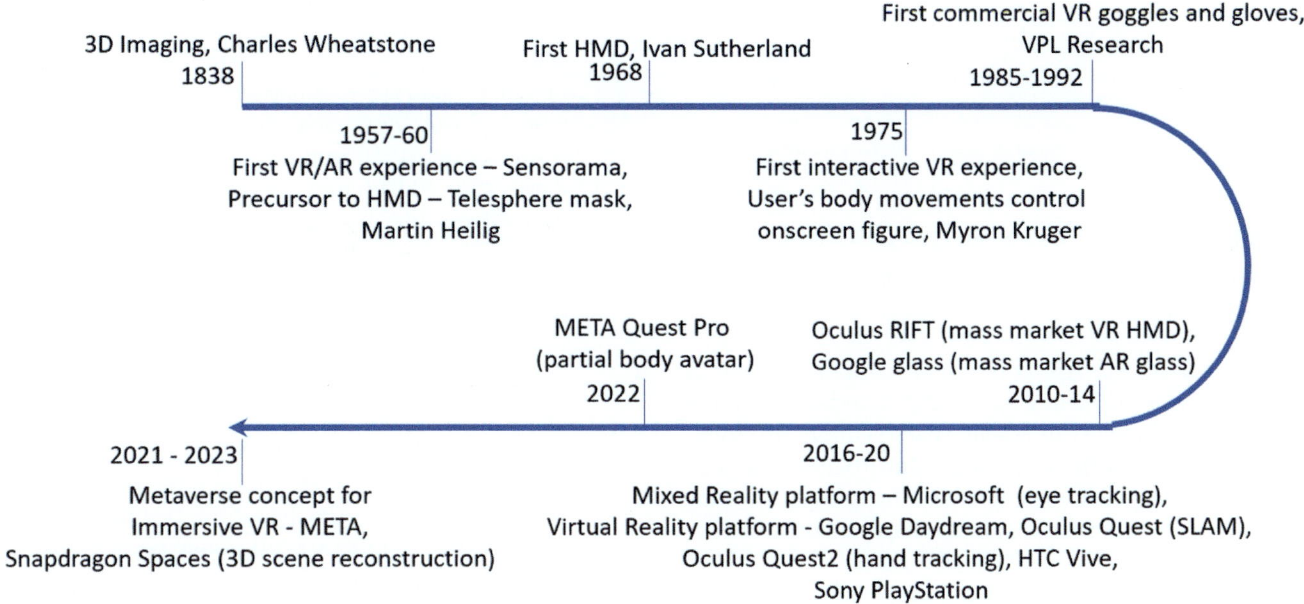

FIGURE 12.1
XR-related technology developments – a timeline.

12.2 Background

There are different ways in which XR devices are realized and some examples of the different device configurations are illustrated in Figure 12.2. Note that MR devices are not shown here since MR functionalities are generally realized using the VR devices. Essentially by having the front-mounted cameras look outwards from the HMD, the scene around the HMD is captured and projected onto the VR device's display and thus it provides augmentation of the virtual content on the VR display with the real-world content captured from the outward-facing cameras. Furthermore, AR standalone devices usually come in two configurations – some vendors provide what is referred to as an all-in-one device, where all the AR-related tasks run on the glass itself. Other vendors are proposing a split perception model wherein each perception workload in VR or AR is split into a portion of the workload that is run on the glass and the rest is run on a companion device, such as smartphone connected to the AR glass or even in the cloud[50]. This type of system partitioning can still provide a good immersive experience due to the advancements in communications technologies such as WiFi and 5G, which result in low latency and high bandwidth communication links between the glass and the untethered companion device. We will discuss in more detail the split perception model of computing that

FIGURE 12.2
Examples of AR and VR devices.

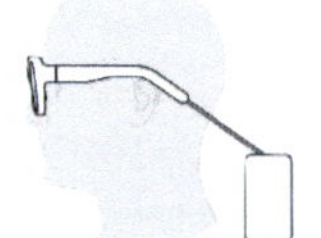
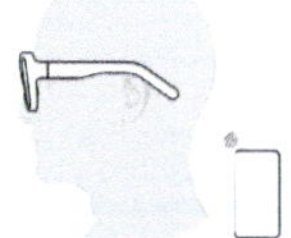

FIGURE 12.3
Evolution of AR glasses from All-in-one to untethered setup.

could be utilized for XR perception workloads later in this chapter. Note that in Figure 12.2, a tethered system is also shown; however, in recent years vendors are moving towards an untethered system due to its enhanced mobility advantages. The progression from the all-in-one system to an untethered XR glass setup is shown in Figure 12.3.

Regardless of the configuration, there are some basic constraints from a user experience perspective that needs to be met for all XR systems. These constraints that directly impact the associated perception workloads in VR and AR include:

1. **Motion-to-photon latency (MTP)**: This measures the time it takes for the VR system to respond to the user's head movement and update the display, which can affect the overall quality and comfort of the experience[56]. In AR systems, this measures the time it takes for the AR system to track the user's movement and overlay digital content onto the physical world in real-time, which can affect the overall quality and usability of the experience. In immersive experiences, the latency can cause discomfort or motion sickness if the movement of the virtual environment does not correspond quickly enough to the user's movement in the real world[53]. Low motion-to-photon latency is crucial for creating a seamless, realistic, and comfortable immersive experience. In VR systems, as the user only sees virtual contents, up to 20 milliseconds of time lag can remain undetected by the user[10]. In AR, optical see-through (OST) systems are commonly used to overlay virtual content in the real world. Since there is no lag in the real world, latency in such AR systems becomes visually apparent in the form of misregistration between virtual content and the real world. Since the human visual system is extremely sensitive to such time lags, 5 milliseconds of latency or less is required in OST systems in order to remain unnoticed by the user[16].

2. **Frame rate**: This measures the number of frames rendered per second, which affects the smoothness and fluidity of the experience. A higher frame rate is generally desirable to provide a more immersive experience. Typically frame rates are around 90–120Hz. Frame rates lower than 90Hz may induce some negative user effects like headache, disorientation, and nausea[37]. Note that for some of the processing workloads, the processing rate might only be 60Hz even if the rendering rate is 90–120Hz.

3. **Display resolution**: This measures the level of detail and clarity in the VR display, which affects the visual quality and sharpness of the experience. Typically, the VR display resolution is around $2k \times 2k$ to $4k \times 2k$ pixels per eye. For example, the latest Meta Quest Pro VR device has 1800×1920 pixels per eye [38].

4. **Field of view**: This measures the extent of the user's visual field in the VR display, which affects the sense of immersion and presence in the experience. There is a separate FOV constraint on the sensor side, which is imposed by the current technology of those sensors, and the sensor FOV impacts the processing specific to a processing workload. The FOV affects the overall performance of the VR system. A wide FOV requires more processing power and can cause a decrease in performance, resulting in slower frame rates and potentially causing motion sickness or discomfort. A narrower FOV, on the other hand, requires less processing power and can improve overall performance, making the VR experience smoother and more comfortable. However, a narrow FOV on the sensor side can impact the tracking accuracy.

5. **Tracking accuracy**: This measures the accuracy of the AR system in tracking the user's movement and positioning digital content in the physical world, which affects the overall quality and realism of the experience. In the context of VR, tracking accuracy is used in individual workloads such as tracking head movements to estimate the pose of the head.

6. **Image recognition accuracy**: This measures the accuracy of the AR system in recognizing and tracking physical objects in the real world, which is important for creating interactive and engaging AR experiences.

7. **Power**: The power constraint depends on the AR/VR configuration as well as whether the overall system is tethered or untethered. In AR,

one would face a severe power constraint as the device is expected to be worn for long periods of time and thus battery life needs to be long. Typically for AR glass, one might have a power requirement not to exceed 0.5 watts to 1 watt for all perception workloads running on the glass. In VR, the power constraints are relatively relaxed because of the larger form factor that can accommodate larger batteries, can dissipate heat easily, and because of the relatively shorter-term use of the VR HMD devices in a session.

Before delving into the specifics of the perception workloads for AR and VR, the mechanisms for enablement of these perception workloads can be illustrated through a representative AR glass setup as shown in Figure 12.4. The tracking camera feeds can be used for different perception tasks including head movement tracking, egocentric hand tracking, and in-depth understanding of the scene. Because of the tight form factor of the AR glasses, only a limited number of tracking cameras can be hosted and therefore it is expected that multiple perception tasks would share the same camera feeds and configurations. In contrast, VR HMDs have more area and therefore can host more cameras to handle such tasks separately. In Figure 12.4, the inward-facing sensors related to eye tracking are also shown; the number of eye-tracking cameras and associated LEDs will vary from vendor to vendor. Processors shown in Figure 12.4 handle the perception workloads including any display and communication functions in a connected environment. Note that when

processors are put on the glass, due to the power and thereby, head dissipation-related constraints concerning the user's safety, the perception workloads might have to be distributed in multiple processors placed strategically on the frame. The placement and number of processors may vary between solution providers.

12.3   Perception Workload

Perception workloads in XR refer to the computational tasks involved in the processing and interpreting of multisensor data of different modalities from the real world to create an immersive XR experience. A typical set of perception workloads that are used in XR are depicted in Figure 12.5. Depending on critical factors such as power, performance, and latency requirements, these perception workloads can be handled by using conventional computer vision and machine learning algorithms or neural network-based models, or a fusion of both. From an implementation perspective, one has to make a trade-off between the algorithm type versus the benefits it offers in terms of lower latency and/or lower power and/or higher accuracy and, in a hardware engine, lower silicon area. A detailed discussion of these trade-offs is outside the scope of this chapter and interested readers can refer to [20] for more insight. Note that the perception workloads shown in Figure 12.5 are not a complete set of perception workloads that are available in any XR system but only the

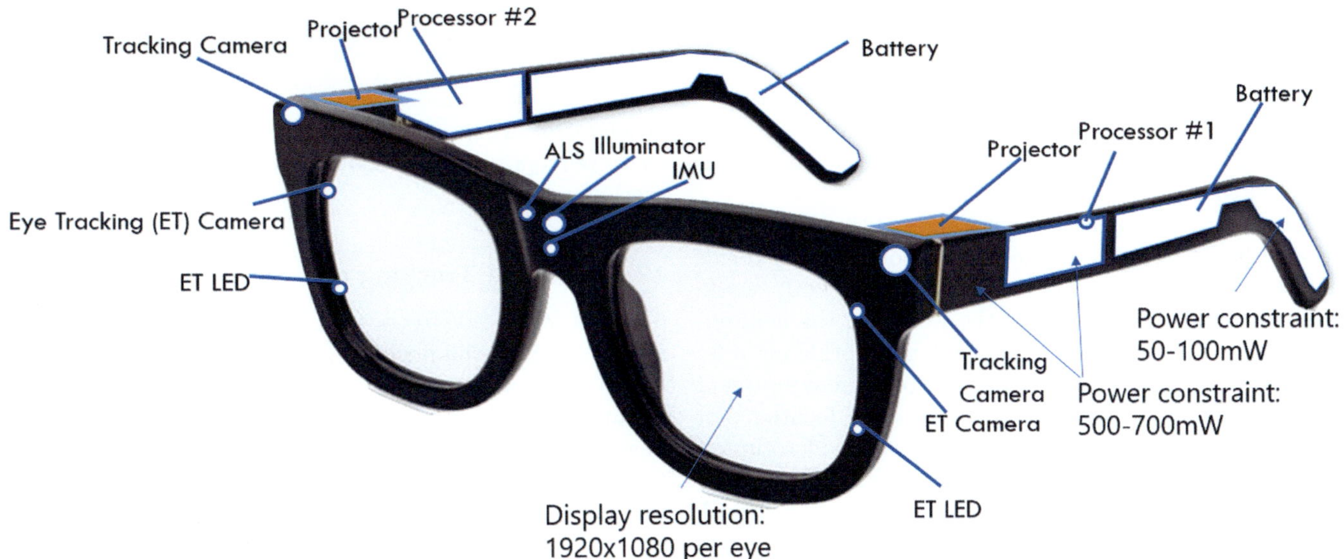

FIGURE 12.4
A representative AR glass with notable components.

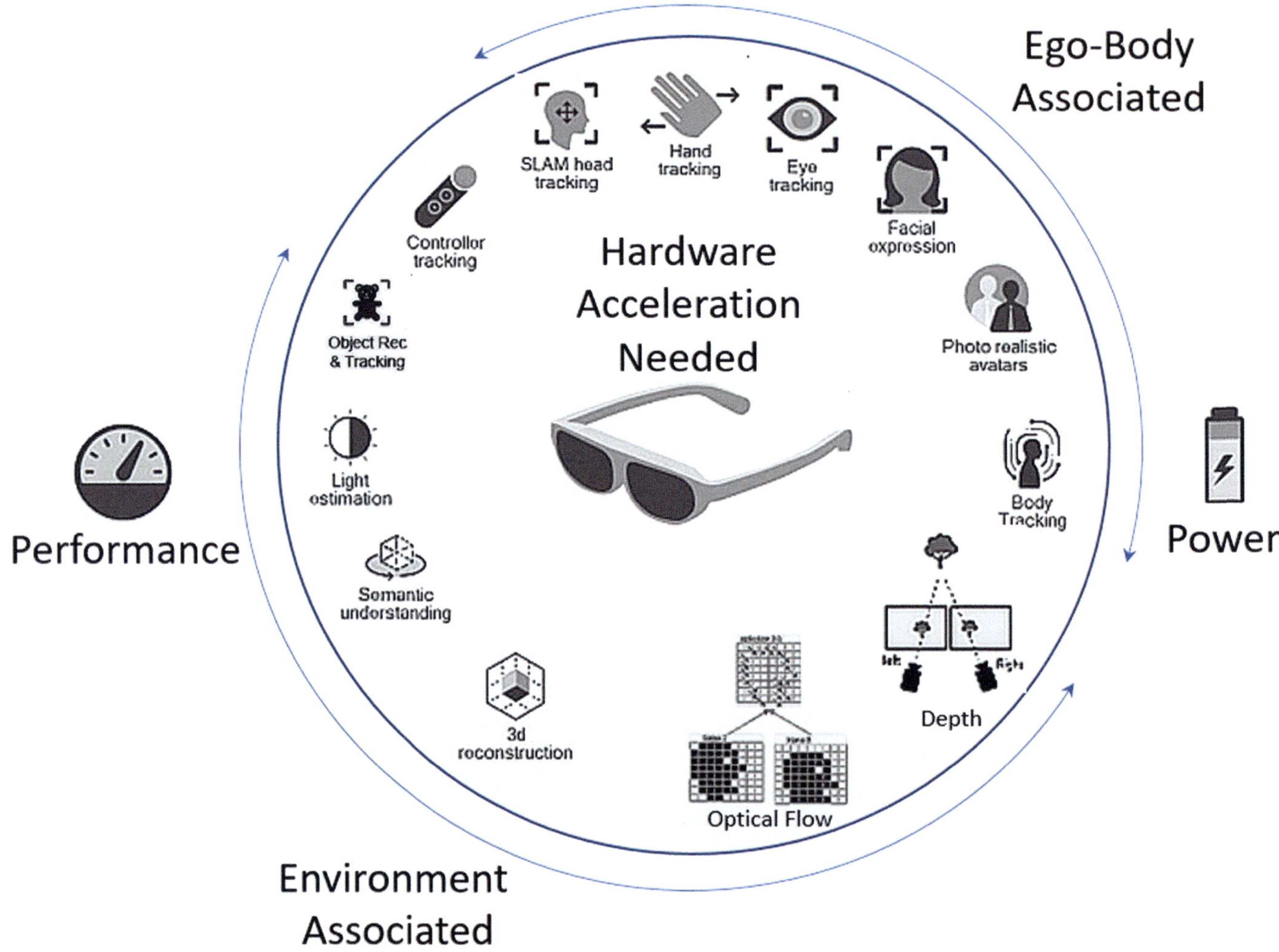

FIGURE 12.5
Typical set of perception workloads in XR.

most notable ones. A VR or AR device may have additional workloads based on the use cases that the device expected to support.

In Figure 12.5, the perception workloads in XR are partitioned into two groups: Ego-body associated workloads and environment-associated workloads. Ego-body workloads refer to tasks specific to the wearer of the headset. On the other hand, environment-related workloads refer to tasks associated with the analysis of the environment in which the wearer/user of the headset is located. For example, if the user is in a room, one may want to image the room and build a 3D model for assisting the user to navigate around the room while avoiding collision with objects. Figure 12.5 also depicts that the computation of these workloads will involve trade-offs in performance and power, and this may lead the system designer to envision offloading some of these workloads to a hardware

accelerator in the XR system. We will discuss each of these perception workloads in some level of detail in this section.

12.3.1 Head Tracking and SLAM

6 DoF (degrees of freedom) head tracking refers to the ability to track the position (x, y, and z) and orientation (pitch, yaw, and roll) of the HMD. In VR, this tracking estimate is used to render the virtual scene relative to the head pose. In a typical XR system, several head tracking technologies including optical tracking and inertial tracking are used jointly to estimate the 6 DoF head pose. Optical tracking uses cameras to track the head position, while inertial tracking uses sensors to measure changes in acceleration and rotation. The 6 DoF pose is used to update the virtual or augmented environment in real time,

allowing the user to look around and interact with their surroundings as if they were there. This technology has become increasingly important in gaming, training, and simulation, where realistic and immersive experiences are highly valued.

In a typical HMD, the Head Tracking perception workload is integrated with the SLAM workload. SLAM refers to the process of creating a map of an environment in real-time while simultaneously tracking the location of the XR device within that environment. This technology is used in AR and VR applications to enable accurate tracking and positioning of virtual objects and characters within the user's physical space. SLAM works by using sensors such as cameras and depth sensors to collect data about the environment, such as the position and orientation of objects. This data is then used to build a 3D map of the environment in real-time. As the user moves around the environment with the XR device, SLAM continually updates the map and tracks the device's position within it. This allows XR applications to accurately place virtual objects and characters within the user's physical space and create a seamless and immersive experience. Head Tracking and SLAM are supported by many HMDs including HTC VIVE, Oculus Quest 2[47], and Hololens[8, 30].

12.3.2　Hand Tracking

Hand tracking in XR refers to the ability of XR devices to track the movements and positions of egocentric hands in real time without the need for physical controllers or other input devices. This technology allows users to interact with virtual objects and environments in the XR setup using natural hand gestures and movements. Hand tracking in XR typically uses a combination of hardware and software to detect and interpret the user's hand movements. The hardware typically includes cameras and/or depth sensors that capture images of the user's hands and surrounding environment. The software then uses computer vision algorithms and machine learning models to analyze the images and track the movements of the user's hands.

Hand tracking can enable a wide range of XR applications, for example, users can manipulate virtual objects, perform complex gestures to control XR interfaces, or use their hands to interact with virtual environments in more natural and intuitive ways. Hand tracking may not be suitable for all types of XR applications, and some users may still prefer physical controllers or other input devices for certain tasks. Support for hand tracking is available in some recent XR devices most notably Oculus Quest 2 and Microsoft Hololens 2. Hand tracking solution is also available on Qualcomm's Snapdragon Spaces [51].

12.3.3　Eye Tracking and Gaze Estimation

Eye tracking is an emerging technology that has the potential to greatly enhance XR experiences. By tracking the movement of a user's eyes, XR devices can determine where the user is looking, which can be used to improve rendering and interaction in several ways. In such an improved rendering scheme for XR usage, referred to as dynamic foveated rendering [26], the rendering resources are focused on the parts of the scene where the user is looking, while areas of the scene that are not being looked at are deprioritized or even hidden. Eye tracking-based dynamic foveated rendering can result in improved performance and reduced battery drain, as well as more immersive and realistic visual experiences.

Eye tracking can also be used to improve interaction in XR. For example, XR devices can use eye tracking to detect when a user is looking at an object and then use gesture or voice commands to interact with it or project additional information on a virtual display about the object. Finally, eye tracking can be used for biometric applications, such as monitoring a user's attention or detecting signs of fatigue. This can be particularly useful in training and simulation applications, where maintaining focus and concentration is important. Some eye-tracking-enabled commercial products are VIVE Pro Eye [42], Pico Nio 3 Pro Eye [40], and Microsoft Hololens 2 [29].

12.3.4　Facial Expression

Facial expression recognition is a technique used in XR to track and interpret the facial expressions of the user. This technology can be used to create more realistic avatars and enhance communication between users, for example, in a telepresence session.

Facial expression recognition systems typically use a combination of computer vision algorithms and machine learning techniques to analyze images of the XR device user's face and extract features such as the position and movement of the eyebrows, mouth, and eyes. The system then interprets these features to determine the user's emotional state and adjust the XR experience accordingly. As the technology continues to improve, it is expected to play an even more important role in enhancing the immersive and interactive experiences offered by XR.

12.3.5　Photorealistic Avatars

Photorealistic avatars are essentially a digital representation of the HMD user. Such avatars currently are restricted to only the face and are starting to emerge in XR environments. Typically, these avatars are created by inward-looking cameras in the HMD wherein the images are stitched together to synthesize the 3D model of the face. As the HMD occludes many areas of the face, it is challenging to create an accurate model, and machine learning techniques are used currently to synthesize the 3D model of the face.

Photorealistic avatars have many potential applications in XR. For example, they could be used to create more immersive virtual environments for gaming or training simulations. They could also be used in telepresence applications, where remote users can communicate and interact with each other in a more realistic way.

However, creating truly convincing photorealistic avatars is a complex task, and there are still many challenges to overcome. One of the biggest challenges is achieving realistic facial expressions and movements, which requires advanced machine-learning algorithms and sophisticated animation techniques. Quest Pro by META introduced facial avatars but these avatars are not photorealistic yet. A full solution for a photorealistic avatar will involve combining the facial avatar with the 3D model of the hand from the hand-tracking perception workload, body-tracking perception workload, and eye-tracking workload. Such a fusion is still a research problem and has not been deployed in a commercial device.

12.3.6   Body Tracking

Body tracking in XR refers to the ability of XR devices, such as VR headsets, to track the movements of a user's body in real-time. This allows for more immersive experiences, as the user's body movements can be reflected in the virtual environment.

There are several ways that XR devices can track the user's body. One common method is using external sensors, such as cameras or depth sensors, which are placed in the room where the user is using the XR device. These sensors capture the user's movements and transmit them to the XR device, which then translates them into corresponding movements in the virtual environment.

Another method is using sensors on the XR device itself, such as accelerometers or gyroscopes, which can track the user's movements as they move their body while wearing the device. This method is less accurate than external sensors, but it allows for more freedom of movement since the user is not restricted by the range of the external sensors. Egocentric body tracking algorithms are still in their infancy and further research is needed to have a commercially viable solution for the mass market.

12.3.7   3D Reconstruction

3D reconstruction is an important tool in creating immersive experiences in XR applications. By using 3D models of real-world objects and environments, XR developers can create realistic and interactive experiences that allow users to explore and interact with virtual environments.

There are several techniques for 3D reconstruction that are commonly used in XR applications including photogrammetry, where multiple views of the scene from different camera poses are synthesized into a 3D model of the scene or using depth from passive stereo or depth from a single camera derived using machine learning techniques to create a 3D point cloud of the environment, which can then be converted to a 3D model of the scene.

Once a 3D model is created, it can be used in a variety of ways in XR applications. For example, a 3D model of a real-world location can be used to create a VR experience that allows users to explore the environment in a realistic and immersive way. A 3D model of a product can be used in an AR application to create an interactive experience that allows users to visualize the product in a real-world setting. 3D reconstruction in commercial XR devices is still in the early stages and may not be delivering a real-time experience due to the high computing resources needed for 3D scene reconstruction.

12.3.8   Semantic Understanding

Semantic segmentation in 3D scene understanding refers to the process of identifying and labeling different objects in a 3D environment with semantic labels, such as chairs, tables, walls, and floors. The goal of semantic segmentation is to provide a comprehensive understanding of the objects and their relationships within the 3D scene, enabling more advanced and context-aware AR and VR applications. One approach to semantic segmentation is to use a 3D point cloud or mesh representation of the scene and apply deep learning algorithms to classify and segment the different objects based on their geometric features and texture. Another approach is to use RGB-D sensors or LiDAR scanners to capture the 3D geometry and texture of the environment and perform semantic segmentation based on the captured data. Semantic segmentation in 3D scene understanding has numerous applications, including AR and VR navigation, object recognition and tracking, and immersive visualization. For example, AR applications can use semantic segmentation to identify and track objects in the real world and augment them with virtual content or information. Similarly, VR applications can use semantic segmentation to create more realistic and immersive virtual environments and enable more natural interactions with virtual objects.

12.3.9   Light Estimation

Light estimation in XR refers to the process of accurately estimating the lighting conditions of the real-world environment and using that information to render virtual objects or environments in a way that appears realistic and consistent with the real world. Accurate light estimation is critical for creating a believable and immersive XR experience, as lighting conditions can significantly impact the appearance and perception of virtual objects and environments. For example, if a virtual object is rendered with incorrect lighting, it may appear out of place or disconnected from the real-world environment. Light estimation can be achieved using various techniques. Some common methods include:

- **Environmental mapping:** This involves capturing the real-world environment using sensors, such as RGB-D cameras or LiDAR, and using that information to estimate the lighting conditions.

- **Physics-based lighting:** This involves using physically based rendering (PBR) techniques to simulate the behavior of light in the virtual environment, based on the estimated lighting conditions.

- **Sensor fusion:** This involves combining data from multiple sensors, such as RGB-D cameras, gyroscopes, and accelerometers, to estimate the lighting conditions.

Accurate light estimation in XR enables more realistic and immersive XR experiences and can be applied to various applications, such as virtual product demonstrations, training simulations, and gaming. It also allows for a more seamless integration of virtual and real-world objects, enabling more natural interactions and transitions between the two.

12.3.10   Controller Tracking

Controller tracking in XR refers to the process of accurately detecting and tracking the movements and orientation of handheld controllers in real time. Handheld controllers are typically used in XR applications to provide users with a more natural and intuitive way to interact with virtual objects and environments. Controller tracking is essential for providing a seamless and immersive XR experience, as any lag or inaccuracies in tracking can result in a disconnect between the user's actions and the virtual environment. Accurate controller tracking also enables more advanced features, such as hand and finger tracking, haptic feedback, and object manipulation. There are various methods for tracking controllers in XR, depending on the specific XR platform and hardware used. Some common techniques include:

- **Optical tracking:** This involves using optical sensors on the XR headset or external cameras to detect and track the position and orientation of the controllers.

- **Inertial tracking:** This involves using built-in sensors in the controllers, such as accelerometers and gyroscopes, to track their movements and orientation.

- **Ultrasonic tracking:** This involves using ultrasonic sensors to detect the position and orientation of the controllers based on the time-of-flight of ultrasonic signals.

Some XR platforms also support hybrid tracking, which combines multiple tracking methods to provide more accurate and robust controller tracking. Most commercial HMDs are equipped with handheld controllers.

12.3.11   3D Object Detection, Recognition, and Tracking

3D object detection, recognition, and tracking[23] are critical components of XR technology and these techniques enable XR systems to interact with real-world objects and create immersive experiences. Object detection in XR involves detecting the presence of objects in a user's environment. This can be achieved using sensors such as cameras or depth sensors. The data collected from these sensors is processed to identify the location, shape, and size of the objects in the user's environment. Object recognition involves identifying objects based on their features, such as shape, color, and texture. This can be achieved using machine learning algorithms that are trained on large datasets of objects. The algorithms can then recognize and categorize objects in real-time, allowing XR systems to interact with them. Object tracking involves following the movement of objects in a user's environment. This can be achieved using techniques such as marker-based tracking, feature-based tracking, or depth-based tracking. The data collected from these techniques is used to track the movement of objects and update their position and orientation in real-time. Combining these techniques allows XR systems to detect, recognize, and track objects in 3D space. This enables users to interact with virtual objects that are placed in their real-world environment, creating immersive and engaging experiences.

12.3.12   Depth and Optical Flow in XR

Depth and optical flow are two key perception workloads in XR and are used by other perception workloads in XR, for example, depth is used as an input in the 3DR perception workload or even in hand tracking perception workload to segment the foreground hand object from the background for more robust tracking of the hands. Optical flow is a key perception workload often used to render the scenes on the HMD and produce a smooth visual experience. Specifically optical flow is used as part of the Adaptive Space Warp (ASW) feature [25] wherein the HMD images are rendered at a higher frame rate than the frame rate at which they are generated and this reduces motion sickness and increases the feeling of immersive experience.

12.4 Current State of the Perception Workloads

This section provides additional detail on some key perception workloads among the many outlined in the previous section. A processing flow for these key workloads will be presented here.

12.4.1 Head Tracking and SLAM

In XR, an essential perception task is that of head tracking and SLAM. This feature is used as an input to many of the other perception workloads depicted in Figure 12.5, such as, 3DR uses the 6 DoF pose generated by this feature, Photorealistic avatars also need the 6 DoF pose. A typical head tracking and SLAM processing flow are as illustrated in Figure 12.6 Other examples of a similar flow that may be implemented in an AR or VR HMD can be found in [39].

The tracking thread estimates camera motion in real-time, whereas mapping predicts the 3D positions of feature points observed in the scene and builds the map interactively or refines a pre-existing map of the environment. The tracking thread could further be split into an IMU thread since IMU data usually comes at 1000Hz and an image processing thread, wherein the vision cameras data comes in at 30Hz. IMU thread can give a pose estimate and this pose estimate is further refined by the processes in the image processing thread based on the visual features detected in the scene. Note that for getting a robust set of visual features, 2–4 cameras data is processed in the feature processing block and feature extraction block. Working in a texture-less environment

with few salient feature points often leads to drift errors in the position and orientation of the head pose. As one of the primary challenges in VSLAM, this error may lead to system failure. Thus, considering complementary scene-understanding methods in feature-based approaches, such as object detection or line features, would be a trendy topic.

The compute-intensive tasks in the tracking thread are mostly in the pose estimation function wherein usually an extended Kalman filter or a sliding window bundle adjustment technique is used. Feature processing and feature tracking are tasks that are easily done in a DSP or dedicated hardware accelerator. In the mapping thread, bundle adjustment is a high computation complexity task given the size of the matrices involved in these computations. Thus, the mapping task in an HMD usually does not run at a high frame rate.

A critical issue in any SLAM system is the drift problem and losing the feature tracks caused by accumulated localization errors. Detection of drifts and loop closures to identify previously visited places contributes to high computation latency and cost as the complexity of loop closure detection increases with the size of the reconstructed map. A counterbalancing strategy to reduce complexity is to do the loop closures at a much lower frame rate relative to the tracking rate.

12.4.2 Egocentric Hand Tracking and Hand Gesture Estimation

Recently, the number of head-mounted devices supporting egocentric hand-tracking features is increasing to keep up with the popular demand for a more immersive XR experience[8]. Hand tracking adds to the degree of immersion and presence, allows more effective hand-based interaction with virtual objects, and enables hand gesture-based communication capabilities [8]. The input to the hand-tracking flow generally varies between monocular/multi-view RGB/Grayscale [3, 22], or RGB/Grayscale with depth [17] along with 6 DoF and camera parameters.

In [5], the authors divided the egocentric hand-related tasks into three categories namely, hand localization, which involves pose estimation, interpretation or gesture analysis, and building real-world applications. Hand localization and interpretation are purely perception tasks and therefore we confine our discussions to those two topics. Also, for most of the current XR use cases, a safe assumption (or even a requirement) can be that the hands are free of any artifacts or objects. A more realistic problem statement should consider hands holding real-world objects, two hands interacting heavily with each other, users wearing rings, gloves, having a tattoo, or wearing ornaments on hands, and the solution may have special

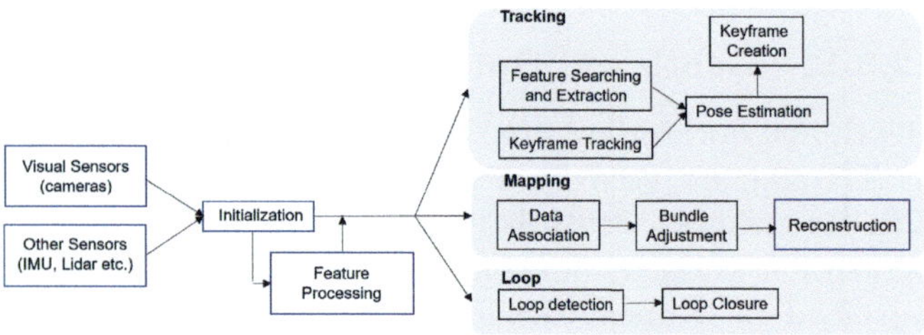

FIGURE 12.6
A high-level processing flow for a typical Visual SLAM approach.

support for anomalous cases like amputated fingers or malformed hands.

The task of egocentric hand localization, that is, pose estimation (HPE) can be divided into three main blocks – hand detection, segmentation, and classification, hand pose estimation (specific tasks such as fingertip detection can be considered as a subcategory for this), and hand tracking [5]. The final output of the hand-tracking flow is the estimated 3D keypoint lists for the two egocentric hands of the HMD's user in the world frame coordinates. In addition, the system may output parametric models of the two hands or full 3D meshes in world coordinates, a segmentation map, fingertip locations, etc. The most common inputs considered for the HPE frameworks are monocular RGB and/or depth images. Some works also take multiple views into account for more robust hand pose estimation.

In Figure 12.7, a sample flow diagram for hand pose estimation is shown. Depending on the accuracy and latency requirements and considering memory and power budgets, the HPE flow may include one or more trackers to track bounding boxes or key points across time. It is more likely for an HPE flow to also output hand classes for each detected bounding box along with confidence scores. Egocentric hand detection can be treated like a regular object detection task where the outputs are bounding boxes and class information. Most popular object detectors are known to estimate multiple boxes at different scales and apply a non-maximum suppression operation to consolidate multiple boxes into one prediction [33]. In [22], the authors customized their solution for the task further by limiting the outputs to two bounding boxes only – one for the left hand and another for the

right. Hence, non-maximum suppression is not needed as a post-processing operation for this solution. An alternative to hand detection can be hand segmentation where segmentation masks are estimated for the scene to differentiate between the contours of hands and the surrounding background. Over the years, researchers proposed an array of pixel-based [31] and patch-based [6, 49] classification methods relying on traditional computer vision techniques to determine ego-hand segmentation masks. More recently, deep learning-based approaches showed great promise to solve the hand and hand+object segmentation problems from an egocentric view [27, 57].

There are many approaches for hand pose estimation – the most popular one being a 2.5D pose estimation followed by an inverse kinematic (IK) hand skeleton model for conversion to 3D. The 2.5D estimation involves estimating 2D key points and a distance measure for the hand in 3D. One such example is reported in [22] where the authors estimate 2D keypoint heatmaps and 1D relative distance heatmaps for each of their 21 predefined key points before fitting to a 3D hand model. A more streamlined approach to directly regressing the 3D hand model (both parametric and nonparametric forms) is proposed in [3] where the key points and vertex locations in world-coordinate are directly regressed from the image features.

The 3D hand pose information can be utilized to render a 3D synthetic hand object that would enable the user to interact with the virtual world. Some additional functionality can be achieved through hand gesture recognition. For example, in META Quest and Microsoft Hololens devices, the user can "pinch" to click a button, and "grab" a virtual object to pick it up. Gesture

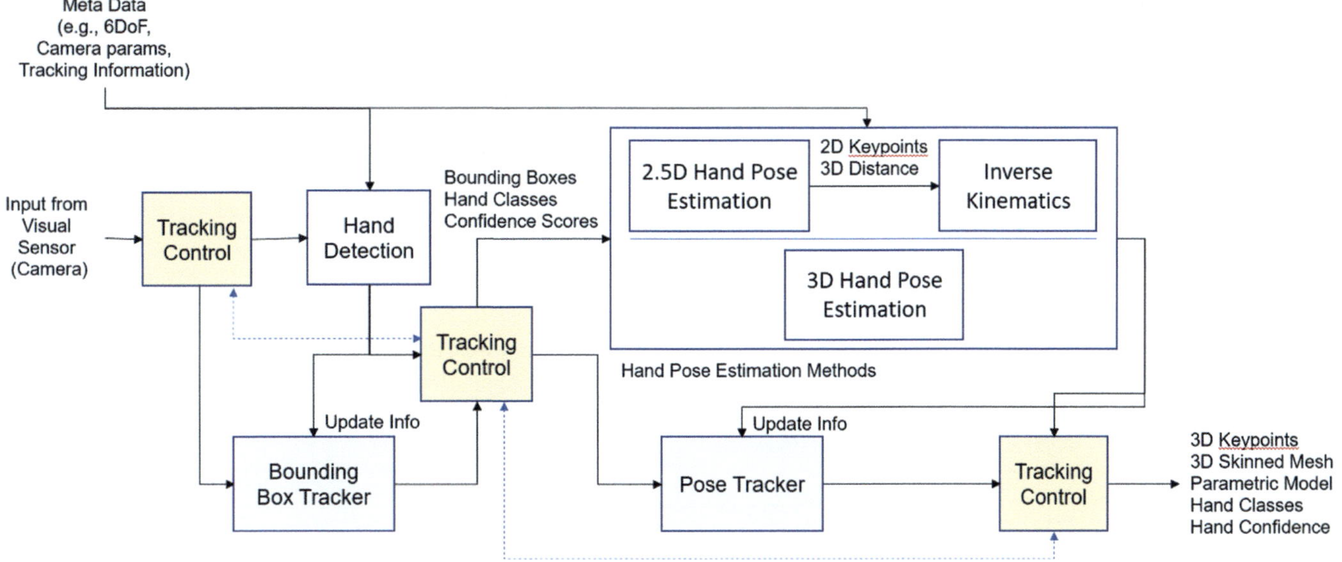

FIGURE 12.7

Egocentric hand pose estimation flow for XR.

recognition from the pose-tracked hand can be rule-based, that is, it can be derived from the relative positions of the key points. Alternatively, hand gestures can also be derived from hand images as part of a multi-class classification task [11]. Directly estimating hand gestures would be specifically useful when only the gesture matters and hand tracking information is not needed. One such practical example would be an ego-hand gesture-based screenshot or photography application that takes a photo when the user performs a predefined gesture.

12.4.3 Eye-Tracking and Gaze Estimation

The gaze estimation task for XR falls under the near-eye-tracking problem domain where a head-mounted device hosts the egocentric eye-facing illuminators and detectors [41]. One of the earliest techniques of near-eye-tracking is based on pupil-centered corneal reflection (PCCR) [24]. To date, variants of PCCR-based gaze estimation methods are utilized by many commercially available eye-tracking solutions for XR such as Tobii [41] and IMOTIONS [15]. In PCCR, near-infrared (NIR) lights from one or more emitters are projected on each eye and the eye image is captured using one or more strategically placed cameras. The reflection of the IR-LEDs on the cornea appears as white dots known as glints in the captured image, whereas the pupil appears as a dark circle. Traditional computer vision-based methods are then applied to the image to determine the pupil, pupil centers, and glint/s locations. A regression-based optimization problem is then solved to estimate the gaze direction for the eye based on the pupil-glint vector and other relevant parameters such as pupil shape, eye openness, and eye shape [32]. A simplified diagram for the eye-tracking flow is shown in Figure 12.8. As can be seen in the diagram, based on the gaze vectors and eye-model parameters additional attributes about the eye state can be predicted such as 3D vergence point, blink/eye-openness, pupil and eye-ball information, and fixation or saccadic movements.

In recent years, eye-tracking using deep learning-based backbones is also getting a lot of attention. In [28], the authors used neural networks for gaze estimation and pupil localization. Depending on the application of gaze tracking, very low latency might be desirable. Along this line, event camera-based eye-tracking is promising extremely fast eye-tracking at up to 10,000Hz frequency [4]. In [4], the authors used a deep learning-based pupil estimator sparsely, while tracking the pupil at a very high frequency using the event camera. This work is a nice example of a hybrid eye-tracking approach where deep learning models are estimated for partial estimation, that is, the pupil, while traditional regression-based modeling is used for final gaze vector estimation. Apart from NIR-illumination and event cameras, researchers are also exploring other sensor technologies such as ultrasound [19] and bright-pupil retinal scanning [7]. Also, instead of multiple fixed IR emitters for PCCR, some works proposed beam scanning [18, 36].

Most gaze tracking systems require one, five, or nine-point calibration for best performance. Auto calibration for different users is still an active research area. Compensating for the slippage of the head-mounted device along the nose bridge is another challenging task in this domain. There are different methods for finding the vergence point in the real-world 3D coordinate, but most produce jittery results – stabilizing the eye gaze estimation for microsaccades is another open research problem. Some gaze tracking use cases such as gaze-based finer control may require high estimation accuracy and precision, whereas other use cases like dynamic foveated rendering or gaze-based attention detection may compromise quantitive performance for power and latency benefits.

12.4.4 3D Reconstruction

In XR, there are many applications that need an understanding of the 3D geometry of the scene, such as, in remote collaboration, plane detection, and occlusion rendering. There are many ways to build a 3D model of a scene by taking different views of the scene or even using

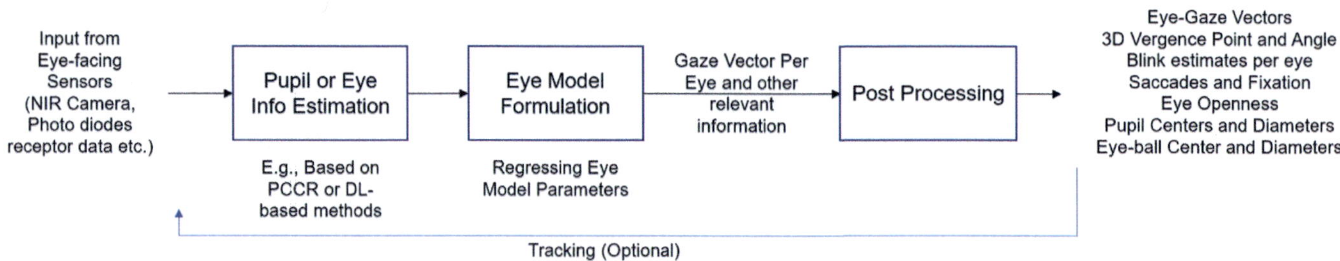

FIGURE 12.8
Gaze estimation and tracking flow for XR.

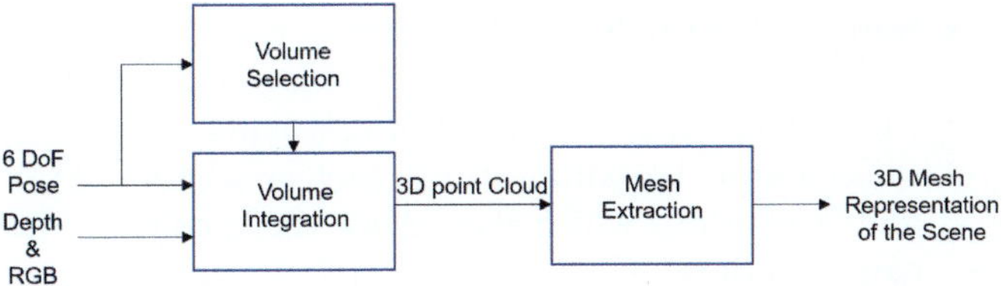

FIGURE 12.9
3D scene reconstruction processing flow.

ML techniques to synthesize a 3D model from a single camera image. In Figure 12.9, we show a representative processing flow for 3D scene reconstruction.

The steps in this process are as follows:

1. **Inputs:** The inputs are depth frames (estimated depth from stereo pair of cameras or time of flight sensors or ML-based depth estimation techniques) and 6 DoF pose of the depth frames (estimated from the head tracking and SLAM perception workload) and optionally the output of an RGB camera aligned to the depth frames.

2. **Volume integration:** There are different techniques for using depth samples from a sequence of camera poses to create a 3D point cloud description of the 3D volume. One approach is to tesselate the scene volume into a set of volume blocks and describe each block by a 3D array of truncated signed distance function (TSDF) volume samples [13]. TSDF for a voxel in the volume block is a function of the depth value as well as the pose and the TSDF value of the voxel is integrated with previous TSDF values of the same voxel derived from another camera pose [55]. After some number of camera frames are integrated in this manner, a 3D point cloud representation of the scene is the result of this volume integration process. Note that when RGB data for each camera pose is also available and assuming it is aligned with the depth data, the color-based integration is done in the same manner as TSDF-based integration [14].

3. **Volume selection:** As the 3D representation is only related to the surface of the 3D objects, it is a sparse representation within the large volume. For instance, if each voxel including color information is 8 bytes and at a resolution of 3cm, a 30m TSDF cube would consume 8GB of memory to store the 3D representation. However, given that the 3D data is sparse, efficient mechanisms need to be devised to access the occupied voxel from a previous time instant in the volume selection process. A hashing strategy such as the

one in [44] would be very efficient in the volume selection process.

4. **Mesh extraction:** The 3D point cloud representation can be used as-is in many XR use cases such as in collision avoidance. However, for some XR use cases, one needs to generate a mesh from the point cloud data. Typically, a marching cubes technique is used for this process [34].

This overview highlights the main processing involved in the 3D scene reconstruction process. There are newer ML-based neural radiance field (NERF) approaches starting to emerge for rendering a 3D scene from camera images directly without the need for depth [9]. However, these methods have high computation complexity, and more research is needed to reduce the complexity of such approaches for an embedded system such as an HMD.

12.5 Challenges and Solution Directions

In this section, we will first enumerate the challenges related to some of the perceptual workloads depicted in Figure 12.5 and then discuss current and future research directions to tackle those challenges.

12.5.1 Head Tracking and SLAM

There are still significant challenges in head tracking and SLAM implementations in XR devices:

- **Reducing latency:** Head tracking latency refers to the delay between the user's head movement and the corresponding change in the virtual environment. High latency can result in motion sickness and reduce the sense of immersion.

- **Operating at high accuracy:** Head tracking accuracy refers to the ability to track the user's head movements with precision. Inaccurate head tracking can result in misalignment between the virtual environment and the user's physical surroundings.

- **Drift reduction:** Drift occurs when the virtual environment gradually drifts away from the user's physical surroundings over time. This can result in misalignment between the virtual environment and the user's physical surroundings.

- **Occlusion handling:** Occlusions occur when objects in the environment obstruct the view of the head tracking sensors. This can result in inaccuracies in the head tracking data and reduce the quality of the user's experience.

- **Operate within computational resource constraints:** XR devices are typically limited in terms of processing power, memory, and battery life. SLAM algorithms require significant computational resources and optimizing them to run efficiently on XR devices can be challenging.

- **Handling lighting variation:** SLAM algorithms rely on visual information to track the user's position and movement. However, changes in lighting conditions, such as shadows and glare, can affect the accuracy of visual tracking and lead to errors.

- **Robust performance in dynamic environments:** XR applications often involve dynamic environments, such as moving objects or changing lighting conditions. SLAM algorithms must be able to adapt to these changes and maintain accurate tracking. Typically, the dynamic objects are detected and masked out of the scene prior to head pose estimations.

- **Overcome sensor limitations:** XR devices use a variety of sensors, such as cameras and gyroscopes, to track the user's head movements. However, these sensors have limitations, such as a limited FOV and sensitivity, which can affect the accuracy of the head tracking.

Addressing these challenges will require continued research and innovation in head-tracking algorithms and sensor technology. Additionally, optimizing algorithms to reduce latency and drift, developing techniques to handle occlusions, and improving the accuracy and sensitivity of head-tracking sensors will be critical to creating high-quality and immersive XR experiences.

12.5.2 Hand Tracking

While hand tracking technology has advanced significantly in recent years, there are still some limitations and challenges:

- **Occlusions:** Occlusions occur when parts of the hand or fingers are obstructed from view, making it difficult to accurately track their position and movement. Rings and gloves on the hand also result in inaccuracies in hand tracking.

- **Articulation:** Hands are complex structures with many degrees of freedom, which can make it difficult to accurately track their movements and poses.

- **Lighting conditions:** Changes in lighting conditions can affect the quality of the captured data and result in inaccuracies in hand tracking.

- **Sensor limitations:** XR devices use a variety of sensors, such as cameras and depth sensors, to capture information about the user's hands. However, these sensors have limitations, such as a limited FOV and resolution, which can affect the quality of the captured data.

- **Computational complexity:** Hand-tracking algorithms are computationally intensive and can require significant resources. This can limit the ability to perform real-time hand tracking on XR devices with limited processing power.

12.5.3 Eye Tracking and Gaze Estimation

Eye tracking technology is still in its infancy in XR devices. There are many challenges to enabling eye tracking on XR. Accurate tracking and prediction of gaze require estimating at very low latency, which is computationally challenging and also may require a special sensor such as an event camera instead of regular eye-tracking cameras. Slippage compensation, that is, retaining gaze estimation performance across the slight movement of the HMD along the nose bridge, is a huge challenge. Another challenge is ensuring that eye tracking is accurate and reliable across different lighting conditions and user profiles. Also, ensuring that eye-tracking does not negatively impact user privacy is quite important, as eye-tracking data can be sensitive and personal.

12.5.4 Facial Expressions

There are some challenges associated with facial expression recognition in XR. One of the main challenges is to estimate the user's facial features as parts of the face are occluded by the HMD. Additional challenges include achieving accurate and reliable recognition in different lighting conditions and environments. Another challenge is dealing with individual differences in facial features and expressions, which can make it difficult to develop a system that works equally well for all users.

12.5.5 Photo-realistic Avatars

Creating photorealistic avatars in XR can be a challenging task. There are several factors that contribute to the complexity of creating photorealistic avatars in XR, including:

- Realistic facial expressions: Creating realistic facial expressions that accurately represent the user's emotions is a significant challenge in creating photorealistic avatars. Subtle changes in facial expressions can significantly impact the perceived emotion and capturing these nuances can be difficult.

- Skin and hair textures: Realistic skin and hair textures are essential to creating photorealistic avatars, but they are challenging to create. Hair and skin textures are complex and require detailed models to capture accurately.

- Motion capture: Motion capture technology is used to capture the movements of real people and animate avatars, but it can be challenging to achieve high-quality results. The movements of the avatar need to be synchronized with the user's movements in real time, which requires accurate and responsive motion capture technology.

- Lighting conditions: Lighting conditions can significantly impact the perceived realism of an avatar. Achieving realistic lighting requires sophisticated algorithms and hardware to simulate natural lighting conditions accurately.

- Real-time performance: Real-time performance is critical in XR applications, where users expect a seamless and responsive experience. Creating photorealistic avatars in real time requires significant computational resources, which can impact the real-time performance of the XR system.

To overcome these challenges, various techniques are being investigated including using machine learning algorithms to create more realistic facial expressions and using sophisticated texture mapping techniques to create realistic skin and hair textures. Realistic lighting conditions can be achieved using advanced rendering techniques, and real-time performance can be improved by using optimized algorithms and dedicated hardware.

12.5.6 Body Tracking

Body tracking enables users to interact with virtual environments using their own body movements. However, body tracking in XR can be challenging due to several factors, including:

- Body pose variability: Real-world bodies can vary widely in shape, size, and pose, making it challenging for XR systems to accurately track body movements.

- Occlusion: Body parts can be partially or fully occluded, making it challenging for XR systems to accurately track body movements in real time.

- Lighting conditions: Changes in lighting conditions, such as shadows or reflections, can make it challenging for XR systems to accurately track body movements.

- Real-time performance: Real-time performance is critical in XR applications, where users expect a seamless and responsive experience. Body tracking can be computationally intensive, which can impact the real-time performance of the XR system.

To overcome these challenges, different techniques have been developed, including using machine learning algorithms to improve tracking accuracy, combining multiple sensors and data sources to improve tracking robustness, and optimizing algorithms for real-time performance. Some XR systems also use markerless tracking approaches to improve tracking accuracy and robustness.

12.5.7 3D Reconstruction

3D scene reconstruction is emerging as a fundamental building block in XR systems. However, this technology has several challenges that limit the immersive experience for the user:

- **Sensor limitations:** XR devices use a variety of sensors, such as cameras and depth sensors, to capture information about the environment. However, these sensors have limitations, such as a limited FOV and resolution, which can affect the quality of the reconstructed scene. When depth is not directly sensed but estimated from mono RGB cameras or stereo cameras, the resulting depth may not be adequate for a high-quality 3D reconstruction of the scene due to incomplete depth in textureless scenes.

- **Occlusions:** Objects in the environment can occlude each other, making it difficult to capture a complete view of the scene. This can lead to missing or incomplete geometry and texture information in the reconstructed scene.

- **Dynamic environments:** XR applications often involve dynamic environments, such as moving objects or changing lighting conditions. This can lead to inconsistencies in the reconstructed scene over time.

- **Lighting conditions:** Changes in lighting conditions, such as shadows and reflections, can affect the quality of the captured data and result in inaccuracies in the reconstructed scene.

- **Computational complexity:** 3D scene reconstruction algorithms are computationally intensive and can require significant resources. This can limit the ability to perform real-time scene reconstruction on XR devices with limited processing power.

Semantic Understanding: Semantic understanding, which refers to the ability of XR systems to recognize and understand the meaning of objects and scenes in the real world, however, semantic understanding in XR, can be challenging due to several issues, including:

- **Object and scene variability:** Real-world objects and scenes can vary widely in shape, size, and appearance, making it difficult for XR systems to accurately recognize and understand them.

- **Environmental variability:** XR applications can be used in a wide range of environments, from indoor to outdoor settings, which can present different challenges for semantic understanding. For example, outdoor environments may contain more complex and dynamic objects, such as trees or moving vehicles, which can be challenging to recognize and understand.

- **Limited data:** Training machine learning algorithms for semantic understanding requires large amounts of data, which can be difficult to obtain in XR applications. This is particularly true for AR applications, where the environment can change rapidly and unpredictably.

- **Real-time performance:** Real-time performance is critical in XR applications, where users expect a seamless and responsive experience. Semantic understanding can be computationally intensive, which can impact the real-time performance of the XR system.

12.5.8 Light Estimation

Light estimation is an important aspect of XR as it enables the virtual and real-world environments to be integrated seamlessly, creating a more immersive and realistic experience for users. However, light estimation in XR can be challenging due to several factors, including:

- **Lighting variability:** Lighting conditions can vary widely across different environments, and changes in lighting conditions can impact the performance of the XR system. For example, a bright light source in the background can cause objects in the foreground to appear dark and underexposed.

- **Color temperature:** Color temperature refers to the warmth or coolness of a light source and can impact the way colors are perceived in the environment. Accurately estimating the color temperature of the environment is critical for creating realistic virtual objects that match the lighting conditions of the real-world environment.

- **Shadows:** Shadows are a critical aspect of lighting that can be difficult to estimate accurately in XR applications. Shadows can be caused by a variety of factors, including the position and intensity of light sources, and the geometry of the objects in the environment.

- **Real-time performance:** Real-time performance is critical in XR applications, where users expect a seamless and responsive experience. Light estimation can be computationally intensive, which can impact the real-time performance of the XR system.

- **Limited hardware capabilities:** XR devices, such as smartphones or AR/VR headsets, may have limited hardware capabilities that can impact the accuracy and speed of light estimation algorithms.

Despite these challenges, light estimation remains a critical aspect of XR applications, enabling a more immersive and realistic experience for users.

12.5.9 Object Recognition and Tracking

Tracking 3D objects in XR applications can be challenging due to various factors, including:

- **Occlusion:** Occlusion occurs when an object is partially or completely hidden from view, which can make it difficult for the XR system to track the object. Occlusion can occur in both AR and VR applications and can be caused by other objects in the environment, user movements, or even the XR device itself.

- **Lighting conditions:** Changes in lighting conditions, such as shadows or reflections, can make it difficult for the XR system to accurately track objects in the environment. This is especially true for AR applications, where the XR system needs to integrate virtual objects seamlessly with the real world.

- **Scale:** Tracking objects of different sizes and scales can be challenging, especially in AR applications where the XR system needs to accurately match the size and position of virtual objects with real-world objects.

- **Real-time performance:** Real-time performance is critical in XR applications, where users expect a seamless and responsive experience. 3D object tracking can be computationally intensive, which can impact the real-time performance of the XR system.

- **Environmental variability:** XR applications are designed to be used in a variety of environments, from indoor to outdoor settings. Variations in environmental conditions, such as weather conditions, can impact the performance of the XR system and make 3D object tracking more challenging.

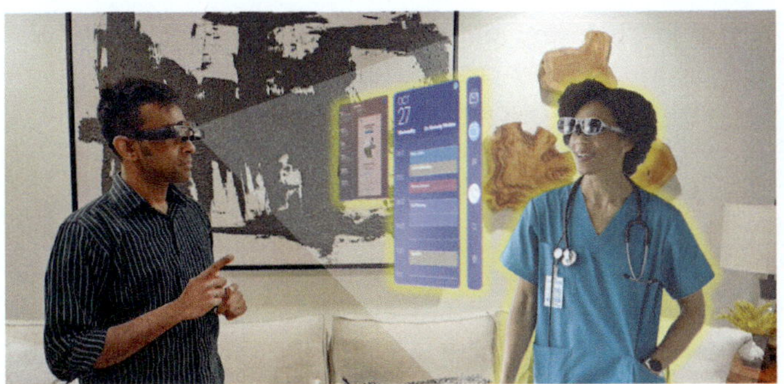

FIGURE 12.10
An illustrative telehealth immersive experience enabled by XR features.

In the above discussion, the focus was on the challenges faced in individual perception workloads. In any immersive experience, there will be a plurality of perception workloads that need to run concurrently. This is illustrated through a few examples below.

A Telehealth appointment scenario is depicted in Figure 12.10. Upal opens his telehealth app to connect with his doctor. A photorealistic avatar of his doctor is mapped in his environment. His doctor can gesture and even annotate in the air where he should apply the topical cream for his affliction. The doctor can also pull up virtual displays to show his medical history and provide treatment.

In this simple example, one can envision the following perception workloads need to run concurrently for an immersive experience: 6 DoF head tracking, eye tracking, hand tracking including gesture tracking, 3DR to create the model of the room he is in and locate the photorealistic avatar of his doctor properly positioned in this environment, and light estimation in case synthetic objects such as virtual displays are mapped into Upal's environment.

Another example one can envision is Guided shopping as illustrated in Figure 12.11. Upal goes to the grocery store to pick up ingredients for dinner. His AR glasses can scan for the items and he can see full-scale models of the items scanned and in another virtual display see possible recipes for the item he scanned. He can even contact his friend and pull up a photorealistic avatar of his friend to discuss the shopping list. This example illustrates several concurrent perception workloads that need to run in this XR environment including 6 DoF, eye tracking, hand tracking, plane detection as part of 3DR to locate the virtual displays properly pinned to the shelf, occlusion rendering to place the virtual objects properly w.r.t the real objects on the shelf.

These concurrent workloads stress the computing resources of the processor in the HMD and also will stress the power limits on these devices. There are various strategies to counter these challenges. One can distribute the workloads to multiple processors and thereby reduce power and also support the concurrent workloads. A commercial example of such a setup is in the AR2 Gen1 Snapdragon platform [46]. Another approach taken to handle the concurrencies is to put more processing on the sensors so that some primitives needed like depth or feature detector tasks, etc. can be run on the sensor itself [20]. This frees up the computing resources on the processor on HMD, so it can handle more tasks and also lowers the bandwidth for some tasks and thus lowers the processor power. Another strategy proposed in XR is to think of the HMD as part of a connected system wherein the HMD is connected tethered or wireless to a computing resource such as the smartphone and also connected to the cloud. In this way, the tasks originally to be run on the HMD can be distributed between HMD, smartphone, and even the cloud and this task distribution can involve trade-offs between power, end-to-end latency, and accuracy in the final results from a systems perspective.

12.6 Conclusion

This chapter outlined the perception workloads typically supported in an XR system to deliver immersive experiences to the user. The nature of these perception workloads and the challenges in the realization of these workloads were also outlined in this chapter. Furthermore, some insights into the processing flow associated with the processing workload were shown. As XR is an emerging area and extensive research is still ongoing for these perception workloads, there will be significant advancements in the coming years. One would also see a hybrid approach to support these perception workloads emerge in terms of using traditional computer vision methods and machine learning methods so as to offer a good trade-off in terms

FIGURE 12.11
An illustrative immersive experience related to augmented grocery shopping.

of silicon area, use case power, and still meet use case requirements on latency and accuracy.

References

[1] Harry Agius and Damon Daylamani-Zad. Guest editorial: Interaction in immersive experiences. *Multimedia Tools and Applications*, 80(20):30939–30942, 2021.

[2] Ian F. Akyildiz and Hongzhi Guo. Wireless extended reality (XR): Challenges and new research directions. volume 3, pages 1–15, 2022.

[3] Ashar Ali, Upal Mahbub, Gökçe Dane, and Gerhard Reitmayr. DIREG3D: Directly regress 3D hands from multiple cameras. In *Fifth Workshop on Computer Vision for AR/VR (CV4ARVR), ICCV*, 2021.

[4] Anastasios N. Angelopoulos, Julien N. P. Martel, Amit P. Kohli, Jörg Conradt, and Gordon Wetzstein. Event-based near-eye gaze tracking beyond 10,000 Hz. *IEEE Transactions on Visualization and Computer Graphics*, 27(5):2577–2586, 2021.

[5] Andrea Bandini and José Zariffa. Analysis of the hands in egocentric vision: A survey. *IEEE Transactions on Pattern Analysis and Machine Intelligence*, 2019.

[6] Lorenzo Baraldi, Francesco Paci, Giuseppe Serra, Luca Benini, and Rita Cucchiara. Gesture recognition in ego-centric videos using dense trajectories and hand segmentation. In *2014 IEEE Conference on Computer Vision and Pattern Recognition Workshops*, pages 702–707, 2014.

[7] Maciej M. Bartuzel, Krystian Wróbel, Szymon Tamborski, Michal Meina, Maciej Nowakowski, Krzysztof Dalasiński, Anna Szkulmowska, and Maciej Szkulmowski. High-resolution, ultrafast, wide-field retinal eye-tracking for enhanced quantification of fixational and saccadic motion. *Biomedical Optics Express*, 11 6:3164–3180, 2020.

[8] Gavin Buckingham. Hand tracking for immersive virtual reality: Opportunities and challenges. *Frontiers in Virtual Reality*, 2, 2021.

[9] Anh-Quan Cao and Raoul de Charette. SceneRF: Self-supervised monocular 3d scene reconstruction with radiance fields. arxiv, 2022.

[10] John Carmack. Latency mitigation strategies. https://danluu.com/latency-mitigation/, 2013.

[11] Tejo Chalasani and Aljosa Smolic. Simultaneous segmentation and recognition: Towards more accurate ego gesture recognition. In *2019 IEEE/CVF International Conference on Computer Vision Workshop (ICCVW)*, pages 4367–4375, 2019.

[12] Pietro Cipresso, Irene Alice Chicchi Giglioli, Mariano Alcañiz Raya, and Giuseppe Riva. The past, present, and future of virtual and augmented reality research: A network and cluster analysis of the literature. *Frontiers in Psychology*, 9, 2018.

[13] Brian Curless and Marc Levoy. A volumetric method for building complex models from range images. *Computer Graphics*, 30(Annual Conference Series):303–312, 1996.

[14] Ivan Dryanovski, Matthew Klingensmith, Siddhartha S. Srinivasa, and Jizhong Xiao. Large-scale, real-time 3D scene reconstruction on a mobile device. *Autonomous Robots*, 41(6):1423–1445, 2017.

[15] B. Farnsworth. What is eye tracking and how does it work? https://imotions.com/blog/learning/best-practice/eye-tracking-work/, 2019.

[16] Heinrich Fink. Optimizing motion-to-photon latency on DAQRI smart glasses. https://xdc2018.x.org/slides/Heinrick_Fink_daqri_optimizing_motion_to_photon_latency.pdf, 2018.

[17] Guillermo Garcia-Hernando, Shanxin Yuan, Seungryul Baek, and Tae-Kyun Kim. First-person hand action benchmark with RGB-D videos and 3D hand pose annotations. In *Proceedings of Computer Vision and Pattern Recognition (CVPR)*, 2018.

[18] Gregory Theodore Gibson and Joshua Owen Miller. Eye tracking using scanned beam and multiple detectors, US Patent 10,303,248 B2, May. 2019.

[19] Andre Golard and Sachin S. Talathi. Ultrasound for gaze estimation—A modeling and empirical study. *Sensors*, 21(13):4502, Jun 2021.

[20] Jorge Gomez, Saavan Patel, Syed Shakib Sarwar, Ziyun Li, Raffaele Capoccia, Zhao Wang, Reid Pinkham, Andrew Berkovich, Tsung-Hsun Tsai, Barbara De Salvo, and Chiao Liu. Distributed on-sensor compute system for AR/VR devices: A semi-analytical simulation framework for power estimation. *CoRR*, abs/2203.07474, 2022.

[21] Simon NB Gunkel, Emmanouil Potetsianakis, Tessa E Klunder, Alexander Toet, and Sylvie S Dijkstra-Soudarissanane. Immersive experiences and XR: A game engine or multimedia streaming problem? *arXiv preprint arXiv:2201.05552*, 2022.

[22] Shangchen Han, Beibei Liu, Randi Cabezas, Christopher D. Twigg, Peizhao Zhang, Jeff Petkau, Tsz-Ho Yu, Chun-Jung Tai, Muzaffer Akbay, Zheng Wang, Asaf Nitzan, Gang Dong, Yuting Ye, Lingling Tao, Chengde Wan, and Robert Wang. Megatrack: Monochrome egocentric articulated hand-tracking for virtual reality. *ACM Transactions on Graphics*, 39(4), aug 2020.

[23] K. He, G. Gkioxari, P. Doll'ar, and R. Girshick. Mask R-CNN. In *Proceedings of the IEEE International Conference on Computer Vision (ICCV)*, pages 2980–2988, 2017.

[24] T.E. Hutchinson, K.P. White, W.N. Martin, K.C. Reichert, and L.A. Frey. Human-computer interaction using eye-gaze input. *IEEE Transactions on Systems, Man, and Cybernetics*, 19(6):1527–1534, 1989.

[25] M. Ivanov, M. Zollmann, and G. Klinker. Real-time adaptive image warping for low-latency head pose correction in virtual reality. *IEEE Transactions on Visualization and Computer Graphics*, 24(4):1625–1634, Apr. 2018.

[26] Susmija Jabbireddy, Xuetong Sun, Xiaoxu Meng, and Amitabh Varshney. Foveated rendering: Motivation, taxonomy, and research directions. *ArXiv*, abs/2205.04529, 2022.

[27] Byeongkeun Kang, Kar-Han Tan, Nan Jiang, Hung-Shuo Tai, Daniel Tretter, and Truong Nguyen. Hand segmentation for hand-object interaction from depth map. In *2017 IEEE Global Conference on Signal and Information Processing (GlobalSIP)*, pages 259–263, 2017.

[28] Joohwan Kim, Michael Stengel, Alexander Majercik, Shalini De Mello, David Dunn, Samuli Laine, Morgan McGuire, and David Luebke. NVGaze: An anatomically-informed dataset for low-latency, near-eye gaze estimation. In *Proceedings of the 2019 CHI Conference on Human Factors in Computing Systems*, CHI '19, page 1–12, New York, NY, USA, 2019. Association for Computing Machinery.

[29] Microsoft Learn. Eye tracking overview - mixed reality. https://learn.microsoft.com/en-us/windows/mixed-reality/design/eye-tracking.

[30] Microsoft Learn. Hand tracking - MRTK 2. https://learn.microsoft.com/en-us/windows/mixed-reality/mrtk-unity/mrtk2/features/input/hand-tracking.

[31] Cheng Li and Kris M. Kitani. Pixel-level hand detection in ego-centric videos. In *2013 IEEE Conference on Computer Vision and Pattern Recognition*, pages 3570–3577, 2013.

[32] Ying li Tian, T. Kanade, and J.F. Cohn. Dual-state parametric eye tracking. In *Proceedings Fourth IEEE International Conference on Automatic Face and Gesture Recognition (Cat. No. PR00580)*, pages 110–115, 2000.

[33] Wei Liu, Dragomir Anguelov, Dumitru Erhan, Christian Szegedy, Scott Reed, Cheng-Yang Fu, and Alexander C. Berg. SSD: Single Shot MultiBox Detector. In Bastian Leibe, Jiri Matas, Nicu Sebe, and Max Welling, editors, *Computer Vision – ECCV 2016*, pages 21–37. Springer International Publishing, 2016.

[34] William E. Lorensen and Harvey E. Cline. Marching cubes: A high resolution 3D surface construction algorithm. In *Proceedings of the 14th Annual Conference on Computer Graphics and Interactive Techniques*, SIGGRAPH '87, page 163–169, New York, NY, USA, 1987. Association for Computing Machinery.

[35] Xinzhong Lu, Ju Shen, Saverio Perugini, and Jianjun Yang. An immersive telepresence system using RGB-D sensors and head-mounted display. *2015 IEEE International Symposium on Multimedia (ISM)*, pages 453–458, 2015.

[36] AdHawk Microsystems. Easy efficient eye racking. https://www.adhawkmicrosystems.com/eye-tracking, 2021.

[37] IrisVR. The importance of frame rates. https://help.irisvr.com/hc/en-us/articles/215884547-The-Importance-of-Frame-Rates.

[38] Meta. This is Meta Quest Pro. https://www.meta.com/quest/quest-pro/tech-specs/#tech-specs, 2023.

[39] Meta AI. Powered by AI: Oculus insight. https://ai.facebook.com/blog/powered-by-ai-oculus-insight/, 2019.

[40] Tobii. VR headset with native Tobii eye tracking - Pico Neo 3 pro eye. https://www.tobii.com/products/integration/xr-headsets/device-integrations/pico-neo-3-pro-eye.

[41] Tobii Connect. How do tobii eye trackers work? https://connect.tobii.com/s/article/How-do-Tobii-eye-trackers-work?, 2021.

[42] VIVE United States. VIVE Pro Eye overview: VIVE united states. https://www.vive.com/us/product/vive-pro-eye/overview/.

[43] Maram Meccawy. Creating an immersive XR learning experience: A roadmap for educators. *Electronics*, 11(21):3547, 2022.

[44] Matthias Nießner, Michael Zollhöfer, Shahram Izadi, and Marc Stamminger. Real-time 3D reconstruction at scale using voxel hashing. *ACM Transactions on Graphics*, 32(6), Nov 2013.

[45] Sang-Min Park and Young-Gab Kim. A metaverse: Taxonomy, components, applications, and open challenges. volume 10, pages 4209–4251. IEEE, 2022.

[46] Qualcomm. Snapdragon AR2 Gen 1 platform. https://www.qualcomm.com/products/mobile/snapdragon/xr-vr-ar/snapdragon-ar2-gen-1-platform.

[47] Ananth Ranganathan. The oculus insight positional tracking system. https://www.aiacceleratorinstitute.com/the-oculus-insight-positional-tracking-system-2/, 2022.

[48] Bruno Simões, Carles Creus, María del Puy Carretero, and Álvaro Guinea Ochaíta. Streamlining XR technology into industrial training and maintenance processes. In *The 25th International Conference on 3D Web Technology*, pages 1–7, 2020.

[49] Suriya Singh, Chetan Arora, and C. V. Jawahar. First person action recognition using deep learned descriptors. In *2016 IEEE Conference on Computer Vision and Pattern Recognition (CVPR)*, pages 2620–2628, 2016.

[50] Jangwoo Son, Serhan Gül, Gurdeep Singh Bhullar, Gabriel Hege, Wieland Morgenstern, Anna Hilsmann, Thomas Ebner, Sven Bliedung, Peter Eisert, Thomas Schierl, et al. Split rendering for mixed reality: Interactive volumetric video in action. In *SIGGRAPH Asia 2020 XR*, pages 1–3, 2020.

[51] Snapdragon Spaces. Introducing hand tracking - Snapdragon Spaces Blog. https://spaces.qualcomm.com/introducing-hand-tracking/, Aug 2022.

[52] Kay M. Stanney, Hannah Nye, Sam Haddad, Kelly S. Hale, Christina K. Padron, and Joseph V. Cohn. *Extended Reality (XR) Environments*, chapter 30, pages 782–815. John Wiley & Sons, Ltd, 2021.

[53] Jan-Philipp Stauffert, Florian Niebling, and Marc Erich Latoschik. Latency and cybersickness: Imapact, causes and measures. A review. *Frontiers in Virtual Reality*, 1, 2020.

[54] Ivan E Sutherland. A head-mounted three-dimensional display. In *Proceedings of the December 9-11, 1968, Fall Joint Computer Conference, Part I*, pages 757–764, 1968.

[55] Diana Werner, Ayoub Al-Hamadi, and Philipp Werner. Truncated signed distance function: Experiments on voxel size. In Aurélio Campilho and Mohamed Kamel, editors, *Image Analysis and Recognition*, pages 357–364. Springer International Publishing, 2014.

[56] Jingbo Zhao, Robert S Allison, Margarita Vinnikov, and Sion Jennings. Estimating the motion-to-photon latency in head mounted displays. In *2017 IEEE Virtual Reality (VR)*, pages 313–314. IEEE, 2017.

[57] Yang Zhou, Bingbing Ni, Richang Hong, Xiaokang Yang, and Qi Tian. Cascaded interactional targeting network for egocentric video analysis. In *2016 IEEE Conference on Computer Vision and Pattern Recognition (CVPR)*, pages 1904–1913, 2016.

13

Multi-camera Bird's Eye View Perception for Autonomous Driving

David Unger, Nikhil Gosala, Varun Ravi Kumar, Shubhankar Borse, Abhinav Valada, and Senthil Yogamani

Most automated driving systems comprise a diverse sensor set, including several cameras, Radars, and LiDARs, ensuring a complete 360° coverage in near and far regions. Unlike Radar and LiDAR, which measure directly in 3D, cameras capture a 2D perspective projection with inherent depth ambiguity. However, it is essential to produce perception outputs in 3D to enable the spatial reasoning of other agents and structures for optimal path planning. The 3D space is typically simplified to the BEV space by omitting the less relevant Z-coordinate, which corresponds to the height dimension.

The most basic approach to achieving the desired BEV representation from a camera image is IPM, assuming a flat ground surface. Surround vision systems that are pretty common in new vehicles use the IPM principle to generate a BEV image and to show it on display to the driver. However, this approach is not suited for autonomous driving since there are severe distortions introduced by this too-simplistic transformation method.

More recent approaches use deep neural networks to output directly in BEV space. These methods transform camera images into BEV space using geometric constraints implicitly or explicitly in the network. As CNN has more context information and a learnable transformation can be more flexible and adapt to image content, the deep learning-based methods set the new benchmark for BEV transformation and achieve state-of-the-art performance.

First, this chapter discusses the contemporary trends of multi camera–based DNN (deep neural network) models outputting object representations directly in the BEV space. Then, we discuss how this approach can extend to effective sensor fusion and coupling downstream tasks like situation analysis and prediction. Finally, we show challenges and open problems in BEV perception.

13.1 Introduction

Scene perception and understanding form the basis for automated driving, and the quality of scene understanding dictates the performance of downstream tasks such as path planning and control. Accurate scene understanding is often challenging in urban environments due to complex interactions with actors and challenging driving scenarios, including construction areas or intersections [4, 18, 48]. Camera-based perception has been a critical component of autonomous driving perception systems including a wide variety of perception tasks like depth estimation [19–21], motion estimation [36, 37], image restoration [49], semantic segmentation [7], and multitask learning [23, 44]. They are frequently used due to their high angular resolution, dense visual representation at high distances, and cheap manufacturing and integration costs. In a multisensor setting consisting of RADARs, LiDARs, and ultra-sonic sensors, a successful multi-sensor camera fusion becomes a key research and engineering problem to focus on [1, 32]. In this chapter, we shall study the various representations used within camera-based perception to better align with 3D-perception outputs obtained from any other sensors mentioned above. In the context of automated driving, two output domain representations are popular in 3D perception systems: 3D detection-based and 2D-occupancy grid-based. Objects are represented by oriented bounding boxes in 3D, while a Cartesian discrete grid is utilized in the latter. Based on the representation utilized, an adapted fusion with other sensors can be achieved. Camera-based systems has depended on two types of image: *Perspective View* (PV) *representation* and *Bird's eye View* (BEV) *representation*.

DOI: 10.1201/9781003328957-13

The *BEV representation* have gained much attention due to their efficacy in different parts of the automated driving pipeline. The PV representation depicts the scene from the viewpoint of a sensor mounted on an automated vehicle. It captures the height dimension of the scene, which makes it extremely useful when the vehicle drives through overhanging regions such as tree canopies, bridges, and tunnels. Further, PV allows for the inference of road rules via traffic light and traffic sign detection, as well as the intention of other agents via blinker and brake-light detection for vehicles and human pose estimation for pedestrians. It also enables the detection of appearance-based social cues, such as a person's age and the presence of a visual disability, which can be used to develop socially aware automated driving systems. The above characteristics make PV extremely suitable for scene-understanding tasks such as segmentation and object detection. The main goal is to analyze and describe the attributes of various elements in the scene. However, it is essential to have these perceived objects in 3D for downstream automated driving tasks.

On the other hand, the BEV representation depicts the scene from the viewpoint of a downward-facing virtual orthographic camera placed above the automated vehicle. Historically, surround vision systems commonly used the IPM principle to generate a BEV image and to show it on display to the driver, as depicted in Figure 13.1. It captures the scene's depth proportional to the metric scale, which allows it to be directly used for distance-sensitive tasks such as collision avoidance. It also can explicitly capture occlusions in the scene, allowing subsequent tasks, such as path planning and control, to handle the ambiguity associated with such regions gracefully. Lastly, being an orthographic projection, BEV representation does not suffer from perspective distortion inherent to

PV, simplifying the representation and processing of lane geometry and road markings. These characteristics make the BEV representation apt for decision-based tasks such as trajectory estimation and control, where the main aim is to safely interact with the static and dynamic elements of the scene. Thus, both PV and BEV representations form an integral part of automated driving and are crucial for an automated vehicle's safe and efficient operation.

PV and BEV representations are typically employed at distinct stages of the perception pipeline, with PV mainly used in the earlier stages for tasks such as segmentation and tracking. In contrast, BEV is employed in subsequent sensor fusion and path planning tasks. Thus, there is a disconnect within the pipeline since the generation of BEV representation requires a depth estimate that cannot be directly obtained from the PV representation. Multiple approaches address this disconnect by either explicitly predicting the scene's depth using depth estimation networks [30] or by extracting depth information from range-based sensors such as LiDARs [35]. These depth estimates are then combined with the intermediate outputs from the PV to generate the required BEV representations. These multistage approaches, however, generate suboptimal BEV representation due to scale inconsistencies in depth estimation networks and the sparsity of range-based sensors. Many recent works have proposed various deep learning-based approaches to generate BEV representation directly from PV images, following an end-to-end learning strategy to alleviate this limitation. These approaches learn the complex characteristics of the PV–BEV mapping using neural networks, thus generating highly accurate representations in the BEV. For instance, VPN [33] uses two MLP layers to learn the PV–BEV mapping, Cam2BEV [39] augments the output of IPM with a learnable transformer, LSS [34] predicts a depth distribution to lift the PV features into the BEV space, PanopticBEV [10] uses dual transformers to independently transform vertical and flat regions in the PV image to BEV, and BEVFormer [25] employs self attention–based transformers to generate the required PV–BEV transformation.

With the development of automotive platforms equipped with multiple cameras generating a 360° view around the vehicle, it has become imperative to address the concerns pertaining to the optimal fusion of data obtained from these surround-view cameras. In automated driving systems, the cameras are designed to have both far-field and near-field 360° view around the vehicle as shown in Figure 13.2. Accordingly two strategies, namely, *late fusion* and *early fusion*, have been proposed in the literature. In late fusion, data from each camera is first processed independently to generate the required output. The outputs are then manually fused into a coherent representation via a post-processing step to generate the

FIGURE 13.1
Vehicle manufacturers offer to surround vision systems to show a top-down representation of the environment, which is helpful for humans but inappropriate for autonomous driving since it does not provide an accurate scale. Further, it only shows color but does not provide information about the image context.

FIGURE 13.2
A typical camera sensor set of an automated vehicle comprises several pinhole cameras and typically four fisheye cameras.

surround-view output. In contrast, approaches following the early fusion paradigm fuse data from multiple cameras within the model and then coherently process the fused data to generate the required surround-view output. Early fusion approaches typically have several advantages over their late fusion counterpart. To name a few: The detection of large objects that often span multiple cameras becomes easier as all necessary information is readily available in the fused feature space. Re-identification of objects across cameras while performing object tracking becomes redundant. Predictions in regions where one camera overlaps another become trivial as no heuristic-based post-processing step is needed to handle contradictory predictions. In this direction, multiple approaches have been proposed that leverage information from surround view cameras to generate coherent predictions in the BEV. However, early fusion comes with an increased computational and memory cost in terms of the extremely large intermediate feature space, which often limits the resolution of the input and output spaces. With more cameras becoming an integral part of an automated vehicle's sensor suite, unifying all cameras into a coherent multicamera pipeline is essential.

In this chapter, we present the various deep learning–based techniques developed for the task of single- and multi-view BEV perception. We first introduce the various tasks pertaining to BEV perception in Section 13.2 and then elaborate on the architectural details of various algorithms developed to address these tasks in Section 13.3. Furthermore, we present an overview of the datasets and metrics used for BEV perception in Section 13.4. Finally, we discuss the various challenges and open problems pertaining to various stages of the BEV perception pipeline in Section 13.5.

13.2 Perception Tasks

Perception and planning methods have developed independently in different communities. They have led to an inherent disconnect between PV and BEV, hindering the quest for developing a coherent end-to-end automated driving pipeline. This has led to work producing BEV representations directly from PV images in the past few years. This development has been fueled by the success of deep learning-based approaches, which enabled more end-to-end learned complex mappings from large amounts of data. Accordingly, several authors have proposed learning-based solutions to transform PV images into BEV. In most approaches, the PV–BEV transformation occurs within the feature space, creating a comprehensive perception framework where multiple tasks, such as segmentation, tracking, and prediction, can be performed in the BEV space. In this section, we present two popular BEV perception tasks, namely, 3D Object Detection and BEV Segmentation.

13.2.1 3D Object Detection in Cameras

3D Object Detection refers to the task of identifying objects in the scene and estimating 3D bounding box parameters around them. This task encompasses two significant schools of thought: Image plane-based object detection and BEV space-based object detection, which differ in the input domain over which the detection head performs object detection. Figure 13.3 illustrates the 3D object detection output by projecting the bounding box predictions onto the corresponding PV image. Image plane-based

FIGURE 13.3
3D object detection has the target to output an object's location, size, rotation, and class.

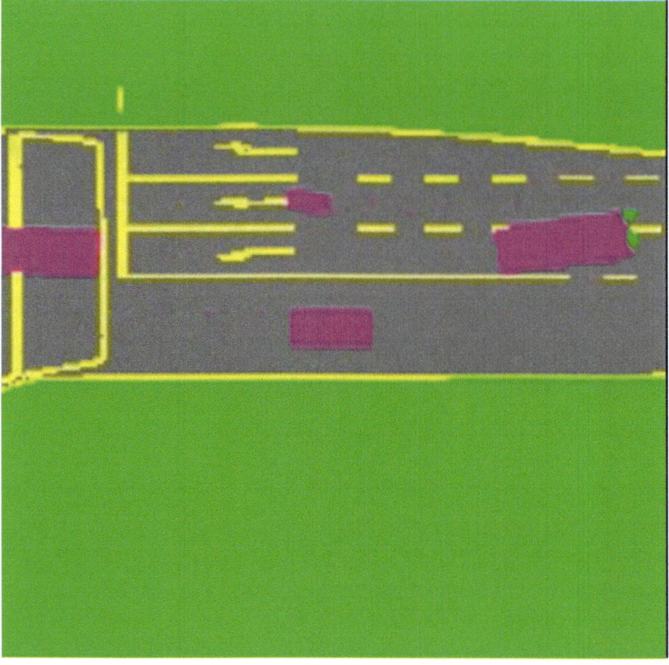

FIGURE 13.4
BEV segmentation provides the occupancy, semantic label, and precise information about which thing or stuff is located in space.

object detection pipelines comprise approaches that directly regress 3D information of the object, such as 3D size, 3D position, and orientation, onto the input PV image [51]. These approaches are typically affected by the perspective distortion of PV images that causes similar objects at different distances to appear at different scales, which prevents the network from learning meaningful object representations [41]. The lack of object consistency also results in poor 3D localization performance, which forces such approaches to rely on explicit depth supervision to improve their performance. Lastly, the detection task is required to be robust to object occlusions.

In contrast, BEV space-based camera-based object detection consists of approaches that perform object detection by estimating 3D bounding box coordinates by evaluating transformed BEV features from the camera front view. The BEV space represents the world using the orthographic projection, which addresses issues on scale inconsistency and object occlusion of perspective projection. Further, the transformation from PV to BEV enables better spatial and geometric reasoning resulting in better object localization and, accordingly, a better overall performance [15, 41].

13.2.2 BEV Segmentation with Cameras

BEV segmentation in cameras is the task of evaluating a metric semantic occupancy grid map using a single monocular image or a cocoon of images around the ego-vehicle. Semantic segmentation on point clouds or LiDARs is easier to define since point clouds provide a native BEV presentation. At the same time, an 2D/3D occupancy representation depends strongly on the projective geometry and camera parameters. The BEV-based representations are useful upstream for planned and scene interpretation models. These frameworks frequently sim-

plify the autonomous vehicle's movement into the XY plane with a single angle representing orientation within the plane, whereas 3D detection provides oriented bounding boxes with more degrees of liberty.

BEV segmentation-based perception systems generally involve (1) BEV semantic segmentation, which assigns a semantic class label to each pixel in the output space; and (2) BEV instance segmentation, which distinguishes between the different instances of an object; and (3) BEV panoptic segmentation, which combines BEV semantic and instance segmentation to generate a coherent output that semantically distinguishes regions in the BEV space while simultaneously discerning between instances of an object [10]. The BEV segmentation output is typically represented using a rasterized grid representing the vehicle's surroundings. The grid cells are often uniformly spaced, and each grid cell contains the segmentation label about the corresponding real-world location. Figure 13.4 graphically represents a typical BEV panoptic segmentation output where the road class is depicted in gray, road markings in yellow, grass and shrubs using green, and distinct instances of the vehicle class are colored differently.

BEV segmentation forms an integral part of the scene understanding pipeline and is critical for the proper functioning of various subsequent tasks. For example, BEV panoptic segmentation facilitates the segmentation of static classes, such as roads and sidewalks. BEV panoptic segmentation, which identifies object instances belonging to dynamic classes, is critical for downstream tasks such as path planning and trajectory estimation. The BEV instance labels also assist in tracking vehicles and pedestrians over time, which can help the model reason about occlusions and can positively influence the decision-making process of the autonomous vehicle. Further, BEV segmentation allows for the creation of instantaneous HD maps, which can be accumulated over time to generate temporally coherent HD maps. These temporally coherent maps can, in turn, be used in creating HD maps for new locations or can assist in the maintenance of previously annotated HD maps.

The BEV segmentation output is characterized by two important design parameters: grid size and grid resolution of the rasterized BEV grid. The grid size defines the number of grid cells and determines the spatial size of the BEV segmentation output. The larger the grid size, the higher the information that can be represented in the segmentation output. However, a larger grid size increases the computational complexity, which subsequently increases the computational requirements as well as the runtime of the model. In contrast, grid resolution determines the real-world size of a single grid cell. Given a fixed grid size, a finer resolution allows the BEV grid to capture finer details, such as elements belonging to lane markings and pedestrians, at the expense of a smaller real-world area. In comparison, a coarser grid resolution increases the real-world area covered by the BEV grid at the cost of a decreased segmentation granularity. Some works in the literature use a grid size of 200×200 cells with each having a resolution of 0.5 m, thus capturing a total real-world area of $100\,m \times 100\,m$ [13, 34, 53]. However, a human covers an area of less than one grid cell at such coarse resolutions, which makes it unsuitable for city driving. Other works use a finer grid resolution of 0.075 m with a $\sim 4\times$ larger grid size (768×704) to segment small objects while capturing a range of 50 m in each direction [10]. Due to the large segmentation output, the method proposed in [10] has a higher runtime than their coarse-resolution counterparts, making it unsuitable for time-sensitive applications. In general, a trade-off must be made between the spatial size and grid resolution of the segmentation output, as well as the runtime of the overall model. Given the large impact of each of these parameters not only on each other but also on the applicability of the overall segmentation model in the real-world, it is crucial to incorporate them into the set of fundamental design decisions for BEV perception. In the ideal case, the segmentation output would have a large grid size, a fine grid resolution, while the segmentation model would have a very low runtime. Since most autonomous driving applications are bounded by strict runtime constraints for

safety purposes, a compromise is made between the real-world area covered and the grid resolution of the segmentation output. We further discuss this trade-off in section 13.5.4.

13.3 Network Architectures for BEV Perception

Having explored the two main tasks in the BEV space, namely, 3D object detection and BEV segmentation, it is important to observe that the BEV perception enables the accumulation of transformed BEV features across multiple sensors and behaves as a unified fused representation. Various algorithms can exploit the spatial and geometric reasoning of the aligned BEV features to generate more accurate outputs than directly using PV features where object positions are not realistic due to perspective distortion introduced by the onboard cameras. Consequently, the fundamental goal of BEV perception is to map the image features in the PV to their corresponding position in the BEV space relative to their real-world 3D locations.

Most recent PV–BEV transformation approaches use an end-to-end deep neural network that takes camera images and processes them to generate the desired output in the BEV space. These PV–BEV transformation networks typically comprise three main components, namely, (1) an image encoder that processes the input RGB image and generates image features in PV, (2) a transformation module that transforms the PV features into BEV, and (3) a task-specific head that processes BEV features and generates the task-specific output in BEV. Figure 13.5 illustrates the three main components of a typical PV–BEV transformation network. In this section, we explore the aforementioned network modules in detail and provide examples of the most common network architectures found in the literature for each component.

13.3.1 Image Encoder

The image encoder, also known as a feature extractor or backbone, forms the first component of a deep neural network. Its main task is to extract features from the input image for use in the downstream modules. This module often consumes a substantial portion of computational resources compared to the whole network. The structure of the network backbone is often nontrivial. It involves engineering several hyper-parameters, such as the number of layers, the size of each layer, and connections between layers, among many others. Consequently, several works have explored various strategies to develop optimal

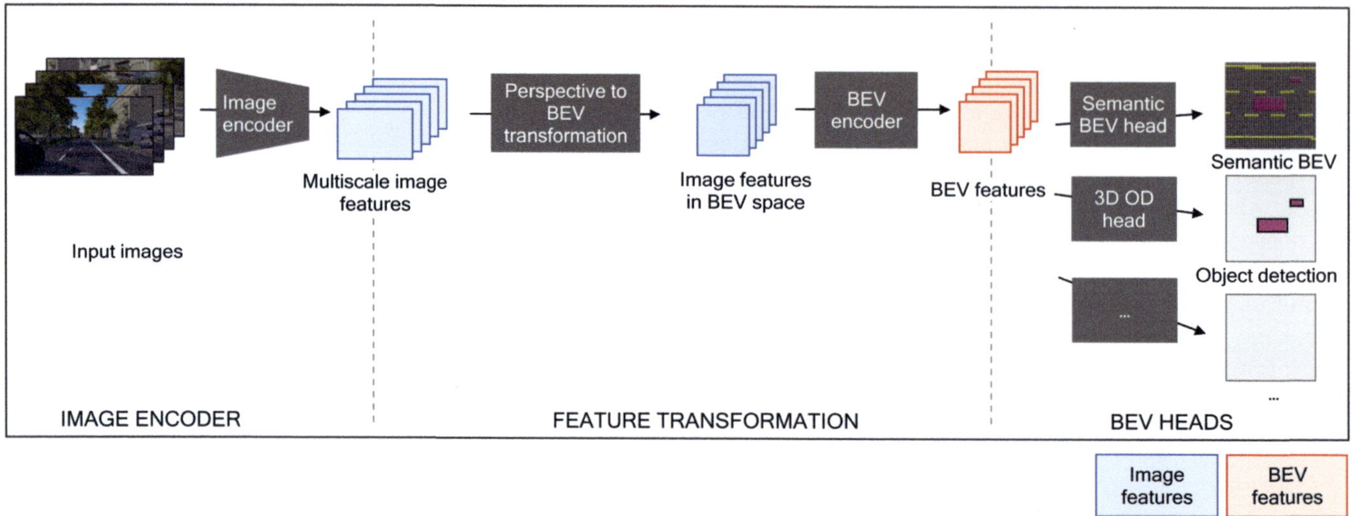

FIGURE 13.5
The architecture of BEV transformation networks consists of typically three parts: the image encoder, the feature transformation, and the task-specific head with an optional BEV encoder.

network backbones [12, 26, 46]. Several works reuse these backbone designs in BEV perception since the feature extraction step is similar in both PV and BEV perception pipelines.

A few of the many available network backbones in the literature have gained wide popularity in BEV perception tasks. The ResNet family of image encoders [12], specifically ResNet-50 and ResNet-101, has been employed in both BEV-based object detection [15] and BEV segmentation [40]. LSS [34], which propose the implicit estimation of depth in the image encoder to transform the features into 3D space for BEV segmentation, prefer the EfficientNet [46] family of backbones due to its computational efficiency and superior performance as compared to ResNet. The same family is used by CVT [57] for their BEV segmentation approach, which applies cross-attention to decide relevant PV features for each BEV grid cell. The EfficientDet [47] image encoder extends EfficientNet with a BiFPN layer for multiscale feature fusion. It is used by PanopticBEV [10] to generate the image features for the task of BEV panoptic segmentation. Another popular class of image encoders is attention-based vision transformers such as Swin [26]. Through their success in image classification [5], attention-based backbones are also employed for BEV perception [15, 27]. It is important to note that switching from one image encoder to another is relatively easy. However, the model runtime and the training convergence rate heavily depend on the choice of the image encoder [34]. This decision becomes even more relevant when multiple cameras are used since the inference time of the model can decide whether a model is deployed in the real world or not. Tables 13.1

and 13.2 present an overview of the image encoder used by multiple popular BEV perception approaches.

13.3.2 BEV Transformation

The BEV transformation module forms the second component of the PV–BEV transformation network and is responsible for transforming the image features from the PV to BEV. Existing BEV transformation modules can broadly be classified into two groups: (1) forward mapping and (2) backward mapping. The classification is based on the BEV features generated from the PV ones.

Forward mapping-based approaches generate the BEV features by lifting the features from the PV to the BEV using various techniques. For example, OFT [41] and M2BEV [53] generate the intermediate 3D voxel grid by assuming a uniform depth distribution and accumulating the image features along the corresponding camera ray. Depth distribution is a discrete probability distribution of depth for each pixel. LSS [34], BEVDet [15], BEVFusion [27], and FIERY [13] learn the PV–BEV mapping by first predicting the depth distribution in the PV and then using it to lift the features from the PV to the 3D space. PanopticBEV [10] uses a dual transformer architecture where each transformer independently attends to vertical and flat regions in the scene and transforms them from the PV to the BEV. The flat transformer employs IPM with an error correction module. In contrast, the vertical transformer implicitly learns the depth of the vertical regions to lift the corresponding features from the PV to the BEV.

TABLE 13.1
Different BEV Segmentation Approaches

Name	Datasets	Multi Camera	Input Resolution	Encoder	Grid Dimension / Cell Size
VED [28]	Cityscapes, KITTI	no	256×512	VGG-16	64×64
VPN [33]	House3D, CARLA, nuScenes	yes		ResNet-18	
PON [40]	nuScenes, Argoverse	no		ResNet-50	200×200 / 0.25m
LSS [34]	nuScenes, Lyft	yes	128×352	EfficientNet-B0	200×200 / 0.5m
TIIM [43]	nuScenes, Argoverse, Lyft	no		ResNet-50	200×200 / 0.25m
PanSeg [10]	KITTI-360, NuScences	no	768×1408, 448×768	EfficientDet	
BEVFormer [25]	nuScenes, Waymo	yes	900×1600	ResNet101-DCN	200×200 / 0.512m
GitNet [9]	nuScenes, Argoverse	no		ResNet-50	200×200 / 0.25m
M2BEV [53]	nuScenes	yes	900×1600	ResNeXt-101	200×200 / 0.5m
CVT [57]	nuScenes, Argoverse	yes		EfficientNet-B4	200×200 / 0.5m

TABLE 13.2
Different 3D Object Detection Approaches

Name	Datasets	Multi Camera	Input Resolution	Encoder	Grid Dimension / Cell Size
OFT [41]	KITTI	no	-	ResNet-18	160×160 / 0.5m
CaDDN [38]	KITTI, Waymo	no	832×1248	ResNet-101	320×336 / 0.16m
BEVFormer [25]	nuScenes, Waymo	yes	900×1600	ResNet-101	200×200 / 0.512m
BEVDet [15]	nuScenes	yes	256×704	ResNet-50, ResNet-101, Swin	128×128 / 0.8m
BEVDet4D [14]	nuScenes	yes	$256 \times 704 \times 2$ (temporal)	ResNet-50	128×128 / 0.8m
M2BEV [53]	nuScenes	yes	900×1600	ResNeXt-101	200×200 / 0.5m
BEVDepth [24]	nuScenes	yes	256×704, 512×1408	ResNet-50	

In contrast, backward mapping-based approaches learn the PV–BEV transformation by querying the value of each destination cell from the image features. These approaches are often computationally expensive since each algorithm loops through every voxel grid location to assign a value from the image features.

Attention-based methods recently became popular for natural language processing [50] and computer vision applications [5]. Attention-based BEV transformation represents the image-to-BEV transformation as a translation problem from one space to another. Since, in general, a column of an image corresponds to a projected ray in BEV space, the transformation can be interpreted as an image column to BEV ray translation [43]. A more general attention-based transformation is to imagine the whole image to BEV space transformation as a

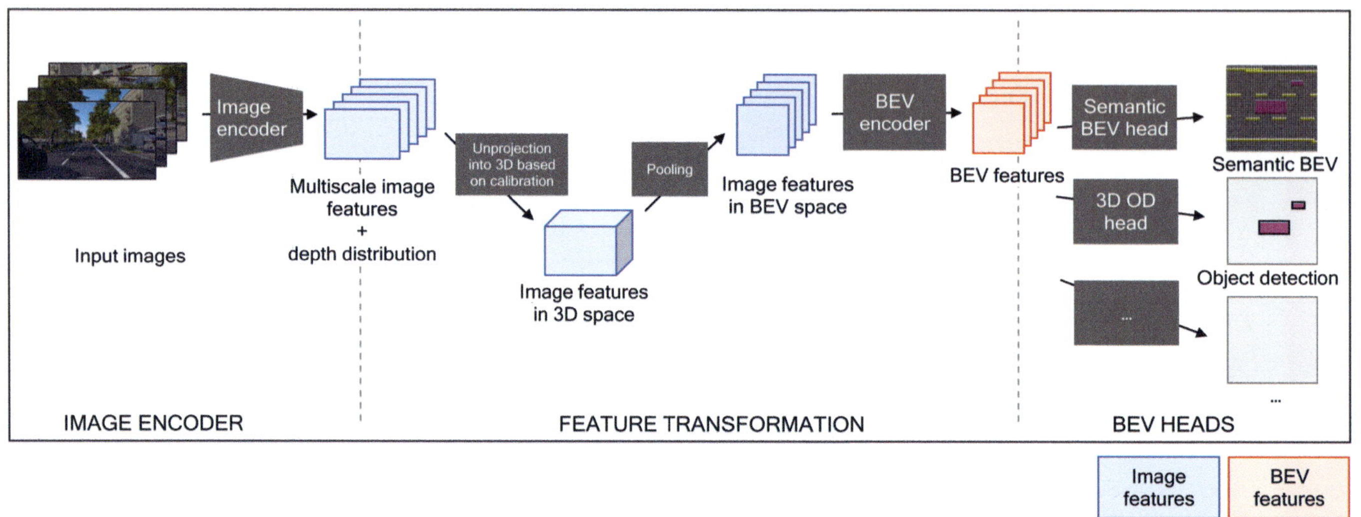

FIGURE 13.6
BEV architecture of LSS' two-step transformation approach: They take the image features and the predicted depth distribution and project them into the 3D space. Then they pool the 3D space in the height dimension to obtain the BEV representation.

translation problem [57]. They use cross-attention to determine which parts of the image space are relevant for a specific area in BEV.

Geometry-based transformation approaches use the camera parameters like position, orientation, or focal length, which are given by the extrinsic and intrinsic camera parameters and are known for cameras used in the autonomous driving context. Based on those parameters, one can calculate the 3D position of the image pixel up to the ambiguous depth of the pixel. Geometry-based transformation approaches use these geometrical constraints to project image features into BEV feature space. Early works like OFT project the image features into 3D space, assuming a uniform distribution for the depth distribution [41]. The seminal work from LSS uses the image encoder backbone to estimate the depth distribution for each image feature additionally. It uses this depth distribution to project the image feature into 3D space. As of now, geometry-based architectures show the most promising overall results, so we limit further architectural discussion to this group of works and elaborate only on those in the following more detailed subsection.

Geometry-based BEV transformation approaches show great performance since they model the real world using a 3D model that is condensed in a subsequent step. Recent architectures use a four-step transformation approach [15, 34]: First, a primarily CNN-based image encoder generates image features from the input images. Second, the view transformation module transforms the image features from the 2D perspective space into the BEV space. Then a BEV encoder processes the features using a convolutional network in the BEV space. Finally, the encoded BEV features are fed into one or several task-specific heads, for example, for semantic segmentation, 3D object detection, or both.

Figure 13.6 shows the transformation approach from LSS [34], which is widely used by subsequent works [15, 27]. The backbone image encoder is shared between all camera images and outputs the image space features and a predicted depth distribution for each image feature. This depth distribution is used to lift the image features into 3D space, as sketched in Figure 13.7. Each image feature is projected into the 3D space along a ray. The ray direction matches the direction where the light ray hits the camera and is determined by the camera intrinsics. The depth estimation part of the network backbone determines how intense the features are projected to each distance step. Each projected 3D feature is assigned to the corresponding 2D BEV grid cell in the splat step. Features that are projected outside the valid grid area are dismissed. After this lift-splat step, the features exist in BEV space and are processed by the BEV encoder. The encoded BEV features are fed into one or several task-specific heads to provide the desired output.

13.3.3 Task-specific Head

The network head forms the last step of a typical neural network pipeline and is responsible for generating the required task-specific output. The BEV head accepts the transformed BEV features from the PV–BEV transformation module as input. It generates the 3D bounding

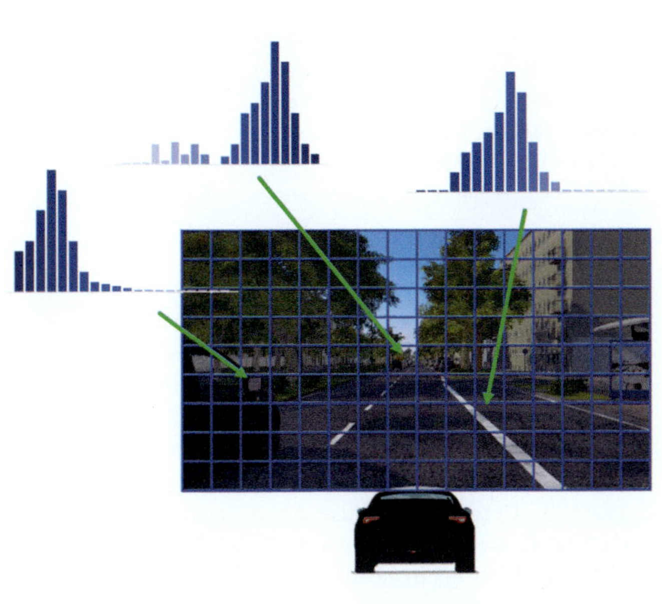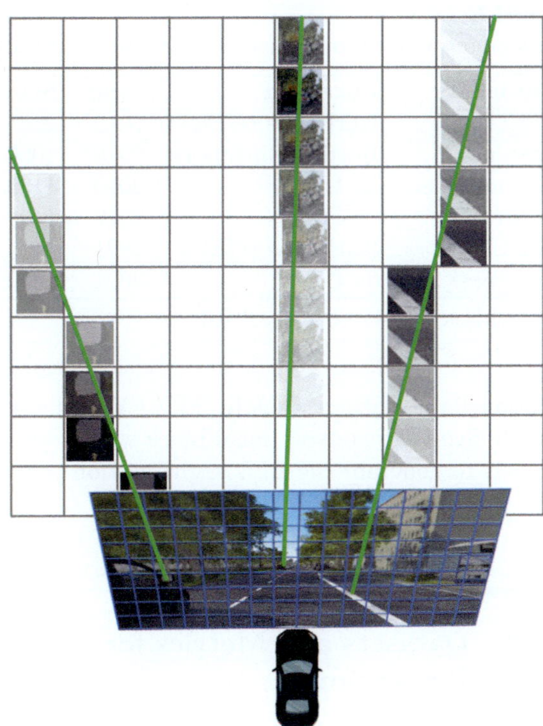

FIGURE 13.7
The network backbone estimates the features and, in addition, the depth distribution for each feature. As a second step, the image features are projected into the 3D space based on the depth distribution and the camera parameters.

box regressions for the 3D object detection task and either BEV semantic, instance or panoptic maps in case of the BEV segmentation task. The task-specific heads in the BEV representation have largely remained unchanged from the existing PV-based and 3D LiDAR-based approaches, apart from minor modifications to adapt them to the characteristics of the novel view.

Focusing on the task of 3D object detection, no consensus is found in the literature, and different authors employ different object detection heads in their proposed frameworks. OFT [41] takes inspiration from DenseBox [16] and predicts a confidence map containing the coarse locations of objects in the scene. This output is augmented with three other heads that predict the position offsets to improve object localization and the size and orientation of the 3D bounding boxes. CaDNN [38] and M2BEV [53] employ the popular 3D LiDAR detection head, Point-Pillars [22], as their object detection head. The former employs the vanilla PointPillar architecture. The latter augments it with a dynamic box assignment strategy that learns to dynamically assign 3D anchors to ground truth boxes instead of using a fixed Intersection over Union (IoU) threshold. BEVFormer [25] adapts the 2D detector deformable DETR [52] to predict 3D bounding boxes and uses it as part of its 3D object detection pipeline. Lastly,

BEVDet [15] and BEVDet4d [14] use the first stage of CenterPoint [54] augmented with a novel scaleNMS strategy as their object detection head. ScaleNMS scales the bounding box predictions of each object based on its class before applying NMS to ensure that the NMS algorithm is effective even in BEV space, where the overlap between the bounding boxes is close to zero.

Similar to 3D object detection, no agreement is found in the literature regarding the architecture of the BEV segmentation head. VED [28] uses six upsampling modules, each consisting of a transpose convolutional layer followed by two convolutional layers. In contrast, VPN [33] uses a pyramid pooling module [56] for generating the BEV semantic predictions from the BEV features. PON [40] employs a stack of eight residual blocks and a transpose convolutional layer to generate the segmentation map. At the same time, LSS [34] uses the PointPillar [22] architecture to flatten the intermediate 3D features into BEV and generate the semantic maps in the BEV space. TIIM [43] employs multiple ResNet blocks with Deep Layer Aggregation [55] to generate the BEV maps. Lastly, PanopticBEV [10] uses DPC and LSFE modules along with depthwise separable convolutions as proposed in EfficientPS [31] to generate the required BEV semantic map from the

TABLE 13.3
Autonomous Driving Datasets Can Be Used for BEV Perception.

Dataset	Location	Year	Scenes	Train	Val	Test	3D Boxes	Maps	Cameras	Lidars
KITTI	Karlsruhe (GER)	2012	22	7k	-	7k	200k	No	2 s	1
nuScenes	Boston, Singapore	2019	1000	28k	6k	6k	1400k	Yes	6 p	1
Waymo Open	3x USA	2019	1150	170k	30k	30k	12,000k	No	5 p	5
Argoverse	2x USA	2019	113	39k	15k	12k	993k	Yes	7 p + 1 s	2
KITTI-360	Karlsruhe (GER)	2021	11	78k	-	-	68k	Yes	1 s + 2 f	2
Argoverse 2	6x USA	2021	1000	-	-	-	-	Yes	7 p + 1 s	2

p: pinhole, s: stereo pair, f: fisheye

intermediate features. Lastly, M2BEV [53] uses a very simple five-layer convolutional block with 3×3 and 1×1 kernels to generate the BEV segmentation predictions.

13.4 Datasets and Metrics for BEV Perception

Publicly available datasets are significant for research because they reduce the barrier to entry and prevent the need for resource and time-intensive data collection and annotation. In addition, public datasets make different approaches comparable with each other since they can be tested using the same data constraints. Considering that BEV perception is a relatively new research area, there are, to our knowledge, no datasets designed explicitly for BEV perception. However, many datasets for autonomous driving contain 3D bounding box annotations and semantic maps that can be used for BEV perception

13.4.1 Datasets

Most works in the BEV field [29] use the nuScenes dataset for their research [2]. The main reason is that it provides full 360° camera field of view and accompanying HD map with Lidar to provide BEV ground truth. The nuScenes dataset comprises six pinhole cameras that yield a combined 360° field of view around the vehicle, in addition to one LiDAR sensor and five Radars. They annotate 3D bounding boxes for 23 classes, such as vehicles and pedestrians. Additionally, they provide a detailed HD map with 11 semantic classes that can be used for BEV segmentation. nuScenes includes 1000 scenes from Boston and Singapore with each 20 s length and provides a strong baseline for BEV perception data. Moreover, the Waymo Open Dataset is like nuScenes but includes five LiDAR sensors and, therefore, better availability of depth information [45]. It contains 1150 scenes of each 20 s, but with more annotated frames, which is why it is larger than the nuScenes. The Waymo Open Dataset cameras

all face to the front or the sides; this is why they do not cover the full 360° for BEV perception, but only around 240°. Therefore, it is unsuitable for 360° BEV perception. Furthermore, some works like TIIM [43] and GitNet [9] additionally report their performance on the Argoverse dataset [3]. In terms of labels, Argoverse provides both 3D bounding boxes and semantic maps and provides even the benefits of having, in addition to the 360° field of view, a pair of front-facing stereo cameras. Argoverse contains fewer scenes than nuScenes or the Waymo Open dataset, which is why it is less used. The KITTI-360 dataset provides 3D bounding boxes and semantic map information for a pair of front-viewing stereo cameras and two fisheye cameras. Fisheye cameras are used by many OEM for their ADAS since fisheye cameras are often utilized for parking visualization. Table 13.3 gives a more detailed overview of the various public datasets that can be used for BEV perception.

In addition to the mentioned public datasets, it can be useful to use a synthetic dataset generated by a 3D simulation tool like CARLA [6]. Using 3D simulation, accurately generating the desired information with minimal labeling errors is simple. In the simulation, a camera can be mounted above the car and follow the car like a drone to record the scene for generating the BEV perception labels. The synthetic BEV camera has the advantage that it does not introduce measurement errors and that most simulation tools can output both per-pixel depth information and per-pixel semantic information, which leads to accurate labels for BEV segmentation. In addition, simulators can output objects as instances with precise position and size, which is more accurate than the labels for real-world datasets generated by LiDAR data, which still have measurement errors. Training models with synthetic data come with a downside called the reality gap [17]. Since the simulation is only a simplified model of reality, the simulated data is generally less complex and has different statistical properties than the real-world data. This gap between simulation and reality often results in substantial degradation when testing a simulation-trained neural network for real-world problems and requires countermeasures.

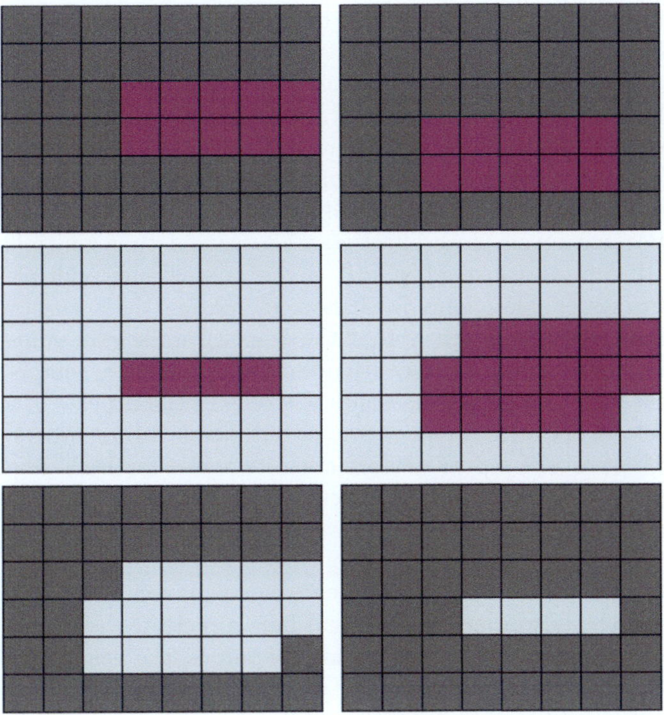

FIGURE 13.8
Top: GT and prediction of a vehicle. Middle: Intersection and union for the class vehicle. $IoU = 4/16 = 0.25$. Bottom: Intersection and union for the background class. $IoU = 32/44 = 0.73$

13.4.2 Metrics

It is crucial to have suitable metrics to evaluate different methodologies for BEV perception and to measure progress during training. The standard metric for BEV segmentation is IoU, which is also standard for image segmentation tasks. The IoU score for each class is defined in Equation (13.1) where y denotes the ground truth and $\hat{y}$ denotes the predicted output.

$$IoU_c(y, \hat{y}) = \frac{|y_c \cap \hat{y}_c|}{|y_c \cup \hat{y}_c|} \qquad (13.1)$$

Naturally, classes with a more significant representation in the real world, both in size and frequency, such as streets, vegetation, and buildings, achieve a higher IoU w.r.t instances like pedestrians. Figure 13.8 visualizes this behavior, where the vehicle class scores lower than the background class. The mean across IoUs of all classes is the mIoU metric and is often used to compare semantic segmentation approaches.

For 3D object detection, the primary metric is mAP, which is explained in the following. The mAP metric compares detected and ground truth objects, and its calculation requires several steps. The first computation step evaluates if an object is detected or not. The KITTI dataset [8] proposes to use the bounding box overlap or IoU per object to determine a correct detection. If a detected object reaches the IoU threshold of 50%, it counts as a TP, otherwise as a FN. Falsely detected objects that do not have an IoU with the ground truth, which lays above the threshold, count as FP. The nuScenes dataset [2] compute the correctness of a detection differently. They use the 2D center distance on the ground plane as a threshold, decoupling the detection rate from the object size. If detection and ground truth are within a distance threshold, the detection counts as a TP, otherwise as a FN. Based on TP, FP, and FN, the intermediate measures precision and recall can be calculated according to formulas 13.2. Precision states how accurate a detector is when detecting a class, whereas recall states the percentage of all objects of a class the detector can detect. A detector can achieve a high precision score when predicting only distinct detections and missing less clear ones. Furthermore, a detector can achieve a high recall score when predicting many objects, even unclear ones. Section 13.5 contrast shows that each metric alone is not meaningful, but combining both is an expressive metric.

$$\text{precision} = \frac{TP}{TP + FP} // \text{recall} = \frac{TP}{TP + FN} \qquad (13.2)$$

Based on precision and recall, a precision–recall curve can be constructed by computing the precision value obtained when assuming a fixed recall value. The final object detection metric AP can be calculated based on the precision-recall curve by computing the AUC by integrating the precision p over the recall r as stated in Equation (13.3).

$$AP = \int_0^1 p(r)dr \qquad (13.3)$$

Ordinarily, mAP is used to measure the final performance. mAP can be calculated from AP by averaging the dataset dependent on the different classes, subclasses, or scenario difficulty levels. mAP only focuses on the position and the overlap of a detected object with the corresponding ground truth but leaves substantial other performance indicators aside. Therefore, Caesar et al. propose the NDS in their work. NDS is a weighted score that is composed of 50% mAP metric. The other 50% of the score takes box location, size, orientation, attributes, and velocity into consideration [2].

13.5 Challenges and Open Problems in BEV Perception

Literature on BEV perception demonstrates good performance on various tasks such as BEV segmentation or 3D

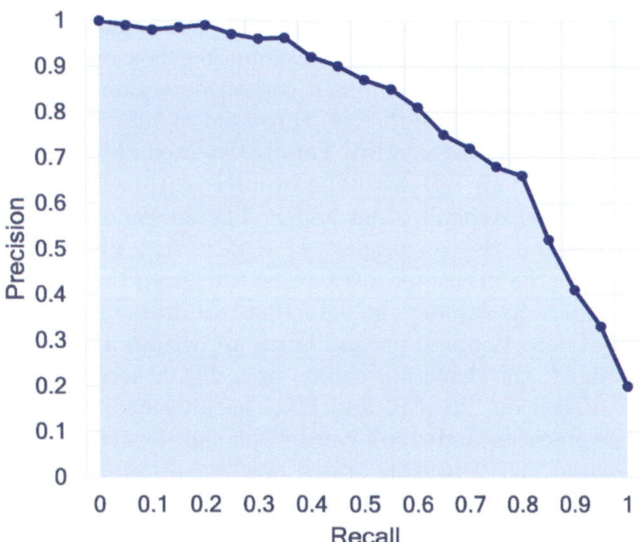

FIGURE 13.9
The PR curve (dark blue) and AUC measure (light blue) for an example detector

object detection. As the success of these algorithms is relatively new, BEV perception is currently in a transitioning process into the industry. There remain unsolved challenges that need to be resolved for BEV perception to become an overarching framework for the usage in commercial deployment projects where a whole fleet of vehicles is involved. This section discusses open topics and encourages new research in those areas.

13.5.1 Extension to Fusion and Downstream Tasks

The autonomous driving stack from sensor, perception, situation analysis, and path planning to actuation and control is quite complex. This subsection elaborates on how BEV perception can be extended and used for downstream tasks from perception.

One significant advantage of the representation of BEV appears in the multimodal sensor fusion or step across Cameras, LiDARs, and Radars. LiDAR and Radar provide 3D measurements that can easily be converted into BEV space. BEVFusion uses camera and LiDAR features and feeds both together in a shared BEV encoder [27] and achieves a state-of-the-art. Another work shows that concatenating camera features in BEV space with Radar detections boosts the performance for BEV segmentation when no LiDAR sensor is available [11]. In addition, a network can learn not only about what it sees but also about what it does not see. It can be achieved by training the network with an additional parameter for occlusion

probability [39], which can help handle situations with hidden road users.

Besides the apparent advantage of multimodal sensor fusion, BEV space representation is also well suited for several other tasks for situation analysis and planning. The situation analysis (future trajectory prediction for actors, collision detection and avoidance, and other tasks) becomes even more effective when considering multiple BEV frames instead of just single ones. Using multiple frames enables temporal fusion, increasing the accuracy because it is less likely that the detector misses the same object for several consecutive frames. In addition, temporal fusion can improve scenes where essential aspects of a set are occluded for some frames but visible during others. Many works report performance increases for BEV segmentation and 3D object detection when enabling temporal fusion [14, 42, 43]. However, not only fusion profits when evaluating the situation based on multiple frames. Optical flow in BEV space can be easily computed and has the advantage that optical flow directly corresponds to velocities in the real world. Therefore, the velocity of the ego vehicle and the other road users can be calculated accurately based on the optical flow in BEV. We want to mention the benefits of mapping and localization to complete the advantages of situation analysis. Maps can be generated from the BEV output by compensating for the ego-motion and combining all parts of the BEV that are usually fixed and do not move between time steps. Further, a vehicle can localize on HD maps based on BEV outputs. HD maps enhance the scene understanding as they provide very accurate environment information.

For trajectory planning, it is not only important to know what the current situation is, but we also need to reason about the plausible future paths of other vehicles. FIERY [13] proposes a model in BEV space, where they predict probabilistic future trajectories of other road users. This information can improve trajectory prediction. Generally, trajectory planning is done by evaluating the current situation using conventional programming methods and then proposing a path based on the result. A neural network can perform trajectory planning for the ego vehicle in BEV space by proposing a trajectory based on trajectories from a training dataset. LSS, for example, proposes a method to predict potential trajectories in BEV space [34].

13.5.2 Perception Challenges

BEV perception using monocular cameras as the only input modality has been a challenging task thus far. This is due to the inability of monocular cameras to provide depth information, which causes inaccuracies in the BEV Transformation illustrated in Section 13.3.2. While some applications, such as 3D object detection for vehicles, achieve reliable results, other tasks, such as Occupancy Flow prediction, are more ambitious. The ambiguity in depth particularly affects small objects in the BEV space. For instance, pedestrians are hard to annotate accurately by the network. This problem worsens in faraway regions, as the errors in transforming distant regions from the Perspective View into the orthographic BEV space are substantially higher.

13.5.3 Practical Challenges

Many changes are required when transforming a computer vision architecture from an image-space-based architecture toward a BEV-space-based architecture. One cannot just replace the PV encoder architectures for segmentation and detection and keep the other parts in a BEV framework unmodified. The perception model itself would require extensive modifications. Moving towards BEV perception also means moving towards a more unified perception architecture where the information from the different cameras is not combined after the different task-specific heads but instead combined at feature level in BEV space. This alignment is learned by the network end-to-end, simplifies the post-processing, and automates the effort to consolidate the information from different cameras into a unified representation. On the other hand, having a data-driven approach for the feature consolidation requires a new set of labels in BEV space to train the new end-to-end architecture. In practice, this means that a large-scale BEV dataset needs to be created, which requires labeling tools to annotate in BEV space.

13.5.4 Computational Challenges

There are computational challenges to using deep learning: during training on the server side and inference on the edge side. Since computations in 3D space, especially 3D convolutions, require many operations and increase the FLOPS for training and inference, it is beneficial to avoid computations in the 3D space as much as possible. The other two main factors determining the computational load are input and output resolution. Some works use an input resolution as low as 0.05 MP per image [34], which has the advantage that the image encoder can be small and does not require many resources. The computational effort is much larger when switching to images with resolutions around 1 MP or above. It becomes challenging to compute this high-resolution camera without adaptions to optimize performance. In industry, there is a trend towards an 8 MP front camera, which makes a compute-optimized architecture for BEV perception essential.

Another challenge is the design of a suitable output format. When predicting BEV segmentation, there is a trade-off between grid resolution and detection range. If one wants to detect objects up to 200 m around the car, which is especially required at high speed, and in addition needs accuracy of 0.1 m to sense lane markings and pedestrians, the grid output would be 200 m × 2/0.1 m = 4000 *pixel* per side of the BEV grid. This grid size would not only be too large to compute but also require too many parameters that would tend to overfit and, therefore, not feasible. On the other hand, the suggested 0.1 m accuracy is mainly required for the near range, for example, 20 m around the car, as precise localization of pedestrians and lane markings is less important further away from the ego vehicle. Therefore, the far range could be sampled with a lower resolution and still provide the relevant information. This reduction in spatial resolution with the distance from the ego vehicle is similar to how we as humans perceive the environment.

There are several potential solutions. One could think about a non-uniform grid cell size. The grid cells close to the ego-vehicle would be small and have a high resolution, while far away cells would have a lower resolution and larger grid cell size. While this is mathematically possible, a non-uniform BEV grid comes with a crucial downside as it is no longer translation equivariant, which makes it harder to create and complicates further processing. A better option is probably to create a near-field and a far-field BEV grid. A possible configuration could be a near-field grid with 0.125 m grid cell size and 25 m range in each direction, which would require a total of 400 × 400 grid cells and would be suitable for near-range reasoning, mainly during low-speed maneuvering and for parking situations. On the other hand, the far-field grid could be resolved with 0.5 m grid cell size and a range of 200 m to the front and rear and 100 m to the sides, resulting in a grid of 800 × 400 pixels. Using this dual grid setup would result in a total output resolution of 0.48 MP, which is still relatively high compared to the current state-of-the-art resolution in research, which is 200 × 200, pixels or 0.04 MP. This increase in output resolution still requires significantly more computing and a rich dataset that comprises near-range and far-range sensors to train the larger network.

13.5.5 Geometric Challenges

Since BEV perception depicts the real-world occupancy information, it needs to model the real world accurately to achieve reliable results. This modeling process is realized as a combination of the BEV feature transformation step and the learning within the model weights. A scene with some vehicles on a flat street can be modeled relatively easily when using the camera parameters to project the features. More complex scenarios encompass, such as, uneven road surfaces, roads with inclines, up and down winding roads, or changing extrinsic camera parameters through roll or pitch angles of the vehicle suspension. When not handled geometrically by the transformation approach or by adapted camera parameters, the BEV perception is susceptible to those situations since the large variety of situations cannot be all part of the training data. Therefore, a pure learning approach cannot handle those situations well. Intuitively, it makes sense to compensate for the camera rotation parameters by the roll and pitch angles of the suspension when they are precisely available. To tackle the different road geometries, an additional road surface estimation that predicts inclines might prove helpful. The transformation model would be required to input this road surface estimation and use the information efficiently to ease the modeling part for the network weights and to reduce the learning effort.

13.5.6 Limitations

The BEV representation is well suited for many tasks around autonomous driving, but the simplification from 3D to BEV comes at the cost of missing height information, which is still required for some tasks. Information like the over-drivable and under-drivable properties of objects can be provided as additional object parameters and are not a direct problem for the BEV representation. However, for information like traffic signs, traffic lights, and pedestrian poses, additional object parameters are very inefficient in BEV space, which is why this information should be handled differently.

Existing methods for the mentioned tasks work well in image space, which is why the tasks should ideally still be handled in the image space. It can be part of a coherent architecture, with heads for traffic signs, traffic lights, and poses after the image encoders to perform the image space tasks. In addition, the encoded image features are fed into the BEV transformation module to enable BEV perception.

13.6 Conclusion

BEV-based representation shows superior performance for 3D semantic segmentation and 3D object detection, especially in accurate 3D localization, based on the reasoning in 3D space. As BEV perception directly learns to combine features of multiple cameras into a coherent representation, it avoids the manual effort of combining outputs from different cameras. In addition, BEV perception shows the potential to perform multisensor fusion between camera, LiDAR, and Radar. This shows why BEV perception can be a foundation for the overall 3D perception of autonomous vehicles. In this chapter, we highlighted the different architectures to implement BEV perception and further elaborated on potential downstream extensions, challenges, and corresponding future research areas. The main motivation of BEV perception is map the pixel-space 2D input to 3D output, which is necessary for downstream tasks in automated driving which operate in 3D. This type of cross-view perception where the input and output domains are different is not a typical problem in computer vision. The tools used in standard dense prediction tasks such as data augmentation and hard negative mining need to be adapted. This is a critical step to improve the maturity of this rapidly progressing research area. However, the fundamental challenge is to learn an optimal feature embedding in 3D using 2D pixels which have depth ambiguity.

Acknowledgments

The synthetic data used in this chapter was generated using a commercial synthetic engine www.cognata.com. The authors thank Ahmed Sadek and Ravi Kiran for the detailed review and feedback.

References

[1] Shubhankar Borse, Marvin Klingner, Varun Ravi Kumar, Hong Cai, et al. X-Align: Cross-modal cross-view alignment for Bird's-eye-view segmentation. In *Proceedings of the IEEE/CVF Winter Conference on Applications of Computer Vision*, pages 3287–3297, 2023.

[2] Holger Caesar, Varun Bankiti, Alex H Lang, Sourabh Vora, Venice Erin Liong, Qiang Xu, Anush Krishnan, Yu Pan, Giancarlo Baldan, and Oscar Beijbom. Nuscenes: A multimodal dataset for autonomous driving. In *Proceedings of the IEEE/CVF Conference on Computer Vision and Pattern Recognition*, pages 11621–11631, 2020.

[3] Ming-Fang Chang, John Lambert, Patsorn Sangkloy, Jagjeet Singh, Slawomir Bak, Andrew Hartnett, De Wang, Peter Carr, Simon Lucey, Deva Ramanan, et al. Argoverse: 3D tracking and forecasting with rich maps. In *Proceedings of the IEEE/CVF Conference on Computer Vision and Pattern Recognition*, pages 8748–8757, 2019.

[4] Mahesh M Dhananjaya, Varun Ravi Kumar, and Senthil Yogamani. Weather and light level classification for autonomous driving: Dataset, baseline and active learning. In *2021 IEEE International Intelligent Transportation Systems Conference (ITSC)*, pages 2816–2821. IEEE, 2021.

[5] Alexey Dosovitskiy, Lucas Beyer, Alexander Kolesnikov, Dirk Weissenborn, Xiaohua Zhai, Thomas Unterthiner, Mostafa Dehghani, Matthias Minderer, Georg Heigold, Sylvain Gelly, et al. An image is worth 16×16 words: Transformers for image recognition at scale. *arXiv preprint arXiv:2010.11929*, 2020.

[6] Alexey Dosovitskiy, German Ros, Felipe Codevilla, Antonio Lopez, and Vladlen Koltun. CARLA: An open urban driving simulator. In *Proceedings of the 1st Annual Conference on Robot Learning*. PMLR, 2017.

[7] Pramit Dutta, Ganesh Sistu, Senthil Yogamani, Edgar Galván, et al. ViT-BEVSeg: A hierarchical transformer network for monocular birds-eye-view segmentation. In *2022 International Joint Conference on Neural Networks (IJCNN)*, pages 1–7. IEEE, 2022.

[8] A. Geiger, P. Lenz, and R. Urtasun. Are we ready for autonomous driving? The KITTI vision benchmark suite. In *2012 IEEE Conference on Computer Vision and Pattern Recognition*, page 3354–3361, Providence, RI, Jun 2012. IEEE.

[9] Shi Gong, Xiaoqing Ye, Xiao Tan, Jingdong Wang, Errui Ding, Yu Zhou, and Xiang Bai. GitNet: Geometric prior-based transformation for birds-eye-view segmentation. *arXiv preprint arXiv:2204.07733*, 2022.

[10] Nikhil Gosala and Abhinav Valada. Bird's-eye-view panoptic segmentation using monocular frontal view images. *IEEE Robotics and Automation Letters*, 7(2), 2022.

[11] Adam W Harley, Zhaoyuan Fang, Jie Li, Rares Ambrus, and Katerina Fragkiadaki. A simple baseline for BEV perception without Lidar. *arXiv e-prints*, pages arXiv–2206, 2022.

[12] Kaiming He, Xiangyu Zhang, Shaoqing Ren, and Jian Sun. Deep residual learning for image recognition. In *Proceedings of the IEEE Conference on Computer Vision and Pattern Recognition*, pages 770–778, 2016.

[13] Anthony Hu, Zak Murez, Nikhil Mohan, Sofía Dudas, Jeffrey Hawke, Vijay Badrinarayanan, Roberto Cipolla, and Alex Kendall. FIERY: Future instance prediction in bird's-eye view from surround monocular cameras. In *Proceedings of the IEEE/CVF International Conference on Computer Vision*, pages 15273–15282, 2021.

[14] Junjie Huang and Guan Huang. BEVDet4D: Exploit temporal cues in multi-camera 3D object detection. *arXiv preprint arXiv:2203.17054*, 2022.

[15] Junjie Huang, Guan Huang, Zheng Zhu, and Dalong Du. BEVDet: High performance multi-camera 3D object detection in bird-eye-view. *arXiv preprint arXiv:2112.11790*, 2021.

[16] Lichao Huang, Yi Yang, Yafeng Deng, and Yinan Yu. Densebox: Unifying landmark localization with end to end object detection. *arXiv preprint arXiv:1509.04874*, 2015.

[17] Nick Jakobi, Phil Husbands, and Inman Harvey. Noise and the reality gap: The use of simulation in evolutionary robotics. In *European Conference on Artificial Life*, pages 704–720. Springer, 1995.

[18] Marvin Klingner, Varun Ravi Kumar, Senthil Yogamani, Andreas Bär, et al. Detecting adversarial perturbations in multi-task perception. In *2022 IEEE/RSJ International Conference on Intelligent Robots and Systems (IROS)*, pages 13050–13057. IEEE, 2022.

[19] Varun Ravi Kumar, Marvin Klingner, Senthil Yogamani, Markus Bach, et al. SVDistNet: Self-supervised near-field distance estimation on surround view fisheye cameras. *IEEE Transactions on Intelligent Transportation Systems*, 23(8):10252–10261, 2021.

[20] Varun Ravi Kumar, Stefan Milz, Christian Witt, Martin Simon, et al. Near-field depth estimation using monocular fisheye camera: A semi-supervised learning approach using sparse LiDAR data. In *CVPR Workshop*, volume 7, page 2, 2018.

[21] Varun Ravi Kumar, Senthil Yogamani, Markus Bach, Christian Witt, Stefan Milz, and Patrick Mäder. UnRectDepthNet: Self-supervised monocular depth estimation using a generic framework for handling common camera distortion models. In *2020 IEEE/RSJ International Conference on Intelligent Robots and Systems (IROS)*, pages 8177–8183. IEEE, 2020.

[22] Alex H Lang, Sourabh Vora, Holger Caesar, Lubing Zhou, Jiong Yang, and Oscar Beijbom. Pointpillars: Fast encoders for object detection from point clouds. In *Proceedings of the IEEE/CVF Conference on Computer Vision and Pattern Recognition*, pages 12697–12705, 2019.

[23] Isabelle Leang, Ganesh Sistu, Fabian Bürger, Andrei Bursuc, et al. Dynamic task weighting methods for multi-task networks in autonomous driving systems. In *2020 IEEE 23rd International Conference on Intelligent Transportation Systems (ITSC)*, pages 1–8. IEEE, 2020.

[24] Yinhao Li, Zheng Ge, Guanyi Yu, Jinrong Yang, Zengran Wang, Yukang Shi, Jianjian Sun, and Zeming Li. BEVDepth: Acquisition of reliable depth for multi-view 3D object detection. *arXiv preprint arXiv:2206.10092*, 2022.

[25] Zhiqi Li, Wenhai Wang, Hongyang Li, Enze Xie, Chonghao Sima, Tong Lu, Qiao Yu, and Jifeng Dai. BEVFormer: Learning bird's-eye-view representation from multi-camera images via spatiotemporal transformers. *arXiv preprint arXiv:2203.17270*, 2022.

[26] Ze Liu, Yutong Lin, Yue Cao, Han Hu, Yixuan Wei, Zheng Zhang, Stephen Lin, and Baining Guo. Swin transformer: Hierarchical vision transformer using shifted windows. In *Proceedings of the IEEE/CVF International Conference on Computer Vision*, pages 10012–10022, 2021.

[27] Zhijian Liu, Haotian Tang, Alexander Amini, Xinyu Yang, Huizi Mao, Daniela Rus, and Song Han. BEVFusion: Multi-task multi-sensor fusion with unified bird's-eye view representation. *arXiv preprint arXiv:2205.13542*, 2022.

[28] Chenyang Lu, Marinus Jacobus Gerardus van de Molengraft, and Gijs Dubbelman. Monocular semantic occupancy grid mapping with convolutional variational encoder-decoder networks. *IEEE Robotics and Automation Letters*, 4(2), 2019.

[29] Yuexin Ma, Tai Wang, Xuyang Bai, Huitong Yang, Yuenan Hou, Yaming Wang, Yu Qiao, Ruigang Yang, Dinesh Manocha, and Xinge Zhu. Vision-centric BEV perception: A survey. *arXiv preprint arXiv:2208.02797*, 2022.

[30] Yue Ming, Xuyang Meng, Chunxiao Fan, and Hui Yu. Deep learning for monocular depth estimation: A review. *Neurocomputing*, 438:14–33, 2021.

[31] Rohit Mohan and Abhinav Valada. EfficientPS: Efficient panoptic segmentation. *International Journal of Computer Vision*, 129(5):1551–1579, 2021.

[32] Sambit Mohapatra, Mona Hodaei, Senthil Yogamani, Stefan Milz, et al. LiMoSeg: Real-time bird's eye view based LiDAR motion segmentation. In *Proceedings of the 17th International Joint Conference on Computer Vision, Imaging and Computer Graphics Theory and Applications (VISIGRAPP 2022)*, 2022.

[33] Bowen Pan, Jiankai Sun, Ho Yin Tiga Leung, Alex Andonian, and Bolei Zhou. Cross-view semantic segmentation for sensing surroundings. *IEEE Robotics and Automation Letters*, 5(3), 2019.

[34] Jonah Philion and Sanja Fidler. Lift, splat, shoot: Encoding images from arbitrary camera rigs by implicitly unprojecting to 3D. In *European Conference on Computer Vision*, pages 194–210. Springer, 2020.

[35] Cristiano Premebida, Luis Garrote, Alireza Asvadi, A Pedro Ribeiro, and Urbano Nunes. High-resolution Lidar-based depth mapping using bilateral filter. In *2016 IEEE 19th International Conference on Intelligent Transportation Systems (ITSC)*, pages 2469–2474. IEEE, 2016.

[36] Hazem Rashed, Ahmad El Sallab, Senthil Yogamani, and Mohamed El Helw. Motion and depth augmented semantic segmentation for autonomous navigation. In *Proceedings of the IEEE/CVF Conference on Computer Vision and Pattern Recognition Workshops*, pages 0–0, 2019.

[37] Hazem Rashed, Mariam Essam, Maha Mohamed, Ahmad Ei Sallab, and Senthil Yogamani. BEV-MODNet: Monocular camera based bird's eye view moving object detection for autonomous driving. In *2021 IEEE International Intelligent Transportation Systems Conference (ITSC)*, pages 1503–1508. IEEE, 2021.

[38] Cody Reading, Ali Harakeh, Julia Chae, and Steven L Waslander. Categorical depth distribution network for monocular 3D object detection. In *Proceedings of the IEEE/CVF Conference on Computer Vision and Pattern Recognition*, pages 8555–8564, 2021.

[39] Lennart Reiher, Bastian Lampe, and Lutz Eckstein. A Sim2Real deep learning approach for the transformation of images from multiple vehicle-mounted cameras to a semantically segmented image in bird's eye view. In *2020 IEEE 23rd International Conference on Intelligent Transportation Systems (ITSC)*, pages 1–7. IEEE, 2020.

[40] Thomas Roddick and Roberto Cipolla. Predicting semantic map representations from images using pyramid occupancy networks. In *2020 IEEE/CVF Conference on Computer Vision and Pattern Recognition (CVPR)*, Seattle, WA, USA, 2020. IEEE.

[41] Thomas Roddick, Alex Kendall, and Roberto Cipolla. Orthographic feature transform for monocular 3D object detection. *arXiv preprint arXiv:1811.08188*, 2018.

[42] Avishkar Saha, Oscar Mendez, Chris Russell, and Richard Bowden. Enabling spatio-temporal aggregation in birds-eye-view vehicle estimation. In *2021 IEEE International Conference on Robotics and Automation (ICRA)*, pages 5133–5139. IEEE, 2021.

[43] Avishkar Saha, Oscar Mendez, Chris Russell, and Richard Bowden. Translating images into maps. In *2022 International Conference on Robotics and Automation (ICRA)*, pages 9200–9206. IEEE, 2022.

[44] Ganesh Sistu, Isabelle Leang, and Senthil Yogamani. Real-time joint object detection and semantic segmentation network for automated driving. *NeurIPSW on ML on the Phone and other Consumer Devices*, 2018.

[45] Pei Sun, Henrik Kretzschmar, Xerxes Dotiwalla, Aurelien Chouard, Vijaysai Patnaik, Paul Tsui, James Guo, Yin Zhou, Yuning Chai, Benjamin Caine, et al. Scalability in perception for autonomous driving: Waymo open dataset. In *Proceedings of the IEEE/CVF Conference on Computer Vision and Pattern Recognition*, pages 2446–2454, 2020.

[46] Mingxing Tan and Quoc Le. EfficientNet: Rethinking model scaling for convolutional neural networks. In *International Conference on Machine Learning*, pages 6105–6114. PMLR, 2019.

[47] Mingxing Tan, Ruoming Pang, and Quoc V Le. EfficientDet: Scalable and efficient object detection. In *Proceedings of the IEEE/CVF Conference on Computer Vision and Pattern Recognition*, pages 10781–10790, 2020.

[48] Michal Uricár, David Hurych, Pavel Krizek, and Senthil Yogamani. Challenges in designing datasets and validation for autonomous driving. In *Proceedings of the 14th International Joint Conference on Computer Vision, Imaging and Computer Graphics Theory and Applications (VISIGRAPP 2019)*, 2019.

[49] Michal Uricár, Jan Ulicny, Ganesh Sistu, Hazem Rashed, et al. Desoiling dataset: Restoring soiled areas on automotive fisheye cameras. In *Proceedings of the IEEE/CVF International Conference on Computer Vision Workshops*, pages 0–0, 2019.

[50] Ashish Vaswani, Noam Shazeer, Niki Parmar, Jakob Uszkoreit, Llion Jones, Aidan N Gomez, Łukasz Kaiser, and Illia Polosukhin. Attention is all you need. *Advances in Neural Information Processing Systems*, 30, 2017.

[51] Tai Wang, Xinge Zhu, Jiangmiao Pang, and Dahua Lin. FCOS3D: Fully convolutional one-stage monocular 3D object detection. In *Proceedings of the IEEE/CVF International Conference on Computer Vision*, pages 913–922, 2021.

[52] Yue Wang, Vitor Campagnolo Guizilini, Tianyuan Zhang, Yilun Wang, Hang Zhao, and Justin Solomon. DETR3D: 3D object detection from multi-view images via 3D-to-2D queries. In *Conference on Robot Learning*, pages 180–191. PMLR, 2022.

[53] Enze Xie, Zhiding Yu, Daquan Zhou, Jonah Philion, Anima Anandkumar, Sanja Fidler, Ping Luo, and Jose M Alvarez. M$\hat{2}$BEV: Multi-camera joint 3D detection and segmentation with unified birds-eye view representation. *arXiv preprint arXiv:2204.05088*, 2022.

[54] Tianwei Yin, Xingyi Zhou, and Philipp Krahenbuhl. Center-based 3D object detection and tracking. In *Proceedings of the IEEE/CVF Conference on Computer Vision and Pattern Recognition*, pages 11784–11793, 2021.

[55] Fisher Yu, Dequan Wang, Evan Shelhamer, and Trevor Darrell. Deep layer aggregation. In *Proceedings of the IEEE Conference on Computer Vision and Pattern Recognition*, pages 2403–2412, 2018.

[56] Hengshuang Zhao, Jianping Shi, Xiaojuan Qi, Xiaogang Wang, and Jiaya Jia. Pyramid scene parsing network. In *Proceedings of the IEEE Conference on Computer Vision and Pattern Recognition*, pages 2881–2890, 2017.

[57] Brady Zhou and Philipp Krähenbühl. Cross-view transformers for real-time map-view semantic segmentation. In *Proceedings of the IEEE/CVF Conference on Computer Vision and Pattern Recognition*, pages 13760–13769, 2022.

The (Computer) Vision of Sports: Recent Trends in Research and Commercial Systems for Sport Analytics

Rikke Gade, Michele Merler, Graham Thomas, and Thomas B. Moeslund

14.1 Introduction

Sports have a very wide impact on society, from private and local activities with focus on personal exercise to billion dollar industries with the largest sports in the world. Common to all is that there is significant interest in measurements and comparisons – How far did I run? Which NHL player had the most successful passes during this match? Which team had the highest ball possession? What is the most commonly used shot from my upcoming badminton opponent?

Some of these things have traditionally been analysed manually, which was only affordable by a few clubs and broadcast companies. Advances in computer vision have allowed automation of many such video analysis tasks today, which opens up a much more widespread use of video data.

This chapter aims to provide an overview of how computer vision is currently used in sports applications and how we envision the near future of this field. In Section 14.2 we outline the historical and current commercial applications of computer vision in sports, including broadcasting, officiating, game analytics, and injury prevention. Section 14.3 discusses what the research community is currently focusing on for team sports and individual athletes. Finally, in Section 14.4 we give our point of view on the future prospects of using computer vision for sports applications.

14.2 Commercial Applications

14.2.1 Visuals for Broadcasting

The job of a sports presenter is to tell the story of the event to the viewer, and graphics have played a key role in this since the mid-1970s, when 'telestrators' [1] started being used to allow a presenter to draw on a still frame of video. These early systems required no computer vision or calibration.

From the mid-1980s, more advanced systems came into use, that relied on careful calibration of the position, orientation and field-of-view of the camera with respect to the area of play, with mechanical sensors on the camera and lens to track camera movement. One of the earliest systems was the 'First and 10' system, used in 1998 to show the first-down line in American Football [2]. This system rendered a virtual line on the pitch, taking account of the pitch curvature. It also used chroma-key to ensure that the rendered line appeared on the green pitch and not on the players.

There are now many companies offering graphics systems for enhancing broadcast sports coverage, with examples including Avid [3], Chyron [4], Dartfish [5], Deltacast [6], Hawk-Eye Innovations [7], Ross [8], RT Software, [9], SMT, [10], and Vizrt [11]. The capabilities of these systems have advanced significantly since the first 'tied-to-pitch' systems of the 1980s, drawing on advances in computer vision and rendering technology:

- The use of additional sensors for camera motion has been augmented and in some cases replaced with vision-based camera tracking. In sports such as soccer, where there are prominent line markings on the pitch at well-defined positions, a line-based tracker is often used [12]. In other sports such as ice hockey or athletics, where lines are less useful, a more general feature point tracker can be used [13]. Eliminating the need for mechanical sensors makes deployment easier, and even where mechanical sensors are still used for high reliability, vision-based processing can be used to refine calibration errors due to problems such as non-rigidity of the camera mount.

- Keying processes have become more flexible to tackle problems such as changing daylight conditions and strong sun/shadow contrast in stadia, and problems in basketball with reflections from the court surface and similarities between the colour of the

DOI: 10.1201/9781003328957-14

court surface and player skin colour [11]. Related problems have been tackled for long jump and beach volleyball where graphics need to be overlaid on sand but not on the athletes, and where the appearance of the sand can change significantly as the surface is raked or jumped on [14].

- Tracking players and objects such as the ball allows more in-depth analysis of games and practices to be performed, for example displaying or reporting precise information about distance and speed of individual tracked entities. Tracking also enables higher-level analysis such as producing heat maps of areas of the field occupied over time, as well as reporting team formation changes over time. Reliable tracking using images from broadcast cameras can be very difficult due to problems including occlusion, limited field-of-view and resolution, so additional sensors are sometimes employed. For example, for ice hockey, a system using infra-red emitters in the puck, viewed by cameras around the rink was introduced in 1996 to allow the puck position to be highlighted in the broadcast images [15]; it met with a mixed response and was only used for a few years. More recently, the use of IR emitters has been explored again for both puck and player tracking [16]. Multiple cameras have been used for tracking cricket balls since 2001 [7], with the technology being further developed for many other sports including tennis (2007) and football (2016) [17]. Technologies such as GPS are used for tracking players in outdoor sports [18] as well as tracking sailing boats [19]. Radar technology has also been recently used in combination with dedicated cameras to track ball and club by the PGA tour golf tournament [20, 21].

- Creating images from novel viewpoints can help explain a key moment in a game. This can be achieved in situations with minimal player occlusion and for a limited range of viewpoints with a single-camera system, by modelling players as flat planes whose 3D position is determined by assuming their feet are in contact with the ground plane [14]. This approach can be extended by switching between two cameras to give a wider range of viewpoints [22]. Extension to multiple cameras and the creation of a full frame-by-frame volumetric model is possible but requires a large number of cameras and lots of computing power, particularly when covering an area as large as a football pitch. For example, Intel's TrueView system was installed in 19 NFL stadiums and several soccer stadiums in Europe as of September 2021 when Intel shut down the division that was supplying it [23]. Other companies including Canon, 8i, Microsoft, Metastage, Dimension, and TetaVi offer polygonal and point cloud capture systems, which could be configured in a way that allows a viewer to interactively change their viewpoint, for example via Unity's Metacast platform [24]. Such approaches may best target high-value sports that do not require such a large capture area, such as boxing. As an alternative to video-based view synthesis, 3D player models can be placed in a virtual stadium to match the position of players at a key moment [8, 25], giving higher-quality images but with less realism.

In addition to graphical enhancements to help explain or visualise aspects of the sports event, there is a significant market for the use of graphical enhancements for advertising. All the systems referred to above that include functionality for rendering augmented reality graphics can be used for rendering advertising, either on the pitch (using chromakey to make adverts appear to be painted on the grass, behind players) or suspended above the pitch (where chromakey is not needed as the adverts are never obscured). Some systems specifically mention an ability to provide several different sets of advert overlays, for video feeds to different territories [4, 9].

Some specialist systems allow the adverts on boards around the pitch to be replaced, with the aim of presenting different adverts to TV viewers in other countries than are shown to people in the stadium. This presents a challenge with keying, as players and other objects will appear in front of the boards. Whilst earlier systems were limited to adverts above the field of play or placed on green advertising boards [26], some specialist systems now use an additional camera to generate the key signal. One approach is to use infra-red illumination on the billboard and an IR camera to detect the visible parts of the billboard [27]. Another approach, aimed specifically at LED billboards, involves a second camera with an optical filter to remove the relatively narrow parts of the optical spectrum that the LEDs emit, so that the billboard appears dark and any brighter objects in front can be distinguished [28].

14.2.2 Officiating

One of the first uses for computer vision technology in sports officiating was for tennis, with Hawk-Eye system passing the International Tennis Federation Electronic Line Calling tests in October 2005 and being used for the US Open Tennis Championships from 2006, and the Wimbledon Championships from 2007 [7]. The introduction of the system provided scope for considerable debate about its accuracy, and indeed about how to represent uncertainty in measurements to the general public [29]. The International Tennis Federation went on to approve systems made by FOXTENN in 2016 and Flightscope in 2020 [30].

FIGURE 14.1
Goal technology system. Several cameras per goal are used to estimate the position of the ball in 3D (left), which then gets rendered and compared to the goal line (center). If the ball is detected as having fully crossed the goal line, a message is sent via wireless to the referee's watch (right).

The use of similar systems for officiating in cricket started with a trial of a Hawk-Eye system in the winter season of 2008/9, building on the system first used for TV coverage in 2001. As of 2017, ball-tracking systems from Hawk-Eye [7] and Virtual Eye [31] were approved by the International Cricket Council [32] as a part of the Decision Review System. Two sound-based technologies were also approved (Real-Time Snickometer from BBG Sports and UltraEdge from HawkEye), which use microphones to detect when the ball contacts the bat. A system from BBG Sports that uses infra-red cameras to detect a hotspot on the bat due to ball contact was also approved.

A version of the tennis system was adapted to solve the related problem in football of whether the ball crossed the goal line. Goal-line technology (GLT) was introduced by FIFA in 2012 after a two-year trial conducted by The Royal Netherlands Football Association (KNVB) [33]. From May 2013, Hawk-Eye's goal-line technology was used for all Premier League football matches [7]. At the time of writing (July 2023), FIFA has certified goal-line technology installations at 107 stadiums, of which 88 use Hawk-Eye technology and the remaining 19 use technology from Vieww GmbH. Thirty-six installations were in the UK, 23 in Germany, 16 in Italy, with 16 other countries having between 1 and 6 installations [34]. Figure 14.1 illustrates the goal-line technology system and process.

Similar technology was chosen by the Badminton World Federation in 2014 for shuttlecock tracking to help officiate across all major BWF events [7].

A more challenging problem in football officiating is determining whether a player is off-side. In December 2020, Semi-Automatic Offside technology was used for the first time at the FIFA Club World Cup Qatar 2020 [7]. As of July 2023, FIFA had certified five virtual offside line systems, from DELTACAST, EVS, Hawk-Eye, SportsMEDIA Technology, and Vieww. These use multi-

FIGURE 14.2
FIFA semi-automated offside system rendering. Ball and players are tracked, and in certain systems their 3D pose is also automatically estimated using a combination of several cameras.

ple calibrated camera feeds to determine the location of the leading goal-scoring attacker and the second-last defender closest to their own goal at the instant when the ball is played. Two virtual offside lines are generated and the offside decision can be made based on the proximity of each line to the goal line, as shown in Figure 14.2. One type of system requires an operator to manually select the players at the key moment; another semi-automated type is capable of tracking multiple skeletal points on each player as well as the ball in order to automatically select the moment the ball is played and the coordinates of both the attacker and defender [35].

In addition to systems relying on computer vision, VAR (video-assistant referee) systems are being increasingly used in many sports to provide a human operator with quick access to multiple video feeds. The International Football Association Board (IFAB) approved trials and

TABLE 14.1

Applications of Ball and Player Tracking Technologies

System	Application	Configuration and Input data	Sports	Adoption
Sportlogiq [37]	Game Analytics, Player Analysis, Scouting	Broadcast videos	🏑 🏈 ⚽	30 NHL teams, NFL Teams, NCAA Football Programs, English Premier League and MLS Clubs
SkillCorner [38]	Game Analytics, Scouting	Broadcast videos	⚽	45 major soccer leagues clubs
SparkCognition [42]	Scouting	Single camera recorded game videos	🏈	Universities
Connexa Sports [43]	Game and Player Analysis	Fixed/portable single camera Six fixed cameras system Broadcast video	🏏 🎾 ⚽	Professional clubs and amateur players
CatapultSports [40]	Game and Player Analysis, Scouting, Athlete Management	Broadcast videos, Training videos	⚽ 🏈 ⚾ 🏀 🏉 🏑 🏏	2,970 national and major league teams across multiple sports, University programs.
STATS Perform [36]	Player analysis, Scouting	Remote (broadcast) Video	🏀	NCAA teams

a pathway to full implementation during its 2016 general meeting [33], partly driven by the easy access to video replays that the spectators could get on their phones using 4G and WiFi.

14.2.3 Game Analytics

The core computer vision technologies which enable game analytics both at individual player and team level including statistics and tactics are object detection and tracking. One major trend enabled by the latest improvements in vision-based tracking accuracy is the emergence of systems able to reliably estimate each player's location, speed, acceleration and body orientation directly from broadcast video [36–39] or single view fixed cameras, as opposed to traditional sets of costly static calibrated multi-camera systems. As reported in Table 14.1 even at professional level, the proliferation of systems not requiring ad-hoc dedicated hardware installation is relevant across sports. For professional athletes and team performance analytics in top tier leagues, analytics companies may still employ dedicated camera systems, but usually in combination with television broadcast as well as data collected from other sensors. Sportlogiq [37], one of the main NHL hockey data-collection and advanced analytics companies, processes broadcast videos to perform action and group activity recognition, body pose estimation, multi-

target tracking, object detection and recognition, person re-identification and semantic segmentation to provide insights such as expected goals, automated time on ice, pass analysis, team formation, and possession. An example is presented in Figure 14.3. CatapultSports [40] automatically aligns video footage analysis with wearable sensor data (GPS and accelerometers) providing insights for the coaching staff. The tool automatically connects every event, phase of play, or game state to a variety of player outputs including speed, intensity, distance covered, and player load. Coaches can view video with every player metric for any action, event, or game state. This allows them to view team positional data on a 2D pitch that is connected to every event to create visualisations, including heatmaps, average positions, trails, and traces.

One of the newest applications of the technology using simpler sources than ad-hoc dedicated systems is scouting and recruiting. Companies like SkillCorner [38] collect broadcast television video from 45 soccer leagues around the world and runs it through an algorithm that tracks individual players' location, speed, and acceleration. It also offers a simulation option in order to estimate player movement off-camera, during replays and close ups. The firm's 65 clients now use the data to scout potential recruits as well [41]. Sportlogiq [37] uses machine-learning algorithms on top of player tracking and pose data to evaluate players' skills and overall potential, classifying

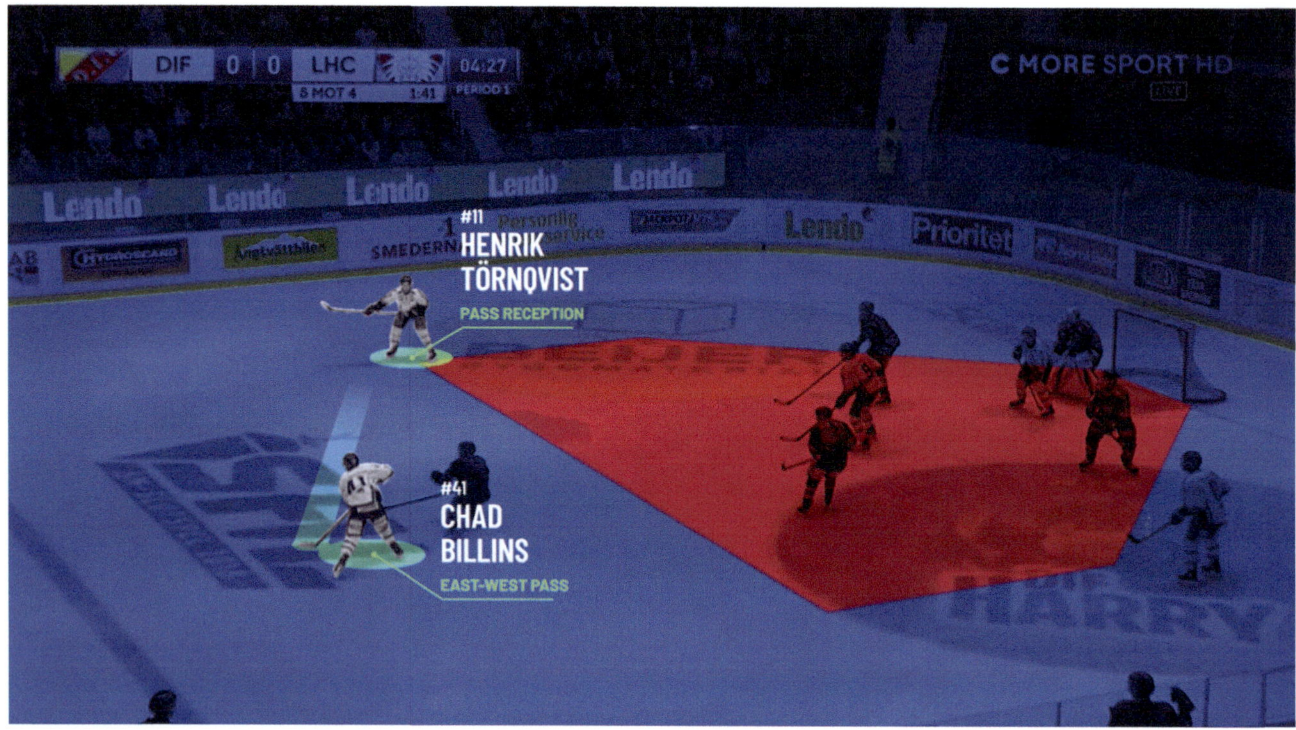

FIGURE 14.3
Sportlogiq (with courtesy from [37]) game analysis system can detect and track hockey players and puck, and automatically recognise passes, team formations and other metrics.

them into what the company calls 'archetypes' — for example, splitting forwards into subcategories like offensive forward, power forward or defensive forward. These types of data-driven player scouting systems are employed not only by professional clubs but also by universities. Companies like SparkCognition [42] are used by university programs to evaluate thousands of applications from student athletes. The system provides insights into players' statistics and their performance in game videos and combines it with academic grades, recommendations, and other relevant information to provide rankings.

14.2.4 Training and Injury Prevention

One of the main advantages of computer vision in sports training is that it can provide objective data that can be used to improve performance. The rapid improvement in the accuracy and reliability of tracking systems for ball trajectory and players, including their body pose, has allowed teams as well as individual players to utilise computer vision technology for enhanced practising, performance assessment, and even injury prevention. Usually computer vision technology is integrated with other sensors, such as radar, to guarantee the highest possible accuracy for such applications [44–46].

Computer vision technology is used to analyze the biomechanics of a player's movements. By analyzing the angles and forces of a player's movements, as well as body pose, coaches can identify areas where players need to improve their technique or reduce the risk of injury.

Such technology is employed across multiple sports, including golf for swing analysis based on club tracking [20], body pose tracking [47], or both [44]. For example, Trackman [20] uses a radar-synchronised high dynamic range camera (HD 720p @60 fps and Full HD 1080p @ 45 fps) to track the player's pose, the club and the ball. For baseball, vision-based systems allow users to upload videos recorded from their phones and perform a comparative analysis with a repository of professional pitchers' moves based on tracking body joints, pose, and bat, thereby enabling the application to recommend specialised training exercises and drills aimed at improving throwing efficiency [46]. Examples of body pose estimation systems for personalised performance analysis and training are depicted in Figure 14.4.

Every NBA team currently uses systems with cameras and other sensors installed in the basketball hoop to track the trajectory of a player's shots. This data is used to provide feedback to the player in real time, in the form of verbal feedback for shot arc, depth, and left/right,

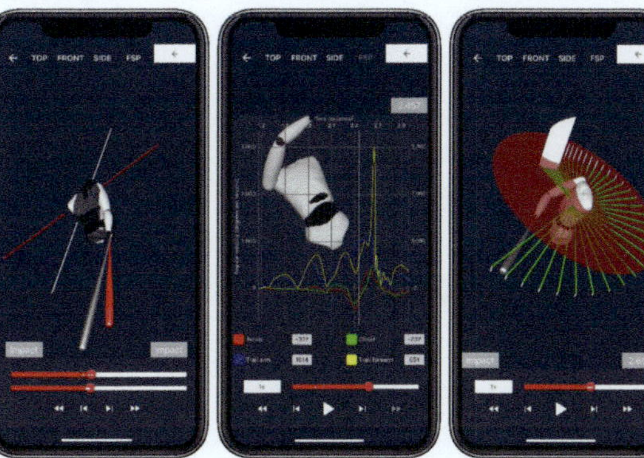

FIGURE 14.4
From left to right: TrackMan system used at the PGA Tour (with courtesy from [21]), 4D Motion (with courtesy from [44]) analysis of golf swing, tracking club as well as body pose, and Mustard app (with courtesy from [46]) for body pose estimation during pitching action.

allowing players to correct their shot in real-time [48]. All the data is also collected for further post analysis to view individual workouts, see unique shot charts and rim maps, and understand exactly which areas need improvement.

Companies have also used computer vision technology in combination with virtual reality to improve training. For example, placing cameras where the batter or receiver stands, researchers filmed cricketers bowling and racket-sport players serving and hitting. They then played these videos to people with varying degrees of skill and stopped the playback at the moment of ball release or racket contact to ask whether the ball would pitch short or long, and go left or right, thus being able to classify pitches as early in the arm's movement as possible. In a pilot test with Duke University baseball team, ball tracking and body pose estimation was used to collect opposing teams pitches and then recreate them in a virtual reality setting, using Win Reality's [49] system, where batters could practise viewing the pitcher throw the ball multiple times, getting familiar with the next opponent's style and pitch types [50].

Computer vision technology is also used for injury prevention in professional sports. Video tracking information is used in combination with on-body sensors, medical devices and other tools to forecast injury risk, tailor prescriptions for training and suggest optimal time to rest for each individual player by clubs across soccer, football and basketball using AI systems [45]. For example, the Wall Street Journal has reported in June of 2022 that "[...]Liverpool Football Club in the U.K. says it reduced the number of injuries to its players by a third over last season after adopting an AI-based data-analytics program" [41].

At the 2022 Winter Olympics in Beijing, ten U.S. figure skaters used a system from 4D Motion [44] to help track fatigue. This was accomplished by using the body pose tracking system to count and record the number of jumps taken during practice, as too many can lead to fatigue.

CatapultSports [40] system combining video footage analysis and wearable sensors data is used by coaching staff to better understand individual players' physical outputs related to their game actions, quickly identify players involved in key moments and analyze performance. Coaches use it to identify each player's maximum intensities and determine the precise physical outputs required to execute coaching strategies and tailor personalised training programs accordingly.

Computer vision-based tracking is used not only for professional performance analysis but also for recreational risk prevention. For example, SwimEye [51] is an AI-based drowning-risk detection system that spots struggling swimmers and raises an alarm when their object recognition software spots signs of distress at the bottom of a swimming pool. It uses sets of double cameras housed underwater with a viewing angle of 180° each, which typically achieve stable detection to a distance of 9m. Object recognition software is employed for detecting and tracking swimmers to spot signs of distress, as shown in Figure 14.5.

Recently, several academic works have presented technology for enhanced practising, performance assessment and injury prevention, including vertical jump[52], baseball pitch [53], ultra-distance runners[54], ACL Injury Risk prediction [55], rehabilitation [56], and even push-up analysis [57]. We're sure commercial applications of such technology are not far away.

14.2.5 Media and Entertainment

Besides the traditional use of computer vision for game specific analysis and broadcast enhancements, two new

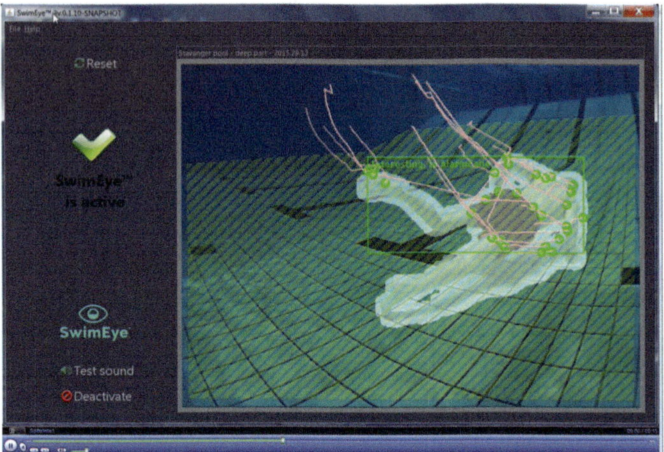

FIGURE 14.5
Swimeye (with courtesy from [51]) tracks swimmers in a pool using underwater cameras to detect signs of distress at the bottom of the pool or possible drowning.

trends have been emerging lately in the use of this technology related to sporting events: crowd monitoring in stadiums and sport journalism.

- **Crowd monitoring in stadiums:** People detection, counting, and tracking technology is used in modern sports stadiums to monitor the flow of patrons, including their movements, dwell times, and paths. Face detection is employed for obscuring personal identities and ensuring privacy. The Los Angeles Rams SoFi Stadium applies custom computer vision models on the feeds from a system of cameras to generate insights around events of interest such as egress bottlenecks, fan engagement, and concession sales trends [58]. The Skyfii [59] system analyses data from various sources including cameras, LiDAR, and WiFi to identify areas of congestion, unauthorised access to areas, busiest times, popular areas, dwell times, heat maps, and fan paths throughout a stadium. Stadiums can deliver targeted content to fans based on their behavior and location in the stadium. Preferential access to sporting arenas is being managed by systems using facial recognition software, in combination with scans of digital tickets. Over 18 venues in the USA use the Clear app [60], the same as used by some airline passengers to speed through security checks by presenting their fingerprints or showing their faces, to manage access to the stadiums via automatic check-in kiosks using facial recognition [61]. A few pilot studies have also been conducted around audience reaction analysis (facial expressions for emotions, crowd mood), with no official product being offered yet.

- **Sports content presentation and journalism:** Highlights of sporting events are now distributed in multiple formats (including length, aspect ratio, and quality) across several platforms. Computer vision is used, in combination with other data sources, for automatic highlight generation across multiple sports by automatically detecting events of interest as well as recognising and tracking players in broadcast videos [62–64]. These systems allow highlights to be ready for consumption within 6–7 minutes [63]. Commercial systems such as Veo [65] and Pixellot [66] can also be used for automatic highlight generation of amateur games. Cloud-based solutions are available for storing recordings of personal games, as well as providing capabilities for fully or semi-automatic editing of video based on ball and player tracking algorithms, as well as targeted sports event detection.

14.3 Current Research Areas

While computer vision research focusing on humans for tasks, such as, detection, tracking, activity recognition, and pose estimation, has come a long way and has been adopted in many commercial applications, robustly analysing sports videos still proves very challenging in many scenarios. The unique challenges in the domain of computer vision for sports can be boiled down to a few main characteristics:

In team sports, players of the same team are often wearing the same clothing, making (re-)identification of individuals difficult. Furthermore, in many team sports close contact and tackles between players is a natural part of a match, resulting in high degrees of occlusions between players. On the other hand, in individual sports, tasks like analysing poses and actions of the athlete is of great interest, but often requires fine-grained recognition and the ability to distinguish between subtle differences for it to be used in applications such as automatic assessment or performance optimisation.

In the following subsections, we will outline some of the latest research in the domains of team sports and individual sports, respectively.

14.3.1 Team Sports Analysis

In this section, we will discuss research aimed at team sports, specifically teams sports where two teams are facilitating the movement of a ball. Examples of these are basketball, football, soccer, hockey, among many others. These sports are often characterised as fast-paced sports,

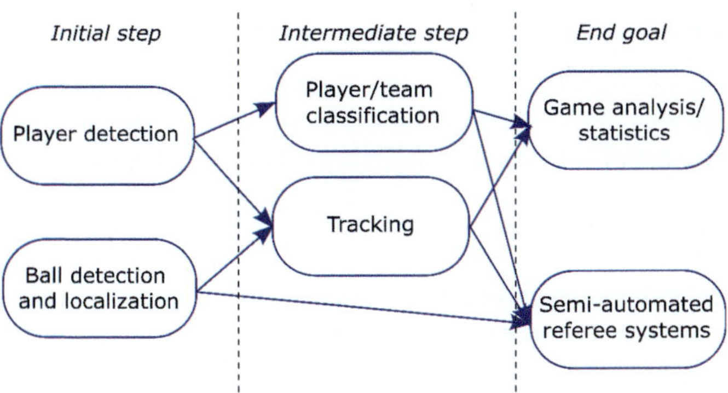

FIGURE 14.6
Sketch of some of the most common building blocks in vision systems for team sports analysis.

with many interactions between players of both the same and the opposing team.

In Figure 14.6 we provide an overview of some of the most common building blocks of which a vision system for teams sports analysis may consist. Active research is still being conducted in all these areas. While the end goal is often either analytics or statistics of the game, such as pass receiver prediction, possession statistics, key actor estimation, action/event spotting, or play phase segmentation, or semi-automated referee systems, such as goal line technology or offside detection, the initial and intermediate tasks are still some of most researched topics. Methods belonging to initial or intermediate steps are impactful, as they feed into several downstream tasks, and important, as their performance directly affects the results of subsequent tasks.

A basis for most analysis is the robust detection and tracking of all players as well as the ball on the field. In early days, player and ball detection algorithms relied on colour/background subtraction, but today all state-of-the-art methods are CNN-based. Player detection differs from general human/pedestrian detection as players may often show more extreme poses, they experience heavy occlusions, and players from the same team usually wear identical clothing. For these reasons, it is necessary to design and train, or at least fine-tune, algorithms specifically for this domain. It has been shown that adapting the networks for specific sports types, and even specific matches, increases the performance of a human segmentation algorithm [67]. However, gathering and annotating enough training data for fine-tuning algorithms for each specific domain/scene/team/etc. is very costly. Recent work instead proposes the use of semi-supervised training, for which the network trains on a small labelled dataset and a large unlabelled dataset [68].

Ball detection and localisation also introduces a number of specific challenges: small object size compared to the total field-of-view of the camera, heavy occlusions by

players, and fast motion in three dimensions. However, the appearance of a ball is often designed to be visible on the field, which may simplify the detection step. High-quality multi-view camera setups on stadia may provide robust ball localisation, as the different viewing angles from the cameras allow the ball to be detected on several cameras despite being occluded from other angles. The 3D localisation is then solved by triangulation. Another solution is to include physics-based constraints on the ball trajectories [69]. Recent work digs into the problem of 3D ball localisation from single images using information on the ball diameter [70].

In the general Multi-Object Tracking (MOT) field, research has been pushed forward by public large-scale annotated datasets. This includes tracking of humans as pedestrians. However, to tackle the domain challenges specific to sports, annotated datasets from sports games are required, and that is a laborious task. In 2022, two new datasets on soccer videos with tracking annotations were published. SoccerNet-Tracking [71] is a collection of 12 complete soccer games, of which one entire half-time and 200 30-seconds clips are annotated with five classes: player, goalkeeper, referee, ball, and other (medical staff or coaches entering the field). The videos in SoccerNet-Tracking are captured by one camera following the main action on the field.

The SoccerTrack dataset [72] consists of 30 30-seconds annotated clips of synchronised videos from a wide-view camera and a bird-view drone camera, and additional GNSS data from body-mounted devices. Both videos of this dataset capture the entire pitch, as opposed to SoccerNet-Tracking.

Many downstream sports analytics tasks may follow after detecting the players and ball on the sports field. One of the valuable intermediate steps is the identification of teams and individual players. Koshkina et al. [73] proposed the use of unsupervised contrastive learning for classifying the team affiliation of players. Due to the unsupervised nature of the method, this is a promising approach that can generalise to novel teams and games without labelling any data. For individual player classification, a common approach is jersey number recognition [74]. New work in this direction exploits the highly popular transformer network on tracklets from ice hockey games [75]. General re-identification of players has also been in focus recently on a new basketball dataset [76] and the papers based on this dataset [77–79].

Many of the sports analytics tasks are rather complicated and have previously been performed by skilled human video analysts. It requires combinations of temporal, spatial, and visual data, and only a few years ago, the technology was still far from being capable of automating such complex tasks, at least without the inclusion of lots of domain-specific knowledge. An example of such a complex downstream task is pass receiver prediction in soccer [80], which is tackled with a 3D CNN for cropped image sequences and LSTM for trajectories, both of which feed into a transformer encoder before the final prediction of pass receiver. Sarkar et al. combine pass event detection and team identification for possession statistics in soccer [81] and Askari et al. proposed first a CNN-RNN-based method with a time-varying attention mechanism [82] and later a self-supervised learning-based method for interaction classification [83], both demonstrated for penalty classification in ice hockey. As these recent papers demonstrate, modern approaches to high-level game analysis and event recognition use multiple modalities of input data and highly complex networks for automatically learning to predict the outcome.

Other research in high level game analysis revolves around temporal action spotting and segmentation of games based on play phases. The quantity of sports footage being captured and broadcast is steadily increasing, and often time-consuming video editing is involved. Automated temporal localisation of actions and phases of a game may lead to efficient automatic production of broadcast videos and highlight summaries. However, this requires the system to have an understanding of the particular sport, and what is important to show the viewer. A substantial amount of work in the domain of understanding soccer games has been carried out in the creation of SoccerNet-v2 [85]. This is a large-scale dataset with annotations for 17 common events in soccer games, replay grounding, and camera shot segmentation. Subsequent work on this dataset aimed to do temporally accurate action spotting and has advanced the state-of-the-art in this area [86–90]. An approach for action spotting is illustrated in Figure 14.7. Another approach to obtaining a time-compressed video of a game is to identify periods of active play and remove any breaks in the game. This has been applied for ice hockey by detecting play phases from visual and auditory cues [91].

14.3.2 Analysis of Individual Athletes

For research focused on individual athletes, we may imagine that we zoom in to take a closer look at the detailed poses and actions of an individual person. These applications often aim to assess the performance of the athlete, used for either automatic scoring or performance optimisation.

In the analysis of human poses and actions, a common first step is localisation of keypoints and body parts. This is a popular research area, also outside the sports domain, with frameworks like OpenPose [92] or HRNet [93] being commonly used. However, networks trained on standard human poses may not react very well to extreme athlete poses, like having the head down, or including high-frequency movements [94]. To improve pose estimation of extreme poses a pre-trained network may be refined using an annotated sports dataset [95, 96]. Example results are shown in Figure 14.8. New research in the sports domain also aims to go beyond the representation of fixed keypoints and allow selection of arbitrary keypoints on the limbs [97, 98] or the entire body [99].

Following keypoint or pose detection, current computer vision research focuses on automated evaluation and feedback on specific poses used for rehabilitation and fitness/strength training [56, 57, 100, 101]. The availability of real-time video-based feedback on poses may eliminate the need for personal trainers, which are not feasible for everyone to hire. Another use-case is fatigue detection by analysing the change in pose over repeated actions, like baseball pitches [53]. Recently, advances in pose estimation have reached a state where analysis of highly complex and occluded poses, as seen in combat sports, can be analysed reliably [102]. The ultimate goal of such an approach may be to automatically score entire matches, eliminating the subjectiveness of human judges. This step towards automatic scoring is also taken in other sports, like ski jumping [103] and diving [104].

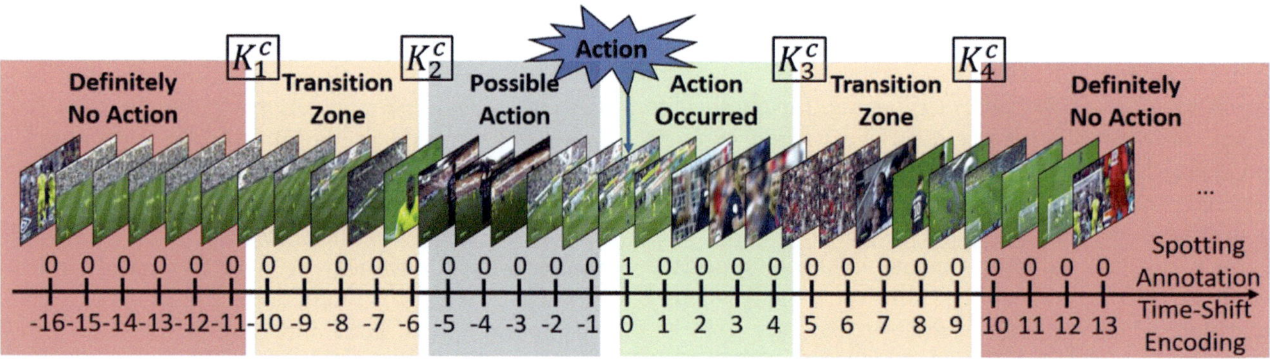

FIGURE 14.7
Action spotting approach from Cioppa et al. © 2020 IEEE. (Reprinted, with permission, from [84].)

FIGURE 14.8
Results from Kitamura et al. for pose estimation on extreme poses after refining on annotated sports dataset. © 2022 IEEE. (Reprinted, with permission, from [95].)

Keypoint estimations over sequences of frames can also lead to action classifications. This is demonstrated by Zhu et al., who used temporal convolutional networks (TCN) on 2D skeleton data for footwork recognition in fencing video [105], and Kulkarni and Shenoy who used a TCN on 2D skeleton data for stroke recognition in table tennis [106]. Martin et al. also tackles the problem of stroke recognition in table tennis, obtaining good results by combining raw image data with optical flow and pose estimations [107].

Traditionally, performance assessment of athletes has utilised special equipment, including both small wearable devices, like heart rate monitors, GPS, and accelerometers, and external devices, like force plates. Since cameras of sufficient quality in terms of both pixel resolution and frame rate are widely available today, computer vision-based methods have become feasible as a cheaper alternative to specialised devices. Blanchard et al. address this idea with a framework for video-based assessment of jumps [52, 55].

While analysis of humans has taken up a lot of focus in the analysis of individual sports, the trajectories of balls/shuttles are also of great importance for fully understanding a game. It is again the case that a calibrated multi-camera setup may help address some of the challenges, like occlusions and 3D localisation of the object. However, in many cases, like at tournaments, it is not possible to install such systems for analysis purposes. In these cases a monocular video may be the only available data source. Liu and Wang worked on 3D shuttlecock trajectory reconstruction for monocular badminton videos [108]. They presented an approach which includes court detection, 2D shuttlecock tracking, 2D human pose tracking, net poles identification, shot segmentation, and physics-based estimations. Hence, it is a highly complex system, and many error sources still need to be reduced before a 3D trajectory with low error can be obtained from a single view.

14.4 Future Prospects

In the previous sections, we have outlined the current state of commercial vision systems for sports and the research in this field. We find it to be a rapidly emerging field, with a close connection between researchers and commercial stakeholders, like broadcast companies, clubs, and sports associations. This is apparent from many mixed co-authorships, and dedicated workshops and conferences with high attendance from both academia and industry. As such, we expect a continued quick transfer from research to applications.

As several hundred different organised competitive sports types exist, they all have their own rule books and ways of assessing the performance of each individual athlete or team of players. The performance assessments can be mapped as a continuum from objective to subjective measurement. Examples of purely objective measurements are time, distance, height, number of goals or faults, while clear subjective measurements are human ratings based on the visual appearance of actions. In between these examples are all types of decisions made by human referees/umpires/line judges according to a sports-specific rule book.

Computer vision-based decision systems have found their way into assessment of sports performance since the first examples of line calling systems by Hawk-Eye in 2005 [7]. Recent trends have shown that we are ready to enter more areas of the sports domain which are closer to the subjective scoring of visual appearance. Examples of these are the scoring of diving, ski jump, gymnastics, and figure skating, which is currently done be a panel of judges.

Sports with very close interactions and occlusions between participants combined with large variations in poses are particularly hard to automatically analyse and score. But despite the difficult nature, we are now starting to also see solutions for combat sports [109] and we expect the recent advances in human pose estimation to open up many applications of action recognition and performance evaluation.

Integration of technology-based systems to substitute for human judges faces a number a challenges, though. Clearly, the technology needs to be ready and sufficiently tested on all possible complex scenarios that are seen in real-world sports. An error margin that is considered small in research communities, for example, 0.5 or 1%, is not acceptable in commercial applications, as it translates to a system that fails for every 200 or 100 participants. The error margin needs to be vanishing, while a back-up solution is still needed if it should fail in rare cases.

Another challenge is the human trust in the system. It may take a long time to build trust in a system, while the trust can easily be broken by just one failure case. Explainable AI (XAI) is a quickly emerging research area within computer vision (among other AI fields), in which the system aims to explain or visualise what the decision of the system is based on. This may help building trust in the system and may really ignite the use of automated decision systems in sports when the system can explain what the decision is based on. With healthcare being one of the big applications of XAI, work on XAI for sports applications is still to be seen, but we believe this will be one of the future directions of research in computer vision for sports.

Another recent trend in both general computer vision and computer vision for sports applications is the use of more unlabelled data. In the sports domain, it is evident that 10–20 years back, domain knowledge for each sports type, like common team formations, was hard-coded into the system. Today neural network-based methods generally require large amounts of annotated data from the specific domain, but besides the few largest sports types with lots of money involved, it is not feasible to do fine-grained annotation of large amounts of data for all sports types, specific teams, etc. However, since lots of sports content is recorded on video for either broadcasting or analytics purposes, the video data is already available, if it could be used directly without annotations. Hence, the trend of self-supervised or semi-supervised methods or domain adaptation with very few labelled data is very useful in sports applications as a way to exploit the large quantity of video data without the expensive annotation process. We expect more research in this direction in the coming years as computer vision methods are spreading to a wider range of different sports types.

14.5 Conclusion

In this chapter, we have provided an overview of the historical and current use of computer vision for commercial applications, like broadcasting, officiating, and performance assessment. Furthermore, we have given our view on the most prominent current research in the human-centred topics of team sports analysis and analysis of individual athletes and pointed out some interesting future research directions. This chapter is not an exhaustive survey on the topic, and many related topics, like camera calibration and play field registration, have not been covered here, although being part of a successful vision system for sport applications.

But as illustrated in this chapter, computer vision for sports applications is a very active research area with close connections to the industry, and there is still a wide range of open problems to work on. The demand for computer vision solutions from the sports industry is huge, and we are excited to follow the progress of computer vision in the coming years.

References

[1] PRWeb. (2005) Telestrator Invention Wins Emmy Just In Time For Super Bowl. [Online]. Available: http://www.prweb.com/releases/2005/2/prweb205940.htm

[2] S. Brannan. How Stuff Works - The First Down Line. [Online]. Available: https://entertainment.howstuffworks.com/first-down-line.htm

[3] Avid. Maestro – For Live Event Production. [Online]. Available: https://www.avid.com/products/maestro/for-live-events

[4] Chyron. Virtual Placement - Virtual Graphics in Any Live Broadcast. [Online]. Available: https://chyron.com/products/virtual-augmented-reality/375/

[5] Dartfish. High-performance Sports Video and Data Analysis Solution. [Online]. Available: https://www.dartfish.com/pro_s

[6] Deltacast. Delta Live. [Online]. Available: https://www.deltacast.com/products/delta-live

[7] Hawk-Eye. Hawk-Eye Innovations. [Online]. Available: https://www.hawkeyeinnovations.com/

[8] Ross. Piero Sports Graphics Analysis. [Online]. Available: https://www.rossvideo.com/products-services/acquisition-production/cg-graphics-systems/piero-sports-graphics-analysis/

[9] RTSoftware. Tactic Live – Graphic Insertion and Virtual Advertising. [Online]. Available: https://rtsw.co.uk/all-products/tactic-live/

[10] SMT. SportsMEDIA Technology. [Online]. Available: https://www.smt.com

[11] Vizrt. Viz arena. [Online]. Available: https://www.vizrt.com/products/viz-arena

[12] G. Thomas, "Real-time camera tracking using sports pitch markings," *Journal of Real Time Image Processing*, November 2007, 2(2-3): 117–132. [Online]. Available: https://www.bbc.co.uk/rd/publications/whitepaper168

[13] R. Dawes, J. Chandaria, and G. Thomas, "Image-based camera tracking for athletics," in *Proceedings of the IEEE International Symposium on Broadband Multimedia Systems and Broadcasting (BMSB 2009)*, 2009. [Online]. Available: https://www.bbc.co.uk/rd/publications/whitepaper181

[14] BBC. Piero Sports Graphics System. [Online]. Available: https://www.bbc.co.uk/rd/projects/piero

[15] R. Cavallaro, "The FoxTrax hockey puck tracking system," in *IEEE Computer Graphics and Applications*, April 1997.

[16] SMT. (2022) How NHL's Sensor-Embedded Puck Allows for Better Broadcast Statistics and Graphics, New Types of Wagers, More. [Online]. Available: https://www.smt.com/articles2019/how-nhls-sensor-embedded-puck-allows-for-better-broadcast-statistics-and-graphics-new-types-of-wagers-more

[17] UEFA. UEFA Euro 2016 to Use Hawk-Eye for Goal-Line Technology. [Online]. Available: https://www.uefa.com/insideuefa/news/022c-0f8e298bb4ce-dae45c5ae038-1000--uefa-euro-2016-to-use-hawk-eye-for-goal-line-technology/

[18] StatSports. Apex Athlete Series GPS Tracker. [Online]. Available: https://statsports.com/apex-athlete-series

[19] SailGPTechnologies. LiveLineFX. [Online]. Available: https://www.sailgptechnologies.com/

[20] TrackMan. TrackMan Golf Simulator. [Online]. Available: https://www.trackman.com/

[21] TrackMan. PGA Tour Selects Trackman Tracking and Tracing Solution Beginning in 2022. [Online]. Available: https://blog.trackmangolf.com/pga-tour-selects-trackman-tracking-and-tracing-solution-beginning-in-2022/

[22] Vizrt. Viz libero. [Online]. Available: https://www.vizrt.com/products/viz-libero

[23] M. Orbach. (2021, August) Intel Shuts Down Sports Division, Sells Part of it to Verizon. [Online]. Available: https://www.calcalistech.com/ctech/articles/0,7340,L-3915635,00.html

[24] H. McLean. All in the Detail: Volumetric Video Set to Hit Touchscreens by End of 2022 Says Unity's Peter Moore. [Online]. Available: https://www.svgeurope.org/blog/headlines/all-in-the-detail-volumetric-video-set-to-hit-touchscreens-by-end-of-2022-says-unitys-peter-moore/

[25] Deltacast. Virtual View. [Online]. Available: https://www.deltacast.com/products/virtual-view

[26] S. Krashinsky. (2011, April) He Shoots - And the Advertisers Score! [Online]. Available: https://www.theglobeandmail.com/report-on-business/industry-news/marketing/he-shoots---and-the-advertisers-score/article578073/

[27] E. Rantalainen, World Patent WO 01/58 147, August, 2001.

[28] E. J. Benolt, Europe Patent EP 3 295 663, March, 2018.

[29] H. Collins and R. Evans, "You Cannot Be Serious! Public understanding of technology with special reference to Hawk-Eye," *Public Understanding of Science*, May 2008, 17(3):283–308.

[30] ITF. (2021, August) ITF Press Release - FlightScope Line-calling System Makes the Grade. [Online]. Available: https://www.itftennis.com/en/news-and-media/articles/flightscope-line-calling-system-makes-the-grade/

[31] Virtaleye. The Story of Virtual Eye Cricket. [Online]. Available: https://virtualeye.tv/the-sports/virtual-eye-cricket

[32] ICC. ICC Men's Test Match Playing Conditions (Effective 28th September 2017). [Online]. Available: https://www.icc-cricket.com/about/cricket/rules-and-regulations/decision-review-system

[33] J. Medeiros. (2018, June) The Inside Story of How FIFA's Controversial VAR System Was Born. [Online]. Available: https://www.wired.co.uk/article/var-football-world-cup

[34] FIFA. FIFA Resource Hub – Goal-line Technology Installations. [Online]. Available: https://www.fifa.com/technical/football-technology/resource-hub?QualityProgram=6Sshn3qiYsRBq6muymEEtY&Category=2xINFjzvbmDbgrU1cRnnaf

[35] FIFA. (2022, October) FIFA Offside Technology Test Manual Version 1. [Online]. Available: https://www.fifa.com/technical/football-technology/standards/goal-line-technology

[36] STATS. Stats Perform. [Online]. Available: https://www.statsperform.com/

[37] Sportlogiq. Swimeye intelligent drowning detection in pool. [Online]. Available: https://sportlogiq.com/

[38] SkillCorner. Skill Corner. [Online]. Available: https://www.skillcorner.com/

[39] R. Vision. Respo Vision. [Online]. Available: https://respo.vision/

[40] C. Sports. Catapult Sports. [Online]. Available: https://www.catapultsports.com/

[41] E. Niilner. (2022) How AI Could Help Predict—and Avoid—Sports Injuries, Boost Performance. [Online]. Available: https://www.wsj.com/articles/how-ai-could-revolutionize-sports-trainers-tap-algorithms-to-boost-performance-prevent-injury-11654353916

[42] SparkCognition. Sparkcognition. [Online]. Available: https://www.sparkcognition.com/

[43] C. Sports. Connexa Sports. [Online]. Available: www.connexasports.com

[44] 4D Motion Sports. 4D Motion Sports. [Online]. Available: https://4dmotionsports.com/

[45] Zone7. Zone7 Performance Hub. [Online]. Available: https://zone7.ai/

[46] Mustard App. Mustard Pitch Analysis. [Online]. Available: https://mustard.ghost.io/

[47] 18birdies. (2022) Golf AI Coach. [Online]. Available: https://18birdies.com/aicoach/

[48] Noah Basketball. Noah Shooting System. [Online]. Available: https://www.noahbasketball.com/

[49] W. Reality. Win reality. [Online]. Available: https://winreality.com/

[50] L. Drew, "How athletes hit a fastball," *Nature*, 592(7852):4–6 , April 2021.

[51] swimeye. Swimeye Intelligent Drowning Detection in Pool. [Online]. Available: https://swimeye.com/products/swimeye/

[52] C. Roygaga, D. Patil, M. Boyle, W. Pickard, R. Reiser, A. Bharati, and N. Blanchard, "APE-V: Athlete performance evaluation using video," in *Proceedings of the IEEE/CVF Winter Conference on Applications of Computer Vision (WACV) Workshops*, January 2022, pp. 691–700.

[53] Y.-W. Ma, J.-L. Chen, C.-C. Hsu, and Y.-H. Lai, "Design and analysis of a pitch fatigue detection system for adaptive baseball learning," *Frontiers in Psychology*, 12, 2021.

[54] D. Freire-Obregón, J. Lorenzo-Navarro, and M. Castrillón-Santana, "Decontextualized I3D ConvNet for ultra-distance runners performance analysis at a glance," in *Image Analysis and Processing – ICIAP 2022*, 2022, pp. 242–253.

[55] N. Blanchard, K. Skinner, A. Kemp, W. Scheirer, and P. Flynn, ""Keep Me In, Coach!": A computer vision perspective on assessing ACL injury risk in female athletes," in *2019 IEEE Winter Conference on Applications of Computer Vision (WACV)*, 2019, pp. 1366–1374.

[56] Y. Qiu, J. Wang, Z. Jin, H. Chen, M. Zhang, and L. Guo, "Pose-guided matching based on deep learning for assessing quality of action on rehabilitation training," *Biomedical Signal Processing and Control*, 72, 2022.

[57] H.-J. Park, J.-W. Baek, and J.-H. Kim, "Imagery based parametric classification of correct and incorrect motion for push-up counter using OpenPose," in *2020 IEEE 16th International Conference on Automation Science and Engineering (CASE)*, 2020, pp. 1389–1394.

[58] Plainsight. (2022) Crowd Analysis NFL Stadium. [Online]. Available: https://plainsight.ai/computer-vision-nfl-stadiums/

[59] Skyfii. Skyfii. [Online]. Available: https://us.skyfii.com/

[60] clear. clear. [Online]. Available: https://www.clearme.com/

[61] P. Olson. Facial Recognition's Next Big Play: The Sports Stadium. [Online]. Available: https://www.wsj.com/articles/facial-recognitions-next-big-play-the-sports-stadium-11596290400

[62] R. AI. Reely AI. [Online]. Available: https://www.reely.ai/

[63] W. Sports. WSC Sports. [Online]. Available: https://wsc-sports.com/

[64] Wimbledon. Wimbledon Cognitive Highlights. [Online]. Available: https://www.wimbledon.com/en_GB/aboutwimbledon/wimbledons_cognitive_highlights.html

[65] VEO. (2022) VEO Analytics. [Online]. Available: https://www.veo.co/en-us/analyze

[66] Pixellot. Pixellot. [Online]. Available: https://you.pixellot.tv/

[67] A. Cioppa, A. Deliège, M. Istasse, C. De Vleeschouwer, and M. Van Droogenbroeck, "ARTHuS: Adaptive real-time human segmentation in sports through online distillation," in *Proceedings of the IEEE/CVF Conference on Computer Vision and Pattern Recognition (CVPR) Workshops*, 2019, pp. 2505–2514.

[68] R. Vandeghen, A. Cioppa, and M. Van Droogenbroeck, "Semi-supervised training to improve player and ball detection in soccer," in *Proceedings of the IEEE/CVF Conference on Computer Vision and Pattern Recognition (CVPR) Workshops*, June 2022, pp. 3481–3490.

[69] P. R. Kamble, A. G. Keskar, and K. M. Bhurchandi, "Ball tracking in sports: a survey," *Artificial Intelligence Review*, 52(3):1655–1705, Oct. 2017. [Online]. Available: https://doi.org/10.1007/s10462-017-9582-2

[70] G. Van Zandycke and C. De Vleeschouwer, "3D ball localization from a single calibrated image," in *Proceedings of the IEEE/CVF Conference on Computer Vision and Pattern Recognition (CVPR) Workshops*, June 2022, pp. 3472–3480.

[71] A. Cioppa, S. Giancola, A. Deliège, L. Kang, X. Zhou, Z. Cheng, B. Ghanem, and M. Van Droogenbroeck, "SoccerNet-Tracking: Multiple object tracking dataset and benchmark in soccer videos," in *Proceedings of the IEEE/CVF Conference on Computer Vision and Pattern Recognition (CVPR) Workshops*, June 2022, pp. 3491–3502.

[72] A. Scott, I. Uchida, M. Onishi, Y. Kameda, K. Fukui, and K. Fujii, "SoccerTrack: A dataset and tracking algorithm for soccer with fish-eye and drone videos," in *Proceedings of the IEEE/CVF Conference on Computer Vision and Pattern Recognition (CVPR) Workshops*, June 2022, pp. 3569–3579.

[73] M. Koshkina, H. Pidaparthy, and J. H. Elder, "Contrastive learning for sports video: Unsupervised player classification," in *Proceedings of the IEEE/CVF Conference on Computer Vision and Pattern Recognition (CVPR) Workshops*, June 2021, pp. 4528–4536.

[74] K. Vats, M. Fani, D. A. Clausi, and J. Zelek, "Multi-task learning for jersey number recognition in ice hockey," in *Proceedings of the 4th International Workshop on Multimedia Content Analysis in Sports*, ser. MMSports'21, 2021, p. 11–15.

[75] K. Vats, W. McNally, P. Walters, D. A. Clausi, and J. S. Zelek, "Ice hockey player identification via transformers and weakly supervised learning," in *Proceedings of the IEEE/CVF Conference on Computer Vision and Pattern Recognition (CVPR) Workshops*, June 2022, pp. 3451–3460.

[76] G. Van Zandycke, V. Somers, M. Istasse, C. D. Don, and D. Zambrano, "DeepSportradar-v1: Computer vision dataset for sports understanding with high quality annotations," in *Proceedings of the 5th International ACM Workshop on Multimedia Content Analysis in Sports*, ser. MMSports '22, 2022, p. 1–8.

[77] K. Habel, F. Deuser, and N. Oswald, "CLIP-ReIdent: Contrastive training for player re-identification," in *Proceedings of the 5th International ACM Workshop on Multimedia Content Analysis in Sports*, ser. MMSports '22, 2022, p. 129–135.

[78] W. Zheng and M. Yuan, "A person re-identification approach focusing on the occlusion problem and ranking optimization," in *Proceedings of the 5th International ACM Workshop on Multimedia Content Analysis in Sports*, ser. MMSports '22, 2022, p. 121–127.

[79] Q. An, K. Cui, R. Liu, C. Wang, M. Qi, and H. Ma, "Attention-aware multiple granularities network for player re-identification," in *Proceedings of the 5th International ACM Workshop on Multimedia Content Analysis in Sports*, ser. MMSports '22, 2022, p. 137–144.

[80] Y. Honda, R. Kawakami, R. Yoshihashi, K. Kato, and T. Naemura, "Pass receiver prediction in soccer using video and players' trajectories," in *Proceedings of the IEEE/CVF Conference on Computer Vision and Pattern Recognition (CVPR) Workshops*, June 2022, pp. 3503–3512.

[81] S. Sarkar, D. P. Mukherjee, and A. Chakrabarti, "Watch and act: Dual interacting agents for automatic generation of possession statistics in soccer,"

in *Proceedings of the IEEE/CVF Conference on Computer Vision and Pattern Recognition (CVPR) Workshops*, June 2022, pp. 3560–3568.

[82] F. Askari, R. Ramaprasad, J. J. Clark, and M. D. Levine, "Interaction classification with key actor detection in multi-person sports videos," in *Proceedings of the IEEE/CVF Conference on Computer Vision and Pattern Recognition (CVPR) Workshops*, June 2022, pp. 3580–3588.

[83] F. Askari, R. Jiang, Z. Li, J. Niu, Y. Shi, and J. J. Clark, "Self-supervised video interaction classification using image representation of skeleton data," in *Proceedings of the IEEE/CVF Conference on Computer Vision and Pattern Recognition (CVPR) Workshops*, 2023, pp. 5228–5237.

[84] A. Cioppa, A. Deliège, S. Giancola, B. Ghanem, M. Van Droogenbroeck, R. Gade, and T. B. Moeslund, "A context-aware loss function for action spotting in soccer videos," in *The IEEE Conference on Computer Vision and Pattern Recognition (CVPR)*, June 2020.

[85] A. Deliege, A. Cioppa, S. Giancola, M. J. Seikavandi, J. V. Dueholm, K. Nasrollahi, B. Ghanem, T. B. Moeslund, and M. Van Droogenbroeck, "SoccerNet-v2: A dataset and benchmarks for holistic understanding of broadcast soccer videos," in *Proceedings of the IEEE/CVF Conference on Computer Vision and Pattern Recognition (CVPR) Workshops*, June 2021, pp. 4508–4519.

[86] S. Giancola and B. Ghanem, "Temporally-aware feature pooling for action spotting in soccer broadcasts," in *Proceedings of the IEEE/CVF Conference on Computer Vision and Pattern Recognition (CVPR) Workshops*, June 2021, pp. 4490–4499.

[87] A. Darwish and T. El-Shabrway, "STE: Spatio-temporal encoder for action spotting in soccer videos," in *Proceedings of the 5th International ACM Workshop on Multimedia Content Analysis in Sports*, ser. MMSports '22, 2022, p. 87–92.

[88] A. Cartas, C. Ballester, and G. Haro, "A graph-based method for soccer action spotting using unsupervised player classification," in *Proceedings of the 5th International ACM Workshop on Multimedia Content Analysis in Sports*, ser. MMSports '22, 2022, p. 93–102.

[89] H. Zhu, J. Liang, C. Lin, J. Zhang, and J. Hu, "A transformer-based system for action spotting in soccer videos," in *Proceedings of the 5th International ACM Workshop on Multimedia Content Analysis in Sports*, ser. MMSports '22, 2022, p. 103–109.

[90] S. Giancola, A. Cioppa, J. Georgieva, J. Billingham, A. Serner, K. Peek, B. Ghanem, and M. V. Droogenbroeck, "Towards active learning for action spotting in association football videos," in *Proceedings of the IEEE/CVF Conference on Computer Vision and Pattern Recognition (CVPR) Workshops*, 2023, pp. 5097–5107.

[91] H. Pidaparthy, M. H. Dowling, and J. H. Elder, "Automatic play segmentation of hockey videos," in *Proceedings of the IEEE/CVF Conference on Computer Vision and Pattern Recognition (CVPR) Workshops*, June 2021, pp. 4585–4593.

[92] Z. Cao, G. Hidalgo Martinez, T. Simon, S. Wei, and Y. A. Sheikh, "OpenPose: Realtime multi-person 2D pose estimation using part affinity fields," *IEEE Transactions on Pattern Analysis and Machine Intelligence*, 2019.

[93] K. Sun, B. Xiao, D. Liu, and J. Wang, "Deep high-resolution representation learning for human pose estimation," in *CVPR*, 2019.

[94] C. K. Ingwersen, J. N. Jensen, M. R. Hannemose, and A. B. Dahl, "Evaluating current state of monocular 3D pose models for golf," in *Proceedings of the Northern Lights Deep Learning Workshop 2023*, vol. 4. Septentrio Academic Publishing, 2023.

[95] T. Kitamura, H. Teshima, D. Thomas, and H. Kawasaki, "Refining OpenPose with a new sports dataset for robust 2D pose estimation," in *Proceedings of the IEEE/CVF Winter Conference on Applications of Computer Vision (WACV) Workshops*, January 2022, pp. 672–681.

[96] C. K. Ingwersen, C. M. Mikkelstrup, J. N. Jensen, M. R. Hannemose, and A. B. Dahl, "Sportspose - A dynamic 3D sports pose dataset," in *Proceedings of the IEEE/CVF Conference on Computer Vision and Pattern Recognition (CVPR) Workshops*, 2023, pp. 5218–5227.

[97] K. Ludwig, D. Kienzle, and R. Lienhart, "Recognition of freely selected keypoints on human limbs," in *Proceedings of the IEEE/CVF Conference on Computer Vision and Pattern Recognition (CVPR) Workshops*, June 2022, pp. 3531–3539.

[98] K. Ludwig, D. Kienzle, J. Lorenz, and R. Lienhart, "Detecting arbitrary keypoints on limbs and skis with sparse partly correct segmentation masks," in *Proceedings of the IEEE/CVF Winter Conference on Applications of Computer Vision (WACV) Workshops*, 2023.

[99] K. Ludwig, J. Lorenz, R. Schön, and R. Lienhart, "All keypoints you need: Detecting arbitrary keypoints on the body of triple, high, and long jump athletes," in *Proceedings of the IEEE/CVF Conference on Computer Vision and Pattern Recognition (CVPR) Workshops*, 2023, pp. 5178–5186.

[100] B. Dittakavi, D. Bavikadi, S. V. Desai, S. Chakraborty, N. Reddy, V. N. Balasubramanian, B. Callepalli, and A. Sharma, "Pose tutor: An explainable system for pose correction in the wild," in *Proceedings of the IEEE/CVF Conference on Computer Vision and Pattern Recognition (CVPR) Workshops*, June 2022, pp. 3540–3549.

[101] M. Deyzel and R. P. Theart, "One-shot skeleton-based action recognition on strength and conditioning exercises," in *Proceedings of the IEEE/CVF Conference on Computer Vision and Pattern Recognition (CVPR) Workshops*, 2023, pp. 5168–5177.

[102] V. Hudovernik and D. Skocaj, "Video-based detection of combat positions and automatic scoring in jiu-jitsu," in *Proceedings of the 5th International ACM Workshop on Multimedia Content Analysis in Sports*, ser. MMSports '22, 2022, p. 55–63.

[103] D. Štepec and D. Skočaj, "Video-based ski jump style scoring from pose trajectory," in *Proceedings of the IEEE/CVF Winter Conference on Applications of Computer Vision (WACV) Workshops*, January 2022, pp. 682–690.

[104] P. Parmar and B. T. Morris, "What and how well you performed? A multitask learning approach to action quality assessment," in *2019 IEEE/CVF Conference on Computer Vision and Pattern Recognition (CVPR)*, 2019, pp. 304–313.

[105] K. Zhu, A. Wong, and J. McPhee, "FenceNet: Fine-grained footwork recognition in fencing," in *Proceedings of the IEEE/CVF Conference on Computer Vision and Pattern Recognition (CVPR) Workshops*, June 2022, pp. 3589–3598.

[106] K. M. Kulkarni and S. Shenoy, "Table tennis stroke recognition using two-dimensional human pose estimation," in *Proceedings of the IEEE/CVF Conference on Computer Vision and Pattern Recognition (CVPR) Workshops*, June 2021, pp. 4576–4584.

[107] P.-E. Martin, J. Benois-Pineau, R. Péteri, and J. Morlier, "Three-stream 3D/1D CNN for fine-grained action classification and segmentation in table tennis," in *Proceedings of the 4th International Workshop on Multimedia Content Analysis in Sports*, ser. MMSports'21, 2021, p. 35–41.

[108] P. Liu and J.-H. Wang, "MonoTrack: Shuttle trajectory reconstruction from monocular badminton video," in *Proceedings of the IEEE/CVF Conference on Computer Vision and Pattern Recognition (CVPR) Workshops*, June 2022, pp. 3513–3522.

[109] Jabbr. (2023, March) Automatic Punch Stats & Content Generation. [Online]. Available: https://jabbr.ai/

15

Spike-Based Neuromorphic Computing for Next-Generation Computer Vision

Md Sakib Hasan, Catherine D. Schuman, Zhongyang Zhang, Tauhidur Rahman, and Garrett S. Rose

15.1 Introduction

Since Gordon Moore's famous prediction in 1965 [1], colloiquially known as Moore's Law, the computing industry has worked relentlessly to achieve exponential increase in speed and functional density over last six decades using progressively advanced fabrication technologies along with architectural innovations. Moore's law along with Dennard scaling [2] resulted in the doubling of performance per joule about every 18 months. Unfortunately, Dennard scaling started breaking down in the early 2000s, resulting in a saturation of single core frequency due to the power constraint popularly known as the power wall [3]. Another major bottleneck of current digital computers is the separation of memory and processor in the conventional von Neumann architecture. The lagging behind of memory speed compared to processors has given rise to von Neumann bottleneck also known as the memory wall, that is, most energy and time is used in trafficking data between these two subsystems instead of the actual computation [3]. Moreover, the scaling of transistor feature size as dictated by Moore's law has also been slowing down and may reach an end in the coming decade due to physical limits [4].

With the ever-increasing demand of computing in this data-intensive world looking forward to a future with billions of smart interconnected devices known as Internet-of-Things (IoT) along with the advent of a new age of artificial intelligence (AI), researchers are looking into novel solutions for sustaining the computing revolution. Diverse approaches, such as optical [5], quantum [6], biomolecular [7], and neuromorphic [8] have been explored to go beyond the traditional paradigm. In this chapter, we focus on spike-based neuromorphic computing, which is a bio-inpsired approach that tries to emulate excellent energy efficiency and superior information processing capabilities that can be observed in biological organisms[9].

The early work in modeling biological neural networks with electrical circuits can be traced back to pioneering work by McCulloch and Pitts in [10]. Another seminal work was Hebbian learning [11] in 1949, which is a classic inspiration for spiking neural network (SNN) researchers. 1952 saw a big breakthrough with the development of Hodgkin-Huxley neuron models [12], which shed light into the complex dynamics of this biological computational primitive followed by subsequent models [13, 14]. Though not the predominant research direction in the AI community (being overshadowed by symbolic AI and other directions), a few notable AI researchers began exploring neural networks as a possible route towards building an intelligent machine which led to seminal works such as perceptron [15], multilayer networks [16], backpropagation training [17], the Hopfield network [18] and self-organizing maps [19], among many others.

However, the dream of neuromorphic computing is to combine brain-inspired algorithm and hardware together to build a different class of intelligent system. This concept was first proposed by Carver Mead at Caltech in the late 1980s [20]. As a pioneer in the VLSI (very large sale integration) digital computer revolution, Mead realized some of its limitations and started exploring an analog/ mixed-signal design paradigm to emulate biological functions with the electronics of an integrated circuit (IC) [8]. Since its inception, vision has been a vibrant field of study in neuromorphic computing as can be seen in the celebrated early works in building the silicon retina in 1994 [21] which followed the famous silcion neuron from 1991[22]. More recently, the world has seen large-scale neuromorphic processors such as BrainScaleS [23], TrueNorth [24], Neurogrid [25], SpiNNaker [26], and Loihi [27]. The research in this field has primarily two thrusts: (1) learning the working principle behind human perception and cognition, a fundamental scientific question of perennial interest, specially to cognitive neuroscientists, (2) building a new class of computing machines overcoming the limitations of traditional von Neumann digital computers, which is of primary interest to researchers working in the frontier of computing technology. Since building such machine requires understanding and

DOI: 10.1201/9781003328957-15

innovation across the entire design hierarchy, namely, materials, device, circuit, system, architecture, communication, and algorithm along with a deep understanding of neuroscience principles, neurmorphic computing has become a vibrant interdisciplinary research endeavor over the last three decades.

The early work on neuromorphic computing [28] explored deep similarity between conduction in electronics and ion-channel dynamics of biological neural networks based on the physics of electronic components such as transistors under special operating conditions. The term has gradually evolved to describe a set of brain-inspired hardware and algorithms for neural networks with varying degrees of biofidelity. Modern digital computers usually store information using 32 or 64 bits and process it using synchronous deterministic architecture with separate memory and processing units made of solid-state electronic devices. On the other hand, our brains use patterns of neuron spikes to represent and process information using stochastic computational elements made of organic materials with collocated memory and processing units. In the literature, designs lying anywhere between traditional digital computer and brain-like architecture have been termed as neuromorphic computing. Since we cannot do justice to this wide variety of works, we focus on SNNs for vision applications in this chapter for the sake of brevity and coherence.

The rest of this chapter is organized as follows: Section 15.2 gives basic background on biological neural networks neuromorphic computing systems. Section 15.3 discusses several traditional and emerging devices and circuits with significant promise for building scalable neuromorphic systems with enhanced functionality followed by a brief summary of state-of-the-art neuromorphic processors in Section 15.4. Section 15.5 discusses several architecture and algorithms using SNNs, which are the principal abstraction of the nervous system that neuromorphic computing systems employ to emulate brain function. In Section 15.6, we discuss several promising vision applications including ongoing research on dancing pose estimation. Finally, Section 15.7 concludes this chapter with a summary along with directions for future research.

15.2 Background on Neuromorphic Computing and Bio-Inspired Spiking Neural Network

In this section, we will first outline the main characteristics of a neuromorphic system followed by a concise overview of the working principle of biological neural network for inspiration. Then we briefly outline the

defining aspects of the spiking neuromorphic computing system to distinguish it from traditional digital computer as well as the artificial neural network (ANN) using neurons with a continuous activation function without temporal dynamics.

15.2.1 Characteristics of Neuromorphic Computing System

There are several brain-inspired characteristics that distinguish a neuromorphic computer from a traditional computer. First, it should have a brain-like, massively parallel operation unlike the traditional von Neumann single core CPU, which was developed using a 'stored program' model for strictly sequential computation. Second, it will have collocated memory and processing, aka CIM (computation in memory) analogous to synapses and neurons in a biological brain overcoming the 'memory wall' resulting from the von Neumann architecture with separate units for storage and computation. For example, synaptic weight serves both as a memory and a computational element where presynaptic spikes produce currents that lead to an increase in the post-synaptic membrane potential. Third, asynchronous and/or analog/mixed-signal computation as opposed to globally synchronized digital computation in traditional platforms. In brain, synapses can store and perform analog operation on the input data; neurons produce binary spikes but they have internal temporal states which are analog along with information encoded in time interval between spikes which is also analog; no global clock. Any one of the four possible combinations between synchronous/asynchronous and analog/digital is possible with its unique trade-offs and different combinations have been used for different neuromorphic processors. Fourth, sparsity in connection and activation, which is essential for energy-efficient computation and large-scale network like brain. Additionally, stochasticity and nonlinear dynamics play a key role in the operation of biological brain, and the learning mechanism is almost certainly not global and mostly unsupervised unlike supervised global algorithm such as backpropagation widely used in ANN. This is an active field of research and a thorough understanding still eludes the research community. Of course, there are other features but achieving these above-mentioned properties through algorithm and hardware innovation remains the primary thrust.

15.2.2 Biological Underpinning

The generation and transmission of action potentials in biological neural networks is a complex dynamical process. Here, we present a significantly simplified account to convey the big picture. Neurons are tiny cells that respond to spiking electrochemical stimuli by generating their own spiking outputs. Biological neural networks are networks of neuron. Figure 15.1(a) shows two neurons connected to each other. The pink neuron (to the left) produces a spiking response that flows into the purple neuron (to the right) via synaptic cleft which is commonly known as the synapse [29]. In this setup, the pink one is the pre-synaptic neuron and the purple one is the post-synaptic neuron. Synapses are responsible for regulating the impact an incoming signal will have on the post-synaptic neuron. This regulation is often modeled as a weight that either allows spiking signals to pass through virtually unobstructed or provides some resistance that essentially dulls the signal strength. Much of our ability to learn has been attributed to this synaptic regulation, as well as the creation or removal of synaptic connections.

A prototypical spiking neuron has four main parts: dendrites, soma, axon, and synapse. Dendrites together with the soma act as a leaky integrator of ionic currents that flow into or out of the neuron. These currents result from the opening and closing of channels in the cell membrane and through ion diffusion or active ion pumps. For example, when a presynaptic neuron creates an action potential, neurotransmitters are released and cause the opening of ligand-gated ion channels at the post-synaptic neuron's dendrites. This causes ions to flow resulting in an increase (in the case of an excitatory pre-synaptic neuron) or a decrease (in the case of an inhibitory presynaptic neuron) in the potential of the post-synaptic neuron's cell membrane. Interestingly, a number of complex computations such as boolean logic functions can take place in the dendritic arbor before information reaches the cell body. If the membrane potential rises (also called depolarization) from its resting value of around -70 mV to a threshold value (typically around -55 mV), then an action potential about 1 ms wide and +40 mV at its peak as shown in Figure 15.1(b) is generated, which travels down the neuron's axon, ending at the synaptic terminals where neurotransmitter is released to the next set of neurons. The length of an axon makes it analogous to a transmission line that will provide some delay and attenuation of signals as they propagate from one neuron to another. After a neuron spikes, there is typically a refractory period of several milliseconds during which it cannot produce another spike. During this time, the neuron's membrane is hyperpolarized below its resting potential.

Neurons communicate with each other using two types of synapses, namely, electrical and chemical synapses. Chemical synapses play a major role in both communication and learning in the central nervous system (Figure 15.1(c)). Generation of action potential in a pre-synaptic neuron causes an influx of calcium ions followed by docking of synaptic vesicles at the cell membrane and releases neurotransmitters into the synaptic cleft. We would like

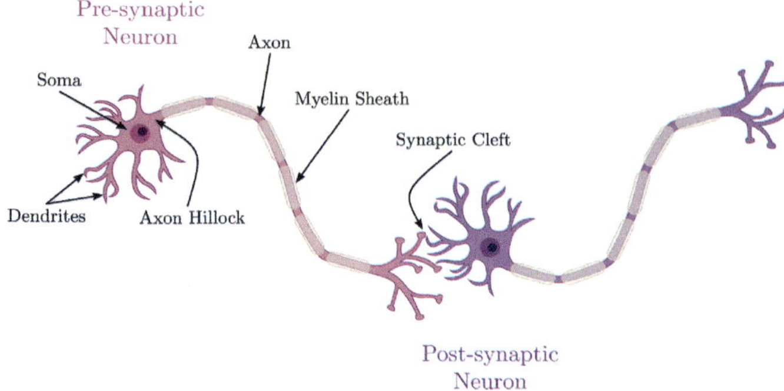

(a)

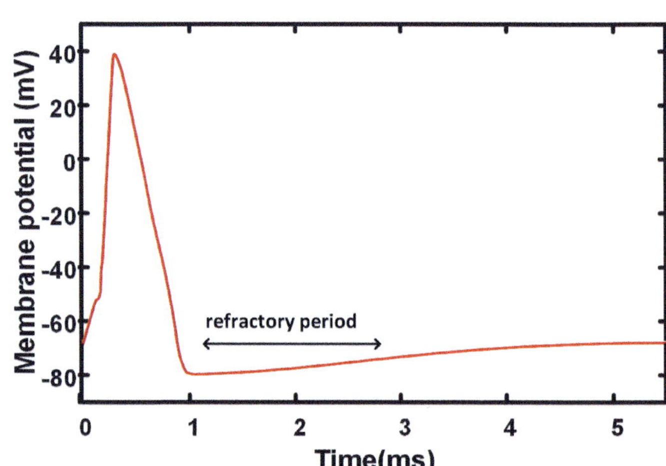

(b)

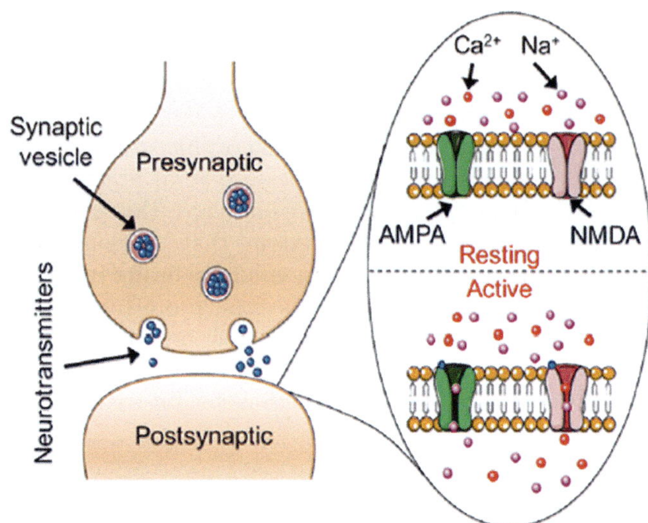

(c)

FIGURE 15.1
(a) Simplified neural communication schematic [29]. (Reproduced from Rose et al., Neuromorphic Computing and Engineering 1.2 (2021). Copyright CC BY 4.0.); (b) Postsynaptic spike generation, (c) Chemical synapse [30]. (Reproduced with permission from Najem et al., ACS nano 12.5 (2018), Copyright 2018 American Chemical Society.)

to mention two neurotransmitters, namely, glutamate and γ-aminobutyric acid (GABA), which exert excitatory and inhibitory effect on the post-synaptic membrane potential, respectively. The released neurotransmitters are received at the post-synaptic plasma membrane by AMPA and NMDA receptors/ions channels, events that trigger ion flux into the post-synaptic neuron that depolarizes the cell [31]. Importantly, these channels remain closed, holding the membrane in an insulating state, until NTs bind to their receptors, at which point the conductance of the post-synaptic membrane increases exponentially. Other additional factors influencing neural dynamics included relative timing of pre- and post-synaptic neuron activity, which can result in long-term strengthening or, weakening of synaptic connections known as LTP (long-term potentiation) and LTD (long-term depression), respectively. One such example is famous Hebbian learning rule popularly phrased as "neurons that fire together, wire together" [11]. Finally, electrical synapse also joins neurons via gap junction ion channels enabling fast, threshold-independent, and bidirectional ion transport and this synaptic strength can be modulated significantly via activity-dependent plasticity with a significant impact on the synchronization in mammalian neural networks [32] [33].

15.2.3   Spiking Neural Network

The seminal paper from Maass [34] divided neural netowrks into three generations based on the neuronal dynamics. The first generation is named McCulloch–Pitt perceptrons, based on the simple McCulloch–Pitt thresholding neuron operation [10]. The second generation neuron has continuous nonlinear activation functions such as tanh, ReLU etc. which enables it to evaluate a continuous set of output values, enables gradient descent based backpropagation due to its differentiable nature, and has propelled the current boom in Artificial Neural Network (ANN) [17]. The third generation of networks use spiking neurons primarily of the 'integrate-and-fire' (IF) type [35], which communicate using spikes and have the promise the create the next paradigm shift in AI.

Biological intelligence emerges from the computation carried out by networks of neurons communicating through spikes. State-of-the-art ANNs significantly abstract the behavior of biological neuronal networks, with several simplifications such as reducing a neuron's behavior to a simple spike rate. However, the performance of ANNs in the field of computer vision tasks is less efficient than their biological counterparts in terms of power and speed [36]. Biological brain processes information in a massively parallel asynchronous manner, whereas deep neural networks (DNN), even in parallel multi-core computing platforms, compute in a essentially sequential form since each layer's computation has to be completed before starting the computation in the next layer and can only parallelize operations in the same layer. The situation is much worse in traditional single core system. As a result, ANN can suffer from significant delay compared to SNN [37]. In 1996, Thorpe et al.[38] showed that the biological brain is able to recognize visual images with one spike propagating through all layers of the visual cortex, and Rolls and Tovee [39] measured the same visual processing speed in the macaque monkey. Theses works highlight the amazing efficiency of the spike-based information encoding technique in brains.

The efficiency of spike-based computation motivated machine learning (ML) and neuroscience researchers to begin exploring SNNs, the third generation of neural networks. Computation in SNNs is event-driven as in the biological brain, so each neuron in the network generates its outputs only when enough spikes indicating the existence of a specific feature or pattern have been detected [37]. This feature gives SNNs the capability to solve complex spatiotemporal tasks and to make use of efficient event-driven sensors, such as event-based cameras.

Table 15.1 shows a comparison of the main aspects between some of the key properties of ANNs and SNNs for vision [40]. As stated in the previous section, synchronous computation in each layer of DNNs can be time consuming. On the other hand, in SNNs, the computation is processed asynchronously in spike form, allowing information to propagate to the next layer before all computation in the current layer is complete. However, this asynchrony combined with the nondifferentiable nature of spikes

TABLE 15.1

Comparison between ANN and SNN

Feature	ANN	SNN
Data	Static Frame-based	Dynamic Event-based
Neuronal activation	continuous valued e.g. ReLU, tanh etc.	discrete-valued spike
Differentiable	Yes	No
Short-term memory	Network	Synapse, neuron and network
Computational Complexity	moderate	high (greatly benefited from compute-in-physics)

complicates the credit assignment problem and limits the use of many popular training algorithms employed in DNNs. On the other hand, the inherent temporal dynamics of SNNs allows them to perform more complex tasks than DNNs [34]. For example, SNNs have neuron-level temporal memory, enabled by the leaky integration of information at the neuron's input. This means that even purely feed-forward SNNs have an inherent short-term memory. Contrast this with DNNs, which can have short-term memory enabled by their network topology (e.g., with recurrent connections), but there is usually no built-in temporal memory at the level of individual neurons. This extra layer of short-term memory in SNNs makes them a good fit for temporal processing of data such as audio and video.

A big challenge in the area of neuromorphic computing is determining how much detail of the physiological processes needs to be modeled to faithfully capture the underlying computational principles. Popular ANN neuron models such as such as tanh and ReLUs lose temporal information and only convey information about the spike rate. Due to their computationally efficiency and differentiable nature, they are employed in most modern DNNs. The spiking neuron models used in neuromorphic computing need to have enough complexity to capture key dynamic properties of biological neurons without having significant computational or hardware overhead. In general, adding more biological features leads to exponential growth in the computational cost. Complex neuron models such as the multi-compartment Hodgkin–Huxley model [12], and the FitzHugh–Nagumo model [13], [14] capture several complex dynamics behavior of spiking neurons. Unfortunately, their exceeding computational cost renders them mostly suitable for research in neuroscience. On the other hand, very simple threshold models such as the McCulloch–Pitts [10] capture only simple neuron behavior, such as spiking above or below a particular rate. It should be noted that calculating implementation cost in the traditional digital computer is not the only way forward. As we will see in Section 15.3.2, to bypass the digital computational cost incurred by complex neuron models, researchers are exploring analog nanoscale low-power scalable emerging neuromophric devices such as memristors, which can natively perform some of these complex neuronal dynamics via intrinsic device physics (known as compute-in-physics), which can reduce these costs by orders of magnitude.

IF or leaky IF (LIF) is one of the most popular spiking neuron models used in neuromorphic computing since it exhibits adequate complexity to capture important temporal information of spike statistics, but abstracted enough to be computationally efficient and suitable for simple hardware implementation. Figure 15.2 illustrates a LIF-based SNN comprising a post-neuron driven by

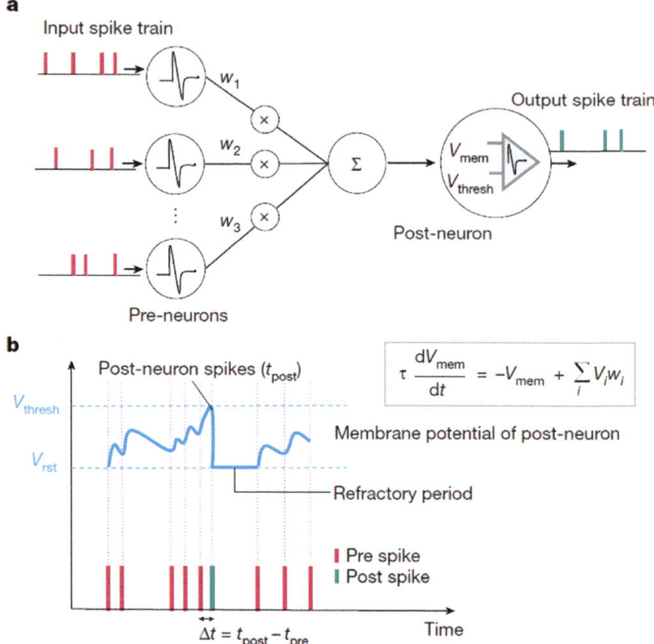

FIGURE 15.2
SNN with LIF (Leaky integrate and fire) neuron, (a) construction and (b) dynamics. [41]. (Reproduced with permission from Roy et al., Nature 575.7784 (2019). Copyright 2019 Springer Nature.)

input pre-neurons. The pre-neuronal spikes, V_i, are modulated by synaptic weights w_i to produce a resultant current, $\sum_i V_i \times w_i$ (equivalent to a dot-product operation) at a given time. The resulting current affects the membrane potential of the post-neuron. Additionally, the dynamics of LIF spiking neurons are shown. The membrane potential, V_{mem}, integrates incoming spikes and leaks with time constant, τ in the absence of spikes. The post-neuron generates an outgoing spike whenever V_{mem} crosses a threshold, V_{thresh}. A refractory period ensues after spike generation, during which V_{mem} of the post-neuron is not affected. However, it is not able to capture some of the more complex biological features of real spiking neurons such as phasic spiking, tonic and phasic bursting, spike frequency adaptation, and accommodation, to name a few. More detailed neuron models such as the Hodgkin–Huxley model [12], which requires numerical solution of four nonlinear differential equations, have significant computational overhead, making it difficult to employ them in low-power neuromorphic systems. Moreover, it is still unclear which biological features are necessary for designing neuromorphic systems with a particular set of desired behaviors. A number of other models have been proposed in the literature that have medium complexity and are a good tradeoff between feature richness and computational cost. These include models such as the adaptive IF model [42], the spike response model [43], and the Izhikivich model [35], which are able to capture more complex dynamics than the LIF model, such as bursting and chattering. Still, other models capture additional behaviors that are important for neural computation, such as stochasticity of spiking and synaptic transmission [44] or energy dependence of neural activity [45]. An excellent review of these features and the associated neuron models can be found in [46].

15.3   Neuromorphic Devices and Circuits

As shown in Table 15.1, SNN model complexity is significantly more than ANN due to dynamic behavior of building blocks such as synapses and neurons. Solving the constituent equations using digital electronics can lead to prohibitive computational cost nullifying the energy and information processing advantages of this emerging paradigm. Hence, analog circuit implementations of synapses and neurons capable of natively performing these computations can be particularly beneficial. Here, we first introduce traditional MOSFET based building blocks and then use memristor as a representative example of an emerging beyond-CMOS device with great promise for building next-generation neuromorphic computing systems.

15.3.1   Synapses and Neurons using MOSFET

1. **Synapses:** Several synaptic circuit using traditional CMOS circuits have been reported in the literature. Indiveri et al. reported synapses in 800 nm process technology with both short and long-term plasticity [47]. Another analog CMOS synapse was reported in [48] with on-chip STDP (spike time dependent plasticity) learning. Similarly there have been several other works such as [49, 50], etc. In recent years, more attention has been focused towards NVM (Non-Volatile Memory) cross-bar architecture, which has great promise for efficient synaptic implementation for in-memory computing.

2. **Neurons:** Researchers have explored several traditional CMOS-based synapse and neuron circuits over the years since the early days of Mead and others [8, 28]. In 2003, Indiveri introduced an IF neuron with spike frequency adaptation and a configurable refractory period with lower power consumption compared to existing axon hillock designs [51]. Later, Indiveri et al. proposed a current-mode conductance-based IF silicon with plastic-

ity for learning [52]. Another IF neuron operating in two asynchronous phases, integration phase followed by firing was reported in [53]. Since the inception of this field, subthreshold conduction in MOSFET has attracted a lot of attention since the low power exponential characteristics have certain similarity with ion-channel dynamics in biological neurons, and the reduced speed is a reasonable trade-off considering the desired biomimetic time scale. Ref. [54] is a valuable resource for the interested reader willing to delve deeper into many designs resulting from this decades-long exploration. Besides, there have been several digital implementations reported in the literature trading cost and complex native dynamics for robust scalable operation [55–57]. A mixed mode neuron has also been reported [58] with on-chip tunability of accumulation rate enabling flexible interfacing with different types of devices.

15.3.2   Synapses and Neurons Using Memristors

During the last 15 years, quite a few emerging devices have been explored as potential candidates for neuromorphic computing. Emerging memory devices capable of non-volatile storage of analog values along with extreme density advantages are highly desirable to be used as programmable weights or synapses. Our brain essentially operates as in-memory computing paradigm which enables it to be highly parallel circumventing the von Neumann bottleneck inherent in conventional digital computers. Building large scale neuromorphic systems using traditional CMOS is inefficient since it takes roughly ten transistors for a synapse and even more for building a neuron with rudimentary functionality. Given our brain has 10^{11} neurons and 10^{14-15} synapses, building a brain-like consuming machine with CMOS transistors will be equivalent to making millions of state-of-the-art chips for emulating a single brain. Hence, scientists have been exploring emerging devices with tiny form factors and energy consumption that has intrinsic properties suitable for emulating synaptic and neuronal functionality for building large scale brain inspired machined on chip. Comprehensive reviews on emerging devices for such applications can be found elsewere [59–62]. Here, as an illustrative example, we discuss memristors, which have garnered a lot of attention in this domain during last 15 years since the first experimental demonstration in 2008 [63].

Most efforts in bio-inspired computing thus far have focused on mimicking primitive lower-order biological complexities. Historically, such complexities are emulated using transistor-based circuits (such as central and graphics processing units, CPUs and GPUs) to simulate multiple dynamical equations [65, 66] but recent advances

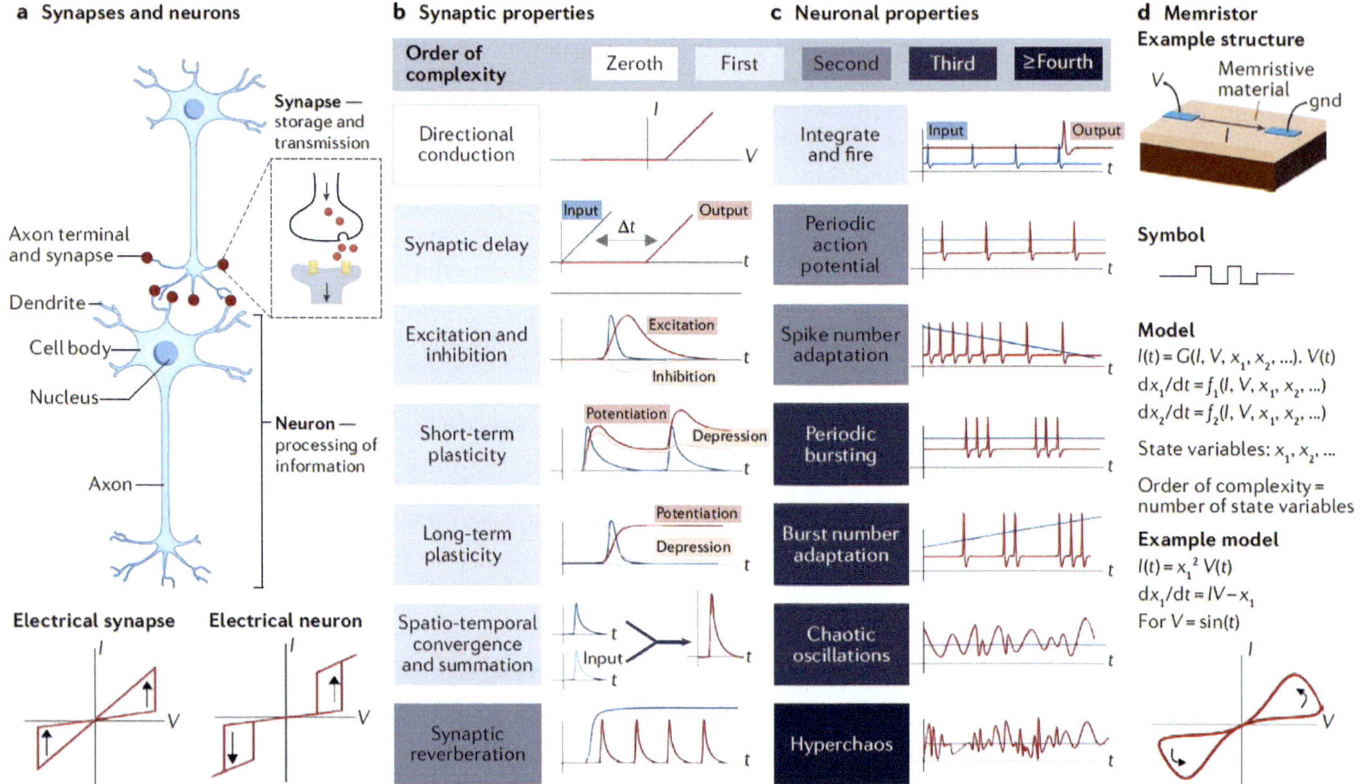

FIGURE 15.3

Concept of memristor and memristive behaviours of various complexity. (a) Illustrations of a biological neuron and synapses, along with the current–voltage characteristics of synaptic and neuronal electrical devices. (b,c) Examples of synaptic (panel b) and neuronal (panel c) behaviours that require different orders of complexity. (d) — Illustration of a memristor and its basic model, depicting with an example how state variables connect currents and voltages with a temporal history dependence. Δt, time delay; G, conductance; gnd, electrical ground; I, current; t, time; V, voltage [64]. (Reproduced with permission from Kumar et al., Nature Reviews Materials 7.7 (2022). Copyright 2022 Springer Nature.)

in memristors have made this approach easier. Memristors, predicted in 1971 [67] and connected to physical devices in 2008 [63], are electrical circuit elements that embody at least one state equation (differential equation of the state variable with respect to time) and, thus, at least first-order complexity (Figure 15.3(d)). The incorporation of state equations necessarily leads to history-dependent behaviours in the current–voltage plot, in either volatile (memory disappears at zero bias) or non-volatile (memory is retained at zero bias) form. Processes such as temperature-driven Mott transitions [68] and field-driven defect generation and recombination [69] lead to volatile or sometimes partially volatile memory effects [70]. By contrast, processes such as electrochemical defect migration [71], spin injection [72], [73], ferroelectric or ferromagnetic switching [74], and crystalline–amorphous phase transitions [75] lead to non-volatile memory effects and are all captured within the memristor framework.

Here, we provide several examples of synaptic and neuronal memristors of different orders of complexity along with their working mechanisms. Several recent reviews have comprehensively covered memristive switching materials, mechanisms, and device-level performance [62, 76, 77], including special material classes, such as 2D materials [78, 79], Mott insulators [80], organic materials [81], and carbon nanomaterials [82]. Here, instead, we focus on higher-complexity memristive materials and devices and discuss how complex computing can be achieved by taking advantage of the intrinsic device dynamics. A few representative examples that illustrate complexity beyond the simple (and often static) functions that can be engineered in memristive electronic devices are summarized in Figure 15.3.

1. **Synaptic Devices:** The non-volatile resistance switching in the first experimentally demonstrated memristor [63] was caused by the movement of

oxygen vacancies in TiO2 under the influence of an electric field. It was a first-order memristor in the sense that the dynamics of the underlying state variable (width of the TiO2 region occupied by the oxygen defects) can be modeled using a first-order differential equation. Most memristors built so far are first-order, that is, one dominant dynamical process that enables memory. Later, memristors based on TaOx ([83]) were designed to be sensitive to thermal effects. These are second-order devices with two state variables, namely, the radius of the filament of oxygen vacancies and temperature. The device exhibited dynamic volatile memory along with the static non-volatile resistance switching property. Second-order effects were also seen in HfO2 memristors [84]. Another second-order memristor dynamics was found in ferroelectric memristors [85] with built-in electric fields and the dynamical response of the interfacial defects being the two state variables. These devices also exhibit temporal memory along with the resistance switching behaviour. In addition to solid state devices, biomolecular volatile memristors with SRDP (spike rate-dependent plasticity) have also been reported [86]. These devices are even more biologically faithful in their composition, ion-channel based conduction, and native time scale of operation and have been used for solving benchmark problems with promising accuracy [87]. The functionality is governed by two independent first-order dynamics, namely, pore generation, and electrowetting. In addition to emulating chemical synapse, biomolecular memristors have also been used to build artificial electrical synapse that exhibits voltage-dependent, dynamic changes in its conductance [88].

2. **Neuronal devices:** First-order neuronal memristors essentially exhibit volatile switching in the current–voltage plane and some simple dynamics (such as a characteristic response time). Volatile switching can be caused by Mott transition, thermal runaway, tunnelling, etc. For example, modelling of threshold switching in TiO2 as a first-order process with internal temperature as the state variable [89] showed that, as temperature increases due to Joule heating, the superlinear temperature dependence of the conductance makes the conductance increase, which, above a threshold, is a runaway process, leading to volatile (reversible) switching. Volatile memristors placed in a relaxation oscillator circuit can exhibit self-sustained oscillations. For example, volatile memristors (exhibiting current-controlled negative differential resistance) with a parallel capacitor can

exhibit oscillations via two alternating dynamical processes: charging–discharging of the capacitor and volatile switching of the memristor, thus exhibiting second-order complexity. The electrode structure of a NbO2 volatile-switching memristor was shown to form a built-in capacitor, which was sufficient to create oscillations without the need for any external capacitor [90]. The device was modelled with a Mott-transition-driven volatile filament (conduction channel) formation process, although later models were based on more realistic and general thermal runaway processes [91].

The only reported third-order memristor [91] was constructed using NbO2 and modelled with three state variables: temperature (representing internal thermal dynamics), charge on the built-in capacitor (representing charge dynamics), and the speed of formation of a metallic region (a volatile filament resulting from the Mott transition dynamics). The devices were carefully designed in structure and material stoichiometry to enable all the above dynamics. When powered by a tunable static voltage input, a single device could produce 15 different neuronal dynamics (including spiking, bursting and chaos). Although third- order complexity can produce many key neuronal behaviours, their usefulness in designing computational system is still a open question that needs rigorous examination. In addition to single neuron, dynamic volatile memristors, and memcapacitors with higher-order behavior have recently found application as a reservoir in the reservoir computing (RC) system, replacing a complex recurrent neural network (RNN) for solving several benchmark temporal, dynamic, and chaotic problems with low cost energy efficient analog hardware implementation [92–95].

15.4 State-of-the-Art Neuromorphic Processors

In this section, we give a brief summary of state-of-the-art neuromorphic processors. Several research groups from academia and industry have reported very promising implementations of neuromorphic processors. For example, a mixed-signal, multi-core neuroprocessor called dynamic neuromorphic asynchronous processor (DYNAP) was reported, which combines the efficiency of analog computational circuits with the robustness of asynchronous digital logic for communications [96] and was implemented in 180 nm CMOS process. Thakur et al. introduced an improved version called DYNAP with scalable and learning devices (Dynap-SEL) containing additional features implemented in a 28 nm FDSOI (Fully Depleted Silicon-On-Insulator) process [97]. There are four cores with each containing 16×16 analog neurons where each neuron has 64 programmable (4-bit) synapses. There is an additional fifth core containing 1×64 analog neurons, 64×128 plastic synapses (on-line learning capability), and 64×64 programmable synapses.

Two other famous neuroprocessors named "SpiNNaker" [26] and "BrainScaleS" [23] came out of the Human Brain Project (HBP) in Europe [98]. The SpiNNaker, developed by researchers at the University of Manchester, contains more than one million parallel ARM processors, which are used to model one billion spiking neurons with biologically realistic synaptic connections in real time [26]. On the other hand, BrainScaleS, a mixed-signal neuromorphic system at wafer-scale with upwards of 40 million synapses and 180 thousand neurons, has been developed from a research collaboration between the University of Heidelberg and the Technische Universität Dresden [23].

TrueNorth, a famous neuroprocessor from IBM [24], consists of 4096 neurosynaptic cores with 1 million digital neurons and 256 million synapses tightly interconnected by an event-driven routing infrastructure consuming 65 mW power. They also introduced a novel hybrid asynchronous–synchronous model along with new CAD tools for the design and verification. Another prominent neuroprocessor is Loihi from Intel [27], which contains 128 neuromorphic cores, three ×86 processor cores, and four communication interfaces that extend the mesh in four directions to other chips. Each neuromorphic core has 1024 primitive spiking neural units grouped into sets of neuronal trees. The mesh protocol can support up to 16,384 chips and 4096 on-chip cores using hierarchical addressing. Loihi introduced several novel features, such as hierarchical connectivity, dendritic compartments, synaptic delays, and programmable synaptic learning rules.

A family of dynamically adaptive neural processors has been developed by the TENNLab neuromorphic research group at the University of Tennessee. The first one is called DANNA (dynamic adaptive neural network array) [99], which was initially designed for FPGA and later adapted for 130 nm CMOS ASIC (application specific integrated circuit) implementation. An improved version called DANNA2 was introduced in 2018 with improved network density, achievable clock speeds, and training convergence rate [100]. In parallel, a mixed-signal extension known as memristive dynamic adaptive neural network array (mrDANNA) was later developed that utilized memristor devices in the synapses to improve the efficiency of the neuromorphic system [101] with an online learning methodology called synchronous digital long-term plasticity (DLTP). Currently, TENNLab is working on developing a convergent and flexible architecture as

part of a reconfigurable and very efficient neuromorphic system or RAVENS [102].

The description above is not exhaustive by any means and meant as in introduction to the exciting world of hardware implementation of a new computing paradigm. Table 15.2 lists several neuromorphic processor implementations from the literature along with key aspects and references for further enquiry.

15.5 Algorithm and Architecture

In this chapter, we focus on neuromorphic systems that implement SNNs. A key ongoing challenge in the field of SNNs is how to do training or learning effectively. Here, we overview some of the common approaches used in training SNNs for vision applications. It is worth noting, many of these algorithms have been evaluated on simple image classification tasks such as MNIST and CIFAR-10, with a few extending on to ImageNet-style classifications.

15.5.1 Mapping Traditional ANNs to SNNs

One of the common approaches used in the field is to train a traditional ANN and then create a mapping from that ANN to an SNN for neuromorphic hardware deployment [104–111] (Figure 15.4). In creating a mapping from a traditional ANN to an SNN on hardware, several issues are encountered. First, some accommodation has to be made to move from a nonlinear activation function like a rectified linear unit (ReLU) or sigmoid to a spiking neuron model. Second, due to hardware limitations, there may be restricted precision on the synaptic weight values, requiring quantization of the weight values. Many deep learning software packages now allow for quantization as part of the training process, but the quantization level required for the hardware must be known a priori to leverage this (i.e., it must be known which hardware system will be targeted). Finally, mapping onto neuromorphic systems with emerging architectures may also require dealing with cycle-to-cycle and device-to-device variation, which can add noise to the operation of the network. Any of these issues may lead to a performance degradation from the original ANN to the SNN on neuromorphic hardware.

15.5.2 Spike-Based Quasi-Backpropagation

Another approach that is taken in training SNNs for neuromorphic hardware is to directly adapt the training procedure to produce an SNN rather than an ANN. In this case, the typical training procedures of backpropagation

TABLE 15.2
State-of-the-art Neuromorphic Processors

Name	Operation	Power/Energy	timescale	on-chip learning
Dynap-SEL [97]	Mixed	260 pJ/spike	ns	STDP
FPAA [103]	Analog	$< 1 \mu$ W	ms to s	STDP
BrainScaleS[23]	Digital	10 pJ/transmit	ns	Configurable plasticity
TrueNorth[24]	Digital	60 mW	ns	none
SpiNNaker[26]	Digital	100 nJ/neuron + 43 nJ/synapse	ns	Configurable
mrDANNA[101]	Mixed	22.31 pJ/neuron/spike + 0.48 pJ/synapse/spike	ns to μ s	DLTP
Loihi[27]	Digital	81 pJ/neuron + 120 pJ/synapse	ns	Configurable STDP

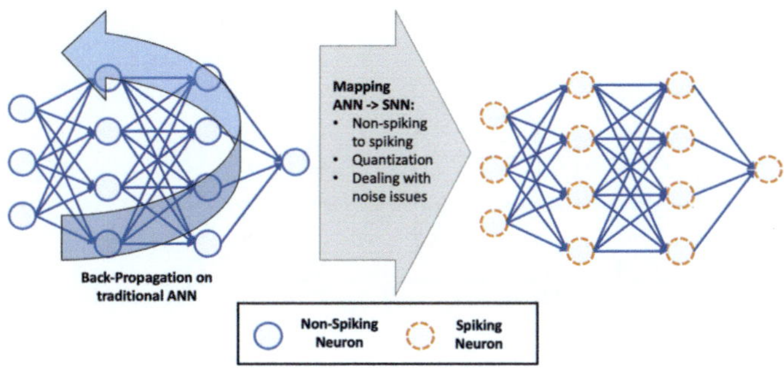

FIGURE 15.4
Mapping procedures take a pre-trained ANN and map it into an SNN suitable for neuromorphic hardware.

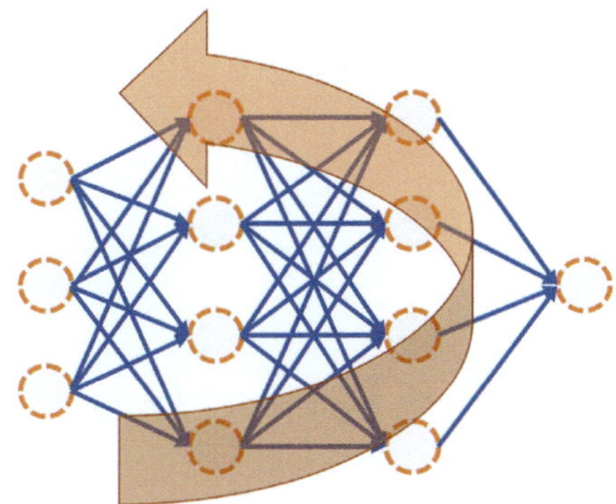

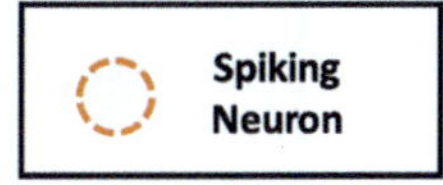

FIGURE 15.5
Spike-based quasi-backpropagation approaches adapt backpropagation to work directly on SNNs.

or stochastic gradient descent are modified so that they will produce an SNN suitable for hardware deployment (Figure 15.5). SpikeProp pioneered a bespoke gradient descent approach for spiking neural networks, specifically focused on first-spike times as a proxy for the neuron's output value and applying a process similar to traditional gradient descent [112].

In [113], an approach for training binary-activated networks suitable for hardware deployment is presented. This approach, called Whetstone, begins with differentiable activation functions such as sigmoid or ReLU and gradually "sharpens" those functions over the course of training to behave like binary activation functions that are suitable for neuromorphic deployment.

Several approaches have adapted back-propagation by using a surrogate gradient. Wu et al. introduce an approach for training both the spatial and temporal aspects of SNNs [114]. They adapt for traditional back-propagation by maintaining an approximated derivative of spike activity throughout. Neftci et al. provide an overview of different approaches for using surrogate gradients to perform backpropagation-based training in recurrent SNNs [115]. Lee et al. treat the membrane potential of the spiking neurons as differentiable signals, which allows the application of backpropagation [116]. Others have adapted backpropagation through time (BPTT) to be amenable to recurrent SNN architectures [117].

Still yet, others have used differentiation on the spike representation to overcome the non-differentiability issue [118]. An overview of other deep learning-like training approaches for SNNs is provided in [119].

15.5.3 Plasticity-Based Training

Spike-timing-dependent plasticity (STDP) is a synaptic-weight plasticity mechanism in which the weight of the synapse is adjusted based on the relative spike times of the pre- and post-synaptic neuron (Figure 15.6). If the post-neuron spikes after the pre-neuron, it leads to potentiation or, increase in the synaptic weight. Conversely, if the post-neruron fires before the pre-neuron, it causes depression or, decrease in the synaptic weight. The weight change usually decays exponentially with respect to time interval between two spikes as shown in Figure 15.6 and A and τ are learning rates and time constants governing the weight change, Δw. Here, we show symmetric behavior for simplicity but in general, the parameters can be different for potentiation and depression. It has been

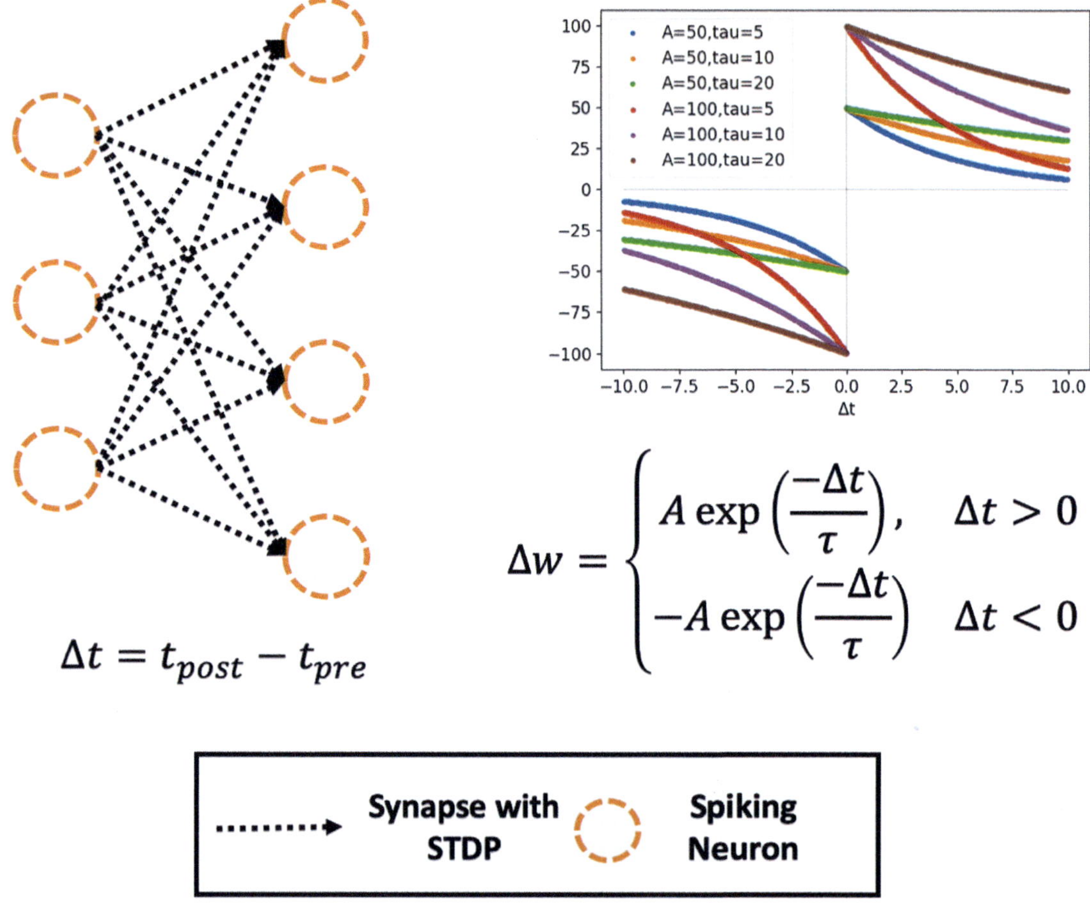

$$\Delta w = \begin{cases} A\exp\left(\dfrac{-\Delta t}{\tau}\right), & \Delta t > 0 \\[2ex] -A\exp\left(\dfrac{-\Delta t}{\tau}\right) & \Delta t < 0 \end{cases}$$

$$\Delta t = t_{post} - t_{pre}$$

FIGURE 15.6

One common example of synaptic plasticity mechanisms is STDP. In this figure, a simple STDP equation is given, along with plots showing how the weight changes are affected by the parameter values.

observed in biological neural systems [120], and it has been implemented in a vast array of neuromorphic hardware implementations [121]. STDP has been demonstrated as a standalone unsupervised learning rule for SNNs on simple vision tasks, such as MNIST classification [122], but STDP alone has been shown to have difficulties scaling to deeper architectures and more complicated tasks. STDP has also been used in combination with traditional learning methods, for example, using STDP for training convolutional filters and developing a supervised STDP-based learning rule that is meant to approximate gradient descent [123]. Others have used a layer-by-layer STDP training approach to leverage STDP in deeper networks for more complicated tasks [124].

15.5.4 Reservoir Computing (RC)

In RC, a sparse, highly recurrent neural network is defined as the "reservoir", and data is passed through the reservoir and then processed by a trained, typically linear "readout" layer that interprets the output of the reservoir (Figure 15.7). Reservoirs are typically well-suited to temporal data tasks. In the context of neuromorphic computing, reservoirs were typically implemented using a recurrent spiking neural network; in this case, the RC approach is referred to as liquid state machines. RC approaches have also been applied to static computer vision tasks as well as event-based camera data and videos. A key issue with RC is defining the appropriate structure of the reservoir. Iranmehr, et al., have demonstrated a spiking reservoir approach on N-MNIST, an event-based version of the MNIST dataset, where they define the reservoir based on ionic density in an ionic environment [125]. Others have used evolutionary approaches to define the reservoir structure [126, 127]. Others have proposed leveraging ensembles of reservoirs to achieve higher accuracy on image classification tasks [128].

The role of the reservoir in RC is to nonlinearly transform sequential inputs into a high-dimensional space such

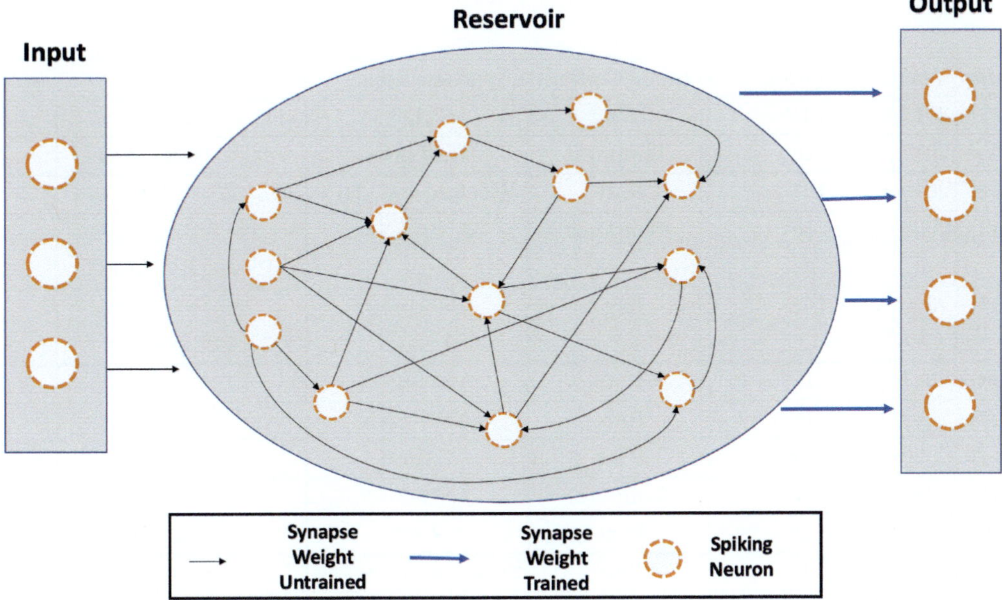

FIGURE 15.7
Reservoir computing or liquid state machines is commonly used as an algorithm for training SNNs. In this case, only the weights between the reservoir (or liquid) and the outputs are trained.

that the features of the inputs can be efficiently read out by a simple learning algorithm. Therefore, instead of RNNs, other nonlinear dynamical systems can also be used as reservoirs. In particular, physical RC using reservoirs based on physical phenomena has recently attracted increasing interest in many research areas. Various physical systems, substrates, and devices have been proposed including electronic [129–132], photonic [133, 134], mechanical [135] [136], biological [137–139], quantum [140] and spintronic [141, 142] RC. A motivation for physical implementation of reservoirs is to realize compact and fast information processing devices with low learning cost.

There are several requirements for a physical reservoir to efficiently solve computational tasks [143]. (1) High dimensionality is necessary to map inputs into a high-dimensional space. This property facilitates the separation of originally inseparable inputs in classification tasks and allows reading out spatiotemporal dependencies of inputs in prediction tasks. The dimensionality is related to the number of independent signals obtained from the reservoir. (2) Nonlinearity is necessary for a reservoir to operate as a nonlinear mapping. This property allows inputs that are not linearly separable to be transformed into those that are linearly separable in classification tasks. It is also useful for effectively extracting nonlinear dependencies of inputs in prediction tasks. (3) Fading memory (or short-term memory) [144] is necessary to ensure that

the reservoir state is dependent on recent-past inputs but independent of distant-past inputs. It is also referred to as the echo state property, indicating that the influence of past inputs on the current reservoir states and outputs asymptotically fades out [145]. Such a property is particularly important for representing sequential data with short-term dependencies. (4) Separation property is required to separate the responses of a reservoir to distinct signals into different classes. On the other hand, a reservoir should be insensitive to unessential small fluctuations, such as noise, so that similar inputs are classified into the same class.

15.5.5 Evolutionary Algorithms

To deal with the highly recurrent and spiking nature of recurrent SNNs, others have turned to evolutionary algorithms to evolve the structure and/or parameters of the networks [146, 147]. Though evolutionary approaches have shown to perform well on a variety of tasks and to outperform deep learning-style SNN training on imitation learning tasks for control [148], there have been limited results on computer vision tasks. We expect that the use of evolutionary algorithms for computer vision tasks for SNNs would be more suited towards neural architecture search [149] and used alongside one of the other training approaches described above.

TABLE 15.3
Static Image Classification with SNN

Problem	**Paper**	Year	Neuron	Input Coding	Algorithm	Architecture	Accuracy (%)
MNIST	[122]	2015	LIF	Rate	STDP	2FC	95
	[150]	2017	IF	Temporal	Backprop	784FC-600FC-10FC	96.98
	[151]	2019	LIF	Rate	Stochastic STDP	36C3-2P-128FC-10FC	98.54
	[152]	2020	IF	Temporal	ANN-SNN	32C5-P2-64C5-P2-1024FC-10FC	99.41
	[153]	2020	LIF	Encoding Layer	Backprop	15C5-P2-40C5-P2-300FC	99.53
CIFAR-10	[108]	2017	IF	Rate	ANN-SNN	4 Conv, 2 FC	90.85
	[107]	2019	IF	Rate	ANN-SNN	VGG16	91.55
	[154]	2020	LIF	Rate	Hybrid	VGG16	92.02
	[155]	2020	IF	Temporal	ANN-SNN	VGG16	93.63
	[156]	2022	LIF	Rate	TET	ResNet-19	94.50
	[157]	2022	LIF	Rate	Surrogate Gradient	ResNet-19	95.60
CIFAR-100	[158]	2020	RMP(soft-reset)	Rate	ANN-SNN	ResNet-20	67.82
	[159]	2021	LIF	Rate	Surrogate Gradient	ResNet-18	74.24
	[156]	2022	LIF	Rate	TET	ResNet-19	74.72
	[157]	2022	LIF	Rate	Surrogate Gradient	ResNet-19	78.76
ImageNet	[107]	2019	IF	Rate	ANN-SNN	VGG16	69.96
	[154]	2020	LIF	Rate	Hybrid	VGG16	65.19
	[155]	2020	IF	Temporal	ANN-SNN	VGG16	73.46
	[160]	2021	IF	Rate	ANN-SNN	ResNet-34	73.45
	[156]	2022	LIF	Rate	TET	SEW-ResNet-34	68
	[157]	2022	LIF	Rate	Surrogate Gradient	SEW-ResNet-34	68.28
	[161]	2023	LIF	Rate	Direct Training	ResNet-104	77.08

15.6 Application

SNNs are promising for analyzing both static and spatio-temporally varying sequential data. In this section, we discuss three promising applications of SNNs, namely static image classification, neuromorphic dataset classification, and dancing pose estimation.

15.6.1 Static Image Classification

Researchers have used SNNs on several benchmark image classification problems as summarized by Rathi et al. [162]. Table 15.3 expands upon the prior work to provide a thorough comparison among the performance of various reported recent SNN models on image classification tasks from frame-based static image datasets such as MNIST [163], CIFAR10 [164], CIFAR100 [164], and ImageNet [165]). Of course, conventional deep ANNs have performed very well in these problems, but as can be seen from this table, innovative SNN implementation has become quite competitive in this domain over the years.

15.6.2 Neuromorphic Dataset (from event-based camera) Classification

Neuromorphic cameras [166], also known as Dynamic Vision Sensors (DVS) or event cameras, have a silicon retina design based on mammalian vision, making them sensitive to moving targets and fluctuating lighting conditions. Each pixel operates asynchronously, independently monitoring logarithmic minute brightness changes, ensuring sensitivity to motion in various lighting conditions. This mechanism inherently filters out static backgrounds, transmitting detailed contents only when an event occurs. Neuromorphic cameras are resilient in situations ranging from nighttime to glaring noon and exhibit reduced sensitivity to skin color and brightness changes [167, 168]. The can offer very high temporal resolution and low

TABLE 15.4
Neuromorphic Dataset Classification with SNN

Problem	Paper	Year	Neuron	Algorithm	Architecture	Accuracy (%)
N-MNIST	[195]	98.88	2018	SRM	Backprop	2FC
	[196]	2018	SRM	Backprop	12C5-2P-64C5-2P-10FC	99.20
	[197]	2019	LIF	Surrogate Gradient	128C3-128C3-AP2-128C3-256C3-AP2-1024FC-Voting	99.53
	[198]	2020	LIF	STBP	128C3-128C3-AP2 -128C3-256C3-AP2-1024FC-10	99.42
	[199]	2021	IF	Surrogate Gradient	5 conv, 2 FC	99.31
	[199]	2021	LIF	Surrogate Gradient	5 conv, 2 FC	99.22
	[200]	2023	IF	Surrogate Gradient	2 conv, 2linear	99.44
DVS-CIFAR10	[197]	2019	LIF	Surrogate Gradient	128C3-128C3-AP2-128C3-256C3-AP2-1024FC-Voting	60.50
	[201]	2020	IF	ANN-SNN	4 Conv, 2 FC	65.61
	[202]	2021	LIF	Surrogate Gradient	5 Conv, 3 FC	63.20
	[199]	2021	IF	Surrogate Gradient	5 Conv, 2 FC	65.59
	[199]	2021	LIF	Surrogate Gradient	5 Conv, 2 FC	63.73
	[159]	2021	**LIF**	Surrogate Gradient	ResNet-18	75.40
	[156]	2022	LIF	TET	VGGSNN	83.17
	[157]	2022	LIF	Surrogate Gradient	VGGSNN	84.90
DVS128 Gesture	[193]	2017	LIF	ANN-SNN	16-layer SNN	94.59
	[196]	2018	SRM	Surrogate Gradient	8-layer-SNN	93.64
	[203]	2021	PLIF	Surrogate Gradient	7B-Net	97.92
	[204]	2021	LIAF	BPTT	Conv-LIAF	97.56
	[205]	2021	LIF	STBP and BPTT	Input-MP4-64C3-128C3-AP2-128C3-AP2-256FC-11	98.61
DVS128 Gait	[194]	2019	NA	BP	6 CNN, 3 FC	89.9
	[206]	2021	NA	BP	6 CNN, 3 FC	94.90
	[205]	2021	LIF	STBP and BPTT	Input-MP4-64C3-128C3-AP2-128C3-AP2-256FC	87.59
	[207]	2021	PLIF	Surrogate Gradient	5 Conv, 3 FC	89.87
	[161]	2023	PLIF	Surrogate Gradient (TCSA)	5 Conv, 3 FC	92.78

latency (order of microseconds), high dynamic range (140 dB versus 60 dB of standard camera), and low power consumption.

Since the first commercial event camera of 2008 [166], there has been a lot of interest in recent years on developing new sensors [169–175] for diverse applications and algorithms for efficient processing the data from these cameras. According to a recent survey [168], applications of event-based cameras may include real-time interaction systems such as robotics and wearable electronics [176], systems requiring low latency and power in uncertain lighting condition [177], object tracking [178, 179], surveillance and monitoring [180], object/gesture recognition lee2014real,orchard2015hfirst, depth esitmation [181, 182], structured light 3D scanning [183], optical flow estimation [184], HDR (high dynamic range) image reconstruction [185, 186], simultaneous Localization and Mapping (SLAM) [187, 188], image deblurring [189], or star tracking [190]. Since, they asynchronously measure per-pixel brightness changes ('events') in contrast to standard cameras measuring absolute brightness at a constant rate, novel methods are required to process their output. Due to its similarity to biological vision and spiking output, we envision that diverse applications related to event-camera would benefit greatly from advances in neuromorphic computing [41, 168].

In Table 15.4, we are showing SNN results from recent works for neuromorphic datasets such as N-MNIST [191], CIFAR10-DVS [192], DVS128 Gesture [193], and DVS128 Gait [194] dataset. N-MNIST is a basic neuromorphic dataset converted from MNIST by a DVS (Dynamic Vision Sensor) camera. Like MNIST, there are 60,000 training samples and 10,000 testing samples in N-MNIST. Each sample in N-MNIST has a pixel size of 34×34, a channel size of two, and a time length of 300 ms. DVS-CIFAR10 is converted from CIFAR10 by a DVS camera, but is more challenging than CIFAR10 due to its larger environmental noise and intra-class variance. DVS-CIFAR10 contains a total of 10,000 samples with 10 labels: "airplane", "automobile", "bird", "ca" "deer", "do" "frog", "horse", "ship", and "truck". DvsGesture contains 11 gestures: "hand-clapping", "right-hand-wave", "left-hand-wave", "right-arm-clockwise", "right-arm-counter-clockwise", "left-arm-clockwise", "leftarm-counter-clockwise", "arm-roll", "air-drums", "air-guitar", and "other gestures". There are 1342 samples from 29 subjects under three types of illumination and the average duration of each gesture is 6 seconds. DVS128 Gesture consists of a series of human hand and arm gestures recorded by a DVS camera. DVS128 Gait dataset has various gaits from 21 volunteers (15 males and 6 females) under two kinds of viewing angles with 4200 recorded samples and the average duration of each gait is 4.4 seconds. With the development of more mature DVSs, researchers are now developing benchmarks for neuromorphic dataset and evaluation metrics. We believe that the advantage of SNN over ANN will be more prominent for such datasets, which seem natively suited for this computational paradigm in contrast to traditional frame-based static datasets.

15.6.3 Dancing Pose Estimation

Technology-mediated dancing (TMD) uses digital systems to enable remote, engaging, and health-promoting dance activities as part of gaming and immersive experiences, increasingly blending digital and physical realities [208–210]. TMD forms range from gaming console games to Virtual Reality (VR) platforms, integrating with users' living spaces. Human pose estimation (HPE) is critical in TMDs, as it identifies users' unique, complex dance poses for computer interaction. TMD requires high-fidelity HPE that functions reliably in diverse, challenging, and realistic indoor environments, including dynamic lighting and background conditions.

Contemporary HPE systems predominantly rely on depth and RGB cameras [211–213], which struggle to generate ultra-fast, high-speed pose inferences due to limited frame rates. This limitation is crucial for applications like VR dance games and high-frequency motion characterization for tremor monitoring. Neuromorphic cameras can achieve a bandwidth of over 10 million events per second with low latency, making them suitable for high-frequency inference. RGB-based HPE struggles in low-light conditions and depth cameras have a limited working depth range, while neither camera inherently distinguishes between static and moving objects, leading to a waste of transmission bandwidth. These issues compromise HPE robustness in dynamic settings, and depth cameras also require significant power consumption.

DVS HPE has gained interest due to its advantages, but current datasets in the field exhibit limitations concerning real-world applicability. They primarily focus on fixed everyday movements and are collected under optimal lighting conditions with static backgrounds, causing models to struggle with intricate movements like those in dance performances. To address this, two novel DVS dancing HPE datasets are introduced: one featuring a real-world dynamic background under various lighting conditions, and another synthetic dataset with variable human models, motion dynamics, clothing styles, and background activities, generated using a comprehensive motion-to-event simulator.

Previous DVS HPE efforts are also constrained by the "missing torso" problem, which occurs when neuromorphic cameras only capture moving components and disregard static body parts. To overcome this, a two-stage system called *YeLan* [214] is proposed, which accurately estimates human poses in low-light conditions with noisy backgrounds. The first stage uses an early-exit-style mask prediction network to eliminate moving background objects, while the second stage employs a BiConvLSTM to facilitate information flow between frames, addressing the missing torso issue. TORE (Time-Ordered Recent Event) volume is also utilized to construct denser input tensors, tackling the low event rate problem in low-light settings. Extensive experiments show that *YeLan* achieves state-of-the-art results on the two proposed new datasets.

YeLan, the first neuromorphic camera-based 3D human pose estimation solution designed for dance moves, functions robustly under challenging conditions such as low lighting and occlusion. It effectively overcomes neuromorphic camera limitations while capitalizing on their strengths. An end-to-end simulator was developed, allowing for precise, low-level control over generated events and resulting in the creation of the first and largest neuromorphic camera dataset for dance HPE, called Yelan-Syn-Dataset. This synthetic dataset surpasses existing resources in both quantity and variability. A human subject study was conducted to collect a real-world dance HPE

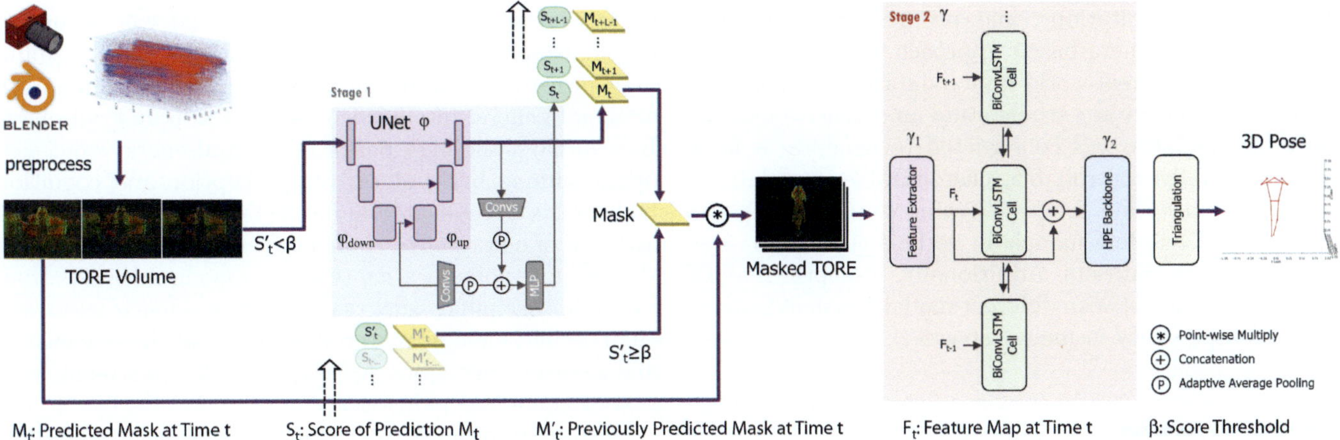

FIGURE 15.8

The pipeline of *YeLan*. It initially processes the event stream into TORE (Time-Ordered Recent Event) volumes, which are subsequently sent to the stage one human body mask prediction network. This network predicts a series of masks for the ensuing frames, accompanied by quality-assessment scores to minimize computation costs. The estimated human mask undergoes point-wise multiplication with the original TORE volume before advancing to the next stage. Stage two encompasses the human pose estimation network, where BiConvLSTM and three hourglass-like refinement blocks are employed to estimate the heatmap of joints' projections on three orthogonal planes. The precise 3D coordinates of these joints are determined through a triangulation method based on these heatmaps.

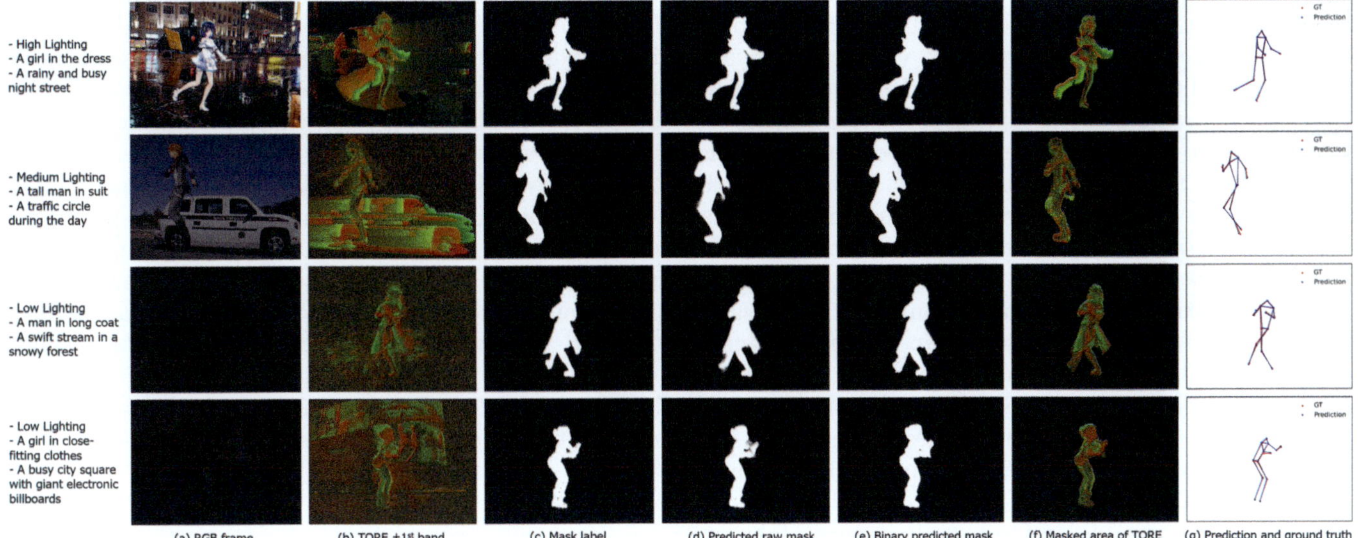

FIGURE 15.9

Sample results from *Yelan-Syn-Dataset* in different lighting conditions with dynamic backgrounds. (a) shows the RGB frames, (b) shows the corresponding event representation TORE, (c-f) shows the different masks generated, (g) shows the ground truth and predicted 3D human pose.

dataset, named Yelan-Real-Dataset, which considers low-light conditions and mobile background content.

YeLan is also compared with RGB-based HPE methods, and the results show that although they behave similarly in high-lighting conditions, *YeLan* performs much better in low-lighting conditions. Also, the comparison results show that *YeLan* is less sensitive to dancers'

depth, while the performance of RGB-Depth camera-based methods is affected more. Lastly, *YeLan* can be trained on one accumulation frame rate, and operates stably in much higher frame rates, enabling its usage in high-frequency reference.

To summarize, the work presented discusses existing 3D HPE techniques used in dance games, evaluating their strengths and limitations, and proposes an innovative neuromorphic camera-based approach to address these shortcomings. A real-world dance dataset was collected through human subject studies, and an extensive motion-to-event simulator was constructed to generate a large amount of fully controllable, customizable, and labeled synthetic dance data for pre-training the model. *YeLan* surpasses all baseline models in various challenging scenarios on both datasets. Additionally, in-depth analysis and comparison between different modalities demonstrate *YeLan*'s superiority in many aspects.

15.7   Conclusion and Future Directions

In this chapter, we have provided an overview of the design stack related to spike-based neuromorphic computing for application in computer vision. The confluence of several factors such as slowing down of traditional computing, the exponential increase in time and energy consumption due to growing complexity ANN models in traditional hardware facing progressively larger dataset and the need for ultra-low power computing for millions of smart edge devices calls for an alternative computing paradigm such as neuromorphic computing to address the challenges of today. With the advent of high performance event cameras and neuromorphic imaging, efficient spatio-temporal image and video processing is turning into a multi-disciplinary vibrant field of research spanning from fabricating new materials and devices to developing new learning algorithms informed by neuroscience.

In recent years, there has been significant progress in SNN algorithms but still scalable supervised and unsupervised learning rules applicable for a wide range of applications remain elusive. Similarly, a host of emerging devices has been reported that exhibit complex dynamic properties to emulate neural activity using intrinsic physical mechanism with tiny form factor and low energy dissipation, yet reliable and robust large scale fabrication with commercially viable yield is still out of reach. More importantly, our knowledge of human perception and cognition along with our mathematical understanding of highly nonlinear large scale complex system such as brain is still at its infancy and as such, nobody is certain which parts of the neural dynamics are beneficial for computation and which parts are historical evolutionary happenstances. Along with computational efficiency, systematic reliability and security framework for ensuring robustness against adversial attack or device variability needs to be developed before deploying these systems for critical applications. In addition, the work in this field so far has focused on neuronal dynamics, although there are dendritic dynamics along with neuroglial cells such as astrocytes, which may play a crucial role in ensuring continual adaptive and flexible lifelong learning and may prove to be indispensable for achieving humanlike artificial general intelligence (AGI), the holy grail of AI. In summary, the high performance and energy efficiency of the human brain at certain perception and cognition tasks waits to replicated in intelligent machines but there are a lot of outstanding challenges and scientific questions that require collaborative research among scientists from multiple disciplines such as computer science, electrical engineering, biology, material science, physics, chemistry, and neuroscience along with significant investment from governments and industries.

Despite its promise for vision applications, it is important to recognize that several issues remain inadequately addressed. Primarily, a significant proportion of the current literature on DVS are still utilizing accumulated representations, eschewing more efficient and apt SNN or GNN-based methodologies which are potentially more suitable for DVS applications. Additionally, given the relatively nascent status of the DVS-based vision field, both the quantity and quality of available datasets are significantly inferior to those in the conventional frame-based vision domain. Rather than undertaking each data collection process and model training de novo, it would be beneficial to devise superior strategies to adapt these datasets and models effectively to the DVS paradigm. We believe that neuromorphic computing will play a pivotal role in the growth and development of the DVS or event camera for everyday applications. More specifically, spike-based neuromorphic computing and real-time high speed processing of event streams can fundamentally unlock new applications in different domains including scientific computing, climate science, human-computer interaction, and medicine. We anticipate that efficient and robust simulation platform and cross-modality training (e.g., from RGBD to DVS) might play an instrumental role in developing the humanlike AGI for event-based vision.

References

[1] G. E. Moore, "Cramming more components onto integrated circuits, reprinted from electronics, volume 38, number 8, april 19, 1965, pp. 114 ff." *IEEE solid-state circuits society newsletter*, vol. 11, no. 3, pp. 33–35, 2006.

[2] R. H. Dennard, F. H. Gaensslen, H.-N. Yu, V. L. Rideout, E. Bassous, and A. R. LeBlanc, "Design of ion-implanted MOSFET's with very small physical dimensions," *IEEE Journal of Solid-State Circuits*, vol. 9, no. 5, pp. 256–268, 1974.

[3] D. A. Patterson and J. L. Hennessy, *Computer organization and design ARM edition: the hardware software interface.* Morgan kaufmann, 2016.

[4] R. S. Williams, "What's next?[the end of moore's law]," *Computing in Science & Engineering*, vol. 19, no. 2, pp. 7–13, 2017.

[5] D. R. Solli and B. Jalali, "Analog optical computing," *Nature Photonics*, vol. 9, no. 11, pp. 704–706, 2015.

[6] J. Preskill, "Quantum computing in the NISQ era and beyond," *Quantum*, vol. 2, p. 79, 2018.

[7] G. Păun, G. Rozenberg, and A. Salomaa, *DNA computing: New computing paradigms.* Springer, 1998.

[8] C. Mead, "Neuromorphic electronic systems," *Proceedings of the IEEE*, vol. 78, no. 10, pp. 1629–1636, 1990.

[9] S. Herculano-Houzel, *The human advantage: A new understanding of how our brain became remarkable.* MIT Press, 2016.

[10] W. S. McCulloch and W. Pitts, "A logical calculus of the ideas immanent in nervous activity," *The Bulletin of Mathematical Biophysics*, vol. 5, pp. 115–133, 1943.

[11] D. O. Hebb, "The organization of behavior; a neuropsychological theory," *The American Journal of Psychology*, vol. 63, no. 4, p. 633, 1949.

[12] A. L. Hodgkin and A. F. Huxley, "A quantitative description of membrane current and its application to conduction and excitation in nerve," *The Journal of physiology*, vol. 117, no. 4, p. 500, 1952.

[13] R. FitzHugh, "Impulses and physiological states in theoretical models of nerve membrane," *Biophysical Journal*, vol. 1, no. 6, pp. 445–466, 1961.

[14] J. Nagumo, S. Arimoto, and S. Yoshizawa, "An active pulse transmission line simulating nerve axon," *Proceedings of the IRE*, vol. 50, no. 10, pp. 2061–2070, 1962.

[15] F. Rosenblatt, "The perceptron: a probabilistic model for information storage and organization in the brain," *Psychological Review*, vol. 65, no. 6, 1958.

[16] P. Werbos, "Beyond regression: New tools for prediction and analysis in the behavioral sciences," *PhD thesis, Committee on Applied Mathematics, Harvard University, Cambridge, MA*, 1974.

[17] D. E. Rumelhart, G. E. Hinton, and R. J. Williams, "Learning representations by back-propagating errors," *Nature*, vol. 323, no. 6088, pp. 533–536, 1986.

[18] J. J. Hopfield, "Neural networks and physical systems with emergent collective computational abilities." *Proceedings of the National Academy of Sciences*, vol. 79, no. 8, pp. 2554–2558, 1982.

[19] T. Kohonen, "Self-organized formation of topologically correct feature maps," *Biological cybernetics*, vol. 43, no. 1, pp. 59–69, 1982.

[20] C. Mead, "How we created neuromorphic engineering," *Nature Electronics*, vol. 3, no. 7, pp. 434–435, 2020.

[21] M. Mahowald and M. Mahowald, "The silicon retina," *An Analog VLSI System for Stereoscopic Vision*, pp. 4–65, 1994.

[22] M. Mahowald and R. Douglas, "A silicon neuron," *Nature*, vol. 354, no. 6354, pp. 515–518, 1991.

[23] J. Schemmel, D. Brüderle, A. Grübl, M. Hock, K. Meier, and S. Millner, "A wafer-scale neuromorphic hardware system for large-scale neural modeling," in *2010 IEEE International Symposium on Circuits and Systems (ISCAS)*. IEEE, 2010, pp. 1947–1950.

[24] P. A. Merolla, J. V. Arthur, R. Alvarez-Icaza, A. S. Cassidy, J. Sawada, F. Akopyan, B. L. Jackson, N. Imam, C. Guo, Y. Nakamura *et al.*, "A million spiking-neuron integrated circuit with a scalable communication network and interface," *Science*, vol. 345, no. 6197, pp. 668–673, 2014.

[25] B. V. Benjamin, P. Gao, E. McQuinn, S. Choudhary, A. R. Chandrasekaran, J.-M. Bussat, R. Alvarez-Icaza, J. V. Arthur, P. A. Merolla, and K. Boahen, "Neurogrid: A mixed-analog-digital multichip system for large-scale neural simulations," *Proceedings of the IEEE*, vol. 102, no. 5, pp. 699–716, 2014.

[26] S. B. Furber, F. Galluppi, S. Temple, and L. A. Plana, "The spinnaker project," *Proceedings of the IEEE*, vol. 102, no. 5, pp. 652–665, 2014.

[27] M. Davies, N. Srinivasa, T.-H. Lin, G. Chinya, Y. Cao, S. H. Choday, G. Dimou, P. Joshi, N. Imam, S. Jain *et al.*, "Loihi: A neuromorphic manycore processor with on-chip learning," *IEEE Micro*, vol. 38, no. 1, pp. 82–99, 2018.

[28] C. Mead and M. Ismail, *Analog VLSI implementation of neural systems.* Springer Science & Business Media, 1989, vol. 80.

[29] G. S. Rose, M. S. A. Shawkat, A. Z. Foshie, J. J. Murray, and M. M. Adnan, "A system design perspective on neuromorphic computer processors," *Neuromorphic Computing and Engineering*, vol. 1, no. 2, p. 022001, 2021.

[30] J. S. Najem, G. J. Taylor, R. J. Weiss, M. S. Hasan, G. Rose, C. D. Schuman, A. Belianinov, C. P. Collier, and S. A. Sarles, "Memristive ion channel-doped biomembranes as synaptic mimics," *ACS Nano*, vol. 12, no. 5, pp. 4702–4711, 5 2018.

[31] K. Krnjević, "Chemical nature of synaptic transmission in vertebrates," *Physiological Reviews*, vol. 54, no. 2, pp. 418–540, 1974.

[32] J. S. Haas, B. Zavala, and C. E. Landisman, "Activity-dependent long-term depression of electrical synapses," *Science*, vol. 334, no. 6054, pp. 389–393, 2011.

[33] J. S. Haas, C. M. Greenwald, and A. E. Pereda, "Activity-dependent plasticity of electrical synapses: Increasing evidence for its presence and functional roles in the mammalian brain," *BMC Cell Biology*, vol. 17, no. 1, pp. 51–57, 2016.

[34] W. Maass, "Networks of spiking neurons: The third generation of neural network models," *Neural Networks*, vol. 10, no. 9, pp. 1659–1671, 1997.

[35] E. M. Izhikevich, "Simple model of spiking neurons," *IEEE Transactions on Neural Networks*, vol. 14, no. 6, pp. 1569–1572, 2003.

[36] C. Frenkel, D. Bol, and G. Indiveri, "Bottom-up and top-down neural processing systems design: Neuromorphic intelligence as the convergence of natural and artificial intelligence," *arXiv preprint arXiv:2106.01288*, 2021.

[37] L. A. Camuñas-Mesa, B. Linares-Barranco, and T. Serrano-Gotarredona, "Neuromorphic spiking neural networks and their memristor-cmos hardware implementations," *Materials*, vol. 12, no. 17, p. 2745, 2019.

[38] S. Thorpe, D. Fize, and C. Marlot, "Speed of processing in the human visual system," *Nature*, vol. 381, no. 6582, pp. 520–522, 1996.

[39] E. T. Rolls and M. J. Tovee, "Processing speed in the cerebral cortex and the neurophysiology of visual masking," *Proceedings of the Royal Society of London. Series B: Biological Sciences*, vol. 257, no. 1348, pp. 9–15, 1994.

[40] H. Hendy and C. Merkel, "Review of spike-based neuromorphic computing for brain-inspired vision: Biology, algorithms, and hardware," *Journal of Electronic Imaging*, vol. 31, no. 1, pp. 010 901–010 901, 2022.

[41] K. Roy, A. Jaiswal, and P. Panda, "Towards spike-based machine intelligence with neuromorphic computing," *Nature*, vol. 575, no. 7784, pp. 607–617, 2019.

[42] R. Brette and W. Gerstner, "Adaptive exponential integrate-and-fire model as an effective description of neuronal activity," *Journal of Neurophysiology*, vol. 94, no. 5, pp. 3637–3642, 2005.

[43] W. Gerstner, R. Ritz, and J. L. Van Hemmen, "Why spikes? Hebbian learning and retrieval of time-resolved excitation patterns," *Biological Cybernetics*, vol. 69, no. 5-6, pp. 503–515, 1993.

[44] W. Gerstner and W. M. Kistler, *Spiking neuron models: Single neurons, populations, plasticity*. Cambridge University Press, 2002.

[45] J. Burroni, P. Taylor, C. Corey, T. Vachnadze, and H. T. Siegelmann, "Energetic constraints produce self-sustained oscillatory dynamics in neuronal networks," *Frontiers in Neuroscience*, vol. 11, p. 80, 2017.

[46] E. M. Izhikevich, "Which model to use for cortical spiking neurons?" *IEEE Transactions on Neural Networks*, vol. 15, no. 5, pp. 1063–1070, 2004.

[47] G. Indiveri, E. Chicca, and R. Douglas, "A VLSI array of low-power spiking neurons and bistable synapses with spike-timing dependent plasticity," *IEEE Transactions on Neural Networks*, vol. 17, no. 1, pp. 211–221, 2006.

[48] T. J. Koickal, A. Hamilton, S. L. Tan, J. A. Covington, J. W. Gardner, and T. C. Pearce, "Analog VLSI circuit implementation of an adaptive neuromorphic olfaction chip," *IEEE Transactions on Circuits and Systems I: Regular Papers*, vol. 54, no. 1, pp. 60–73, 2007.

[49] I. E. Ebong and P. Mazumder, "CMOS and memristor-based neural network design for position detection," *Proceedings of the IEEE*, vol. 100, no. 6, pp. 2050–2060, 2011.

[50] G. M. Tovar, E. S. Fukuda, T. Asai, T. Hirose, and Y. Amemiya, "Analog CMOS circuits implementing neural segmentation model based on symmetric STDP learning," in *Neural Information Processing: 14th International Conference, ICONIP 2007,*

Kitakyushu, Japan, November 13-16, 2007, Revised Selected Papers, Part II 14. Springer, 2008, pp. 117–126.

[51] G. Indiveri, "A low-power adaptive integrate-and-fire neuron circuit," in *Proceedings of the 2003 International Symposium on Circuits and Systems, 2003. ISCAS'03.*, vol. 4. IEEE, 2003, pp. IV–IV.

[52] G. Indiveri, F. Stefanini, and E. Chicca, "Spike-based learning with a generalized integrate and fire silicon neuron," in *Proceedings of 2010 IEEE International Symposium on Circuits and Systems.* IEEE, 2010, pp. 1951–1954.

[53] X. Wu, V. Saxena, and K. A. Campbell, "Energy-efficient stdp-based learning circuits with memristor synapses," in *Machine Intelligence and Bio-inspired Computation: Theory and Applications VIII*, vol. 9119. SPIE, 2014, pp. 33–39.

[54] S.-C. Liu, T. Delbruck, G. Indiveri, A. Whatley, and R. Douglas, *Event-based neuromorphic systems.* John Wiley & Sons, 2014.

[55] M. Skrbek, "Fast neural network implementation," *Neural Network World*, vol. 9, no. 5, pp. 375–391, 1999.

[56] H. Hikawa, "A digital hardware pulse-mode neuron with piecewise linear activation function," *IEEE Transactions on Neural Networks*, vol. 14, no. 5, pp. 1028–1037, 2003.

[57] A. Muthuramalingam, S. Himavathi, and E. Srinivasan, "Neural network implementation using FPGA: Issues and application," *International Journal of Electrical and Computer Engineering*, vol. 2, no. 12, pp. 2802–2808, 2008.

[58] S. Sayyaparaju, M. M. Adnan, S. Amer, and G. S. Rose, "Device-aware circuit design for robust memristive neuromorphic systems with STDP-based learning," *ACM Journal on Emerging Technologies in Computing Systems (JETC)*, vol. 16, no. 3, pp. 1–25, 2020.

[59] J. Zhu, T. Zhang, Y. Yang, and R. Huang, "A comprehensive review on emerging artificial neuromorphic devices," *Applied Physics Reviews*, vol. 7, no. 1, p. 011312, 2020.

[60] R. Islam, H. Li, P.-Y. Chen, W. Wan, H.-Y. Chen, B. Gao, H. Wu, S. Yu, K. Saraswat, and H. P. Wong, "Device and materials requirements for neuromorphic computing," *Journal of Physics D: Applied Physics*, vol. 52, no. 11, p. 113001, 2019.

[61] D. Ielmini and S. Ambrogio, "Emerging neuromorphic devices," *Nanotechnology*, vol. 31, no. 9, p. 092001, 2019.

[62] Z. Wang, H. Wu, G. W. Burr, C. S. Hwang, K. L. Wang, Q. Xia, and J. J. Yang, "Resistive switching materials for information processing," *Nature Reviews Materials*, vol. 5, no. 3, pp. 173–195, 2020.

[63] D. B. Strukov, G. S. Snider, D. R. Stewart, and R. S. Williams, "The missing memristor found," *Nature*, vol. 453, no. 7191, pp. 80–83, 2008.

[64] S. Kumar, X. Wang, J. P. Strachan, Y. Yang, and W. D. Lu, "Dynamical memristors for higher-complexity neuromorphic computing," *Nature Reviews Materials*, vol. 7, no. 7, pp. 575–591, 2022.

[65] R. A. Nawrocki, R. M. Voyles, and S. E. Shaheen, "A mini review of neuromorphic architectures and implementations," *IEEE Transactions on Electron Devices*, vol. 63, no. 10, pp. 3819–3829, 2016.

[66] Y. He, L. Zhu, Y. Zhu, C. Chen, S. Jiang, R. Liu, Y. Shi, and Q. Wan, "Recent progress on emerging transistor-based neuromorphic devices," *Advanced Intelligent Systems*, vol. 3, no. 7, p. 2000210, 2021.

[67] L. Chua, "Memristor-the missing circuit element," *IEEE Transactions on Circuit Theory*, vol. 18, no. 5, pp. 507–519, 1971.

[68] M. D. Pickett, G. Medeiros-Ribeiro, and R. S. Williams, "A scalable neuristor built with mott memristors," *Nature materials*, vol. 12, no. 2, pp. 114–117, 2013.

[69] M. Zhu, K. Ren, and Z. Song, "Ovonic threshold switching selectors for three-dimensional stackable phase-change memory," *MRS Bulletin*, vol. 44, no. 9, pp. 715–720, 2019.

[70] Z. Wang, S. Joshi, S. E. Savel'ev, H. Jiang, R. Midya, P. Lin, M. Hu, N. Ge, J. P. Strachan, Z. Li *et al.*, "Memristors with diffusive dynamics as synaptic emulators for neuromorphic computing," *Nature Materials*, vol. 16, no. 1, pp. 101–108, 2017.

[71] S. Kumar, C. E. Graves, J. P. Strachan, E. M. Grafals, A. L. D. Kilcoyne, T. Tyliszczak, J. N. Weker, Y. Nishi, and R. S. Williams, "Direct observation of localized radial oxygen migration in functioning tantalum oxide memristors," *Advanced Materials*, vol. 28, no. 14, pp. 2772–2776, 2016.

[72] Y. V. Pershin and M. Di Ventra, "Spin memristive systems: Spin memory effects in semiconductor

spintronics," *Physical Review B*, vol. 78, no. 11, p. 113309, 2008.

[73] S. Manipatruni, D. E. Nikonov, and I. A. Young, "Beyond CMOS computing with spin and polarization," *Nature Physics*, vol. 14, no. 4, pp. 338–343, 2018.

[74] T. Endoh, H. Koike, S. Ikeda, T. Hanyu, and H. Ohno, "An overview of nonvolatile emerging memories—Spintronics for working memories," *IEEE Journal on Emerging and Selected Topics in Circuits and Systems*, vol. 6, no. 2, pp. 109–119, 2016.

[75] S. Raoux, F. Xiong, M. Wuttig, and E. Pop, "Phase change materials and phase change memory," *MRS Bulletin*, vol. 39, no. 8, pp. 703–710, 2014.

[76] H. Wang and X. Yan, "Overview of resistive random access memory (RRAM): Materials, filament mechanisms, performance optimization, and prospects," *physica Status Solidi (RRL)–Rapid Research Letters*, vol. 13, no. 9, p. 1900073, 2019.

[77] X. Hong, D. J. Loy, P. A. Dananjaya, F. Tan, C. Ng, and W. Lew, "Oxide-based RRAM materials for neuromorphic computing," *Journal of Materials Science*, vol. 53, pp. 8720–8746, 2018.

[78] M. M. Rehman, H. M. M. U. Rehman, J. Z. Gul, W. Y. Kim, K. S. Karimov, and N. Ahmed, "Decade of 2D-materials-based RRAM devices: A review," *Science and Technology of Advanced Materials*, vol. 21, no. 1, pp. 147–186, 2020.

[79] X. Feng, X. Liu, and K.-W. Ang, "2D photonic memristor beyond graphene: Progress and prospects," *Nanophotonics*, vol. 9, no. 7, pp. 1579–1599, 2020.

[80] Y. Wang, K.-M. Kang, M. Kim, H.-S. Lee, R. Waser, D. Wouters, R. Dittmann, J. J. Yang, and H.-H. Park, "Mott-transition-based RRAM," *Materials Today*, vol. 28, pp. 63–80, 2019.

[81] S. Goswami, S. Goswami, and T. Venkatesan, "An organic approach to low energy memory and brain inspired electronics," *Applied Physics Reviews*, vol. 7, no. 2, p. 021303, 2020.

[82] E. C. Ahn, H.-S. P. Wong, and E. Pop, "Carbon nanomaterials for non-volatile memories," *Nature Reviews Materials*, vol. 3, no. 3, pp. 1–15, 2018.

[83] S. Kim, C. Du, P. Sheridan, W. Ma, S. Choi, and W. D. Lu, "Experimental demonstration of a second-order memristor and its ability to biorealistically implement synaptic plasticity," *Nano Letters*, vol. 15, no. 3, pp. 2203–2211, 2015.

[84] A. Rodriguez-Fernandez, C. Cagli, L. Perniola, E. Miranda, and J. Suñé, "Characterization of HFO2-based devices with indication of second order memristor effects," *Microelectronic Engineering*, vol. 195, pp. 101–106, 2018.

[85] V. Mikheev, A. Chouprik, Y. Lebedinskii, S. Zarubin, Y. Matveyev, E. Kondratyuk, M. G. Kozodaev, A. M. Markeev, A. Zenkevich, and D. Negrov, "Ferroelectric second-order memristor," *ACS Applied Materials & Interfaces*, vol. 11, no. 35, pp. 32 108–32 114, 2019.

[86] J. S. Najem, G. J. Taylor, R. J. Weiss, M. S. Hasan, G. Rose, C. D. Schuman, A. Belianinov, C. P. Collier, and S. A. Sarles, "Memristive ion channel-doped biomembranes as synaptic mimics," *ACS Nano*, vol. 12, no. 5, pp. 4702–4711, 2018.

[87] M. S. Hasan, C. D. Schuman, J. S. Najem, R. Weiss, N. D. Skuda, A. Belianinov, C. P. Collier, S. A. Sarles, and G. S. Rose, "Biomimetic, soft-material synapse for neuromorphic computing: from device to network," in *2018 IEEE 13th Dallas Circuits and Systems Conference (DCAS)*. IEEE, 2018, pp. 1–6.

[88] S. Koner, J. S. Najem, M. S. Hasan, and S. A. Sarles, "Memristive plasticity in artificial electrical synapses via geometrically reconfigurable, gramicidin-doped biomembranes," *Nanoscale*, vol. 11, no. 40, pp. 18 640–18 652, 2019.

[89] A. Alexandrov, A. Bratkovsky, B. Bridle, S. Savel'Ev, D. Strukov, and R. Stanley Williams, "Current-controlled negative differential resistance due to joule heating in TiO2," *Applied Physics Letters*, vol. 99, no. 20, p. 202104, 2011.

[90] M. D. Pickett, J. Borghetti, J. J. Yang, G. Medeiros-Ribeiro, and R. S. Williams, "Coexistence of memristance and negative differential resistance in a nanoscale metal-oxide-metal system," *Advanced Materials*, vol. 23, no. 15, pp. 1730–1733, 2011.

[91] S. Kumar, R. S. Williams, and Z. Wang, "Third-order nanocircuit elements for neuromorphic engineering," *Nature*, vol. 585, no. 7826, pp. 518–523, 2020.

[92] C. Du, F. Cai, M. A. Zidan, W. Ma, S. H. Lee, and W. D. Lu, "Reservoir computing using dynamic memristors for temporal information processing,"

Nature Communications, vol. 8, no. 1, pp. 1–10, 2017.

[93] J. Moon, W. Ma, J. H. Shin, F. Cai, C. Du, S. H. Lee, and W. D. Lu, "Temporal data classification and forecasting using a memristor-based reservoir computing system," *Nature Electronics*, vol. 2, no. 10, pp. 480–487, 2019.

[94] Y. Zhong, J. Tang, X. Li, X. Liang, Z. Liu, Y. Li, Y. Xi, P. Yao, Z. Hao, B. Gao *et al.*, "A memristor-based analogue reservoir computing system for real-time and power-efficient signal processing," *Nature Electronics*, vol. 5, no. 10, pp. 672–681, 2022.

[95] M. R. Hossain, A. S. Mohamed, N. X. Armendarez, J. S. Najem, and M. S. Hasan, "Energy-efficient memcapacitive physical reservoir computing system for temporal data processing," *arXiv preprint arXiv:2305.12025*, 2023.

[96] S. Moradi, N. Qiao, F. Stefanini, and G. Indiveri, "A scalable multicore architecture with heterogeneous memory structures for dynamic neuromorphic asynchronous processors (dynaps)," *IEEE Transactions on Biomedical Circuits and Systems*, vol. 12, no. 1, pp. 106–122, 2017.

[97] C. S. Thakur, J. L. Molin, G. Cauwenberghs, G. Indiveri, K. Kumar, N. Qiao, J. Schemmel, R. Wang, E. Chicca, J. Olson Hasler *et al.*, "Large-scale neuromorphic spiking array processors: A quest to mimic the brain," *Frontiers in Neuroscience*, vol. 12, p. 891, 2018.

[98] K. Amunts, C. Ebell, J. Muller, M. Telefont, A. Knoll, and T. Lippert, "The human brain project: Creating a European research infrastructure to decode the human brain," *Neuron*, vol. 92, no. 3, pp. 574–581, 2016.

[99] M. E. Dean, C. D. Schuman, and J. D. Birdwell, "Dynamic adaptive neural network array," in *Unconventional Computation and Natural Computation: 13th International Conference, UCNC 2014, London, ON, Canada, July 14-18, 2014, Proceedings 13*. Springer, 2014, pp. 129–141.

[100] J. P. Mitchell, M. E. Dean, G. R. Bruer, J. S. Plank, and G. S. Rose, "Danna 2: Dynamic adaptive neural network arrays," in *Proceedings of the International Conference on Neuromorphic Systems*, 2018, pp. 1–6.

[101] G. Chakma, M. M. Adnan, A. R. Wyer, R. Weiss, C. D. Schuman, and G. S. Rose, "Memristive mixed-signal neuromorphic systems: Energy-efficient learning at the circuit-level," *IEEE Journal on Emerging and Selected Topics in Circuits and Systems*, vol. 8, no. 1, pp. 125–136, 2017.

[102] A. Z. Foshie, N. N. Chakraborty, J. J. Murray, T. J. Fowler, M. S. A. Shawkat, and G. S. Rose, "A multicontext neural core design for reconfigurable neuromorphic arrays," in *2021 IEEE Computer Society Annual Symposium on VLSI (ISVLSI)*. IEEE, 2021, pp. 67–72.

[103] J. Hasler, "Large-scale field-programmable analog arrays," *Proceedings of the IEEE*, vol. 108, no. 8, pp. 1283–1302, 2019.

[104] P. U. Diehl, D. Neil, J. Binas, M. Cook, S.-C. Liu, and M. Pfeiffer, "Fast-classifying, high-accuracy spiking deep networks through weight and threshold balancing," in *2015 International Joint Conference on Neural Networks (IJCNN)*. IEEE, 2015, pp. 1–8.

[105] P. U. Diehl, G. Zarrella, A. Cassidy, B. U. Pedroni, and E. Neftci, "Conversion of artificial recurrent neural networks to spiking neural networks for low-power neuromorphic hardware," in *2016 IEEE International Conference on Rebooting Computing (ICRC)*. IEEE, 2016, pp. 1–8.

[106] E. Hunsberger and C. Eliasmith, "Training spiking deep networks for neuromorphic hardware," *arXiv preprint arXiv:1611.05141*, 2016.

[107] A. Sengupta, Y. Ye, R. Wang, C. Liu, and K. Roy, "Going deeper in spiking neural networks: VGG and residual architectures," *Frontiers in Neuroscience*, vol. 13, p. 95, 2019.

[108] B. Rueckauer, I.-A. Lungu, Y. Hu, M. Pfeiffer, and S.-C. Liu, "Conversion of continuous-valued deep networks to efficient event-driven networks for image classification," *Frontiers in Neuroscience*, vol. 11, p. 682, 2017.

[109] D. Neil, M. Pfeiffer, and S.-C. Liu, "Learning to be efficient: Algorithms for training low-latency, low-compute deep spiking neural networks," in *Proceedings of the 31st Annual ACM Symposium on Applied Computing*, 2016, pp. 293–298.

[110] S. K. Esser, R. Appuswamy, P. Merolla, J. V. Arthur, and D. S. Modha, "Backpropagation for energy-efficient neuromorphic computing," *Advances in Neural Information Processing Systems*, vol. 28, 2015.

[111] Y. Cao, Y. Chen, and D. Khosla, "Spiking deep convolutional neural networks for energy-efficient

object recognition," *International Journal of Computer Vision*, vol. 113, pp. 54–66, 2015.

[112] S. M. Bohte, J. N. Kok, and J. A. La Poutré, "Spikeprop: Backpropagation for networks of spiking neurons." in *ESANN*, vol. 48.   Bruges, 2000, pp. 419–424.

[113] W. Severa, C. M. Vineyard, R. Dellana, S. J. Verzi, and J. B. Aimone, "Training deep neural networks for binary communication with the whetstone method," *Nature Machine Intelligence*, vol. 1, no. 2, pp. 86–94, 2019.

[114] Y. Wu, L. Deng, G. Li, J. Zhu, and L. Shi, "Spatio-temporal backpropagation for training high-performance spiking neural networks," *Frontiers in Neuroscience*, vol. 12, p. 331, 2018.

[115] E. O. Neftci, H. Mostafa, and F. Zenke, "Surrogate gradient learning in spiking neural networks: Bringing the power of gradient-based optimization to spiking neural networks," *IEEE Signal Processing Magazine*, vol. 36, no. 6, pp. 51–63, 2019.

[116] J. H. Lee, T. Delbruck, and M. Pfeiffer, "Training deep spiking neural networks using backpropagation," *Frontiers in Neuroscience*, vol. 10, p. 508, 2016.

[117] G. Bellec, D. Salaj, A. Subramoney, R. Legenstein, and W. Maass, "Long short-term memory and learning-to-learn in networks of spiking neurons," *Advances in Neural Information Processing Systems*, vol. 31, 2018.

[118] Q. Meng, M. Xiao, S. Yan, Y. Wang, Z. Lin, and Z.-Q. Luo, "Training high-performance lowlatency spiking neural networks by differentiation on spike representation," in *Proceedings of the IEEE/CVF Conference on Computer Vision and Pattern Recognition*, 2022, pp. 12 444–12 453.

[119] A. Tavanaei, M. Ghodrati, S. R. Kheradpisheh, T. Masquelier, and A. Maida, "Deep learning in spiking neural networks," *Neural Networks*, vol. 111, pp. 47–63, 2019.

[120] N. Caporale and Y. Dan, "Spike timing–dependent plasticity: A Hebbian learning rule," *Annu. Rev. Neurosci.*, vol. 31, pp. 25–46, 2008.

[121] C. D. Schuman, T. E. Potok, R. M. Patton, J. D. Birdwell, M. E. Dean, G. S. Rose, and J. S. Plank, "A survey of neuromorphic computing and neural networks in hardware," *arXiv preprint arXiv:1705.06963*, 2017.

[122] P. U. Diehl and M. Cook, "Unsupervised learning of digit recognition using spike-timing-dependent plasticity," *Frontiers in Computational Neuroscience*, vol. 9, p. 99, 2015.

[123] A. Tavanaei, Z. Kirby, and A. S. Maida, "Training spiking convnets by STDP and gradient descent," in *2018 International Joint Conference on Neural Networks (IJCNN)*.   IEEE, 2018, pp. 1–8.

[124] S. R. Kheradpisheh, M. Ganjtabesh, S. J. Thorpe, and T. Masquelier, "STDP-based spiking deep convolutional neural networks for object recognition," *Neural Networks*, vol. 99, pp. 56–67, 2018.

[125] E. Iranmehr, S. B. Shouraki, and M. M. Faraji, "ILS-based reservoir computing for handwritten digits recognition," in *2020 8th Iranian Joint Congress on Fuzzy and Intelligent Systems (CFIS)*.   IEEE, 2020, pp. 1–6.

[126] J. J. Reynolds, J. S. Plank, and C. D. Schuman, "Intelligent reservoir generation for liquid state machines using evolutionary optimization," in *2019 International Joint Conference on Neural Networks (IJCNN)*.   IEEE, 2019, pp. 1–8.

[127] C. Tang, J. Ji, Q. Lin, and Y. Zhou, "Evolutionary neural architecture design of liquid state machine for image classification," in *ICASSP 2022-2022 IEEE International Conference on Acoustics, Speech and Signal Processing (ICASSP)*.   IEEE, 2022, pp. 91–95.

[128] P. Wijesinghe, G. Srinivasan, P. Panda, and K. Roy, "Analysis of liquid ensembles for enhancing the performance and accuracy of liquid state machines," *Frontiers in Neuroscience*, vol. 13, p. 504, 2019.

[129] M. C. Soriano, S. Ortín, L. Keuninckx, L. Appeltant, J. Danckaert, L. Pesquera, and G. Van der Sande, "Delay-based reservoir computing: Noise effects in a combined analog and digital implementation," *IEEE Transactions on Neural Networks and Learning Systems*, vol. 26, no. 2, pp. 388–393, 2014.

[130] B. Schrauwen, M. D'Haene, D. Verstraeten, and J. Van Campenhout, "Compact hardware liquid state machines on FPGA for real-time speech recognition," *Neural Networks*, vol. 21, no. 2-3, pp. 511–523, 2008.

[131] Y. Zhang, P. Li, Y. Jin, and Y. Choe, "A digital liquid state machine with biologically inspired learning and its application to speech recognition," *IEEE Transactions on Neural Networks and Learning Systems*, vol. 26, no. 11, pp. 2635–2649, 2015.

[132] N. Soures, L. Hays, and D. Kudithipudi, "Robustness of a memristor based liquid state machine," in *2017 International Joint Conference on Neural Networks (IJCNN)*. IEEE, 2017, pp. 2414–2420.

[133] G. Van der Sande, D. Brunner, and M. C. Soriano, "Advances in photonic reservoir computing," *Nanophotonics*, vol. 6, no. 3, pp. 561–576, 2017.

[134] D. Brunner, B. Penkovsky, B. A. Marquez, M. Jacquot, I. Fischer, and L. Larger, "Tutorial: Photonic neural networks in delay systems," *Journal of Applied Physics*, vol. 124, no. 15, p. 152004, 2018.

[135] H. Hauser, A. J. Ijspeert, R. M. Füchslin, R. Pfeifer, and W. Maass, "Towards a theoretical foundation for morphological computation with compliant bodies," *Biological Cybernetics*, vol. 105, no. 5, pp. 355–370, 2011.

[136] K. Caluwaerts, J. Despraz, A. Işçen, A. P. Sabelhaus, J. Bruce, B. Schrauwen, and V. SunSpiral, "Design and control of compliant tensegrity robots through simulation and hardware validation," *Journal of the Royal Society Interface*, vol. 11, no. 98, p. 20140520, 2014.

[137] M. R. Dranias, H. Ju, E. Rajaram, and A. M. VanDongen, "Short-term memory in networks of dissociated cortical neurons," *Journal of Neuroscience*, vol. 33, no. 5, pp. 1940–1953, 2013.

[138] D. Sussillo and L. F. Abbott, "Generating coherent patterns of activity from chaotic neural networks," *Neuron*, vol. 63, no. 4, pp. 544–557, 2009.

[139] B. Jones, D. Stekel, J. Rowe, and C. Fernando, "Is there a liquid state machine in the bacterium Escherichia coli?" in *2007 IEEE Symposium on Artificial Life*. Ieee, 2007, pp. 187–191.

[140] K. Fujii and K. Nakajima, "Harnessing disordered-ensemble quantum dynamics for machine learning," *Physical Review Applied*, vol. 8, no. 2, p. 024030, 2017.

[141] J. Torrejon, M. Riou, F. A. Araujo, S. Tsunegi, G. Khalsa, D. Querlioz, P. Bortolotti, V. Cros, K. Yakushiji, A. Fukushima *et al.*, "Neuromorphic computing with nanoscale spintronic oscillators," *Nature*, vol. 547, no. 7664, pp. 428–431, 2017.

[142] G. Bourianoff, D. Pinna, M. Sitte, and K. Everschor-Sitte, "Potential implementation of reservoir computing models based on magnetic skyrmions," *AIP Advances*, vol. 8, no. 5, p. 055602, 2018.

[143] G. Tanaka, T. Yamane, J. B. Héroux, R. Nakane, N. Kanazawa, S. Takeda, H. Numata, D. Nakano, and A. Hirose, "Recent advances in physical reservoir computing: A review," *Neural Networks*, vol. 115, pp. 100–123, 2019.

[144] W. Maass, T. Natschläger, and H. Markram, "Fading memory and kernel properties of generic cortical microcircuit models," *Journal of Physiology-Paris*, vol. 98, no. 4-6, pp. 315–330, 2004.

[145] I. B. Yildiz, H. Jaeger, and S. J. Kiebel, "Re-visiting the echo state property," *Neural Networks*, vol. 35, pp. 1–9, 2012.

[146] C. D. Schuman, J. P. Mitchell, R. M. Patton, T. E. Potok, and J. S. Plank, "Evolutionary optimization for neuromorphic systems," in *Proceedings of the Neuro-inspired Computational Elements Workshop*, 2020, pp. 1–9.

[147] C. D. Schuman, J. S. Plank, A. Disney, and J. Reynolds, "An evolutionary optimization framework for neural networks and neuromorphic architectures," in *2016 International Joint Conference on Neural Networks (IJCNN)*. IEEE, 2016, pp. 145–154.

[148] C. Schuman, R. Patton, S. Kulkarni, M. Parsa, C. Stahl, N. Q. Haas, J. P. Mitchell, S. Snyder, A. Nagle, A. Shanafield *et al.*, "Evolutionary vs imitation learning for neuromorphic control at the edge," *Neuromorphic Computing and Engineering*, vol. 2, no. 1, p. 014002, 2022.

[149] Y. Liu, Y. Sun, B. Xue, M. Zhang, G. G. Yen, and K. C. Tan, "A survey on evolutionary neural architecture search," *IEEE Transactions on Neural Networks and Learning Systems*, 2021.

[150] H. Mostafa, B. U. Pedroni, S. Sheik, and G. Cauwenberghs, "Fast classification using sparsely active spiking networks," in *2017 IEEE International Symposium on Circuits and Systems (ISCAS)*. IEEE, 2017, pp. 1–4.

[151] G. Srinivasan and K. Roy, "Restocnet: Residual stochastic binary convolutional spiking neural network for memory-efficient neuromorphic computing," *Frontiers in Neuroscience*, vol. 13, p. 189, 2019.

[152] M. Zhang, Z. Gu, N. Zheng, D. Ma, and G. Pan, "Efficient spiking neural networks with logarithmic temporal coding," *IEEE Access*, vol. 8, pp. 98 156–98 167, 2020.

[153] W. Zhang and P. Li, "Temporal spike sequence learning via backpropagation for deep spiking neural networks," *Advances in Neural Information Processing Systems*, vol. 33, pp. 12 022–12 033, 2020.

[154] N. Rathi, G. Srinivasan, P. Panda, and K. Roy, "Enabling deep spiking neural networks with hybrid conversion and spike timing dependent backpropagation," *arXiv preprint arXiv:2005.01807*, 2020.

[155] B. Han and K. Roy, "Deep spiking neural network: Energy efficiency through time based coding," in *Computer Vision–ECCV 2020: 16th European Conference, Glasgow, UK, August 23–28, 2020, Proceedings, Part X.* Springer, 2020, pp. 388–404.

[156] S. Deng, Y. Li, S. Zhang, and S. Gu, "Temporal efficient training of spiking neural network via gradient re-weighting," *arXiv preprint arXiv:2202.11946*, 2022.

[157] C. Duan, J. Ding, S. Chen, Z. Yu, and T. Huang, "Temporal effective batch normalization in spiking neural networks," *Advances in Neural Information Processing Systems*, vol. 35, pp. 34 377–34 390, 2022.

[158] B. Han, G. Srinivasan, and K. Roy, "RMP-SNN: Residual membrane potential neuron for enabling deeper high-accuracy and low-latency spiking neural network," in *Proceedings of the IEEE/CVF Conference on Computer Vision and Pattern Recognition*, 2020, pp. 13 558–13 567.

[159] Y. Li, Y. Guo, S. Zhang, S. Deng, Y. Hai, and S. Gu, "Differentiable spike: Rethinking gradient-descent for training spiking neural networks," *Advances in Neural Information Processing Systems*, vol. 34, pp. 23 426–23 439, 2021.

[160] Y. Li, S. Deng, X. Dong, R. Gong, and S. Gu, "A free lunch from ANN: Towards efficient, accurate spiking neural networks calibration," in *International Conference on Machine Learning.* PMLR, 2021, pp. 6316–6325.

[161] M. Yao, G. Zhao, H. Zhang, Y. Hu, L. Deng, Y. Tian, B. Xu, and G. Li, "Attention spiking neural networks," *IEEE Transactions on Pattern Analysis and Machine Intelligence*, 2023.

[162] N. Rathi, I. Chakraborty, A. Kosta, A. Sengupta, A. Ankit, P. Panda, and K. Roy, "Exploring neuromorphic computing based on spiking neural networks: Algorithms to hardware," *ACM Computing Surveys*, vol. 55, no. 12, pp. 1–49, 2023.

[163] Y. LeCun, L. Bottou, Y. Bengio, and P. Haffner, "Gradient-based learning applied to document recognition," *Proceedings of the IEEE*, vol. 86, no. 11, pp. 2278–2324, 1998.

[164] A. Krizhevsky, G. Hinton *et al.*, "Learning multiple layers of features from tiny images," 2009.

[165] J. Deng, W. Dong, R. Socher, L.-J. Li, K. Li, and L. Fei-Fei, "Imagenet: A large-scale hierarchical image database," in *2009 IEEE Conference on Computer Vision and Pattern Recognition.* IEEE, 2009, pp. 248–255.

[166] P. Lichtsteiner, C. Posch, and T. Delbruck, "A 128 × 128 120 db 15 *mus* latency asynchronous temporal contrast vision sensor," *IEEE Journal of Solid-state Circuits*, vol. 43, no. 2, pp. 566–576, 2008.

[167] C. Posch, D. Matolin, and R. Wohlgenannt, "An asynchronous time-based image sensor," pp. 2130–2133, 2008.

[168] G. Gallego, T. Delbrück, G. Orchard, C. Bartolozzi, B. Taba, A. Censi, S. Leutenegger, A. J. Davison, J. Conradt, K. Daniilidis *et al.*, "Event-based vision: A survey," *IEEE Transactions on Pattern Analysis and Machine Intelligence*, vol. 44, no. 1, pp. 154–180, 2020.

[169] C. Posch, D. Matolin, and R. Wohlgenannt, "A QVGA 143 db dynamic range frame-free PWM image sensor with lossless pixel-level video compression and time-domain cds," *IEEE Journal of Solid-State Circuits*, vol. 46, no. 1, pp. 259–275, 2010.

[170] C. Brandli, R. Berner, M. Yang, S.-C. Liu, and T. Delbruck, "A 240× 180 130 db 3 μs latency global shutter spatiotemporal vision sensor," *IEEE Journal of Solid-State Circuits*, vol. 49, no. 10, pp. 2333–2341, 2014.

[171] T. Finateu, A. Niwa, D. Matolin, K. Tsuchimoto, A. Mascheroni, E. Reynaud, P. Mostafalu, F. Brady, L. Chotard, F. LeGoff *et al.*, "A 1280x720 back-illuminated stacked temporal contrast event-based vision sensor with 4.86 um pixels, 1.066 geps readout, programmable event-rate controller and compressive data-formatting pipeline," in *IEEE International Solid-State Circuits Conference*, 2020.

[172] B. Son, Y. Suh, S. Kim, H. Jung, J.-S. Kim, C. Shin, K. Park, K. Lee, J. Park, J. Woo *et al.*, "4.1 a 640× 480 dynamic vision sensor with a 9μm pixel and 300meps address-event representation," in *2017 IEEE International Solid-State Circuits Conference (ISSCC).* IEEE, 2017, pp. 66–67.

[173] Y. Suh, S. Choi, M. Ito, J. Kim, Y. Lee, J. Seo, H. Jung, D.-H. Yeo, S. Namgung, J. Bong *et al.*, "A 1280× 960 dynamic vision sensor with a 4.95-μm pixel pitch and motion artifact minimization," in *2020 IEEE International Symposium on Circuits and Systems (ISCAS).* IEEE, 2020, pp. 1–5.

[174] S. Chen and M. Guo, "Live demonstration: Celex-v: A 1m pixel multi-mode event-based sensor," in *2019 IEEE/CVF Conference on Computer Vision and Pattern Recognition Workshops (CVPRW).* IEEE, 2019, pp. 1682–1683.

[175] "Insightness event-based sensor modules," *Available: http://www.insightness.com/technology/,* 2020.

[176] T. Delbruck, "Neuromorophic vision sensing and processing," in *2016 46Th European Solid-state Device Research Conference (ESSDERC).* IEEE, 2016, pp. 7–14.

[177] S.-C. Liu, B. Rueckauer, E. Ceolini, A. Huber, and T. Delbruck, "Event-driven sensing for efficient perception: Vision and audition algorithms," *IEEE Signal Processing Magazine*, vol. 36, no. 6, pp. 29–37, 2019.

[178] T. Delbruck and M. Lang, "Robotic goalie with 3 ms reaction time at 4% cpu load using event-based dynamic vision sensor," *Frontiers in Neuroscience*, vol. 7, p. 223, 2013.

[179] A. Glover and C. Bartolozzi, "Event-driven ball detection and gaze fixation in clutter," in *2016 IEEE/RSJ International Conference on Intelligent Robots and Systems (IROS).* IEEE, 2016, pp. 2203–2208.

[180] M. Litzenberger, B. Kohn, A. N. Belbachir, N. Donath, G. Gritsch, H. Garn, C. Posch, and S. Schraml, "Estimation of vehicle speed based on asynchronous data from a silicon retina optical sensor," in *2006 IEEE Intelligent Transportation Systems Conference.* IEEE, 2006, pp. 653–658.

[181] P. Rogister, R. Benosman, S.-H. Ieng, P. Lichtsteiner, and T. Delbruck, "Asynchronous event-based binocular stereo matching," *IEEE Transactions on Neural Networks and Learning Systems*, vol. 23, no. 2, pp. 347–353, 2011.

[182] H. Rebecq, G. Gallego, E. Mueggler, and D. Scaramuzza, "EMVS: Event-based multi-view stereo—3D reconstruction with an event camera in real-time," *International Journal of Computer Vision*, vol. 126, no. 12, pp. 1394–1414, 2018.

[183] N. Matsuda, O. Cossairt, and M. Gupta, "Mc3d: Motion contrast 3d scanning," in *2015 IEEE International Conference on Computational Photography (ICCP).* IEEE, 2015, pp. 1–10.

[184] R. Benosman, C. Clercq, X. Lagorce, S.-H. Ieng, and C. Bartolozzi, "Event-based visual flow," *IEEE Transactions on Neural Networks and Learning Systems*, vol. 25, no. 2, pp. 407–417, 2013.

[185] H. Kim, A. Handa, R. B. Benosman, S.-H. Ieng, and A. J. Davison, "Simultaneous mosaicing and tracking with an event camera," in *British Machine Vision Conference*, 2014. [Online]. Available: https://api.semanticscholar.org/CorpusID:14402590

[186] H. Rebecq, R. Ranftl, V. Koltun, and D. Scaramuzza, "High speed and high dynamic range video with an event camera," *IEEE Transactions on Pattern Analysis and Machine Intelligence*, vol. 43, no. 6, pp. 1964–1980, 2019.

[187] H. Kim, S. Leutenegger, and A. J. Davison, "Real-time 3D reconstruction and 6-DOF tracking with an event camera," in *Computer Vision–ECCV 2016: 14th European Conference, Amsterdam, The Netherlands, October 11-14, 2016, Proceedings, Part VI 14.* Springer, 2016, pp. 349–364.

[188] A. R. Vidal, H. Rebecq, T. Horstschaefer, and D. Scaramuzza, "Ultimate slam? combining events, images, and IMU for robust visual slam in HDR and high-speed scenarios," *IEEE Robotics and Automation Letters*, vol. 3, no. 2, pp. 994–1001, 2018.

[189] L. Pan, C. Scheerlinck, X. Yu, R. Hartley, M. Liu, and Y. Dai, "Bringing a blurry frame alive at high frame-rate with an event camera," in *Proceedings of the IEEE/CVF Conference on Computer Vision and Pattern Recognition*, 2019, pp. 6820–6829.

[190] T.-J. Chin, S. Bagchi, A. Eriksson, and A. Van Schaik, "Star tracking using an event camera," in *Proceedings of the IEEE/CVF Conference on Computer Vision and Pattern Recognition Workshops*, 2019, pp. 0–0.

[191] G. Orchard, A. Jayawant, G. K. Cohen, and N. Thakor, "Converting static image datasets to spiking neuromorphic datasets using saccades," *Frontiers in Neuroscience*, vol. 9, p. 437, 2015.

[192] H. Li, H. Liu, X. Ji, G. Li, and L. Shi, "Cifar10-dvs: An event-stream dataset for object classification," *Frontiers in Neuroscience*, vol. 11, p. 309, 2017.

[193] A. Amir, B. Taba, D. Berg, T. Melano, J. McKinstry, C. Di Nolfo, T. Nayak, A. Andreopoulos,

G. Garreau, M. Mendoza *et al.*, "A low power, fully event-based gesture recognition system," in *Proceedings of the IEEE Conference on Computer Vision and Pattern Recognition*, 2017, pp. 7243–7252.

[194] Y. Wang, B. Du, Y. Shen, K. Wu, G. Zhao, J. Sun, and H. Wen, "EV-gait: Event-based robust gait recognition using dynamic vision sensors," in *Proceedings of the IEEE/CVF Conference on Computer Vision and Pattern Recognition*, 2019, pp. 6358–6367.

[195] Y. Jin, W. Zhang, and P. Li, "Hybrid macro/micro level backpropagation for training deep spiking neural networks," *Advances in Neural Information Processing Systems*, vol. 31, 2018.

[196] S. B. Shrestha and G. Orchard, "Slayer: Spike layer error reassignment in time," *Advances in Neural Information Processing Systems*, vol. 31, 2018.

[197] Y. Wu, L. Deng, G. Li, J. Zhu, Y. Xie, and L. Shi, "Direct training for spiking neural networks: Faster, larger, better," in *Proceedings of the AAAI Conference on Artificial Intelligence*, vol. 33, no. 01, 2019, pp. 1311–1318.

[198] L. Deng, Y. Wu, X. Hu, L. Liang, Y. Ding, G. Li, G. Zhao, P. Li, and Y. Xie, "Rethinking the performance comparison between snns and anns," *Neural Networks*, vol. 121, pp. 294–307, 2020.

[199] J. Wu, Y. Chua, M. Zhang, G. Li, H. Li, and K. C. Tan, "A tandem learning rule for effective training and rapid inference of deep spiking neural networks," *IEEE Transactions on Neural Networks and Learning Systems*, 2021.

[200] G. Qiao, N. Ning, Y. Zuo, P. Zhou, M. Sun, S. Hu, Q. Yu, and Y. Liu, "Batch normalization-free weight-binarized snn based on hardware-saving if neuron," *Neurocomputing*, p. 126234, 2023.

[201] A. Kugele, T. Pfeil, M. Pfeiffer, and E. Chicca, "Efficient processing of spatio-temporal data streams with spiking neural networks," *Frontiers in Neuroscience*, vol. 14, p. 439, 2020.

[202] Y. Kim and P. Panda, "Revisiting batch normalization for training low-latency deep spiking neural networks from scratch," *Frontiers in Neuroscience*, p. 1638, 2021.

[203] W. Fang, Z. Yu, Y. Chen, T. Huang, T. Masquelier, and Y. Tian, "Deep residual learning in spiking neural networks," *Advances in Neural Information Processing Systems*, vol. 34, pp. 21 056–21 069, 2021.

[204] Z. Wu, H. Zhang, Y. Lin, G. Li, M. Wang, and Y. Tang, "Liaf-net: Leaky integrate and analog fire network for lightweight and efficient spatiotemporal information processing," *IEEE Transactions on Neural Networks and Learning Systems*, vol. 33, no. 11, pp. 6249–6262, 2021.

[205] M. Yao, H. Gao, G. Zhao, D. Wang, Y. Lin, Z. Yang, and G. Li, "Temporal-wise attention spiking neural networks for event streams classification," in *Proceedings of the IEEE/CVF International Conference on Computer Vision*, 2021, pp. 10 221–10 230.

[206] Y. Wang, X. Zhang, Y. Shen, B. Du, G. Zhao, L. Cui, and H. Wen, "Event-stream representation for human gaits identification using deep neural networks," *IEEE Transactions on Pattern Analysis and Machine Intelligence*, vol. 44, no. 7, pp. 3436–3449, 2021.

[207] W. Fang, Z. Yu, Y. Chen, T. Masquelier, T. Huang, and Y. Tian, "Incorporating learnable membrane time constant to enhance learning of spiking neural networks," in *Proceedings of the IEEE/CVF International Conference on Computer Vision*, 2021, pp. 2661–2671.

[208] M. M. López-Rodríguez, M. Fernández-Martínez, G. A. Matarán-Peñarrocha, M. E. Rodríguez-Ferrer, G. G. Gámez, and E. A. Ferrándiz, "Efectividad de la biodanza acuática sobre la calidad del sueño, la ansiedad y otros síntomas en pacientes con fibromialgia," *Medicina Clínica*, vol. 141, no. 11, pp. 471–478, 2013.

[209] S.-L. Cheng, H.-F. Sun, and M.-L. Yeh, "Effects of an 8-week aerobic dance program on health-related fitness in patients with schizophrenia," *Journal of Nursing Research*, vol. 25, no. 6, pp. 429–435, 2017.

[210] D. X. Marquez, R. Wilson, S. Aguiñaga, P. Vásquez, L. Fogg, Z. Yang, J. Wilbur, S. Hughes, and C. Spanbauer, "Regular latin dancing and health education may improve cognition of late middle-aged and older latinos," *Journal of Aging and Physical Activity*, vol. 25, no. 3, pp. 482–489, 2017.

[211] T. Chen, C. Fang, X. Shen, Y. Zhu, Z. Chen, and J. Luo, "Anatomy-aware 3D human pose estimation with bone-based pose decomposition," *IEEE Transactions on Circuits and Systems for Video Technology*, vol. 32, pp. 198–209, 2022.

[212] Y. Chen, Y. Tian, and M. He, "Monocular human pose estimation: A survey of deep learning-based methods," *Comput. Vis. Image Underst.*, vol. 192, p. 102897, 2020.

[213] M. Hassan, V. Choutas, D. Tzionas, and M. J. Black, "Resolving 3D human pose ambiguities with 3D scene constraints," *2019 IEEE/CVF International Conference on Computer Vision (ICCV)*, pp. 2282–2292, 2019.

[214] Z. Zhang, K. Chai, H. Yu, R. Majaj, F. Walsh, E. Wang, U. Mahbub, H. Siegelmann, D. Kim, and T. Rahman, "Neuromorphic high-frequency 3D dancing pose estimation in dynamic environment," *Neurocomputing*, p. 126388, 2023.